Sulfur, Energy, and Environment

Sulfur, Energy, and Environment

BEAT MEYER

Chemistry Department,
University of Washington,
Seattle, WA, U.S.A.

ELSEVIER SCIENTIFIC PUBLISHING COMPANY
Amsterdam – Oxford – New York 1977

ELSEVIER SCIENTIFIC PUBLISHING COMPANY
335 Jan van Galenstraat
P.O. Box 211, Amsterdam, The Netherlands

Distributors for the United States and Canada:

ELSEVIER NORTH-HOLLAND INC.
52, Vanderbilt Avenue
New York, N.Y. 10017

Library of Congress Cataloging in Publication Data

Meyer, Beat.
Sulfur, energy, and environment.

Bibliography: p.
Includes index.
1. Sulphur. 2. Sulphur compounds.
3. Sulphur--Environmental aspects. I. Title.
QD181.S1M42 546'.723 77-23017
ISBN 0-444-41595-5

Printed in The Netherlands

Preface

In 1612 Rulando listed sixteen types of sulfurs which he assigned to two classes: "The lifegiving sulfurs," and those "which are and remain enemies of all metals." Today we know far more about sulfur, but the basic division remains between the viable sulfurs needed in agriculture and industry, and the corrosive and polluting sulfurs which constitute a nuisance for both the electric power industry and the public.

In talking with researchers and practitioners in any of the twenty-five fields listed in Table 1.1, one might easily gain the impression that sulfur is twenty-five different materials. It is the goal of this book to describe the properties of elemental sulfur and its three more important compounds, and to review the production, use, and recovery of sulfur in relation to energy production and environmental protection.

It is the purpose of this book to serve as a guide to the literature in fourteen areas which the author considers most important. Each chapter contains a short review, references to recent specialist reviews, and many key references to original research papers. Thus, the reader can choose the depth of his involvement, and, hopefully, the text is equally useful to both specialists and non-specialists. During the last fifteen years some 15,000 articles have dealt with sulfur. Despite Chemical Abstracts, it has become increasingly difficult to find access to basic new facts, not only because of the diversity of the areas involved, but because of changes in the publication styles. Obviously, a full coverage of the field would fill fifteen volumes rather than fifteen chapters, and a full set of references alone would fill an entire book. There is a definite need for such an encyclopedia on sulfur research, use, and recovery. In contrast, this book is intended to serve merely as a reference guide so that readers can quickly establish whether material is relevant to their work. Thus, several fields had to be omitted or neglected. For example, the entire field of pulp and paper chemistry has been totally ignored. Sulfur production, shipment, and other equally important fields receive abbreviated and incomplete treatment. Flue gas desulfurization is only described from a basic chemical viewpoint, and engineering considerations—no matter how important—have been neglected. Thus, this book does not contain one single process diagram, even though they constitute important information and are valuable aids. Likewise, only one out of ten references could be included in the bibliography to keep it manageable. This necessitated some quite arbitrary cuts in order to keep perspectives in a very large field which is still quickly developing.

The Introduction correlates traditions and trends in the fields of sulfur, energy use, and environmental attitude, and delineates the role of chemistry. Chapter 2 sketches a short history of sulfur chemistry. Chapter 3 reviews the chemical properties of the element and some of its most important compounds. Chapter 4 indicates common analytical methods and some problems encountered in their use. Chapter 5 provides a short list of sulfur sources and reserves. Chapter 6 reviews recent work on the four most important sulfur cycles. Chapter 7 summarizes sulfur production methods, and Chapter 8 describes coal combustion chemistry and flue gas desulfurization efforts. Chapter 9 describes some of the problems which occur at the interfaces of science, industry, and society. Chapters 10 through 14 deal with the use of sulfur in medicine, agriculture, chemical industry, the plastics industry, and in other applications. The final Chapter discusses possible future trends in sulfur production, use, and recovery, and the role of chemistry, government, and education in these areas.

This book reflects personal views based on my training and experience. My interests in sulfur were stimulated by Prof. E. Schumacher in Zürich, who showed me a paper by F. O. Rice on trapped sulfur vapor, and guided my Ph.D. research. Prof. Leo Brewer in Berkeley introduced me to high temperature and combustion chemistry, and showed me the challenges of conducting independent research in an applicable field. Furthermore, he brought me into contact with many leading scientists. There is not sufficient room to thank all those in academe, industry, and government who have been instrumental in my learning about sulfur. Dr. J. R. West and Dr. R. Coleman introduced me to many outstanding people in the industry, and to Mr. P. N. Kokulis, who practices a special art which makes it possible to efficiently translate basic scientific and innovative ideas into practical applications while protecting the interests of all sides.

The preparation of this book was greatly aided by Prof. L. Brewer who has regularly provided key references in basic and applied chemistry. My wife and daughter helped search, copy, and collect references; Harry

Weeks traced several dozen difficult to find references in the UC Berkeley library; Christal Shaskey Rosenlund helped research the history chapter and collected references dealing with the sulfur cycle. Phyllis Ayer helped with the literature search, the bibliography, the preparation of figures, and the indices. Carey Julian typed and edited Chapters One and Nine and helped with the bibliography. Special thanks belong to Mrs. Marilee Kapsa who coordinated the production and assembly of the text. She assisted in searching the literature, typed the text, edited and proofed the entire manuscript, and prepared the camera copy in Aldine Roman type.

This book is dedicated to all those in industry, government, and at universities who work with sulfur and help sound scientific methods find acceptance. If they should ever get discouraged, they should remember that Hermann Frasch struggled for many long years, and that, shortly before his sulfur production method finally succeeded, a prominent man promised him that he would eat every ounce of sulfur produced by his method.

Beat Meyer

Seattle, Washington
March 1977

Contents

Chapter 1

Introduction

This book deals with elemental sulfur. It describes the chemical properties of sulfur and their relation to the production, use, and recovery of the element. Both basic chemistry and industrial use of sulfur influence energy production and the extent to which the environment can be protected. Thus, sulfur is an important parameter in the choice of new technology and in determining practically enforceable environmental laws. The purpose of this book is to delineate the role of sulfur at a time when technological transitions and social adjustments are setting the framework of society for the next hundred years.

In 1976 almost fifty million tons of sulfur were produced worldwide. Most of this was used to manufacture sulfuric acid, which is unchallenged as the leading industrial chemical. Its production steadily increased from 5 tons in 1750 to 110 million tons in 1976. However, this smooth overall growth does not properly reflect development, which was marked by stagnant as well as dynamic periods. In the 18th and 19th Centuries sulfur production was centered in Sicily. Frasch's development of a clean and cheap sulfur production process made the Sicilian industry obsolete, and within ten years Louisiana and Texas had become the world centers for sulfur production. They retained this position unchallenged for 60 years, until 1960, when production of sour natural gas made Canada and France major producers of sulfur. With this development, an initially unwanted by-product became the major source of the leading industrial chemical. Currently, and for at least the next ten years, oil production in the Middle East will yield important quantities of sulfur, probably equalling the stepped-up production in Poland, which is implementing increasingly modern techniques based on traditional production methods.

Sulfur production from sour gas and oil is based on the Claus process, which was patented in 1882. Even before that time, the manufacture of sulfuric acid had reached essentially the present stage of chemical art. Since then sulfur chemistry and industrial inorganic chemistry have remained almost stagnant in comparison to other fields of basic chemistry and other sciences which have proceeded through a period of unique growth. Only during the last fifteen years has sulfur chemistry slowly revived. Recent progress has indeed been surprising: twelve new solid allotropes of the element have been synthesized, some 20 components of elemental sulfur vapor have been identified, and substantial progress has been made in the understanding of liquid sulfur.

At the same time, energy consumption has increased at a similarly spectacular rate. During the dynamic 1960's, when all progress and values were re-examined, the public suddenly became aware that involuntary release of sulfur dioxide from coal burning power plants–mainly into the air of the northern hemisphere–equaled the intentional world sulfur production. The public became fearful of the enormous quantities of efferent sulfur, because it did not realize that the total sulfur involved in all of man's recorded activities is smaller than the sulfur dioxide emission resulting from any of the large volcanic eruptions of Mt. Katmai in Alaska in 1912, Mt. Hekla in Iceland in 1947, or Mt. Agnug in Bali in 1963. The conflict between consumption and conservation brought into the open the emotional origin and the political potential of technical terms such as power, waste, and pollution, and an increasing government effort to regulate industries and abate pollution was demanded.

Today, a large number of highly competent and skilled specialists in industry, government, and education struggle to translate the results of a hundred years of progress in diverse fields of science into technology acceptable under the new standards of society. It is the goal of this book to help these researchers find their way through the increasingly incoherent literature of the last ten years, and gain access to work related to their fields. It is too early at this time to gain a full overview, but this book constitutes an effort to make a modest start and point out the direction of developments which are under way. It is hoped that it can support those working in the field as well as newcomers in their momentous task, which will determine the chemical basis for the large scale technology to be used during the next hundred years. This task will also influence the role of sulfur, and the nature and quality of human life. Since the work involves coordinating results from a variety of divergent sciences, it seems reminiscent of the task of salvaging the Tower of Babel.

Changing Traditions

A hundred-fifty years ago, sulfur was widely used in industry and in the chemistry lab, and could be found in the medicine cabinet of every home. Sulfur alone filled a third of the inorganic chemistry texts. During the following decades it maintained its position as the leading industrial chemical, but it disappeared from public view, from chemistry labs, and finally from college chemistry textbooks. One of the three leading U.S. college general chemistry textbooks of 1970 contains only six sentences on sulfur, of which two are incorrect. During the same time academic interest in sulfur also waned; ten years ago at the 20 academically highest ranking U.S. universities, a total of only three inorganic or physical chemistry professors conducted research focussing on sulfur. Most of the progress resulted from basic research which was conducted as a side activity by a small group of little-noticed but outstanding industrial scientists whose main responsibility was supervising plant production.

In the meantime oil, gas and sulfur production became cheaper, and abundant energy became available to exploit technology developped during recent wars for mass production. This, and the rapidly growing influence of media and international communication fostered a hunger for consumption of manufactured goods, which in turn caused quick economic growth. Sputnik caused demand for quick implementation of new technology and instant mass education. This proved to be socially disruptive, because democracy and free enterprise depend on equilibrating forces which require more time than was available.

At the time when manufacturers, in response to earlier trends, used linear extrapolation to compute anticipated needs, perfected better process methods, and finished building larger manufacturing facilities, products became obsolete. Industry responded by diversifying and intruding into each other's established fields in order to survive. By this time, education had caught up with the fifties, and had geared up for large scale improvement of traditional structures. Mass education in obsolete academic and scientific fields began. This caused frustration among graduates and insecurity among students. The unleashed anxiety and fears increased awareness of previously ignored complaints beyond the threshhold of political inactivity. The sudden change in the public attitude caught industry at an economically unfavorable time. Universities abolished traditional paths of study, and everybody rushed to explore new laws, policies and academic fields.

Table 1.1
ACADEMIC FIELDS WITH PROGRAMS OF SULFUR STUDY

Field of Study	Subject
Law	Air quality legislation
Business management	Impact of SO_2 control
Economics	Electric rate structure
Political science	Distribution of abatement burden
Social science	Impact of SO_2 abatement
Medical epidemiology	Urban excess of morbidity & mortality
Clinical physiology	Symptoms of SO_2 exposure
Oceanography	Sedimentary sulfur cycle
Atmospheric science	Atmospheric sulfur cycle
Meteorology	Effect of sulfate particulates on climate
Soil science	Fertility of sulfur soils
Agricultural science	Effect of sulfur on crop yields
Veterinary science	Inorganic sulfur in feed of non-ruminants
Forestry	Effect of SO_2 on forest growth
Geochemistry	Artesian sulfur and pyrite
Environmental science	Effect of excess sulfur on climate
Biochemistry	Sulfur-containing proteins
Combustion chemistry	Sulfur chemistry in flame
Metallurgy	Corrosion of high temperature steels
Mining	Mechanical properties of high sulfur coals
Civil engineering	Sulfur dioxide monitoring
Mechanical engineering	Sulfur in construction materials
Physics	Optical properties of sulfur dioxide
Chemistry	Analytical chemistry, kinetics, etc.

Today, in response to the political and social events of the last 15 years, interest in pollution abatement is well established, and every major U.S. university has sulfur specialists in at least 20 different academic fields (table 1.1).

Every government has a branch dealing with sulfur emissions, each government branch has sulfur specialists, and even businesses which neither buy nor sell any sulfur or sulfur-containing materials have pollution specialists.

Needless to say, academic chemists have also rediscovered sulfur; about a third of all research proposals from chemistry departments now deal with sulfur in connection with energy or pollution. Furthermore, technical, semi-popular and general literature is full of articles on subjects connected with sulfur. Even today, however, those who deal professionally with sulfur find it difficult to acquire basic chemical facts. As a rule, their specialized educations did not permit study of chemistry beyond the freshman level, and chemical education has adjusted slowly. Most college freshman classes deal with sulfur superficially, if at all, and often discuss the social impact of man on the environment instead of explaining basic sulfur chemistry. Introductory and general chemistry textbooks still focus on computer modeling, atomic physics and quantum theory, and ignore experimental chemistry and chemical facts in applicable fields. Even those who major in chemistry have a hard time learning about sulfur, as it is only superficially treated even in advanced texts, and less than one in six inorganic chemistry professors in North America knows the properties of gypsum, the structure of the bisulfite ion, or the composition of elemental sulfur vapor. This is a result of the fact that the subject was not taught when they acquired their own general chemical knowledge, and information explosions in their own specialties have prevented them from following progress in the field of sulfur. The burden of basic sulfur chemistry, then, rests with only a few professors, and primarily with industrial chemists and engineers, who have continued through the confusion to provide sulfur as needed. They are now aided in their research by highly competent government scientists.

Those attempting to gain access to the field are confronted with the results of the information explosion. Chemical Abstracts currently quotes over 10 articles connected with sulfur per day, and more than one patent per day. Anyone trying to read the original literature finds it difficult, as an increasing fraction of important data is in government reports or other articles which are published outside the traditional literature and may be difficult to procure. The traditional U.S. literature clearly suffers from lack of research funds; it is rich in general discussions and computer modeling, which is cheaper and quicker than experimental work. The most interesting work is frequently buried in unusual journals which don't levy publication charges, or in long and difficult to read papers derived from Ph.D. theses. Furthermore, professional literature increasingly uses a "modern" technical language in which structure and grammar are replaced with strings of often incomprehensible and artificial technical nouns, which undoubtedly would have aroused jealousy among the masters of the trade guilds of the middle ages.

Table 1.2

SULFUR CONTENT OF VARIOUS MATERIALS

Material	Wt. %	Material	Wt. %
Cosmos	0.002	Colza cabbage	0.98
Crust of the earth	0.052	Alfalfa	0.50
Coal	1-14	Oats	0.41
Oil	0.1-14	Barley	0.30
Gas	0.1-40	Beef meat	0.1
Gypsum	18.6	Cow milk	0.08
Soil	0.01&0.05	Human brain tissue	1.1
Human body	1.1	Human dietary need	0.5[a]

a) Grams per day.

It is the goal of this book to help specialists cross interdisciplinary barriers and increase awareness of the relationship and importance of their own work to that being done in other fields.

Sulfur, Chemistry and Engineering

Table 1.2 illustrates that sulfur is more ubiquitous than is generally assumed. If the concentration of sulfur involved is modest, the total amount is certainly not trivial. Sulfur participates in numerous important reactions. Here are some examples which reflect its diverse role.

Phosphate rock and sulfuric acid yield fertilizer:

$$3Ca_3(PO_4)_2 \quad + \quad 6H_2SO_4 \quad \rightarrow \quad 3Ca(H_2PO_4)_2 \quad + \quad 6CaSO_4$$

This reaction constitutes the largest use of sulfur. Sulfur acts as an acid to solubilize the rock, and part of it remains in the product, serving as a plant nutrient. This process is fully described in Chapter 2. The quantities and values involved are described in the next section.

Sulfate is essential for plant growth:

$$\text{Soil} \quad + \quad CaSO_4 \quad \rightarrow \quad \text{Cysteine}$$

Sulfur is an important plant nutrient, as Chapter 11 explains, because sulfur is a vital component of several proteins. Sulfur shortage stunts the growth of plants and reduces crop yield.

Smelting of copper ore yields copper and fertilizer:

$$CuS \quad + \quad Air \quad + \quad NH_3 \quad \rightarrow \quad Cu \quad + \quad (NH_4)_2SO_4$$

In this reaction, sulfur is a byproduct. Depending on the chemical path chosen, sulfur acts as an air pollutant, as solid waste, or as sellable fertilizer.

Oil and gas refining produce fuel and elemental sulfur, as described in Chapter 8:

$$RSH \quad \rightarrow \quad RH \quad + \quad H_2S \quad \rightarrow \quad RH \quad + \quad S$$

Sulfur has to be removed from oil and gas. Originally, sulfur was considered a waste product; today it constitutes a valuable by-product which accounts for half of all sulfur sold.

Roasting of pyrite (Chapter 7) is used to produce elemental sulfur:

$$FeS \quad \rightarrow \quad Fe_2O_3 \quad + \quad S_x$$

At one time, pyrite was the most important source of sulfur worldwide. Today, pyrite is a major source of sulfur and acid in only Spain and Scandinavia.

Combustion of coal yields both energy and pollution:

$$100\ C \quad + \quad 1100\ Air \quad \rightarrow \quad 100\ CO_2 \quad + \quad 1\ SO_2 \quad + 9500\ kcal/mole$$

On the average, coal contains only 1-4% sulfur, but this sulfur is nonetheless responsible for much of the present controversy concerning the use of coal for electricity generation.

Stack gases and urban air can yield smog (Chapter 6):

$$SO_2 \quad + \quad SO_3 \quad + \quad O_3 \quad \rightarrow \quad H_2SO_4$$

In certain areas, urban air contains ozone and hydrocarbons from automobile exhaust. These chemicals interact synergistically and form particulates which aggravate the detrimental effects of smog on health.

Combustion gases and acid rain destroy marble:

$$CaCO_3 \quad + \quad H_2SO_4 \quad \rightarrow \quad CaSO_4 \quad + \quad CO_2$$

In many countries, and especially in Italy, invaluable outdoor art and architecture is made of marble which is vulnerable to acid. During just the last 50 years, coal emissions have inflicted more damage on some of these treasures than 2000 years of previous exposure to the elements.

Hydrogen sulfide corrodes hot iron:

$$Fe + H_2S \rightarrow Fe_2S$$

In Claus furnaces and coal gasifiers, sulfur can react with iron and form iron sulfides. Some of these sulfide phases form eutectics in which, at normal operating temperatures, one component can become liquid. Under these conditions corrosion is extremely rapid.

In each of the above examples, the impact of sulfur is the result of purely chemical reactions. Chemistry determines the products, the yields of products, the reaction paths, and the speed of the reactions. An understanding of basic chemistry is necessary to predict and influence whether the effect of sulfur will be beneficial or detrimental. It determines whether sulfur will form valuable products or merely waste. For example, in the case of copper smelting, by chemical manipulation of the original gases one can determine whether sulfur is obtained as elemental sulfur, a generally valuable commodity, as ammonium sulfate, which is a valuable fertilizer, or as sulfur dioxide gas, an air pollutant.

The basic chemical properties of sulfur remain the same regardless of the manner in which it is used. Thus, sulfur chemistry is the link between the various applications and uses of the element, and establishes the intrinsic limits of any process in which sulfur is involved. If the basic sulfur chemistry is resulting in undesirable products, or the thermodynamics or kinetics are unfavorable, all efforts, including legislation, political pressure, money or engineering are in vain. On the other hand, chemical inventions and innovation can induce basic changes in industry, superseding carefully planned management predictions or economic programs. Hence, this book will concern itself primarily with chemistry. Engineering constitutes another basic limitation, because it determines whether chemical opportunities can be fulfilled. However, even when engineering efforts fail, a process may become viable at a later time, when new techniques or materials become available.

Sulfur, GNP and Living Standards

In 1843, in his letters on chemistry quoted in Chapter 2, Justus Liebig concluded, "We can fairly judge of the commercial prosperity of a country from the amount of sulfuric acid it consumes." Table 1.3 shows the sulfur production in the U.S. since 1850. The growth is partly due to increasing population, and partly due to increase in the per capita

Table 1.3

U.S. SULFUR PRODUCTION

Year	Native	Production (1,000 t) Gas & Coal	Smelter	Pyrite	Total	Posted Price $/ton
1800	0.5	–	–	0.9	1.8	–
1890	0	–	–	45	58	–
1900	3.2	–	–	92	122	21.15
1910	247	–	–	109	460	22.00
1920	1,255	300	279	120	1,660	23.85
1930	2,560	2.5	269	124	2,955	18.00
1940	2,732	4	191	262	3,205	16.00
1950	5,193	142	216	393	5,985	19.02
1960	5,037	767	345	416	6,660	23.50
1970	7,092	1,590	560	320	9,360	25.00

After Hazleton (1970) and Merwin (1975).

use of sulfur. In 1975 about half of the 100 million tons of sulfuric acid used world-wide were produced in the U.S. Table 1.3 also shows the low price of sulfur during that time, which was possible because of Frasch's revolutionary invention. His inexpensive clean method for producing sulfur and oil was an important factor in helping the U.S. to gain industrial leadership. This illustrates that judicial choice of a chemical process can profoundly influence the wealth of a nation, and that under fortuitous conditions an individual can contribute more than a decade of well-financed work by hundreds of gifted, government sponsored scientists working under less favorable conditions.

Figure 1.1 shows that sulfur is involved in several economic sectors. A large fraction (about 80%) goes into the production of 38 million tons of fertilizers, with a value of over $2 billion. This amounts to almost half of the value of U.S. agriculture, which in turn constitutes 3.1% of the GNP. Sulfur participates to an important extent in the mining industry, which makes up 6.2% of the GNP: elemental sulfur constitutes about $200 million, while the fuel industry amounts to $34 billion. This sum is comparable to the U.S. Government defense budget. Figure 1.1 shows that sulfur enters a large number of industrial manufacturing products. These products amount to 27.6% of the GNP. Fuel desulfurization greatly affects utilities, which account for 4% of the GNP.

Figure 1.2 shows the correlation between sulfur use and GNP for several nations. Most nations not specifically cited fall equally close to the curve which shows only international sulfur production. In the U.S.,

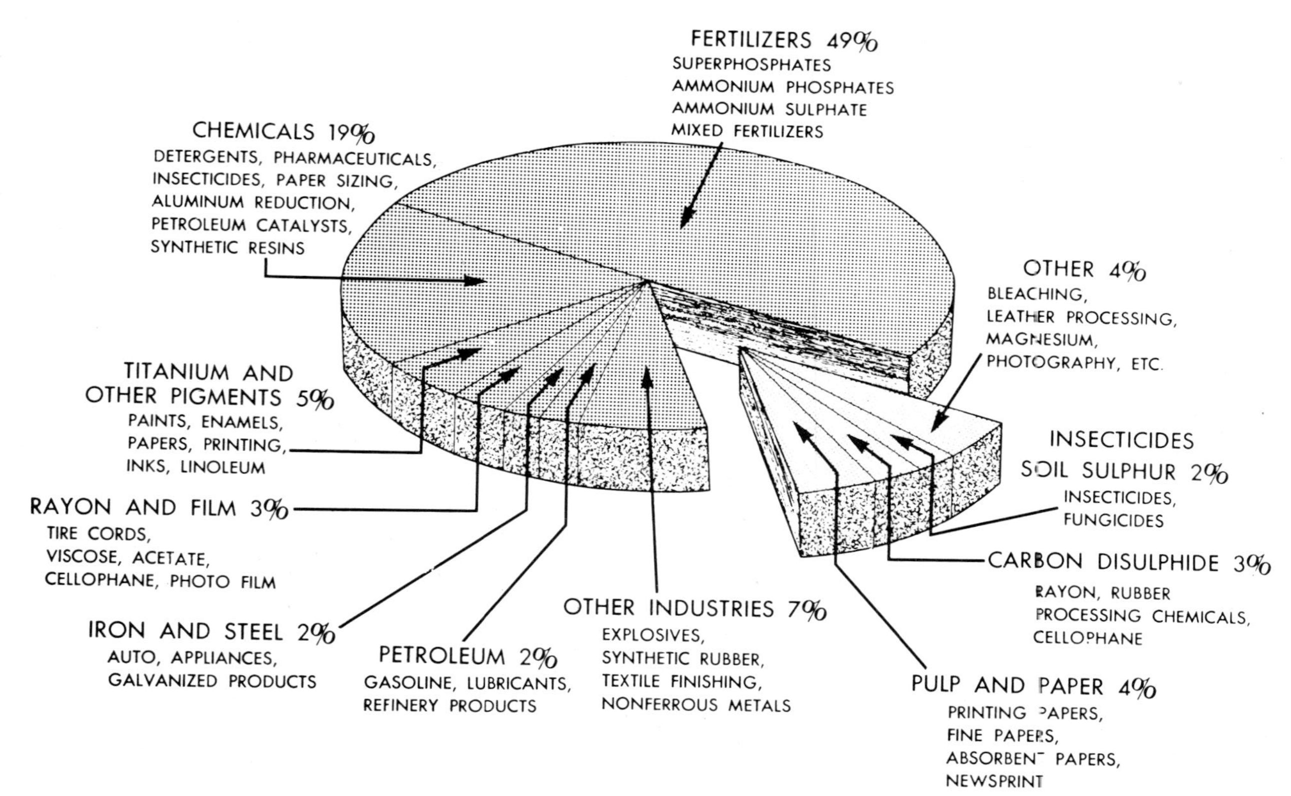

Fig. 1.1. End uses of sulfur.

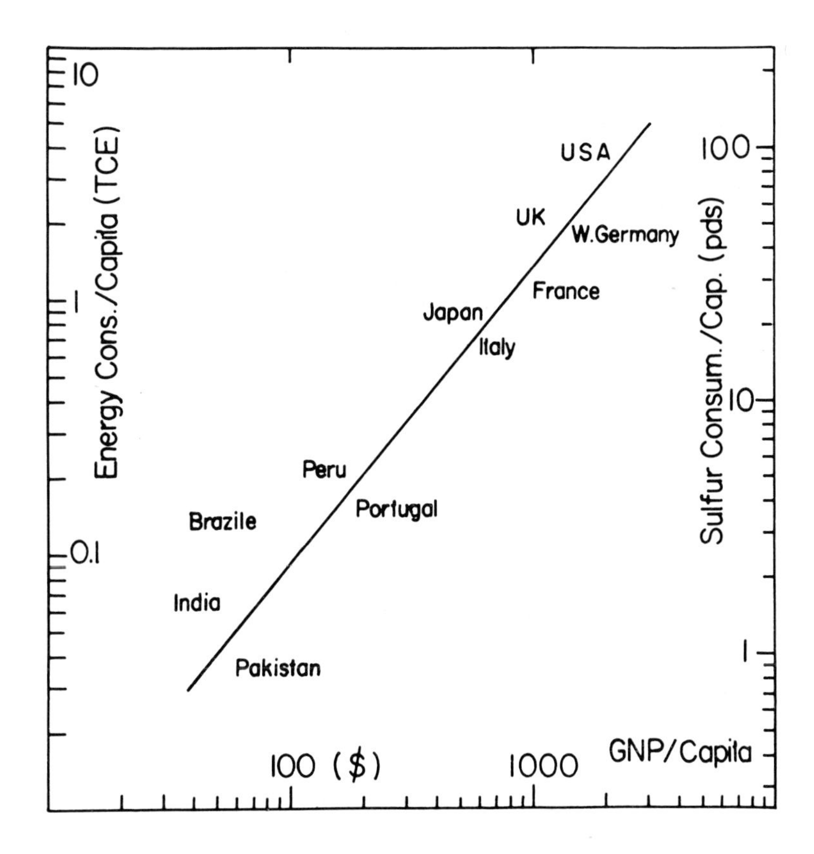

Fig. 1.2. Energy consumption per capita and sulfur consumption per capita in eleven countries, as a function of per capita GNP

per capita use of sulfur is almost 100 pounds per person per year. If all nations equalled the U.S. in sulfur use, world sulfur production would have to be increased to four times today's rate. Unintentional sulfur production is directly correlated to the per capita energy use, also shown in Figure 1.2. Figure 1.3 shows a correlation between industrial production and energy use. It is evident that intentional and unintentional sulfur use are correlated.

It is widely believed that per capita GNP and living standard are identical, and this may indeed be true for some developing countries. However, there is a point beyond which wealth leads to an apparently asymptotic drop in man's efficiency in striving for further fulfillment of his dreams, unless education prepares him for improving his methods

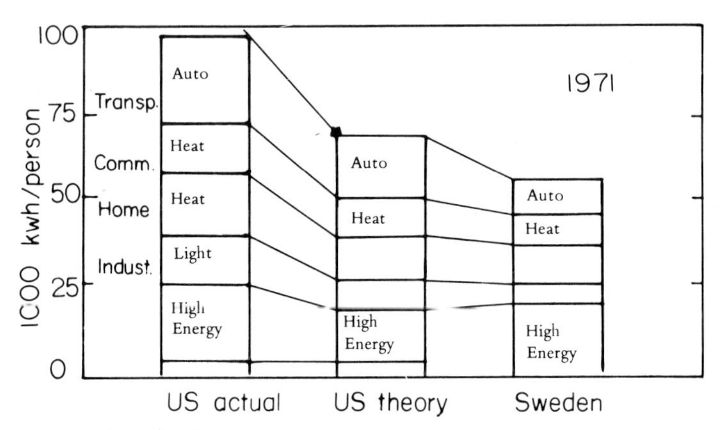

Fig. 1.3. U.S. and Swedish energy use in 1971, and theoretical U.S. energy use based on Swedish efficiency in industry, automobiles, etc. Life-style factors (such as numbers of appliances and passenger miles) were not considered. (After Schipper, 1976.)

and goals. Misunderstood science has often seriously hampered progress (Hildebrand, 1976). Modern agribusiness, for example, requires a subsidy of 15 calories to produce 1 calorie of protein; in 1910 the same product was obtained with an effort of 1 calorie. Rice cultures used 0.1 calorie, and the migrant hunter consumed 0.01 calorie for 1 calorie of food (Steinhart, 1976). It is argued in Chapter 9 that several highly industrial nations have passed a critical point; increased wealth is not used for tangible improvement of living quality, but simply goes to waste and causes pollution. Economic history has shown that in such situations, stimulating the economy leads merely to inflation. However, this is not the place for dwelling on this controversial issue which presently lies outside scientific definition.

Sulfur Production

The history of sulfur production is described in the next chapter and in Chapter 7. Figure 1.4 shows that the incredible growth in production was obtained not by continual improvement and expansion of the original method, but was a result of change of production method and raw materials. The original manual mining method was inefficient and could not remain competitive; it was replaced everywhere except in Sicily by the roasting of pyrites around 1850. Frasch's mining method of 1896 constituted one of the major inventions of the century, and

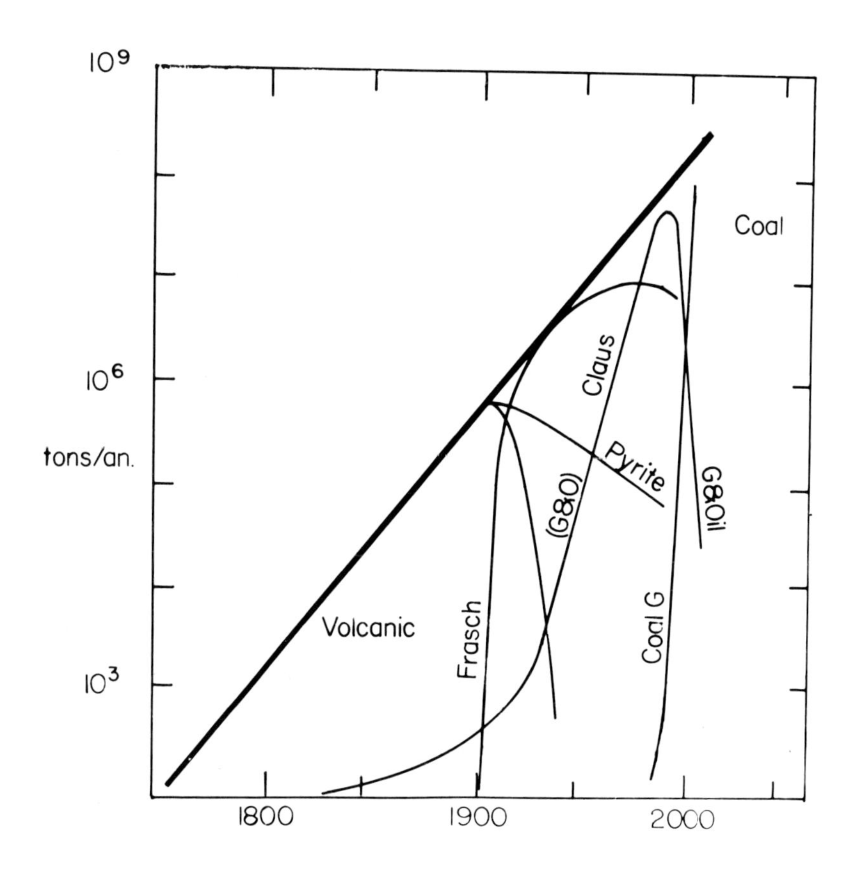

Fig. 1.4. Sulfur production, 1750–2000 (after U.S. historic statistics; Bureau of Census, 1975; Hazleton, 1970; K. Schmidt, 1974).

influenced the U.S. economy as well as world politics for several decades. However, pyrite roasting survived in certain countries. With the incredible increase in energy use, sulfur recovered from oil and gas has steadily gained importance, until it has exceeded that of smelter gases. Since 1970, recovered sulfur has constituted half of all sulfur produced, and the entire sulfur production has been equalled by unrecovered sulfur emission. Chapter 15 deals with the prediction of the future balance between supply and demand for sulfur which changes in irregular cycles, and causes periodic regional strains.

Energy and Environment

Each year, worldwide, 5 x 10^{12} kwhr of electricity is produced by the chemical reaction:

$$C + Air \rightarrow CO_2 + 95 \text{ kcal}$$

Of this energy, about half is lost as heat during generation. While the history of sulfur production has been characterized by innovative changes and continued technological improvement, that of energy production has followed a far choppier path. The history of the coal mining industry has been well chronicled by Jevons (1865), Devine (1925), Rubin (1927), Reid (1973), and many others. When oil originally displaced coal in ship and other mobile engines, experts were confident that it would be only a matter of a few years before coal would regain the market. When it was finally realized that the English coal exports would suffer permanently, the industry had already been weakened by archaic practices and conflicts of vested interests. The industry quickly became a "battlefield of ideologies" (Reid, 1973), and the situation caused the government negotiator, Lord Birkenhead, to state in his 1926 report: "It would be possible to say without exaggeration that the miners' leaders were the stupidest men in England if we had not had frequent occasion to meet the owners." English coal production reached its highest value of 287 mt in 1913, when England accounted for ¼ of the world production. At that time, however, productivity had already dropped from the 1881 level of 400 tons per person employed to an output of 257 tons per person per day. "This fall was a result of three main factors: shorter working hours, initiated 1908, difficulties in extracting coal from seams, and the slow development of new or improved extracting methods." (Lubin, 1927.) The industry began to experience difficulties in attracting skilled young workers. In 1947 England nationalized the industry by forming the National Coal Board, because it was felt that the scope of the capital requirement and the time scale necessary for renewal exceeded the capacity of private industry. As Lubin (1927) stated: "...to say that if each owner seeks his own interest the general interest will be served is like saying that in case of panic or fire the general welfare demands that each individual struggle as vigorously and independently as he can for his own safety." Figure 8.4 shows trends in the English coal production; its decline after 1924 is paralleled by

Table 1.4
AVERAGE COST OF ELECTRICITY

Country	Cost, U.S. cents per kwh	
	Industrial	Domestic
Norway	0.4	0.7
South Africa	0.8	2.1
Sweden	0.8	1.3
United States	0.9	2.4
United Arab Republic	1.0	3.6
France	1.3	4.3
United Kingdom	1.6	2.1
Belgium	1.7	4.7
West Germany	1.8	3.6
Luxemburg	1.8	4.9
Romania	2.5	3.3
Burma	4.1	12.2

After Guyol (1969).

a decline in sulfur production in Sicily, which is reflected by an overall decline in English industrial production. At the same time, Germany increased coal production. The curve shows that the German production was only insignificantly slowed after the war, but that it was stopped by the influx of the Middle Eastern oil in 1950. U.S. coal production never boomed during this century, because oil development had already started at the turn of the century. Thus, the use of coal was restricted to the steel and electric power industries. In Russia and Asia, coal consumption has steadily increased, and post-war energy management reflects the stability of a government controlled industry. The main use of coal is now in the electric power industry, which grew from 5×10^9 kwh in 1900 to 2×10^{10} kwh in 1960 (Guyol, 1969). The increase in energy use from 1900 to the anticipated use in 1990 in the U.S. is shown in Figure 1.5a. This increase is due to the availability of new cheap fuels; Figure 1.5b shows the relative importance of the different fuels used in the U.S. in the period from 1800 to 2100. Table 1.4 lists the price of industrial and domestic electricity in selected countries during 1966. The U.S. energy consumption was 1.5×10^{11} kwh in 1970, i.e. about one-millionth of the energy transfer from the sun. The total world consumption was three times that of the U.S. The per capita use corresponded to an equivalent of 12 tons of coal per capita per year. Canada produced the equivalent of 10 tons of coal per capita, Denmark and Czechoslovakia 7, Belgium, Germany, Sweden and England 6, and Australia

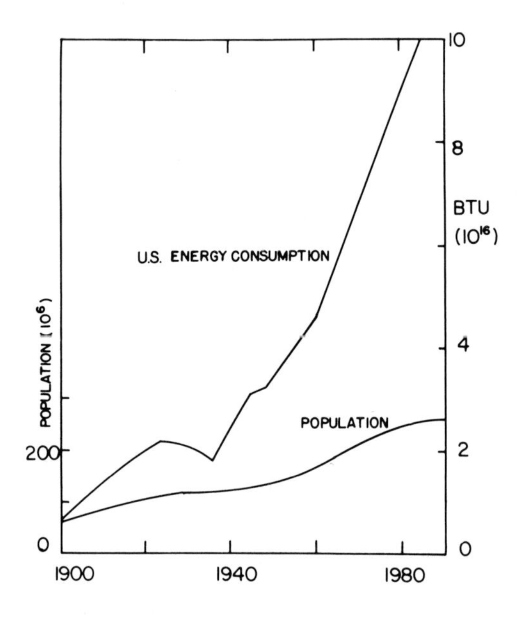

Fig. 1.5a. U. S. energy consumption, 1900-1990.

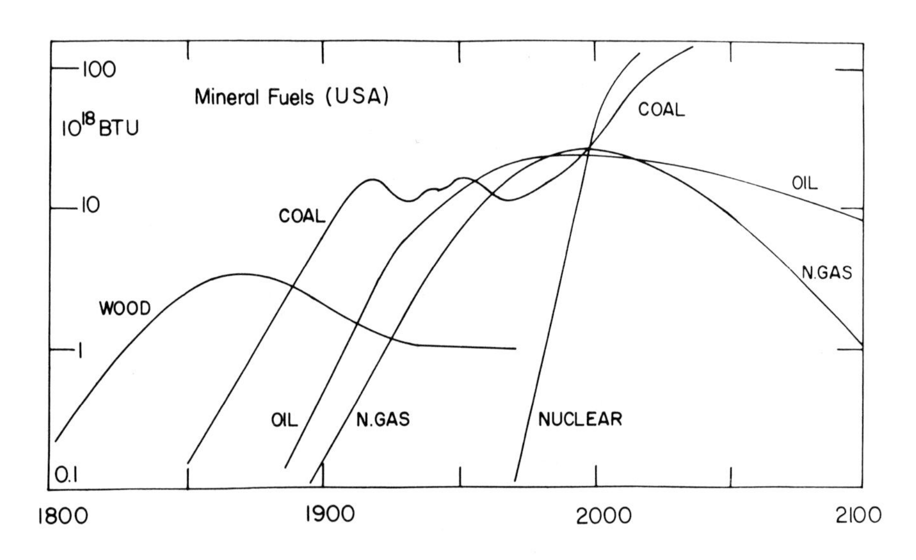

Fig. 1.5b. U. S. mineral fuel sources, 1800-2100.

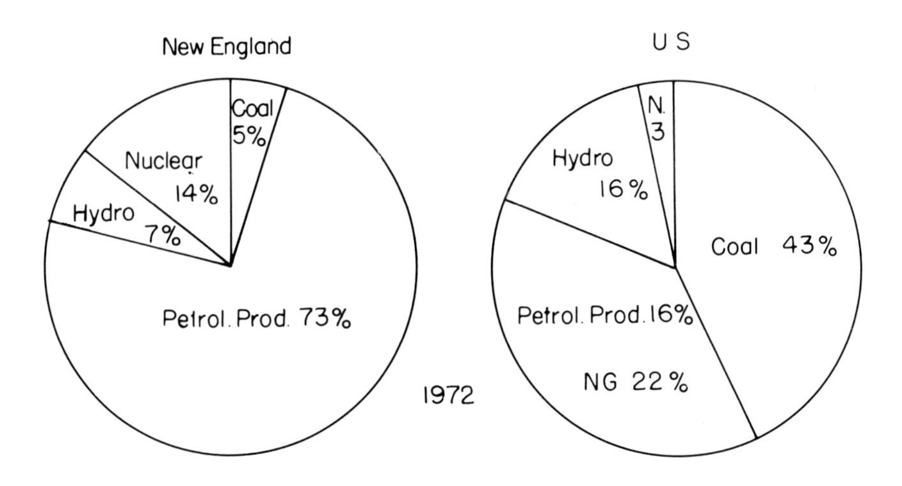

Fig. 1.6. Fuel sources for electricity generation in U.S. and N.E. sector (ERDA, 1975)

five. Switzerland used the equivalent of 3 tons of coal per capita in 1970, the same quantity the U.S. had used in 1850, and the same as the present U.S. per capita use for transportation alone. In Bolivia, Equador and the United Arab Republic, total per capita use of energy corresponds to 0.2 tons per capita per year; in the U.S. this amount of energy goes to air conditioning alone. Total per capita energy use in Burma is one-third of the preceeding. The abundance of cheap energy helped the U.S. gain industrial dominance, and the generous "American" style of life is based on the premise that inefficient use of resources helps increase employment and industrial production; this implies that inefficiency is a political necessity and that waste helps the economy (Schipper, 1976). This present U.S. situation is reminiscent of that in England circa 1913. However, the source of energy is drastically different, as seen in Figure 1.6: only 17% of all energy is deduced from coal, and even in electricity generation coal constitutes merely 43% of all fuel used. In several populated states and population centers, the situation is far more extreme. In New England only 5% of all electricity is produced by coal; the bulk, 73% is produced by burning residual oil. The reason for this is visible in Table 1.5, which shows the approximate 1970 cost of different fuel materials, and their heating value per unit weight. The cost of coal is increased because coal leaves ashes, and solid wastes cannot be ignored, but must be removed for a price. Coal consumers are responsible for coping with their own

Table 1.5
ENERGY SOURCES

Fuel	Price ($/unit)	Heat Content (kcal/g)	U.S. Consumption Btu x 10^{15}	World Reserves 10^{12} ton coal equivalent	Sulfur content (tons)
Wood	6/t	3-4	1	renewable	-
Bituminous coal	5/t	3-7	12	2.1	10^9
Anthracite	6/t	8	2	6.8	2×10^9
Tar & shale oil	1/t	6-10	-	.6	3×10^8
Crude oil	3/bl	10	22.5	.1	4×10^7
Natural gas	$.2/10^3$ cft	12	21.9	.1	4×10^7
Uranium oxide	30/t	4×10^5	.3	.1	-
Nuclear fission (D_2O)	1000/t	10^6	-	35,000	-
Solar energy	-	-	-	125/yr	-

wastes. Oil and gas producers desulfurize their fuel at their own expense and sell clean fuel. In the case of nuclear energy, the U.S. Government presently absorbs this cost and stores the wastes with no long range plan for protecting future generations against its poisonous effects (Rochlin, 1976).

Sulfur and Pollution

Pollution from combustion gases is as old as fire. However, until fifteen years ago the sight of smoke above a home evoked a feeling of human warmth, stability and peace, and the black plume belching from an industrial stack was taken as a symbol of opportunity, power and a prospering economy. In 1969 the Small Business Administration removed the factory plume (Figure 1.7) from its seal; since then industries which have opaque plumes are fined, and in several countries, Switzerland for example, domestic oil burners are inspected biannually, and home owners whose combustion gases color filter paper gray must buy new burners and pay a fine. When asked the reason for the sudden action, an inspector stated that it had to do with the high sulfur content of modern oils. He explained that, somehow, a cleaner flame emitted less sulfur. This is a common misconception, and is discussed in Chapters 8 and 9.

Frasch sulfur production involves steam and elemental sulfur. There is no waste, nor any residue. Frasch production helped drain the swamps of Texas and Louisiana, and the process is technologically and chemically mature. Pyrite roasting for recovering sulfur aims at selling sulfur. Losing the product to the environment constitutes waste of the primary saleable

Fig. 1.7. Seal of the United States Small Business Administration: a) prior to 1968; and b) since 1968.

product. Copper smelters produce sulfur as a byproduct. The chemical basis for this process was developed before Agricola's time, around 1580. The two remaining U.S. custom smelters operate, and look, very much like woodcuts from his time. However, all modern smelters are equipped to recover all but traces of SO_2, and the recovery trains are operating to the extent that the economy of the area is able to absorb sulfuric acid, sulfur dioxide, or ammonium sulfate. The history of smelter abatement is demonstrated in Chapter 9 with the example of COMINCO, which was forced to pioneer the field, and succeeded in making a profit.

The oil industry and the gas industry discovered quickly that their materials were suitable neither as lamp fuels nor in other applications unless sulfur was removed. Frasch (1912) demonstrated that desulfurization handsomely paid off in increased product value. Standard Oil and all other producers promptly followed suit. The oil industry has since learned that the recovered byproduct can be sold for profit as well, and the gas industry has been so successful at recovering and selling sulfur that it has outproduced Frasch companies, and shaken the international sulfur market until it has finally acquired an orderly marketing technique. The basic technology of desulfurizing gas and oil was not invented by these industries, but was adapted from the coal gasification industry described in Chapter 8. Ironically, this industry, which prepared the market for oil and gas, was abandoned; it is defunct, and its technology and know-how are now lost, at least in the United States. The reason for the regression in coal gas technology is contained in Table 1.5. Coal gas has 30% less heat content than natural gas; thus, it costs 30% more to ship. Furthermore, coal gas is expensive, as it requires large capital investment and the operating costs are substantial, while natural gas can be simply drawn from a well, and was originally a byproduct of oil. It is explained in Chapter 8 that the design and construction of a new coal gasification plant in the U.S. is estimated to cost about $1 billion and take 20 years. Thus coal, which was used 150 years ago to produce valuable coal tar products, will probably not be used to produce chemicals and fuel gas for some time to come. Instead, sulfur, which was commercially removed from coal gases between 1815 and 1900, has been emitted into the air during the last fifteen years.

Thus, it can be stated that the present public concern over sulfur pollution is due to regression in the technology of coal use. Unfortunately, the large international effort at abating pollution must be initially concentrated on recovering sulfur dioxide after fuel use. This procedure is purely palliative, and does not tackle the cause of the problem. Furthermore, the present method of recovering sulfur dioxide gas in the form of an aqueous slurry of calcium sulfite-sulfate merely transforms gaseous waste into liquid waste, and does so in a questionable manner, because the process itself wastes energy. The liquid wastes are estimated to take approximately 20 years to become harmless, far longer than sulfur dioxide gas, of which 90% settles within less than a day and within 10 miles of its emission source. This book discusses these problems from a chemical viewpoint and explains why such an undesirable stop-gap measure became necessary; it also discusses the problem of producing an economic, closed-loop sulfur removal cycle. The future situation will depend on the relation between application of presently available chemistry and government influence on the balance between consumption and conservation. Each of the following 14 chapters highlights one aspect of the problem, and Chapter 15 gives a short summary of the ways in which the present developments may affect future supply and demand of sulfur and energy.

Chapter 2

History

Sulfur has been known for thousands of years; the first reference to it being found in the Old Testament. Genesis XIX, 24. states: "Then the Lord caused to rain upon Sodom and upon Gomorrah brimstone and fire from the Lord out of heaven."

צֹעַר׃ הַשֶּׁמֶשׁ יָצָא עַל־הָאָרֶץ וְלוֹט בָּא צֹעֲרָה׃ וַיהוָה 23 24
הִמְטִיר עַל־סְדֹם וְעַל־עֲמֹרָה גָּפְרִית וָאֵשׁ מֵאֵת יְהוָה מִן
הַשָּׁמָיִם׃ וַיַּהֲפֹךְ אֶת־הֶעָרִים הָאֵל וְאֵת כָּל־הַכִּכָּר וְאֵת כה
כָּל־יֹשְׁבֵי הֶעָרִים וְצֶמַח הָאֲדָמָה׃ וַתַּבֵּט אִשְׁתּוֹ מֵאַחֲרָיו 26

For the alchemists, sulfur was one of the four basic ingredients of chemistry. Everything combustible was believed to contain sulfur and new substances were assumed to constitute new forms of sulfur compounds. Martin Rulando, in his *'Lexicon Alchemiae sive Dictionarium Alchemisticum,'* published in Frankfurt in 1612, listed 16 different types of sulfur, and stated: "Sulphur ist zweyerley: ein aeusserlicher und innerlicher...Der Schwefel oder Sulphur ist und bleibt ein Feindt aller Metalle...aber der Philosophische Sulphur ist lebensmachend." (There are two types of sulfur: internal and external...Sulfur gives life... but sulfur is and remains an enemy of all metals.) Many outstanding scientists, such as Watson at Cambridge, adhered to such philosophical thoughts, and to the phlogiston theory, until 1800, despite Boyle's statement on the front page of Volume II of his *'The Sceptical Chymist or Chymico-Physical Doubts and Paradoxes Touching the Spagyrist's Principles Commonly Called Hypostatica,'* published in London in 1661: "...Vulgar Spagyrists are wont to endeavour to evidence their salt, sulphur and mercury to be the true principle of things."

The first book fully devoted to sulfur was Michala Sedziwoja's *'Tractatus de Sulphure,'* (published in 1616 in Cologne), the title page of which is shown in Figure 2.1. This book started a tradition of sulfur research which is now again alive in modern Poland. In 1674 an article was published in the *Philosophical Transactions* 9, 66, with the title: "Concerning the Nature of the Salt in Brimstone, and Whence it is Derived." The author claimed:

> Brimstone consists of Mineral Sulphur, and an Acid Salt...Mineral Sulphur derives...from the bowels of the Earth. The Saline Principle of Sulphur...is the foundation of all saline substances in the Universe.

Ten years later, Martin Lyster, M.D., in the *Philosophical Transactions* 14 (1684) 512 wrote an article entitled, "Of the Nature of Earthquakes; more particularly of the Origine of the matter of them, from the Pyrites alone," and stated: "I take all pure Sulphur to have been produced by the fire...from Pyrites...which are the cause of Earthquakes, Thunder and Lightning."

In 1697, G. E. Stahl, in his famous *'Zymotechnica Fundamentalis,'* (published in Frankfurt) explained that sulfur consisted of phlogiston and an acid. In 1718 he published a treatise with the title, *'Gedanken und Bedenken von dem Sulphure,'* in which he discussed elemental sulfur, sulfuric acid, sulfurous acid and other oxyacids of sulfur. Between 1700 and 1710, Homberg published an extensive series of paper on sulfur. He concluded in his "Essays de Chimie," in *'Histoire de l'Academie Royale des Sciences,'* 1702, pg. 33-52 (published in 1720): "Le Souffre bitumineux est attribue aux acides de la seconde classe (vitriol et alum)...le souffre metallique est plus fixe que le souffre vegetal ou animal." (Bituminous sulfur belongs to the second type of acids, i.e. vitriols and alums; sulfur in metals is more strongly bonded than that in plant and animal compounds.)

Then in his "Sur l'Analise du Souffre Commun," in *'Histoire de l'Academie Royale des Sciences,'* 1703, pg. 47 (published in 1720), he reported:

> Le Souffre commun est visiblement un Mixte...Il a decouvert enfin le secret de les separer. Il a vu que c'etoit: un sel acide, une terre, une matiere grasse, bitumineuse, inflammable, et un peu de metal...Une Huile epaisse et rouge comme du sang, prend une consistence de Gomme en refroidissant...n'a point l'odeur desagreable.

TRACTATVS
DE SVLPHVRE
ALTERO NATVRÆ
PRINCIPIO,
AB AVTHORE EO, QVI ET
primum conscripsit principium.

Non nobis Domine non nobis, sed nomini tuo da gloriam.

ANGELVS DOCE MIHI IVS.
Vt possim dijudicare inter verum & falsum,

COLONIAE,
Apud Ioannem Crithium sub signo galli.
ANNO M. D. CXVI.

Fig. 2.1. Michala Sedziwoja's "Tractatus de Sulphure," Cologne, 1616.

F. August von Wasserberg
chemische
Abhandlung
vom
Schwefel.

Wien, bey Johann Paul Krauß 1798.

Fig. 2.2. F. A. Wasserberg's "Chemische Abhandlungen vom Schwefel," Vienna, 1798.

(Common sulfur is visibly a mixture...(I) have finally discovered the secret for separating its components. I discovered that it contains an acid salt, an earth, a matter which is greasy, bituminous, and inflammable, and some metal...flowers of sulfur can be purified by sublimation; (liquid sulfur is a) thick and red oil, similar to blood, which assumes a gluey consistency while cooling and it lacks any disagreeable odor.) The last observation shows that Homberg used purer sulfur than most chemists had available during the following 230 years.

In his "Sur les Matieres Sulphureuses et sur la facilite de les changer d'une espece de souffre en une autre," in *'Histoire de l'Academie Royale des Sciences,'* 1710, pg. 225-234 (published in 1730), Homberg finally states:

> ...j'ai suppose, que le Souffre principe n'est autre chose que la matiere de la lumiere...J.ai divise les Matieres sulphureuses en trois classes: le souffre bitumineux, le souffre dans une matiere aqueuse, et le souffre metallique.

(I have concluded that the principal sulfur is nothing else but the substance of light...I have divided sulfurous matter in three classes: bituminous sulfur, aqueous sulfur, and metallic sulfurs.) These observations formed the basis for the work of Lavoisier. His views are described in an article dealing with a meeting of the Histoire de l'Academie Royale des Sciences, "Sur la Combustion du Phosphore," 1777, pg. 25 (published in 1780). It reports:

> Les soufres ne font-ils donc en general qu'un acide prive de cet air, et l'acide un soufre a qui on l'a rendu? ou le phlogistique du soufre et du phosphore, pendant que leur acide se combinoit avec une partie de l'air atmospherique s'est-il combine avec le reste de cet air? M. Lavoisier penche pour la premiere opinion; mais un grand nombre de chimistes paroissent tenir a la seconde.

(Are the sulfurs nothing but an acid deprived of its air, and the acid a sulfur to whom one has returned it (air); or does the phlogiston of sulfur recombine with air when the acid combines with a portion of the atmospheric air? Mr. Lavoisier tends to hold the first opinion, but a large number of chemists seem to hold the second.)

In 1783, in the *'Histoire de l'Academie Royale des Sciences,'* (1786) Lavoisier published an article on the weight gain of sulfur and phosphor during their combustion and the causes of this weight gain, and observed, "...j'ai deduites que le soufre absorbe, en brulant, de l'air vital..." (I have shown that during combustion sulfur absorbs vital air.) Thus, he confirmed his long held belief that sulfur was an element. Despite the excellent international communication between scientists, R. Watson of Oxford University, in his *'Chemical Essays,'* (London, 1789) still reported, "The constituent parts of sulfur are two; an inflammable principle and an acid." Likewise, Wasserberg in his book, *'Chemische Abhandlung vom Schwefel,'* published in Vienna in 1788 (Figure 2.2) did not report any of the modern chemical views reflected in the contemporary French literature.

J. B. Richter, in *'Anfangsgründe der Stöchiometrie oder Messkunst chymischer Elemente,'* (Introduction to Stoichiometry or the Art of Measuring the Chemical Elements), published in Bresslau and Hirschberg in 1792, lists sulfur among the elements: "Chymische Elemente: Salze (Säuren und Alkalien), Erden (metallische, alkalische und Kieselerde), Andere: Luft, Phlogiston, Wasse, Elementarfeuer, und Schwefel." (Chemical elements: salts (acids and alkalies), earths (metallic, alkaline and silicon), and others: air, phlogiston, water, fire and sulfur.) However, his selection was obviously not very discriminating. Chaptal, a member of the French Academy, handled the matter in his authoritative, five volume work, *'Elemens de Chymie,'* (Montpellier and Paris, 1790) simply by adding sulfur to the traditional list of chemical elements in the Table of Contents: "Substances simples ou elementaires: du feu, du calorique et de la chaleur, de la lumiere, du soufre, du carbone." (Simple or elementary substances: fire, heat, light, sulfur, and carbon.)

In his famous book, *'Elemens de Chimie,'* which is still widely available in the various English translations prepared by R. Kerr (*Elements of Chemistry in a Systematic Order, Containing all the Modern Discoveries,* Translated by R. Kerr, Edinburgh, 1793), Lavoisier lists the following elements; Light, Caloric, Oxygen, Azot, Hydrogen, Sulphur, Phosphorus, Carbon, 17 Metals, Limes, Magnesia, Barytes, Argil, Silex and Strontite. His table of 29 binary combinations of sulfur with simple substances is shown in Figure 2.3.

Lavoisier's reasoning and fame did not discourage L. A. Emmerling in his *'Lehrbuch der Mineralogie,'* (Giessen, 1796) from stating that sulfur contained 40% phlogiston and 60% vitriolic acid; nor C. Girtanner in *Annales de Chimie et Physique* 23 (1799) 229, from claiming that sulfur contains oxygen and hydrogen; nor Mr. Curaudau, *Annales de Chimie* 67 (1808) 72, who reported on sulfur and its decomposition, from concluding, "De l'apparition de cette couleur (bleu) je souconnai que le soufre pouvoit

TABLE

OF THE BINARY COMBINATIONS OF SULPHUR WITH SIMPLE SUBSTANCES.

Simple Substances.	*Resulting Compounds.*	
	NEW NOMENCLATURE.	OLD NOMENCLATURE.
Caloric	Sulphuric gas	
Oxygen	Oxyd of sulphur	Soft sulphur.
	Sulphurous acid	Sulphureous acid.
	Sulphuric acid	Vitriolic acid.
Hydrogen	Sulphuret of hydrogen	Unknown Combinations.
Azot	azot	
Phosphorus	phosphorus	
Carbon	carbon	
Antimony	antimony	Crude antimony.
Silver	silver	
Arsenic	arsenic	Orpiment, realgar.
Bismuth	bismuth	
Cobalt	cobalt	
Copper	copper	Copper pyrites.
Tin	tin	
Iron	iron	Iron pyrites.
Manganese	manganese	
Mercury	mercury	Ethiops mineral, cinnabar.
Molybdena	molybdena	
Nickel	nickel	
Gold	gold	
Platina	platina	
Lead	lead	Galena
Tungstein	tungstein	
Zinc	zinc	Blende.
Potash	potash	Alkaline liver of sulphur with fixed vegetable alkali.
Soda	soda	Alkaline liver of sulphur with fixed mineral alkali.
Ammoniac	ammoniac	Volatile liver of sulphur, smoking liquor of Boyle.
Lime	lime	Calcareous liver of sulphur.
Magnesia	magnesia	Magnesian liver of sulphur.
Barytes	barytes	Barytic liv. of sulph.
Argil	argil	Yet unknown.

2 I

Fig. 2.3. 29 Binary combinations of sulfur from "Elements of Chemistry in a Systematic Order, Containing all the Modern Discoveries," by A. Lavoisier, English translation by R. Kerr, Edinburgh, 1792.

bien etre une combinaison de carbone et d'hydrogene." (From the blue color I suspect that sulfur might well be a combination of carbon and hydrogen.)

In the meantime, on May 8, 1794, at the age of 50, Lavoisier had been executed by the revolutionairies, who misspelled his name on the death certificate. His name vanished from the list of founders on the front page of the *Annales de Chimie* and remained unmentioned in the French literature for almost 100 years. Most of his colleagues, however, fully recognized the value of his scientific findings and adopted them. For example, in 1804, in *'A General System of Chemical Knowledge and its Application to the Phenomena of Nature and Art,'* (translated by W. Nicholson, London), A. F. Fourcroy lists simple bodies: light, caloric, oxygen, air, azote, hydrogen, carbon, phosphorus, sulfur, diamond and metals. Furthermore, the previously mentioned report of Curaudau led to a detailed formal inquiry. In 1808 Vauquelin and Berthollet in *Annales de Chimie* 67, 151, published a "Rapport sur un Memoire presente par M. Curaudau," and summarized their findings: "Nous avons repete cette operation...par laquelle M. Curaudau preten avoir decompose le soufre... ses resultats n'offrent rien qui puisse appuyer ses pretentions." (We have repeated the operations by which Mr. Curaudau claims to have decomposed sulfur; the results contain nothing whatsoever that could support his claims.) And after a visit to his laboratory they stated:

> ...en examinant les matieres (de M. Curaudau) nous reconnumes un melange de rapure de corne, des morceaux d'ivoire, et de different autres substances...Nous avons lieu de presumer que M. Curaudau qui reconnu son erreur ne sera plus tente de revenir sur les questions tres-difficiles...

(While examining Mr. Curaudau's materials we found a mixture of corn shells, ivory bits and other items) and concluded (we have reason to believe that Mr. Curaudau, who has recognized his error, will not be tempted in the future to reconsider these very difficult questions.)

Although it was now widely accepted that "sulfur is an element," as J. Dalton stated in his *'New System of Chemical Philosophy,"* (Vol. 1, pg. 239, 1808), some famous people still held out: Among them was the brilliant Humphrey Davy. In the *Philosophical Transactions of the Royal Society of London* (pg. 39, 1809), "The Bakerian Lecture; an Account of some new analytical Researches on the Nature of certain Bodies, particularly the Alkalies, Phosphorus, Sulphur, Carbonaceous Matter, and the Acids hitherto undecompounded; with some general Observations on

Chemical Theory," Davy stated:

> In considering the analytical powers of the Voltaic apparatus, it occurred to me that though sulphur, from its being a non-conductor, could not yield its elements to the electrical attractions...(it) might possibly (experience) some alteration...A most brilliant spark, which appeared orange coloured through the sulphur, was produced... and a globule equal to about the tenth of an inch in diameter was obtained, which, when examined, was found to be sulphuretted hydrogene. The experiments upon the union of sulphur and potassium...prove...that sulphuretted hydrogene is evolved. The existence of hydrogene in sulphur is fully proved, and we have no right to consider a substance, which can be produced from it in such large quantities, merely as an accidental ingredient. From the general tenour of these various facts, it will not be, I trust, unreasonable to assume, that sulphur, in its common state, is a compound of small quantities of oxygene and hydrogene with a large quantity of a basis that produces the acids of sulfur in combustion.

In 1809 Davy presented supplementary experiments in order to support his claims. Gay-Lussac and Thenard immediately picked up the subject in France, and after long, detailed experimental study, concluded in 1810 in the *Annales de Chimie* 73, pg. 229, "Memoire en reponse aux recherches analytiques de M. Davy, sur la nature du soufre et du phosphore,"

> Les experiences precendentes prouvent...que le soufre lui-meme n'en contient pas...Que le soufre et le phosphore ne contienent point d'oxigene; qu'ainsi on doit toujours continuer a regarder comme simples ou indecomposes ces deux combustibles que M. Davy veut assimiler pour la nature ou la composition, aux substances vegetables.

(The preceding experiments prove that sulfur contains none of the above mentioned (other elements); sulfur and phosphorus contain no oxygen; these two combustibles must be regarded as simple, indecomposeable elements; and they are not among the vegetable substances to which Mr. Davy wants to assign them.)

Shortly thereafter, Gay-Lussac and Thenard published another paper, *Annales de Chimie* 75 (1810) 290, "Observations sur les trois precedens Memoires de M. Davy," and wrote, "...nous avons repete plus de cinquantes foi nos experiences sur le soufre, le gaz hydrogene sulfure et le potassium...Nous assurons de nouveau que ces resultats sont certains." (We repeated our experiments more than 50 times, and ascertain anew that our results are correct.) This caused Davy in *Philosophical Transactions of the Royal Society of London*, 1810, pg. 231, to retract. In "Researches on the oxymuriatic Acid, its Nature and Combinations; etc." he writes, "...new and more minute enquiries have enabled me to correct: The able researches of Dr. Thomson have shewn that sulphur, in its usual state, contains small quantities of acid matter...All the late experiments induce me to suspect a notable proportion of oxygene in Sicilian sulphur."

In 1811 J. Berzelius entered the dispute, and in the process discovered his second law of stoichiometry. In *Annalen der Physik* 37, pg. 208, under the title, "Schreiben über einige Gegenstände, welche zwischen Davy und den H.H. Gay-Lussac und Thenard streitig sind, und über ein zweites neues Gesetz, welches er im Verfolge seinerUntersuchungen aufgefunden hat," (Report on matters which are the content of a dispute between Davy and Gay-Lussac, and on a second novel law which was discovered during these researches), he writes:

> Die Abhandlungen der HH Thenard und Gay-Lussac über Davy's Untersuchungen ...sind sehr interessant; noch haben sie mich aber nicht uberzeugt...Die Versuche... sind ein wenig dunkel, und scheinen mir nicht ganz richtig, habe ich sie anders nicht misverstanden. Sie sehen, dass ich diese so schwierige Sache mit Kühnheit angegriffen habe; die Zeit wird uns belehren, was in diesen Ansichten wahr ist, und worin ich vielleicht geirrt haben kann.

(The treatise of Thenard and Gay-Lussac is very interesting, but it does not yet convince me. The experiments appear a little dark, and do not seem to me quite right, if I did not misunderstand them...You see that I have tackled this difficult matter with courage. Time will show what is correct, and in what matters I might have erred.)

In his next paper, Berzelius summarized the entire history. In his "Versuch, die bestimmten und einfachen Verhältnisse aufzufinden, nach welchen die Bestandtheile der unorganischen Natur mit einander verbunden sind," in *Annalen der Physik* 37 (1811) 249, he says:

> Mehrere Chemiker besonders Klaproth, Bucholz und Richter haben sich bemüht die Menge des Schwefels in der Schwefelsäure zu bestimmen. Davy hat mit grossem Scharfsinne so viele Umstände zusammengestellt...dass Schwefel und Phosphor eigene, bisher unbekannte, metallische Körper, mit geringen Mengen Wasserstoff und Sauerstoff verbunden enthalte...dass sie nicht ganz unwahrscheinlich erscheinen. Die von angestellten Versuche entsprechen indessen der Vermuthung Davy's nicht...nie habe ich irgendeine Spur von Schwefel-Wasserstoffgas oder von Wasserdünsten entdecken können...Der von Davy in dem Schwefel aufgefundene Sauerstoff und der Wasserstoff rührten daher von Feuchtigkeit her.

(Many chemists, especially Klaproth, Bucholz and Richter tried to determine the sulfur content of sulfuric acid. Davy, with great brilliancy, has collected circumstantial evidence which made it appear not impossible that sulfur and phosphorous might contain some unknown, metallic compounds and traces of hydrogen and oxygen. However, my own experiments do not support Davy's suspicions; I have never found a trace of water or H_2S. Thus, Davy's oxygen and hydrogen sulfide must have come from moisture in his samples.) Since Davy reported a boiling point of 280°C; his sulfur must have contained more than merely water. In contrast, the French value for the boiling point, reported by Dumas, was 430°; an amazingly good value for the then very difficult measurement of high temperature.

Despite the efforts of J. W. Döbereimer, who claimed that sulfur contained metal and hydrogen (Schweigger's Journal 13 (1815) 76), and Ch. F. Schönbein, who argued that sulfur is dimorphous and thus could not be an element (Berichte der Naturforchenden Gesellschaft, Basel (1835) 60), the scientific world accepted sulfur as an element. In 1835 in his *Theorie des Proportions Chimiques et Tables Synoptique des Poids Atomiques*, (2nd Ed., Paris), J. J. Berzelius lists the atomic weight of sulfur as 201.17 on the scale O = 100, and its atomic weight as 16.12 on the scale H = 1. But the impurity problem plagued sulfur chemists for another century, until Bacon and Fanelli wrote their paper on the viscosity of liquid sulfur (1943).

Several sources contain good summaries of the status of sulfur chemistry during the above-mentioned time. In addition to the books of Boyle, Richter, Lavoisier, Fourcroy, Chaptal, and Watson, the books, *Tractatus de Sulphure* by Sedziwoja (1616) and *Chemische Abhandlung vom Schwefel* by Wasserberg (1788) are entirely devoted to sulfur. The first edition of Gmelin was translated into English in 1848 as the first project of the newly founded Chemical Society. Highlights of the history of sulfur were reviewed by Kopp (1845), and the progress of sulfur chemistry in the last 150 years has been documented with several hundred key references by Shaskey (1976).

Sulfur Production

Figure 2.4 shows how sulfur was mined and refined from ores during Agricola's time (1556). These methods survived for several hundred years, in combination with the furnaces developed in Sicily. These are described in Section7. The revolution of the sulfur industry caused by Frasch is described in Section 7. His own drawing of the original patent is shown in Figure 2.5.

Fig. 2.4. Sulfur production, G. Agricola, "De re metallica," 1556.

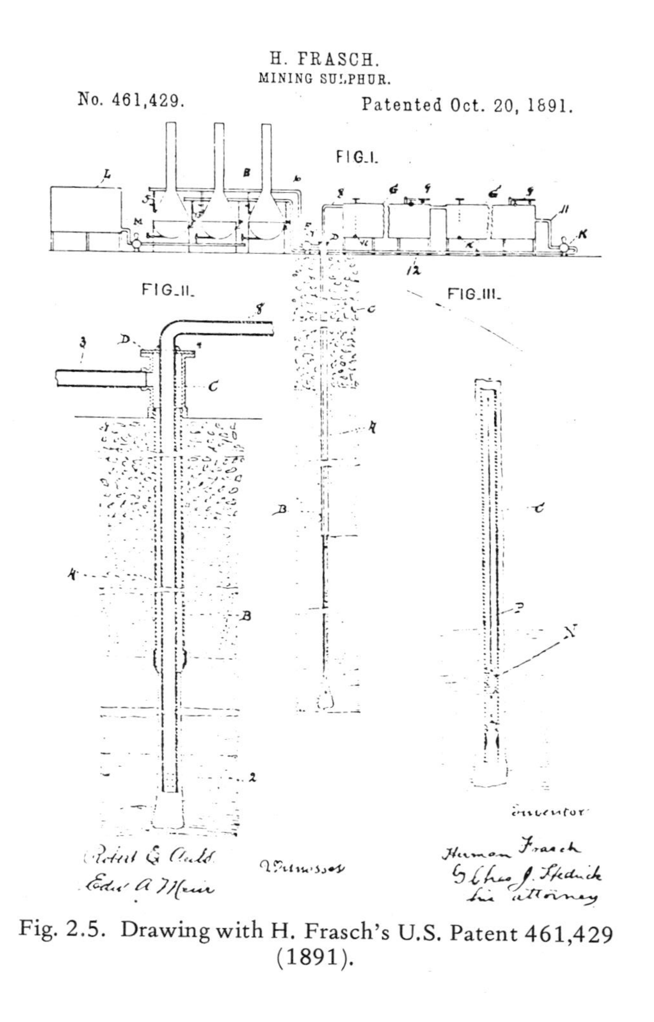

Fig. 2.5. Drawing with H. Frasch's U.S. Patent 461,429 (1891).

Elemental Sulfur

As early as 3000 years ago, Dioskotides distinguished natural and melted sulfur. Milk of sulfur was prepared by Geber; and Valentinus described flowers of sulfur around 1500. Mitscherlich recognized two crystalline forms in 1823 which he called alpha and beta sulfur. Early work was reviewed by Bertholet. Engel and Aten prepared eta and rho sulfur, and though the Greek alphabet was exhausted in labelling new or alleged allotropes for over an hundred years, the existence of none of these was confirmed. However, during the last 30 years Engel and Aten were vindicated, and during the last 15 years, Schmidt has synthesized 12 new allotropes using old reagents, new methods, and ingenious chemical thinking.

Liquid Sulfur

The properties of liquid sulfur are astonishingly well described in a paper by Dumas published in 1834. The history of the natural *vs* the ideal melting point makes a long, but fascinating story (Meyer, 1976). The properties of the viscous liquid were unambiguously described by Eisenberg and his teacher, the late Arthur Tobolsky, who had entered the field under the guidance of Henry Eyring. It has since been found that hot liquid sulfur is a very reactive mixture, which contains S_3, S_4 and other small molecules.

Sulfur Vapor

The properties of the dark brown sulfur vapor were early recognized to be peculiar. Dumas, Mitscherlich in 1833, and Deville and Troost in 1843 conducted some most skillful experiments which indicated that the vapor contained molecular weight 4.5 to 7.6 depending on conditions. Salet was the first to study the spectrum of S_2 and Louis D'Or was the first to propose S_4. Both molecules and other vapor components are fascinating species which were studied by Goldfinger, Marquart, Berkowitz, and Chupka.

Binary Compounds

Lavoisier listed 23 simple combinations in his book in 1789. We now know the binary compounds of 84 elements with sulfur (Meyer, 1973). Some 350 key references documenting the history of binary compounds of sulfur have been collected by Shaskey (1976). The oldest commercial sulfur compounds were the *polysulfides* called 'liver of sulfur;' Scheele called them 'lac of sulfur.' The first pure sulfanes were prepared by Block in 1909. Today, following the work of Feher, they are called sulfanes, in analogy to alkanes. Schmidbaur's NMR work made the analysis of the sulfane mixtures easier, and thus made possible the exciting chemistry described in Chapter 3.

Sulfuric Acid

The commercial preparation of sulfuric acid goes back to B. Valentinus in the 15th Century. Lemery noted in his *Cours de Chymie* in 1675 that iron vitriol gave "clean" acid, while copper vitriol produced a pungent vapor. The commercial production was described by Bernhardt (1755) in his *Chymischen Versuchen und Erfahrungen*. The acid was frequently prepared by burning sulfur in the presence of water in glass retorts by apothecaries in small batches, as needed. LeFebre and Lemery in Paris in 1666 increased the yield substantially be adding KNO_3; and in 1746 in Richmond, Joshua Ward started a factory using a glass balloon. Roebuck in 1746 and Garbett in 1749 in Birmingham built lead chambers which could product about 5 tons of acid per year. In 1774 Hoelker introduced the lead chamber in Rouen, France, where de la Folie injected steam. Clement and Desormes started to blow air into the reaction vessel in 1793. Rollox started the first continuous factory in 1807 in Glasgow. In Germany, fuming concentrated acid was produced intermittently in Pilsen in 1526, by distillation of vitriol prepared from iron and aluminum sulfate, as practiced in Nordhausen until 1858. In 1810 the first lead chambers, with dimensions of 8 x 8 m, were built in Schwemsel bei Leipzig. Sulfur and saltpeter in the ratio 5:1 were reacted in 50 lb batches, using injection of steam. The nitric oxides were discharged into the air. That the exhaust gases were accepted as an inevitable part of industry and progress is evidenced in the case from Modena, Italy, cited in Chapter 9. In 1827 Gay-Lussac introduced the use of a wash tower. The first such tower was installed in Chauncy by St. Gobin.

Justus Liebig described the uses and importance of sulfuric acid in a public letter "written for the purpose of exciting the attention of governments and an enlightened public, to the necessity of establishing Schools of Chemistry." He personally supervised the English translation of the following letter which was printed in 1843 in Philadelphia for his American Audience:

In order to prepare the soda of commerce (which is the carbonate) from common salt, it is first converted into Glauber's salt (sulphate of soda.) For this purpose 80 pounds weight of concentrated sulphuric acid (oil of vitriol) are required to 100 pounds of common salt. The duty upon salt checked, for a short time, the full advantage of this discovery; but when the government repealed the duty, and its price was reduced to its minimum, the cost of soda depended upon that of sulphuric acid.

The demand for sulphuric acid now increased to an immense extent; and, to supply it, capital was embarked abundantly, as it afforded an excellent remuneration. The origin and formation of sulphuric acid was studied most carefully; and from year to year, better, simpler, and cheaper methods for making it were discovered. With every improvement in the mode of manufacture, its price fell; and its sale increased in an equal ratio.

Sulphuric acid is now manufactured in leaden chambers, of such magnitude that they would contain the whole of an ordinary sized house. As regards the process and the apparatus, this manufacture has reached its acme–scarcely is either susceptible of improvement. The leaden plates of which the chambers are constructed, requiring to be joined together with lead, (since tin or solder would be acted on by the acid,) this process was, until lately, as expensive as the plates themselves; but now, by means of the oxyhydrogen blow-pipe, the plates are cemented together at their edges, by mere fusion, without the intervention of any kind of solder.

And then, as to the process; according to theory, 100 pounds weight of sulphur ought to produce 306 pounds of sulphuric acid; in practice, 300 pounds are actually obtained; the amount of loss is therefore too insignificant for consideration.

Again; saltpetre being indispensable in making sulphuric acid, the commercial value of that salt had formerly an important influence upon its price. It is true that 100 pounds of saltpetre only are required to 1000 pounds of sulphur; but its cost was four times greater than an equal weight of the latter.

Travellers had observed, near the small seaport of Yquiqui, in the district of Atacama, in Peru, an efflorescence covering the ground over extensive districts. This was found to consist principally of nitrate of soda. Advantage was quickly taken of this discovery. The quantity of this valuable salt proved to be inexhaustible, as it exists in beds extending over more than 200 square miles. It was brought to England at less than half the freight of the East India saltpetre, (nitrate of potassa;) and as, in the chemical manufacture, neither the potash nor the soda were required, but only the nitric acid, in combination with the alkali, the soda-saltpetre of South America soon supplanted the potash-nitre of the East. The manufacture of sulphuric acid received a new impulse; its price was much diminished without injury to the manufacturer; and, with the exception of fluctuations, caused by the impediments thrown in the way of the export of sulphur from Sicily, it soon became reduced to a minimum, and remained stationary.

Potash-saltpetre is now only employed in the manufacture of gunpowder; it is no longer in demand for other purposes; and thus, if government effect a saving of many hundred thousand pounds annually in gunpowder, this economy must be attributed to the increased manufacture of sulphuric acid.

We may form an idea of the amount of sulphuric acid consumed, when we find that 50,000 pounds weight are made by a small manufactory, and from 200,000 to 600,000 pounds by a large one, annually. This manufacture causes immense sums to flow annually into Sicily. It has introduced industry and wealth into the arid and desolate districts of Atacama. It has enabled us to obtain platina from its ores at a moderate and yet remunerating price; since the vats employed for concentrating this acid are constructed of this metal, and cost from 1000*l.* to 2000*l.* sterling. It leads to frequent improvements in the manufacture of glass, which continually becomes cheaper and more beautiful. It enables us to return to our fields all their potash—a most valuable and important manure—in the form of ashes, by substituting soda in the manufacture of glass and soap.

It is impossible to trace, within the compass of a letter, all the ramifications of this tissue of changes and improvements resulting from one chemical manufacture; but I must still claim your attention to a few more of its most important and immediate results. I have already told you, that in the manufacture of soda from culinary salt, it is first converted into sulphate of soda. In this first part of the process, the action of sulphuric acid produces primary muriatic acid to the extent of one and a half the amount of the sulphuric acid employed. At first, the profit upon the soda was so great that no one took the trouble to collect the muriatic acid; indeed it had no commercial value. A profitable application of it was, however, soon discovered: it is a compound of chlorine, and this substance may be obtained from it purer than from any other source. The bleaching power of chlorine has long been known; but it was only employed upon a large scale after it was obtained from residuary muriatic acid, and it was found that in combination with lime it could be transported to distances without inconvenience. Thenceforth it was used for bleaching cotton; and, but for this new bleaching process, it would scarcely have been possible for the cotton manufacture of Great Britain to have attained its present enormous extent—it could not have competed in price with France and Germany. In the old process for bleaching, every piece must be exposed to the air and light during several weeks in the summer, and kept continually moist by manual labour. For this purpose, meadow land, suitably situated, was essential. But a single establishment near Glasgow bleaches 1,400 pieces of cotton daily, throughout the year. What an enormous capital would be required to purchase land for this purpose! How greatly would it increase the cost of bleaching to pay interest upon this capital, or to hire so much land in England! This expense would scarcely have been felt in Germany. Beside the diminished expense, the cotton stuffs bleached with chlorine suffer less in the hands of skilful workmen than those bleached in the sun; and already the peasantry in some parts of Germany have adopted it, and find it advantageous.

Another use to which cheap muriatic acid is applied, is the manufacture of glue from bones. Bone contains from 30 to 36 per cent of earthy matter—chiefly phosphate of lime, and the remainder is gelatine. When bones are digested in muriatic acid, they become transparent and flexible like leather, the earthy matter is dissolved, and after the acid is all carefully washed away, pieces of glue of the same shape as the bones remain, which are soluble in hot water and adapted to all the purposes of ordinary glue, without further preparation.

Another important application of sulphuric acid may be adduced, namely, to the refining of silver and the separation of gold, which is always present in some proportion in native silver. Silver, as it is usually obtained from mines in Europe, contains, in 16 ounces, 6 to 8 ounces of copper. When used by the silversmith, or in coining, 16 ounces must contain in Germany 13 ounces of silver, in England about 14½. But this alloy is always made artificially, by mixing pure silver with the due proportion of the copper; and for this purpose the silver must be obtained pure by the refiner. This he formerly effected by amalgamation, or by roasting it with lead; and the cost of this process was about 2*l.* for every hundred weight of silver. In the silver so prepared, about 1/1200 to 1/2000th part of gold remained. To effect the separation of this by nitric hydrochloric acid was more expensive than the value of the gold; it was therefore left in utensils, or circulated in coin, valueless. The copper, too, of the native silver was of no use whatever. But the 1/1000th part of gold, being about one and one-half per cent of the value of the silver, now covers the cost of refining, and affords an adequate profit to the refiner; so that he effects the separation of the copper, and returns to his employer the whole amount of the pure silver, as well as the copper, without demanding any payment: he is amply remunerated by that minute portion of gold. The new process of refining is a most beautiful chemical operation. The granulated metal is boiled in concentrated sulphuric acid, which dissolves both the silver and the copper, leaving the gold nearly pure, in the form of a black powder. The solution is then placed in a leaden vessel containing metallic copper; this is gradually dissolved, and the silver precipitated in a pure metallic state. The sulphate of copper is also a valuable product, being employed in the manufacture of green and blue pigments.

Other immediate results of the economical production of sulphuric acid, are the general employment of phosphorus matches, and of stearine candles—that beautiful substitute for tallow and wax. Twenty-five years ago, the present prices and extensive applications of sulphuric and muriatic acids, of soda, phosphorus, &c., would have been considered utterly impossible. Who is able to foresee what new and unthought-of chemical productions, ministering to the service and comforts of mankind, the next twenty-five years may produce?

After these remarks *you will perceive that it is no exaggeration to say, we may fairly judge of the commercial prosperity of a country from the amount of sulphuric acid it consumes.* Reflecting upon the important influence which the price of sulphur exercises upon the cost of production of bleached and printed cotton stuffs, soap, glass, &c., and remembering that Great Britain supplies America, Spain, Portugal, and the East, with these, exchanging them for raw cotton, silk, wine, raisins, indigo, &c., &c., we can understand why the English government should have resolved to resort to war with Naples, in order to abolish the sulphur monopoly, which the latter power attempted recently to establish. Nothing could be more opposed to the true interests of Sicily than such a monopoly; indeed, had it been maintained a few years, it is highly probable that sulphur, the source of her wealth, would have been rendered perfectly valueless to her. Science and industry form a power to which it is dangerous to present impediments. It was not difficult to perceive that the issue would be the entire cessation of the exportation of sulphur from Sicily. In the short period the sulphur monopoly lasted, fifteen patents were taken out for methods to obtain back the sulphuric acid used in making soda. Admitting that these fifteen experiments were not perfectly successful, there can be no doubt it would ere long have been accomplished. But, then, in gypsum (sulphate of lime) and in heavy-spar (sulphate of barytes) we possess mountains of sulphuric acid; in galena (sulphate of lead) and in iron pyrites we have no less abundance of sulphur. The problem is, how to separate the sulphuric acid, or the sulphur, from these native stores. Hundreds of thousands of pounds weight of sulphuric acid were prepared from iron pyrites, while the high price of sulphur consequent upon the monopoly lasted. We should probably ere long have triumphed over all difficulties, and have separated it from gypsum. The impulse has been given, the possibility of the process proved, and it may happen in a few years that the inconsiderate financial speculation of Naples may deprive her of that lucrative commerce. One country purchases only from absolute necessity from another, which excludes her own productions from her markets. Precisely analogous is the combination of workmen against their employers, which has led to the construction of many admirable machines for superseding manual labour. In commerce and industry every imprudence carries with it its own punishment; every oppression immediately and sensibly recoils upon the head of those from whom it emanates.

This letter touches more than one subject which is still of acute interest. Liebig's work induced wide and lasting interest in agricultural science, and in the use of fertilizer. This increased the demand for sulfuric acid further than even Liebig had foreseen. Therefore, more and bigger sulfuric acid plants had to be built. From this experience, new technology developed. In 1859 Glover introduced in England the contact tower. The first commercial effort to recover acid from smelter gases is credited to Oker, in Freiberg. A yet more dramatic increase in acid production began in 1864 with the manufacture of superphosphate. Ever since, sulfuric acid has been the leading industrial chemical.

Chapter 3

Properties

A. ELEMENTAL SULFUR

Pure elemental sulfur is a bright yellow solid and has no odor. It is sold in many different commercial forms and grades which differ in physical and chemical form, and in purity. The purity of commercial sulfur is often described by its color. Bright sulfur is 99+% pure. Frasch and Claus sulfur contain traces of organic compounds which, upon heating, slowly react, evolving hydrogen sulfide. Such reactions can take place in transit, for example on liquid sulfur barges. Mineral sulfur is usually less pure. It often contains As, Se, and oil fractions (Churbanov, 1967). For shipping solid sulfur, traces of additives are sometimes mixed with sulfur to improve the mechanical handling and reduce dust. These prilled, flaked, granulated, slated, and other grades often contain organic amines. For research purposes, sulfur is now commercially available in 99.999% (5N) purity. Ultra-pure sulfur is obtained by zone refining (Pavlov, 1969) and other methods (Suzuki, 1974; Bor, 1975), and residual impurities can be accurately determined (Grünert, 1971; Tuller, 1970), Chapter 4C.

Today, 90% and better enriched isotopes 32-S and 34-S are available in gram quantities. They are prepared by thermo-diffusion of liquid carbon disulfide, and by the exchange between sulfur dioxide and bisulfite (Rutherford, 1975; Stachewski, 1975). It is possible that in the future IR laser enriched isotopes of SF_6 will provide yet cheaper sulfur-34 (Hancock, 1976; Ambartsumyan, 1975; Grinenko, 1975; Moiseyer, 1976). These isotopes will make it possible to study the reaction path and kinetics of many reactions which form the key to new large scale applications of sulfur and its compounds.

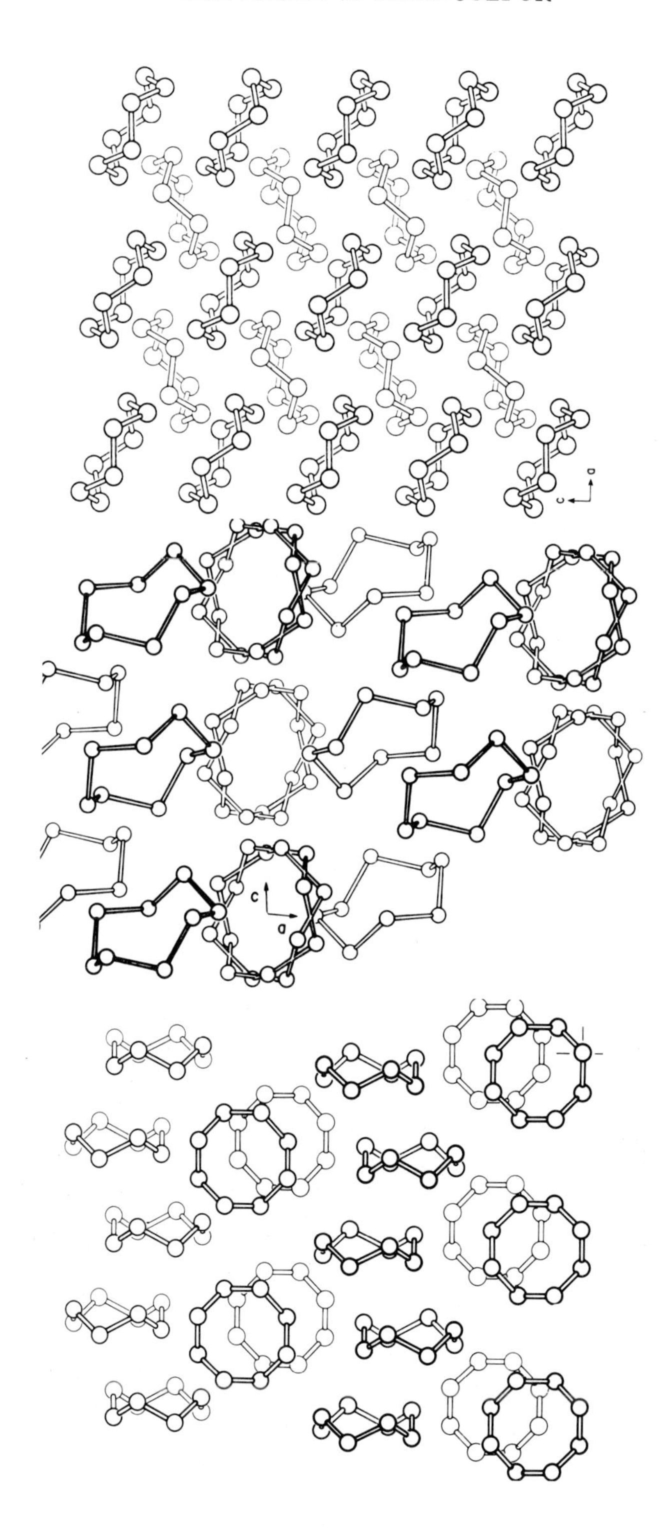

Fig. 3.1. a) Orthorhombic alpha-sulfur, b) monoclinic beta-sulfur, and c) monoclinic gamma-sulfur. After Donohue (1965).

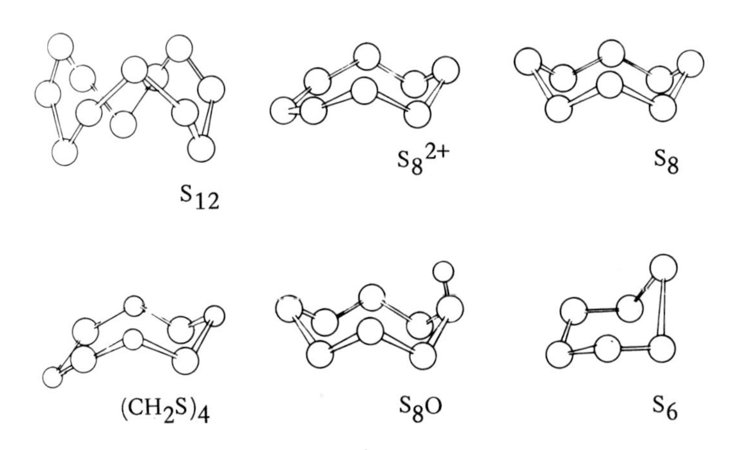

Fig. 3.2. The structure of S_8, S_6, S_{12}, S_8^{2+}, and S_8O.

1. Solid Elemental Sulfur

Sulfur can exist in at least twelve different forms at room temperature. These and other allotropes and polymorphs have been recently reviewd (Meyer, 1976). Only three are presently of practical interest: Orthorhombic sulfur is the most stable form at room temperature; polymeric sulfur, obtained by quenching hot liquid sulfur, is insoluble in carbon disulfide; and monoclinic sulfur forms above 96°K from orthorhombic sulfur and is the stable form at the melting point. The structure of these three forms is now well established (Donohue, 1974). It is shown in Fig. 3.1. The orthorhombic and monoclinic forms contain crown shaped S_8 molecules, Fig. 3.2, which are stacked in a complex array, involving 'crank-case' and other off-set positions. The stable room temperature form was conclusively established by Abrahams (1955) and Caron (1965). The high temperature monoclinic form was recently established by Watanabe (1974) and Templeton (1976). In contrast, polymeric sulfer contains long helices (Zuckerman, 1963; Geller, 1969; Lind, 1969) in which 10 atoms form three full turns. These helices can have random orientation. In Crystex, the polymer is stabilized with 0.5% additive. The X-ray structure shows that upon stretching of freshly quenched polymeric sulfur, long fibres can be formed (Trillat, 1931; Meyer, 1934). The

Table 3.1
THE STRUCTURES OF NINE SULFUR ALLOTROPES

Molecule	Space group	Density (gm/cm³)	Decomp. or melting pt. (°C)	Unit cell (Å) a	b	c	Angle β (deg)	Space cell[a]
S_6	$\bar{R}3-C_{3i}^{2}$	2.209	50–60	10.818	(c/a = 0.3956)	4.280 ± 0.001	—	3; 18
S_7	?	2.090	39	21.77	20.97	6.09	—	16; 112
$S_{8-\alpha}$	$Fddd-D_{2h}^{24}$	2.069	94(112)	10.4646	12.8660	24.4860	—	16; 128
$S_{8-\beta}$	$P2_1/a-C_{2h}^{5}$	1.94	133	10.778	10.844	10.924	95.80	6; 48
$S_{8-\gamma}$	$P_2/c-C_{2h}^{4}$	2.19	~20	8.442	13.025	9.356	124.98	4; 32
S_{12}	$Pnnm-D_{2h}^{12}$	2.036	148	4.730	9.104	14.574	—	2; 24
S_{18}	$P2_12_12_1-D_2^{4}$	2.090	128	21.152	11.441	7.581	—	4; 72
S_{20}	$Pbcn$	2.016	124–125	18.580	13.181	8.600	—	4; 80
S_∞	$Ccm2_1-C_{2v}^{12}$	2.01	104	13.8	4 × 8.10	9.25	85.3	160[b]
S_8O	$Pca2-C_{2v}^{5}$	2.13	20–78	13.197	7.973	8.096	—	4; 32
S_7TeO_2	$Pmnb-D_{2h}^{16}$	2.65	—	8.82	9.01	13.28	—	4; 28

[a] First number = number of molecules; second number = number of atoms in unit cell.
[b] Ten atoms for three turns.

structural data for these common sulfur forms and for other well established forms are listed in Table 3.1. The mechanical properties of sulfur depend on the allotropic state of the element. Unfortunately, orthorhombic sulfur is quite frangible. Except for rare special applications, for example practice skeet targets (Dale, U.S. P. 3,840,232, 1974), this is a serious draw-back which prevents many otherwise attractive uses of pure, unmodified sulfur. Thermal and mechanical properties have been recently analyzed by Hyne (1976), who studied the aging of samples containing mixtures of allotropes. Freshly quenched, pure sulfur is extremely plastic; it can be stretched to twenty times its original length (Meyer, 1934). The conversion to brittle orthorhombic sulfur occurs within less than a month. This conversion is enhanced by mechanical stress, but can be delayed by 'working' the polymer. The mechanical properties of sulfur can be drastically altered by the addition of 1-3% plasticizer or modifier. With 10% 'impurities', regular polymers can be synthesized, as described in Chapter 13. These properties of mixed allotropes of pure sulfur and of modified sulfur are still poorly explored, and the tremendous potential for basic research and large scale application is not widely recognized.

The mechanical properties have been reviewed and studied by Dale; their values are summarized in Table 3.2. ASTM tests designed for concrete were used. The consistently higher tensile strength of quick quenched samples and of modified sulfur is due to the presence of polymeric chains. The Shore B-2 hardness of solid sulfur is 90 +2. The properties of aggregates, foams, coatings, and other sulfur products are

Table 3.2
MECHANICAL PROPERTIES OF SULFUR

Property	Bulk	Cast	Fibrous
Tensile strength (psi)	180-280	160	
quenched		49-620	
Compression (psi)		3,300	
Modulus of rupture (psi)		200	2,600-14,000
Notch impact strength			
(lb/inch)	17-36		
(kg/cm)	3-6.5		
Modulus of elasticity (psi)			28-8,500

After Dale (1965, 1967, 1974), Crow (1970), Rennie (1971).

discussed in Chapter 14. The linear thermal expansion coefficient of orthorhombic sulfur is 4.6 x 10^{-5} at 10°C, 7.4 at 25°C, and increases rapidly from 20 x 10^{-5} at 80°C to 100 x 10^{-5} at 110°C. Looman (1969) has measured the longitudinal thermal expansion of fibrous sulfur. The linear expansion is 20 x $10^{-6}(K^{-1})$ while the axial expansion is about 80 x $10^{-6}(K^{-1})$. The compressibility is 13 x 10^{-6}. The ignition temperature of sulfur in air at standard pressure lies between 478 and 511°F (248-261°C). The heat of combustion is $1/8\ S_8 + O_2 = SO_2 + 72$ kcal/mole (3982.2 Btu/lb at 77°F). The density of sulfur is 2.07 g/cm^3 (129 lb/cft) for orthorhombic, 1.96 g/cm^3 (122 lb/cft) for monoclinic, and 1.92 g/cm^3 (120 lb/cft) for polymeric sulfur.

On certain surfaces, for example on metal edges, unusual sulfur species can be formed. Cocke (1976) found molecules with 2 to 22 atoms on a tungsten needle; Davis (1974) found similar species.

Among the allotropic sulfur forms which are still merely of academic interest is an orange-red rhombohedral material, first prepared by Engel (1891) and Aten (1914), which consists of six-membered rings. It can be stored at room temperature for a few days, but it immediately begins a slow conversion to the more stable orthorhombic form. Table 3.1 also lists solids which consist of rings with 7, 12, 18, and 20 atoms. It is amazing that an element can exist in such a variety of forms, and it is startling that so many of these rings can be crystallized as pure substances. Progress in this field is largely due to the research of Schmidt and his coworkers (1965-1972), who synthesized most of these substances by continuous combination of dilute reagents in dry ether at -40°C:

$$H_2S_n \quad + \quad S_mCl_2 \quad \rightarrow \quad \text{cyclo-}S_{n+m} \quad + \quad 2HCl$$

For this purpose, they used sulfanes, H_2S_n, and chlorosulfanes, S_mCl_2, prepared by Feher's method (1956, 1957). For the preparation of pure substances it was essential that the purity of the reagents could be checked with a method developed by Schmidbaur (1964). Some of the allotropes in Table 3.1 were prepared by elaborate routes. The starting compounds were recently reviewed (Meyer, 1976). The most noteworthy of these forms is S_{12}, which forms a very stable ring and is found in quenched liquid sulfur (Schmidt, 1973), Fig. 3.2. Among the recently established allotropes is also gamma-monoclinic sulfur, which had already been proposed (Muthmann, 1890), rejected, and reconsidered periodically until pure crystals were discovered quite accidentally by Watanabe (1974), who was working with an organometallic compound. He confirmed the 'sheared-penny roll' structure proposed by De Haan (1958). The structure of these elemental compounds has been reviewed by Meyer (1976) and Donohue (1974). The latter provides an excellent historic perspective and reflects upon the progress and contribution of the many outstanding workers who have tested their skills on sulfur. The preparation of allotropes has been reviewed by Meyer (1976) and Schmidt (1973). Work up to 1953 is summarized thoroughly in Gmelin; progress between 1953 and 1964 was reviewed by Meyer (1964, 1965). The latter reference also contains a summary on the transformation of orthorhombic into monoclinic sulfur (Thackray, 1965, 1970; Hampton, 1973, 1974). The crystallization rate of sulfur modified with plasticizing additives has been measured by Currell (1975, 1976). The half life ranges from one month for styrene and thiokols to 10 months for dicyclopentadiene. The kinetics of the degradation of polymeric sulfur by X-ray radiation has been studied by Dorabialska (1973).

Insoluble sulfur, as used in rubber vulcanization, can be prepared by various methods. Monsanto (U.S. P. 2,569,375, 1947) and Ruhrgas (1954) prepared it by quenching liquid sulfur. Stauffer (U.S. P. 2,419,309 and 2,419,310, 1944; 2,419,324 and 2,460,365, 1945; 2,513,524 and 2,524,063, 1948; 2,757,075, 1954) produces it by quenching sulfur vapor in carbon disulfide, and via other paths. The former method has been extremely successful. Insoluble sulfur, in mixture with soluble sulfur, results during the Wackenroder reaction, i.e. the aqueous reaction of hydrogen sulfide with sulfur dioxide. Insoluble sulfur can also be stabilized by addition of unsaturated organic compounds, such as terpenes.

Table 3.3
EQUILIBRIUM VAPOR PRESSURE OF SULFUR[a]

P, Torr	T, °C	P, atm	T,[b] °C
10^{-5}	39.0	1	444.61
10^{-4}	58.8	2	495
10^{-3}	81.1	5	574
10^{-2}	106.9	10	644
10^{-1}	141	20	721
1	186	40	800
10	244.9	50	833
100	328	100	936
760	444.61	200	1035

a) After Jensen (1972); b) Rounded average values; see Baker (1971), Rau (1973).

The equilibrium vapor pressure of elemental sulfur, shown in Table 3.3, has been reviewed by Jensen (1973). The modern data is largely based on West (1959), Rau (1973), and Baker (1971). At low temperature, the vapor consists mainly of S_8 molecules. The vapor pressure of super-cooled droplets was measured by Ford (1950). The structure of these was described by Bolotov (1971). The high temperature vapor is very complex, but is now well understood, as discussed below. The vapor pressure at room temperature is about 10^{-5} torr, but sulfur sublimes quite readily. It is believed that the vapor pressure at room temperature is sufficient to account for the fungicidal action and the toxic action of sulfur towards leaves (McCallan, 1931). The vaporization process depends on the structure and composition of the solid (Berkowitz, 1963), and on the influence of light. The latter has been long recognized (Wigand, 1911), but no one has yet checked this effect with modern analytical tools, even though the vaporization and vapor-phase transformation are of great practical importance in all outdoor processes, for example in agriculture and in the decay of sulfur coatings.

Single crystals of orthorhombic sulfur do not readily convert to monoclinic sulfur; instead they melt at 112°C (Currell, 1974). The self-diffusion rate of S_8 in orthorhombic sulfur was measured by Hampton (1973, 1974). The electric conductivity was measured by Spear (1965), Fittipaldi (1966), Kuramoto (1969), and Watanabe (1968). The electric conductivity is caused by two effects: Hole mobility, which has a value

Table 3.4
SPECIFIC HEAT OF SOLID, LIQUID, AND GASEOUS SULFUR[a]

T, °K	Sulfur Species			
	alpha-S(s)	beta-S(s)[b]	Liquid	Vapor
10	0.103	0.163		
15	0.348	0.412		
20	0.608	0.666		
25	0.868	0.906		
40	1.465	1.490		
50	1.795	1.808		
60	2.089	2.101		
100	3.090	3.077		
150	3.990	4.072		
200	4.650	4.817		
298.15	5.430	5.551		5.659
368.54	5.778	5.913	7.579[c]	
388.36		6.053	7.579	5.569
400			7.712	
420			8.190	
433			11.930	
440			10.800	
460			9.925	
717.75			7.694	5.252
1000				5.137

a) After Jensen (1972); b) After Montgomery (1974); c) Feher (1964) gives 7.423.

of about 10 $cm^2/V.s.$, with a negative temperature coefficient, and electron transport in the vibrationally strongly interacting electronic band, which contributes approximately 10^{-4} $cm^2/V.s.$ (Gibbons, 1970). The electrical resistance of orthorhombic sulfur is 10^{17} ohm/cm at 20°C. The potentially photo conducting properties of sulfur allotropes have not been well explored (Meyer, 1960), but industrial work in this area has been attempted (Eiichi, Japan P 76 67,126). The first patent for dry photocopy was based on sulfur, and not on the selenium system which was so successfully commercialized by Xerox-Battelle. The magnetic susceptibility of solid sulfur is about 5 x 10^{17} cgs units at 20°C.

The thermal conductivity changes from 11 W/m·deg at 4.2°K to 0.29 W/m·deg at 0°C and 0.15 W/m·deg at 95°C (Ericks, 1971; Powell, 1966; Mogilewskii, 1968). The insulating properties of sulfur are comparable to those of asbestos. The specific heat is given in Table 3.4. The heats of transition are summarized in Table 3.5.

Table 3.5
THERMAL DATA FOR PHASE TRANSITIONS

Transition	Process or reaction	T, K	H, kcal/g-atom[a]	S, cal/deg·g-atom	Ref.
alpha, beta	alpha-S_8(s) – beta-S_8(s)	368.46 ± 0.1	0.096	0.261	Miller, 1971
Sublimation alpha	alpha-S_8(s) – cyclo-S_8(g)	368.5	2.979	8.191	Thackray 1965
beta	beta-S_8(s) – cyclo-S_8(g)	368.5	2.883	7.93	Gernez, 2876
eta	eta-S_6(s) – cyclo-S_6(g)	300	4.02	8.38	Meyer, 1964
Fusion alpha	alpha-S_8(s) – cyclo-S_8(l) + ?[b]	383[c]	0.507		Thackray 1970 Currell, 1974
beta	beta-S_8(s) – cyclo-S_8(l) + ?[b]	392.9[c]	0.3842	0.75	Pacor, 1967 Feher, 1971
Ring scission	cyclo-S_8(l) – atena-S_8(l)	432	4.1	2.88	Tobolsky, 1964 Klement, 1974
Polymerization	catena-S_8(l) + cyclo-S_8(l) – catena-S_8(l)	442.8	0.396[d]	0.58	Tobolsky, 1965
Vaporization	S_i(l) – S_i(g)	717.824[e] = 444.674 °C	2.5	3.5	Meyer, 1964

a) 1 g-atom of sulfur = 32.066 g; b) the composition of the melt is not known; c) see also Table 3.9; d) see Tobolsky (1965); e) sulfur is a secondary temperature reference point on the International Practical Temperature Scale (Rossini, 1970).

During the last twenty years an immense wealth of new theoretical and experimental chemical tools has been developed. Most of these have not yet been used to study sulfur, but their application is only a matter of time. We already have a fairly detailed understanding of the structure of the S-S bond. Cruickshank (1968) conducted wave mechanical calculations; Keeton (1976) reviewed the atomic wave-function. The role of d-electrons was explored by Craig (1965). Fogleman (1969), Cusachs and Miller (1968-1971), Spitzer (1972), and Meyer (1977). All concluded that d-orbitals contribute little to ground state properties of divalent sulfur, but that d-orbitals participate in excited state interactions. Carlson (1975) performed an *ab initio* calculation. Semi-empirical calculations were performed by Thompson (1966), Sinencio (1969), Palma (1970), Müller (1967), Clark (1963), Salaneck (1975), Richardson (1975), and Meyer (1972, 1977) who determined the electronic energy levels from which the color of molecules can be determined, and the charge distribution, from which the relative reactivity of various molecular groups and functions can be predicted. Sakai (1975) investigated the correlation between molecular orbitals and the electronic spectra of divalent sulfur compounds. The bond geometry of some common sulfur species is

Table 3.6
TYPICAL S–S BOND DISTANCES

Molecule		S-S bond length, A	Reference
S_2		1.887	Barrow, 1970
S_2F_2		1.89	Kuczkowski, 1964
$S_2O_3^{2-}$		2.00	Csordas, 1969
Diphenyl disulfide		2.03	Lee, 1969
alpha-Cystine		2.03	Oughton, 1959
Me_2S_2		2.038	Sutter, 1965
$S_3(CF_3)_2$		2.065	Bowen, 1954
S_8O		2.04-2.20	Steudel, 1973
S_8^{2+}		2.04	Bali, 1975, Davies, 1971
S_{12}		2.053	Kutoglu, 1966
H_2S_2		2.055	Winnewasser, 1968
S_6		2.057	Donohue, 1974
S_8		2.060	Donohue, 1974
S_n		2.066	Tuinstra, 1966,67
$S_nO_6^{2-}$	n=3	2.15	Zachariasen, 1934
	n=6	2.04; 2.10	Foss, 1965
$S_2O_6^{2-}$		2.15	Stanley, 1956
$S_2O_5^{2-}$		2.209	Lindqvist, 1957
$S_2O_4^{2-}$		2.389	Dunitz, 1956
S_4N_4		2.58	Sharma, 1963

For all references in this table, see Meyer (1976).

summarized in Table 3.6. The 'normal' S-S bond distance is about 2.05 A. Typical S-S-S bonds have a torsion angle of 85.3° and a bond angle of 106.5°, Fig. 3.2 (Kao, 1977). Torsion around the bond has an activation of about $\Delta E = 6$ kcal/mole, but depends on the substituents (Semlyen, 1967, 1968). The different bond distances between 1.887 A for S_2 and 2.389 A for $S_2O_4^{2-}$ are due to differences in the electronic structure, which can be expressed as a ratio of s/p hybridization (Lindqvist, 1958). The bond properties are reflected in the Raman and IR spectra, which are important for analytical work (Anderson, 1969; Strausz, 1965; Meyer, 1977; Gautier, 1974, 1976; Ozin, 1969; Ward, 1968, 1972). The spectrum is pressure dependent (Zallen, 1974). The strong band at 1510 cm^{-1} is not a fundamental, as had been generally assumed, but is due to the strong C-S bond, which becomes dominant even at low impurity

Table 3.7

SOLUBILITY OF SULFUR IN 80 SOLVENTS

Solvent	wt %	T(°C)	Solvent	wt %	T(°C)
acetone	2.65	25	ethyl ether		
acetylene	1.26	25	(alpha-S)	0.080	0
dichloride				0.283	25
ammonia*	38.6	-20	(beta-S)	0.113	0
(anhydrous)	21	30		0.253	25
ammonium	7.3	-9		2.8	100
polysulfide*	25	0	ethyl formate		
	35	50	(alpha-S)	0.019	0
amyl alcohol	1.48	95	(beta-S)	0.028	0
	5.03	131	ethylene chloride	0.826	25
aniline	1	0		9.97	97.5
	46	130	ethylene dibromide	1.2	0
benzene	2.1	25		2.8	25
	4.3	50		36.5	100
	17.5	100	glycerol	0.14	15.5
	36	130	heptane	0.124	0
bromoform	3.64	5.6		0.362	54
carbon disulfide	2	-110		0.926	54
	4	-80	hexane	0.07	-20
	19	0		0.25	20
	34	25		2.8	100
	56	60		8.2	180
	90	98	hydrogen sulfide	0.14	0
carbon	0.15	-24		0.005	-60
tetrasulfide	0.37	0		1.3	80
	0.87	25	iodoform	42	85
	2.0	60	lanolin	0.38	45
chloroform	1.22	24	linseed oil	0.6	30
(alpha-S)	0.788	0	*	9.1	160
(beta-S)	1.101	0	methanol	0.028	18.5
cyclohexane	1.02	22	methylene iodine	9.1	10
	4.39	71	beta-naphthol	25.4	118
dibutyl phthalate	2.27	61	*	75.0	163
*	14.0	198	nicotine	9.6	100
dichlorethylene	1.26	25	olive oil	2.2	15
ethanol	0.05	15		30.1	130
	0.07	25	pentachloroethane	1.183	25
	0.42	79	phenol	8.3	89.5
ethyl bromide				26.7	175
(alpha-S)	0.611	0	pyridine	10.05	84.5
	1.307	25		13.35	80
(beta-S)	0.852	0		98.0	110
	1.676	25		91.94	127.0

Continued...

Table 3.7 (continued)

Solvent	wt %	T(oC)	Solvent	wt %	T(°C)
quinoline	13.8	74.5	trichloroethylene	1.60	25
	85.2	101	turpentine	1.33	10
	24.0	88.75	m-xylene	1.969	25
	97.8	111.5		10.29	80
	27.35	72	p-xylene		
	90.4	72	(alpha-S)	17.85	100
	46.8	91		34.6	148
	49.5	91	(beta-S)	16.3	100
disulfur dichloride	7.3	-9		91.0	148
	17	21	*	91	205
	97	110	*	44	235
sulfur dioxide	0.0078	25	hydrogen sulfide	0.005	0
	0.039	60		0.14	-60
	0.46	140		1.3	80
tetrachloroethane	1.214	25	rubber		
tetrachloro			natural	5	100
ethylene	1.507	25		0.8	50
toluene	0.079	-58.25	SBR 1006	7	100
	1.827	20		0.5	50
	11.64	83			

*indicates chemical interaction

concentrations (Srb, 1962). The Auger spectrum of various simple sulfur compounds has been published by Asplund (1976).

The UV-spectrum which determines the color of sulfur and its sensitivity to sun-light has been discussed by Bass (1953), Oommen (1973), Emerald (1976), and Clark (1963). The low lying triplet state which is expected to be in the blue-green region and might be responsible for the photosensitivity of the S-S bond, is not yet known. Sulfur has a parachor of 49.2 $\pm$ 1$(G^{1/4}cm^3)/sec^{1/2}$. The index of refraction is 1.929 at 110°C (230°F). The polarization is 0.245 cm^3/g. The X-ray spectra will be discussed in Chapter 4 (Barrie, 1975).

The bond energy of the sulfur allotropes has been discussed repeatedly. Kao (1977) and Allinger (1976) used a force field method to calculate apparent bond energies. The computed values are 62.62 kcal/mole for S_8, 61.5 for S_6, 63.0 for S_{12}, 63.1 for S_{18}, 63.1 for S_{20}, and similar values for all molecules with more than 6 atoms. The bond energy of S_2 is 100.69 $\pm$ 0.01 kcal/mole (Ricks and Barrow, 1969), 55.8 kcal/mole for S_3, 58 for S_4, and 60 for S_5 (Rau, 1973). The heat of formation for elemental molecules with 4 to 20 atoms can be approximately represented by a formula developed by Kao (1977). Proposed values vary between 31.2 for S_2 and 24.32 for S_8 (Rau, 1973).

2. Solutions

The solubility of sulfur in 80 solvents is listed in Table 3.7. Typical solubility curves are shown in Fig. 3.3. Sulfur is insoluble in water. It reacts with several inorganic solvents at room temperature, for example with liquid ammonia, and forms a mixture of reaction products, among them tetranitrogen tetrasulfide which is discussed later. With chlorosulfanes sulfur forms polysulfurdichlorides; with hydrogen sulfide it forms sulfanes. These reactions are discussed in Sections 3B-D. Above 150°C, sulfur reacts with every known inorganic and organic solvent; in some cases the reactions are very slow, but the odor and color of these systems change quickly and prove reactivity. Systems in which reactivity is substantial are indicated with an asterisk in Table 3.7. More details are discussed in later sections.

The cryoscopic constant of sulfur is 686°C/g·atom (E.D.West, 1959). In many solvents, chemical reactions between the sulfur and the solvent are observed: Larkin (1967) described the binary equilibria between sulfur and CS_2, CCl_4, benzene, toluene, alpha-xylene, naphthalene, biphenyl, triphenylmethane, and cis- and trans-decalin. Wigand (1909) observed that sun light reduces the solubility of orthorhombic sulfur in organic solvents, and that the precipitate redissolves only very slowly. This well established fact has not yet been satisfactorily explained. Nishijma (1976) studied the UV-spectrum by flash photolysis in EPA at -200°C, and proposes that the intermediates formed in ethanol contain a linear S_8 chain and a dimer. Above 150°C, all these solutions slowly turn pink or brown, indicating chemical reactions. Wiewiorowski (1966) obtained the phase diagram of S_8-CS_2 and determined the effect of pressure on the solubility, which at atmospheric pressure changes almost linearly with temperature. The solubility of oxygen in sulfur is about 3 x 10^{-4}% at 120°C in standard air. The adsorbance of argon, xenon, and SF_6 was studied and interpreted by Stoeckli (1975). The ternary system, sulfur-naphthalene-n-octadecane, was studied by Wiewiorowski (1968) at 113°C. The reaction of liquid sulfur with H_2S, Se, As, I_2, and organic solvents is discussed below. Sulfur readily dissolves in liquid ammonia (Kerouanton, 1973; Lautenbach, 1969; Jander, 1966). Slurries of sulfur

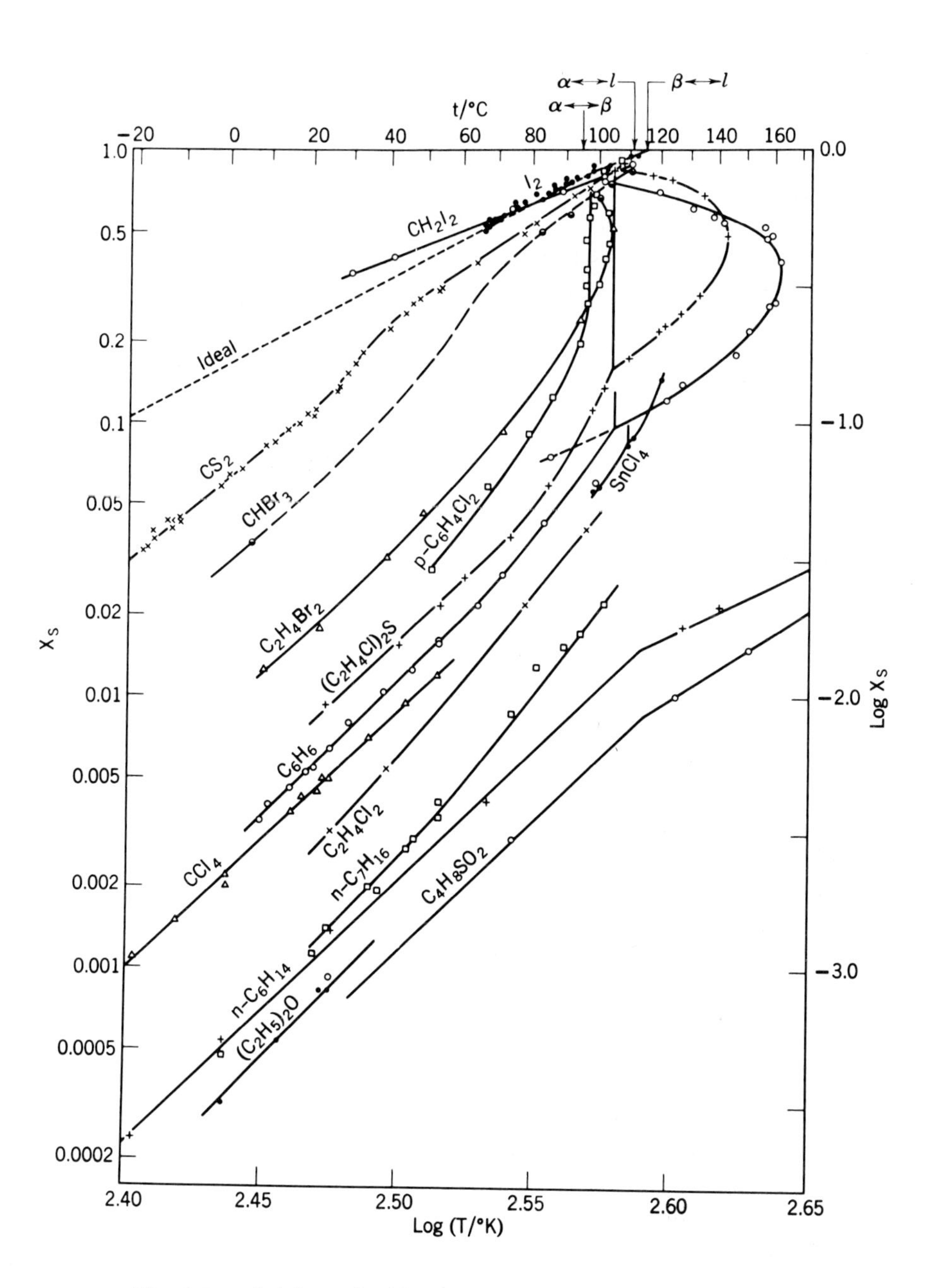

Fig. 3.3. Solubility of sulfur in 16 solvents (after Scott, 1965).

Table 3.8
ELEMENTAL SULFUR SPECIES

Species	Solvent	Nature	n
S_n Rings	Organic, non-polar	Stable, full electron shells	5 to 20
S_n^{2-} Chains	Aqueous	Stable, full electron shells	1 to 400
S_n Chains	Liquid sulfur	Di-radicals	1 to 100,000
S_n^- Chains	Ionic salts, amines	Radicals	2 to 8 (?)
S_n^{2+} Rings	Oleum, super-acids	Di-radicals	4;8;16;24 (?)

in ammonia have been occasionally used as fertilizer. Upon standing, the solution reacts and various compounds containing S-N bonds are formed. Among them are SNH_3, S_2NH_3, and, ultimately, S_4N_4. These solutions are still very poorly understood.

In oleum, sulfur forms S_8^{2+} and similar ions, described below (Stillings, 1971; Gillespie, 1971; Symons, 1972; Low, 1976). The blue species was first described by Weber (1975), who thought that an oxyacid was being formed. The structure of S_8^{2+} is shown in Fig. 3.2. The nature of the other species is not yet conclusively known.

In amines, singly and doubly charged negative ions are formed (Daly, 1976). These species are identical with those found in molten salt (Green, 1971), and many other highly ionizing systems. Work on these ubiquitous S_x^- radical ions has been reviewed by Chivers (1971-1974), Giggenbach (1971, 1973), Merritt (1970), Martin (1973), Holzer (1969), and Bernard (1973). These species also occur in some natural ultramarines (Seel, 1972, 1973). The electronic structure of the ions has been discussed by Meyer and Spitzer (1973, 1977): The charge distribution of these species is shown in Fig. 3.4. The elemental species now known are shown in Table 3.8.

3. Liquid Sulfur

The complexity of liquid sulfur was clearly recognized by Dumas (1827), and Deville (1856), who gave an excellent physical description of the system. The appearance and the molecular composition of liquid sulfur differ in three distinct temperature ranges.

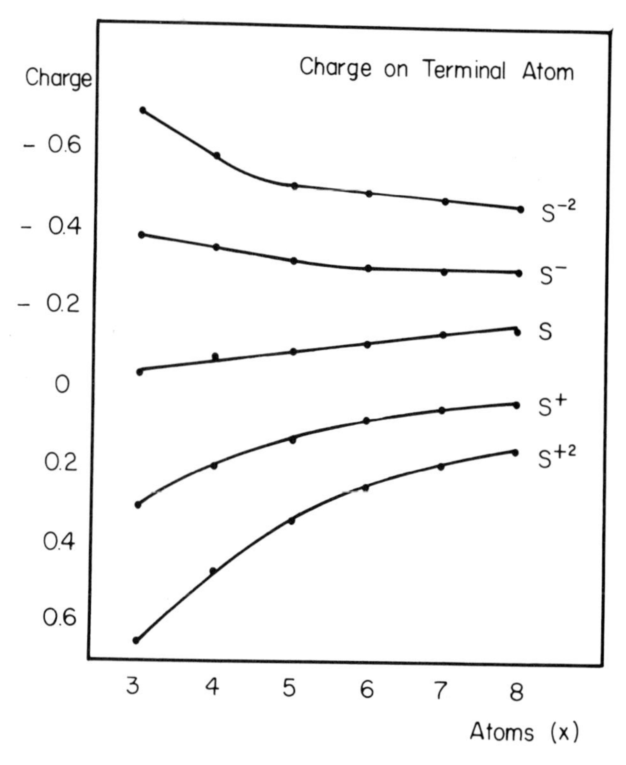

Fig. 3.4. Calculated charge distribution in sulfur chains (Meyer, 1977).

The Melt Below 150°C

Table 3.9 shows the melting points for sulfur. The wide range is due to the fact that sulfur can occur in several different molecular forms which can co-exist. The melting and freezing points are determined by the history of a sample, i.e. the relative concentration of the various metastable species which can act as solutes in S_8 and cause a freezing point depression. This freezing point depression, first observed by Gernez (1876), has been calculated by Wiewiorowski (1968), Semlyen (1971), Harris (1970), Feher (1971), and Schenk (1957). The high pressure melting curve was measured by Vezzoli (1970) and Block (1973). The best value for the heat of melting is $\Delta H_m = 384.2 \pm 1.9$ cal/g·atom at 119.6°C (Feher, 1971; Pacor, 1967). Despite 150 years of research, the composition of the liquid is not yet known. Schmidt (1973) extracted S_{12} from the frozen melt; Krebs (1967) isolated S-pi, which could be separated into three fractions with molecular weights of 6, 8, and 9.2, respectively.

Table 3.9
MELTING POINTS OF SULFUR

Allotrope	Mp, °C	Remarks	Reference
alpha-S	112.8	Single crystal	Currell, 1974
	115.11	Microcrystal	Thackray, 1970
beta-S	114.5	'Natural'	Gernez, 1876
	119.6[a]	'Ideal' and obsd	Feher, 1971
			Pacor, 1967
	120.4	Microcrystal	Thackray, 1970
	133	'Ideal' calcd	Schmidt, 1973
gamma-S	106.8	Classic	Meyer, 1964
	108	Optical, DTA	Miller, 1971
	108.6	Microcrystal	Thackray, 1970
insoluble-S	77; 90; 160	Optical, TDA, DTA	Miller, 1971
	104		Currell, 1974
S_n	75	Optical	Miller, 1971
	104	Classic	Gmelin, 1953
S_6	(50–)	Decomposition	Schmidt, 1973
S_{12}	148	Decomposition	Schmidt, 1973
S_{18}	128	Decomposition	Schmidt, 1974
S_{20}	124	Decomposition	Schmidt, 1974

a) Thermodynamic melting point.

Wiewiorowski (1968), Semlyen (1971), and Harris (1970) also analyzed the situation, but the answers are not yet conclusive. It is quite easy to obtain supercooled solutions. The kinetics of nucleation of small droplets were measured by Hamada (1970) at -20°C and down to -50°C. Calculated weight and mole fractions of the various purported components of liquid sulfur are shown in Fig. 3.5. A more detailed description of the systems which might contain S_8 ring/S_8 chain charge-transfer complexes is given by Meyer (1976), Koningsberger (1971), and Wiewiorowski (1968). The specific heat of the liquid is shown in Table 3.4; it is 7.02 cal/g·atom degree at 120°C according to Feher (1964). The density is 1.901 - 8.00 x 10^{-4} Tg/mole (120-160°C) (Ono, 1957). The electric conductivity has been very carefully studied because the Li-S battery is high on the list of candidates for bulk power storage batteries. It decreases from 2 x 10^{-13} per Ohm cm to 8 x 10^{-9} at the boiling point (444°C). In the batteries, described in Chapter 14B, sulfide acts as a conductor. The surface tension of sulfur has been calculated and measured by Ono (1957), who obtained values of 60.95 dyne/cm at 120°C and 57.64 dyne/cm at 150°C. It can be described by:

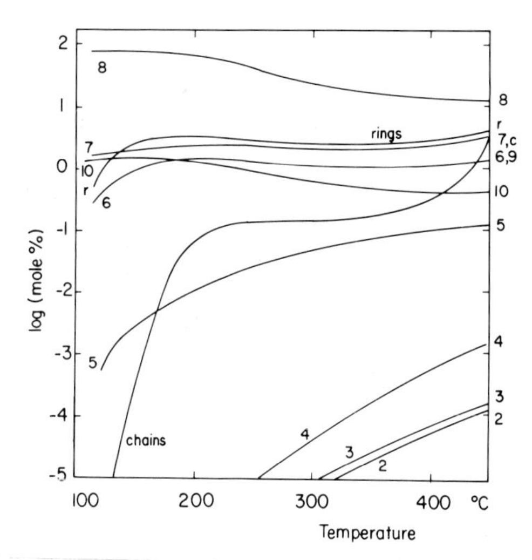

Fig. 3.5. Mole fraction of components of liquid sulfur.

$$T = 73.69 - 0.1066T \quad (120\text{-}160^{o}C)$$
$$T = 66.2 - 0.060T \quad (160\text{-}420^{o}C)$$

Three percent arsenic increases the surface tension over the entire temperature range by 8%. Ten percent selenium increases the viscosity by 5%.

The interfacial behavior of liquid sulfur was studied by Matsushima (1958). He measured a contact angle of 60° in the 120-160°C range between liquid sulfur and fused silica, and of 58.5° between sulfur and pyrite in the same temperature range. In the presence of water, the adhesional work of sulfur is reduced by -31.0 dyne/cm at 120°C. This explains why water helps removal of sulfur from pyrite in the autoclave. The surface tension between liquid sulfur and water was determined:

$$T_{S\text{-}H_2O} = 70.99 - 0.1329T \quad (110\text{-}160^{o}C)$$
$$T_{S\text{-}H_2O} = 57.25 - 0.0470T \quad (160\text{-}180^{o}C)$$

(Matsushima, 1958). The thermal expansion has been measured by Fanelli, 1945).

The viscosity of liquid sulfur attracts the attention of every young science student, because of the unexpected maximum at 196°C. The most careful low temperature values stem from Bacon and Fanelli (1943), Feher (1964), and Doi (1965, 1967).

Polymerization at 159.4°C

At 159.4°C almost all properties of sulfur suffer a discontinuity. The density minimum is shown in Fig. 3.6 (Patel, 1971). The viscosity curve, first described by Eötvös (1886), is shown in Fig. 3.7. Bacon and Fanelli (1943) recognized the influence of impurities, such as iodine. Touro (1966) and Rubero (1964) studied the influence of hydrogen sulfide; Koningsberger (1971) that of selenium; Ward (1969) that of arsenic; Doi (1965) shows plots for iodine, bromine, chlorine, dibenzothiazyl disulfide, disulfur dichloride, naphthalene, p-dichlorobenzene, diphenyl, and m-cresol, and a group of slowly reacting compounds, among them pyrogallol, p-benzoquinone, alpha-naphthol, tetramethylthiuram disulfide, 2-mercapt-benzothiazole, and diphenyl guanidine. Their effect on viscosity is shown in Fig. 3.8. The inorganic halides and dibenzothiazyl disulfide immediately react or form polysulfides, which are in equilibrium. The unreactive group of aromatics forms normal solutions with sulfur; thus, the viscosity is merely dependent on concentration and not on the chemical character of the substance, as seen in Fig. 3.8. Schmidt (1973) showed that 2% S_6 lowers the polymerization temperature by 10° for fifteen minutes, while $(CH_2\text{-}S)_3$ added at 200°C reduces the chain length. According to Eisenberg (1964) and Tobolsky (1965), the polymerization reaction can be represented by two reaction steps; an initiation reaction,

$$S_8 \text{ ring} \rightarrow S_8 \text{ chain, and } [Ch]/[R] = K_1$$

and a propagation reaction:

$$(S_8 \text{ chain})_n + S_8 \text{ ring} \rightarrow (S_8)_{n+1}, \text{ with } K_3$$

If K_1 and K_3 are known at two temperatures, their enthalpy and entropy can be deduced from Van't Hoff's equations:

$$\ln K_1 = -(\Delta H^o/RT) + (\Delta S^o/R), \text{ and}$$
$$\ln K_3 = -(\Delta H_3^o/RT) + (\Delta S_3^o/R)$$

from experimental observation Tobolsky obtained

$$\Delta H_1^o = 32.8 \text{ kcal/mole} \qquad \Delta S_1^o = 23 \text{ cal/deg·mole}$$
$$\Delta H_3^o = 3.17 \text{ kcal/mole} \qquad \Delta S_3^o = 4.63 \text{ cal/deg·mole}$$

The concentration of the various species, and the average chain length in S_8 units (P) are shown in Table 3.10 and Figures 3.5 and 13.2. Koningsberger (1972) computed the equilibrium using Flory parameters.

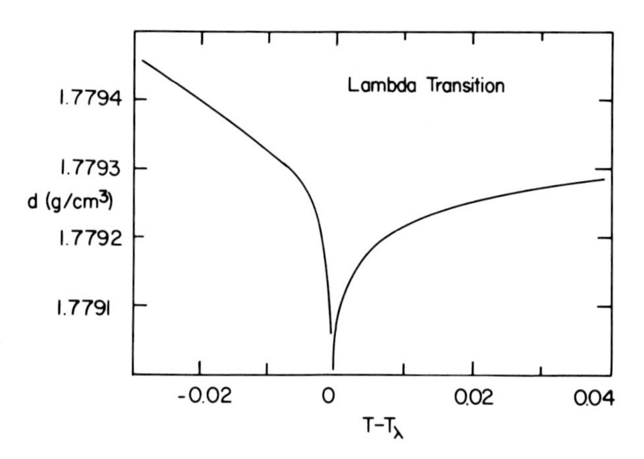

Fig. 3.6. Density of liquid sulfur at lambda point
(after Patel, 1971)

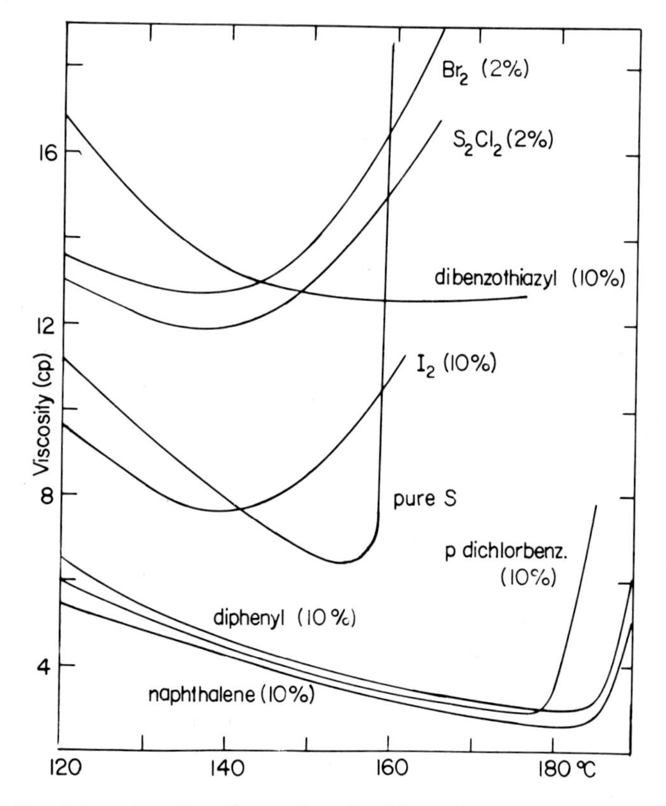

Fig. 3.7. Viscosity of pure liquid sulfur (after Bacon, 1943), and
Fig. 3.8. Viscosity of sulfur containing additives
(after Doi, 1965).

Table 3.10
AVERAGE CHAIN LENGTH, P (IN MULTIPLES OF S_8), MONOMER CONCENTRATION, M, AND EQUILIBRIUM CONSTANTS, K_1 AND K_3, FOR LIQUID SULFUR BETWEEN 112°C AND 305°C

Temp., °K	Polymer	Monomer	log K_1	log K_3
385	2.21		13.546	-.8532
400	3.38		12.959	-.7435
410	5.02		12.572	-.6882
420	9.44		12.054	-.6395
425	16.4		11.729	-.6188
428	27.6		11.699	-.6073
430	57.9		11.641	-.5977
440	112,300	3.65	11.275	-.5628
450	113,900	3.36	10.914	-.5272
460	94,500	3.14	10.568	-.4968
470	75,800	2.89	10.216	-.4609
490	46,000	2.52	9.587	-.4005
510	28,400	2.21	9.025	-.3454
540	13,870	1.86	8.245	-.2703
580	5,750	1.52	7.568	-.1852

From Tobolsky (1965).

The kinetics of the polymerization have been investigated by Klement (1974). Such studies are very difficult because of the poor thermal conductivity of liquid sulfur. Fig. 3.9 shows that equilibrium cannot be efficiently reached below 160°K. The effect of pressure on the polymerization equilibrium was reported by Doi (1967) and Bröllos (1974). Since the polymer chains have radical character, liquid sulfur should have an ESR spectrum with a signal size corresponding to the concentration of polymer. Gardner and Fraenkel (1956) measured far too small signals, and so did Koningsberger (1971), who explained this fact by involving charge-transfer complexes between chain ends and residual rings; his postulate quantitatively fits Wiewiorowski's model. This model also explains the corresponding temperature dependence of the magnetic susceptibility (Poulis, 1965).

What species other than polymer and S_8 rings occur in liquid sulfur? Investigation of the UV absorption edge and the visible spectrum of thin films led Oommen (1971) and Meyer (1971) to the discovery of S_3, S_4, S_5, and S_6 in liquid sulfur. Since polymeric sulfur is yellow at room temperature, and its absorption does not penetrate into the green, the color of hot liquid sulfur must be due to the increasingly larger concentration of S_3, S_4, and other small molecules. However, the color of the

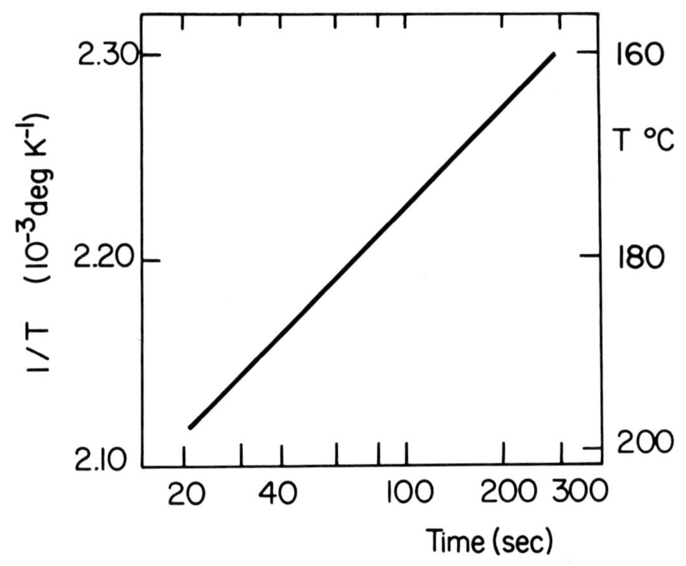

Fig. 3.9. Equilibrium kinetics at 160°C.

polymer still constitutes a puzzle, because a theoretical model for the color of long free radical chains would require them to be black (Meyer, 1972). Once more, the best present explanation is provided by Wiewiorowski's charge-transfer complex model.

Liquid Sulfur Above 250°C

At high temperature, the mole fraction of polymer increases, but the average chain length decreases, as shown in Fig. 3.5. Hot liquid sulfur contains substantial concentrations of short, reactive chains which account for its incredibly high reactivity towards traces of organic compounds. The dark color of the reaction products reflects the presence of impurities so reliably that it can be used for commercial grading. The triple point is 1040°C and 180 atm (Baker, 1971). Rau (1973) reported 200 atm. The density is 0.563 g/cm^3 and the average molecular size is 2.8 atoms/molecule. The normal boiling point of sulfur, once believed to be 298°C by Davy (1809) and recorded at 440°C by Dumas (1811), is 444.64°C. It served for many years as a primary high temperature reference point for IUPAC (Rossini, 1970), but is now only a secondary point. The livid, dark boiling sulfur has not yet been explored for its potential to conduct a large number of selective, direct organic synthetic reactions.

4. Sulfur Vapor

The thermodynamic properties of sulfur vapor are shown in Table 3.5. Fig. 3.10 shows the partial pressure of the various species as a function of temperature for six total pressures. The data was computed by Rau (1976) using his own experimental data (Rau, 1973). Obviously, the vapor is far more complex than had been assumed by early workers. The first conclusive vapor analysis was undertaken by Chupka and Berkowitz (1965), Berkowitz (1969), and Drowart (1967). At low temperature, the vapor is green, due to the S_8 which makes up a large fraction of it. At an intermediate pressure and temperature, the vapor is deep cherry red, due to S_3 and S_4 (Meyer, 1971). At high temperature and low pressure, the vapor is dark violet due to S_2.

Berkowitz also contributed much to the determination of the dissociation energy of S_2, which was determined to be 100.69 ± 0.01 kcal/mole by Ricks and Barrow (1969). Earlier work has been reviewed by Meyer (1976). The electronic energy levels of S_2 are still widely investigated; S_2 is isovalent with O_2 and has a similar basic electron structure. Barrow contributed much of our present knowledge, and reviewed the situation in 1965 and 1970. The ground state is triplet sigma, i.e. the molecule is paramagnetic. The first spin forbidden state, singlet delta, is at 4500 cm^{-1}; chemical reaction with these two states should yield selective and specific products. More details on S_2 and other vapor phase molecules can be found in recent reviews (Barrow, 1970; Berkowitz, 1976; Meyer, 1976). Leone (1977) has demonstrated that S_2 can lase. A sulfur laser could be used to conduct selective reactions which are not energetically feasible with ground state molecules. S_3, thiozone, can be isolated in low temperature glasses. S_4 can be prepared by photolysis of S_4Cl_2 in rare matrices (Meyer, 1971). The spectrum of S_4 has been observed by Gardner (1973); the electronic properties of these species have been reviewed by Berkowitz (1969, 1975, 1976). The other vapor components are not yet well explained, except the sulfur atom, which, however, occurs only at very high temperature and low pressure. The electronic energy levels of the atom are well known (Moore, 1958). Recently, a thermogravimetric method has been described with which the composition of the sulfur vapor above alloys can be continuously monitored (Rilling, 1975). This type of work is important during the study of the formation of phases of mixed alloys. Such knowledge will eventually help reduce corrosion,

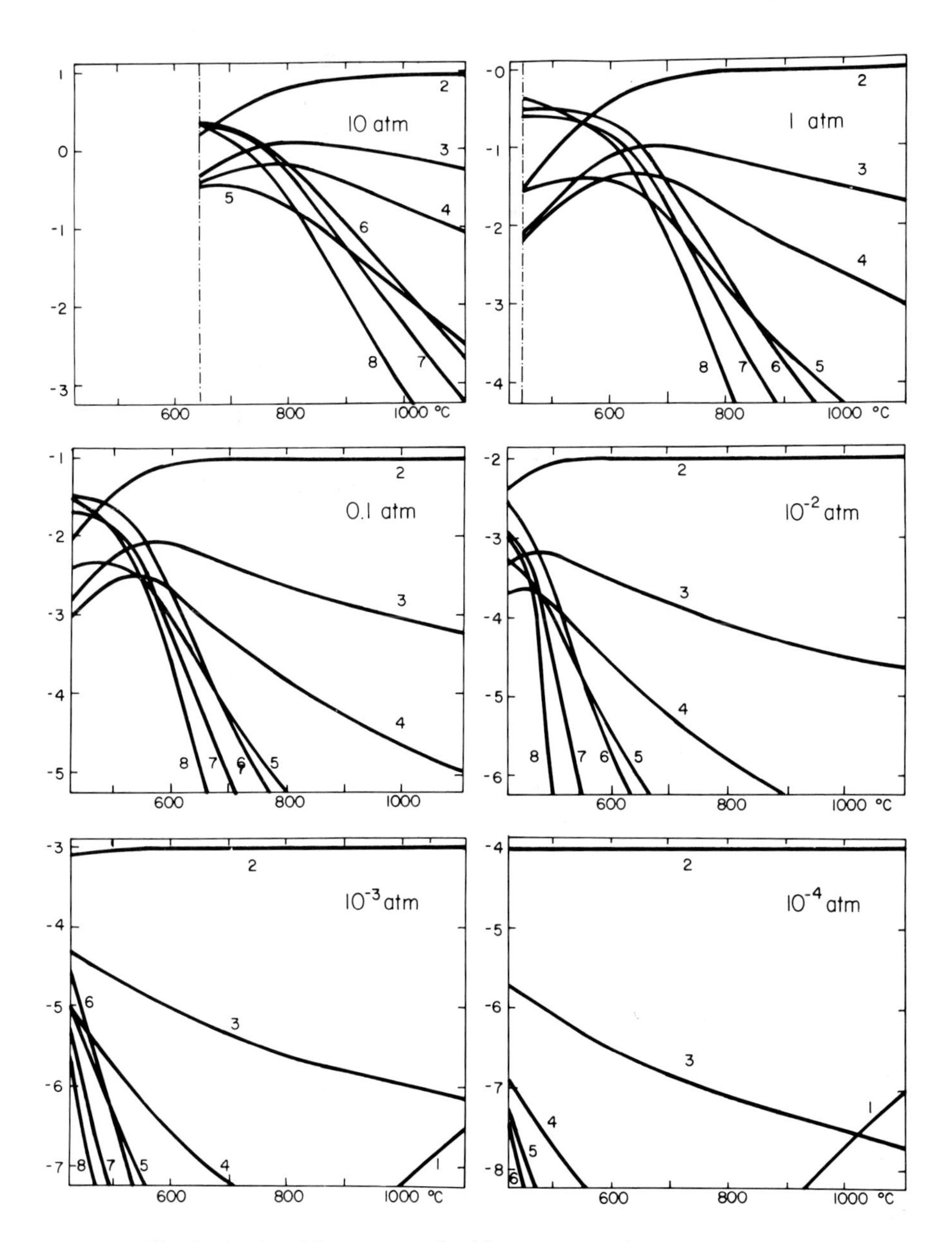

Fig. 3.10. Partial pressure of sulfur vapor species at 6 pressures (after Rau, 1976).

described in Section 3E. Sulfur atoms, both in the ground state and the final excited state, can be produced by photolysis of COS. Strausz and Gunning (1965-1972) have developed an entire branch of sulfur chemistry by reacting atoms with various organic substances. This work has greatly contributed to our understanding of intermediates and the kinetics of reactions of sulfur.

5. Sulfur Hydrosols and Colloids

Colloidal sulfur is not commercially available. Commercial grade 'colloidal' sulfur has either a particle size of about 2-6 micron or is chemically modified. Colloidal sulfur is an unpredictable and poorly defined material; if it is intentionally prepared as a fungicide or as an additive to medicines and toilettries, such as shampoos, it often coagulates and cakes; if it is not wanted–for example during the recovery of tail gases from the Claus process–it occasionally forms in large quantities, survives heating, and stubbornly resists any attempts to precipitate it. Colloidal sulfur usually contains a mixture of different chemical species. Oden's work (1911-1912) and Weitz (1956) showed that colloidal sulfur can contain polythionates of the formula $SO_3^- - S_x - SO_3^-$, with an x value of 10-25. Deines (1924) postulated chain lengths of up to 350. Colloidal sulfur is obtained when elemental sulfur forms in aqueous solution, for example in Wackenroder's (1846) reaction:

$$H_2S + SO_2 \rightarrow {}^-S\text{-}S_x\text{-}SO_3^-$$
$$SO_2 + {}^-S\text{-}S_x\text{-}SO_3^- \rightarrow {}^-SO_3\text{-}S_x\text{-}SO_3^-$$
$${}^-SO_3\text{-}S_x\text{-}SO_3^- \rightarrow S_x + 2SO_3$$

Neutral elemental sulfur and large polysulfide ions, S_x^{2-} or HS_x^-, are not soluble in water, and eventually precipitate. This process is accelerated by light, which induces S-S scission. Thus, such solutions must be stabilized with additives. Barnes (1947) and La Mer (1946) studied colloids prepared from thiosulfate, and interrupted growth at the desired stage by titrating the excess of thiosulfate with iodine. Kaplina (1968) prepared a colloid containing 95.5% sulfur from residues of coke-oven gas by treating it with lye. It is said to be stable for three years. Table 3.11 lists some commercial preparation methods; in most of these, a polysulfide intermediate is decomposed, and ring scission takes place: In the course of this reaction, long sulfur chains are synthesized or degraded.

Table 3.11
COMMERCIAL PREPARATION OF COLLOIDS

Method	Author	Reference	Year
S + gelatin, casein, milled	Szegvari	U.S. 1,969,242	1934
S-glycol, milled	Hoessle	Ger. 588,964	1934
S + alkanes, heated, quenched, milled	Chambers	U.S. 2,111,552	1938
S in water; ultrasonic energy	Sollner	B.P. 489,610	1938
Liquid S + modifier, quenched in water	Du Pont	B.P. 431,134	1935
Liquid S + modifier, quenched in water	Williams	Can. 348,002	1935
Liquid S + modifier, quenched in water	Grindrod	U.S. 1,992,611	1935
Liquid S + modifier, quenched into sulfite liquor	McDougall	Fr. 881,226	1943
Polysulfide + acid, with proteins	Hashimoto	U.S. 2,060,311	1936
Polysulfide + SO_2 + colloid	Wirth	Ger. 707,756	1941
Polysulfide + SO_2 + colloid	H. Kaufmann	U.S. 1,949,797	1934
$(NH_4)_2S$ + formaldehyde + oleate	Schwartz	U.S. 2,439,147	1948
$(NH_4)_2S$ + formaldehyde + dextrin	Geller	Ger. 584,042	1933
Polysulfide + ammonium nitrate	Petroff	Fr. 747,795	1933
Polysulfide + acid + silicates	Brunel	Fr. 940,884	1948
Dithionate + phosphoric acid	Soc. du Soufre	Fr. 800,007	1936
S + NaOH + SO_2	Boulogne	B.P. 465,574	1937
H_2S + SO_2	Ehman	U.S. 2,201,124	1940
Na_2S + SO_2 + dextrin	Torigian	U.S. 1,980,236	1934
S_2Cl_2 + H_2O + vaseline	Kaufmann	Ger. 563,010	1932
S + amines + H_2O	Nietsche	Ger. 527,326	1936
S + glycerine	Hoessle	Ger. 604.118	1934
Chlorobenzene + S + rubber	I.G.Farber	Ger. 536,075	1931

The mechanism of this reaction has been reviewed by Meyer (1977). It involves stepwise chain length changes with subsequent intramolecular ring formation or degradation, as discussed in the section below. Sulfur colloids are also discussed by Alyea (1969). Various colloidal grades of commercial sulfur, for example Colloidal Heyden Sulfur, are discussed in connection with pharmaceutical sulfur, Chapter 10, and agricultural grades, Chapter 11. In the latter field, insecticidal and fungicidal sulfurs have smaller particle sizes, around 200+ mesh, than do agricultural grades, which are often quite coarse.

6. Chemical Reactions of Elemental Sulfur

Sulfur forms ions and radicals with equal ease; thus, it can participate in all kinds of nucleophilic, electrophilic, and free radical reactions. In an aqueous medium, sulfur forms negative hydrogen sulfide and disulfide ions, and reactions are catalyzed by nucleophiles. In liquid ammonia or organic amines both sulfur ions and radicals are present. In liquid sulfur neutral radicals prevail, in oleum and other strong acids doubly charged positive radical ions are found, and in non-polar solvents neutral S_8 rings are abundant. According to Chemical Abstracts, each year over 4,000 papers dealing with new aspects of sulfur chemistry are published. Thus, this book cannot possibly offer a thorough review of sulfur chemistry. Only reactions of present or presently envisioned large scale applications are mentioned here, and even those are only superficially touched.

Several excellent summaries of sulfur chemistry exist. Inorganic sulfur chemistry is periodically reviewed in Gmelin (1959), of which a new edition is in preparation. Recent work on inorganic chemistry has been covered by Meyer (1964, 1965, 1976), Nickless (1968), Senning (1971-1973), Miller and Wiewiorowsky (1972); some excellent reviews have appeared in a book series edited by Emeleus (1977) and in several journals. Analytical procedures are reviewed in the following Chapter. Classical preparative inorganic work has been summarized by Feher (1963) in Brauer's Handbook. An international list of people active in sulfur chemistry as of 1974 is available from The Sulphur Institute (Fike, 1974).

The organic reactions of elemental sulfur have been thoroughly reviewed in a series of volumes authored by Reid (1958-1963). Pryor (1962), Price (1962), Kharasch (1961-1963), Wegler (1961), Plattner (1943), Senning (1971-1973, 1976), and Reid (1975) have given valuable reviews of this field. The International Journal of Sulfur Chemistry, edited by N. Kharasch, has provided both reviews and original contributions. Proceedings of bi-annual international conferences on organic sulfur chemistry sponsored by The Sulphur Institute, edited by Janssen (1967), Tobolsky (1968), Lozac'h (1970), Mangini (1971), Gronowitz (1972), Sterling (1975), and Walter (1977), provide much useful and detailed information. The organic photochemistry of sulfur has been reviewed by Gunning (1966), Strausz (1970), Kharasch (1961), and recently by Mayo (1976). The biochemistry of sulfur has been reviewed by Benesch (1959). Some excellent short reviews have been provided by Wegler (1958), Neumann (1974), Juraszyk (1974), Mayer (1973, 1976), and Schwalm (1972).

Organic Reactions of Elemental Sulfur

The reaction of elemental sulfur with organic compounds can be initiated by heat, light, or catalysts.

Elemental sulfur does not react at room temperature with hydrocarbons or aromatic compounds; however, it reacts with some amines and other nucleophilic reagents which break the sulfur ring and form intermediate ions. Vineyard (1967) showed how n-butylamine and other amines react with thiols; he prepared various di-, tri-, and tetrasulfides and discussed the importance of the polarity of the solvent. Disulfides can also be formed in liquid ammonia, using sodium, sulfur, and organic iodides. Ansinger (1956) showed that sulfur can be made to react with aliphatic ketone at room temperature in the presence of ammonia. The intermediate products are alpha-thiols. Toland (1953) showed that sulfur, thiosulfate, sulfite, and ammonium sulfate can be used to oxidize alkylaromatics, such as p-xylol, p-tolunitriles, in excellent yield. Compounds of the type R-CH=CH-CH_3 react with sulfur, even without amine, to form cyclic trithiones. The best known reaction of sulfur is probably the Willgerodt reaction, discovered in 1887, in which ketones are oxidized to carbonic acid by aqueous ammonium polysulfide at 200°C in an autoclave. A summary of work up to 1958 is given by Wegler (1958).

Monosubstituted acetylenes react with sulfur according to:

$$R - C \equiv CH + S \rightarrow R - C \equiv C - SH$$

which enter further reactions. In dimethyl formamid and similar solvents and in amines, intermediate ions form which can be used to guide product formation (Mayer, 1973, 1976).

Many reactions can be induced by alkali in aqueous solution, because sulfur disporportionates in alkali to thiosulfate and sulfide ions. The sulfide ions can undergo a variety of reactions. With formaldehyde a cyclic trithiane, $(CH_2\text{-}S)_3$, or tetrathiane, $(CH_2\text{-}S)_4$, is formed. Depending on temperature, pH, and reaction time, long polymeric chains, $\text{-}(CH_2\text{-}S\text{-})_n$, can be formed which constitute oils, rubber, or solids. The reaction yielding polymers are discused in Chapter 13.

Room temperature reactions can be induced by photolysis. This type of chemistry forms an entire field all by itself (Kharasch, 1971; Strausz, 1971, etc.). It is not yet clear whether olefins undergo ionic or radical reactions with sulfur. Up to 140°C, intermolecular polysulfides are formed without attack on double bonds; above 140°C continued reaction will induce secondary changes in the moleular structure of reagents. Weitkamp (1959) reacted isomeric terpenes with elemental sulfur at 170°C for two hours and obtained 44% polysulfide and 10-50%

p-cymene. The yield depends on the position, i.e. reactivity, of double bonds. The vulcanization of polybutadiene or natural rubbers has been studied thoroughly by many people (Kharasch, 1961; Bateman, 1961; Pryor, 1962; Nordsiek, 1965), but it is not yet even known whether the reactions involve ions or radicals and what chemical role the excess sulfur plays, which remains in the product and might act as plasticizer. Double bonded chains, such as alkyles, R-CH = CR - CH_3, form 1,2 dithiol-3-thianes, i.e. five membered rings containing two sulfur atoms as ring members, and a third sulfur forming a thioketone. Above 200°C hydrogen sulfide is formed. Butanes undergo ring closure and form heterocyclic thiophene; cyclic compounds are dehydrated and form double bonds. For example, some olefines convert to cyclohexene, alpha-tetralon to alpha-naphthol, cyclohexene to benzene, and by analogous processes other compounds yield propenes. Aromatics such as phenol form diphenyl disulfide. This reaction will be discussed in connection with the polymers, Chapter 13. Cyclic aliphatic compounds can be dehydrated by sulfur according to methods developed by Ruzicka (1921) and summarized by Plattner (1943). Grignard reagents form mercaptans.

Hydrocarbons in residual oils eventually convert to tar (Juraszyk, 1974). The sulfur chemistry in such systems determines whether the residues are useful in chemical reactions as fuel oils or must be used as road and roofing tars. Bocca (1973) studied the reaction of various petroleum fractions with 5% sulfur and found that the reactivity decreases from asphalts to aromatics, to branched alkanes, and, finally, to normal alkanes. He analyzed the reaction products densitometrically and found an average composition compatible with those of Quarles (1965), Tucker (1965), and Petrossi (1972). With the recent development and refinement of analytical techniques for separating natural products, much progress can be expected in this area. It will have a profound impact on the availability of various oil fractions as chemicals and fuels. If tars, for example, can be desulfurized, they might become valuable fuels and and alternate materials must be found to replace their function in road construction. Above 500°C, almost all organic compounds react with sulfur and yield carbon disulfide and hydrogen sulfide.

Inorganic Chemistry of Sulfur

Inorganic reactions, like organic sulfur reactions, can occur by either of three paths: Homolytic scission (cyclo octa-sulfur converts to catena octa-sulfur), nucleophilic degradation ($R^- + S_8$ converts to RS_8^-), and electrophilic degradation, (S_x + 2xHI converts to $xI_2 + xH_2S$.)

The homolytic scission can be induced by light or heat. The thermal homolytic scission causes the formation of polymeric sulfur in the liquid, and, thus, the high viscosity of liquid sulfur at 160-190°C. The inorganic photochemistry of sulfur is not yet well explored (Nishijima, 1976; Meyer, 1976). Nucleophilic attack is the well established reaction path chosen by sulfide ions, cyanide ions, thiocyanate, arsenite, Grignard reagents, and sulfite. The initial step is:

$S_8 + HSO_3^- \rightarrow HSO_3\text{-}S_8^-$, followed by degradation:
$$HSO_3S_x^- + HSO_3^- \rightarrow HS_2O_3^- + HSO_3S_{x-1}$$

i.e. sulfite reacts with sulfur to form thiosulfate and a sulfane-sulfonic acid. The latter compounds have been prepared and identified by Schmidt (1959, 1973), who prepared the free acids in ether at -78°C. The reverse reaction:

$$HS_2O_3^- + S_2O_3^{2-} \rightarrow HSSSO_3^- + SO_3^{2-}$$
$$HS_xSO_3^- + S_2O_3^{2-} \rightarrow HS_{x+1}\text{-}SO_3^- + SO_3^{2-}$$

accounts for the chain formation which occurs in acid systems where elemental sulfur forms and precipitates because it is not soluble in water:

$$HS_nSO_3^- \rightarrow S_n + HSO_3^-$$

The ring size, n, depends on temperature and concentration. S_6, S_8, S_{12}, and polymers are common products. The reaction mechanism for ring formation and degradation has been reviewed by Meyer (1977), who computed the charge distribution in the intermediate chains.

B. HYDROGEN SULFIDE, POLYSULFIDES AND SULFANES

1. Hydrogen Sulfide

Large quantities of hydrogen sulfide are released from swamps and geothermal sources, and enter the atmospheric sulfur cycle (Chapter 6). Equally large quantities are separated from natural gas during production; furthermore, hydrogen sulfide is produced during oil refining. In the future, coal gasification plants are expected to contribute significantly to sulfur in this form.

Hydrogen sulfide is a colorless gas with a density of 0.00153 g/cm^3. The bond angle is 92.5^o. The characteristic 'rotten egg' odor is similar to that of mercaptans found in swamps. This gas is viciously poisonous, as discussed in Chapter 10. The intensity of the smell is not correlated to its concentration, and the heavy gas can accumulate in pockets in closed areas, such as railroad cars, barges, and storage lines and overcome workers by surprise. Hydrogen sulfide is very reactive. In mixture with 4.3 - 46% air, it can explode, and it can react with most metals readily at modest temperature. Because of these properties, hydrogen sulfide must be separated from natural gas at the well head. Only little of this is used in chemical industry; most of it is converted to elemental sulfur. Hydrogen sulfide is shipped in tank cars at a pressure of 17 atm at 20°C.

In the laboratory, pure hydrogen sulfide can be prepared from calcium sulfide with concentrated hydrogen chloride; excess hydrogen chloride is absorbed in barium hydroxide, and the gas is dried by passage through a phosphorus pentoxide column. Carefully dried gas does not react with iron or with mercury. The purity of hydrogen sulfide is best established by chromatography. The chemical reactions of hydrogen sulfide are well known. The use of the differing solubilities of metal sulfides in analytical chemistry is known to anyone who went through a wet chemistry freshman laboratory course. The reaction of dry hydrogen sulfide with metals will be discussed in the section on corrosion (3D).

Hydrogen sulfide reacts with many organic materials. The polymer reactions are discussed in Chapter 13. Hydrogen sulfide adds to double bonds. With epoxides hydroxythiol is formed, with aldehydes geminal olthiols are formed. Formaldehyde reacts and yields trithiane or long linear chains containing the $(CH_2\text{-}S\text{-})_n$ group. With halides additive reactions can be introduced. Hydrogen sulfide has been proposed as an industrial reagent, for example to pretreat wood before pulping (Proctor, 1969).

Liquid Hydrogen Sulfide

The properties of liquid hydrogen sulfide have been reviewed by Jander (1949), Gmelin (1959), Feher (1949), and Meyer (1969). The boiling point of hydrogen sulfide is listed in Table 3.12. The viscosity of liquid hydrogen sulfide is about half that of water at 22°C. The surface tension changes from 33.418 dyne/cm at the melting point to 28.783 at the boiling point. The molar heat capacity is about 16.25 cal/mole·deg

Table 3.12
PHYSICAL PROPERTIES OF H_2S, SCl_2, S_2Cl_2, SO_2, SO_3, AND H_2SO_4

Property	H_2S	SCl_2	S_2Cl_2	SO_2	SO_3	H_2SO_4
Molecular Weight	34	102.97	135.05	64.06	80.10	98.30
Melting point (°C)	-85.6	- 77.8	- 76.1	- 75.5	16.77 (alpha) 32.55 (beta) 62.2 (gamma)	10.371
Boiling point (°C)	-60.75	58.9	137.8	- 10.2	44.8	279.6
Density (l) (g/ml)	0.993	1.621	1.683	1.46	1.9924	1.8269
Critical T (°C)	100.4	-	-	157.12	218.3	-
Critical p (atm)	88.9	-	-	77.7	83.8	-

for the entire liquid range. The vapor pressure increases from 173.7 atm at 85.6°C, the melting point, to 10.2 atm at 0°C, and 17.7 atm at 20°C to 88.9 atm at thc critical point, 100.4°C. The cryoscopic constant is 3.83°C. The electric conductivity of pure liquid hydrogen is 3.7 x 10^{-11} per ohm·cm for direct current and 1.17 x 10^{-9} per ohm·cm for alternate current at -78.5°C. Water reduces the conductivity. The dipole moment is 0.88D at 195°K; the dielectric constant is 8.99, and the polarizability is 3.78 cm^3.

The thermodynamic properties of hydrogen sulfide and the sulfanes, H_2S_n, are listed in Table 3.13. If hydrogen sulfide could be converted to hydrogen and elemental sulfur, substantial quantities of hydrogen could be produced which would be sorely needed in refineries, where much of the hydrogen sulfide occurs. Hyne and Raymont (1975) studied the system, which is endothermic, i.e. consumes 38 kcal/mole, and proposed a step-wise cracking to upset the unfavorable equilibrium by use of a semi-permeable membrane. Hydrogen could be continuously withdrawn at, say 700°C. Suitable membranes are not yet known, but they can probably be developed, because it is well known that sulfides, unlike the oxides, corrode reactive metals. Thus, they form permeable surface films rather than passive films such as those of the oxides. The reaction would proceed by either of the following methods:

$$H_2S + Me \rightarrow MeS + H_2 \qquad I$$
$$MeS \rightarrow Me + S, \text{ or}$$
$$2H_2S + 2CO \rightarrow 2H_2 + 2COS \qquad II$$
$$2COS + SO_2 \rightarrow 2CO_2 + 3/2\ S_2$$

where the sulfur dioxide would be produced by burning sulfur.

Table 3.13
THERMODYNAMIC PROPERTIES OF SULFANES[a]

Species	Temperature (°K)				
	750	800	900	1000	1200
H_2S_2	-9.0	-7.8	- 5.2	- 2.7	2.3
H_2S_3	-6.0	-3.9	0.4	4.7	13.1
H_2S_4	-3.9	-0.8	5.2	11.2	23.0
H_2S_5	-2.1	1.9	9.7	17.5	32.7
H_2S_6	-0.3	4.5	14.1	23.6	42.2
H_2S_7	0.5	6.3	17.7	28.9	50.9
H_2S_8	0.7	7.4	20.6	33.5	59.0
H_2S_9	1.0	8.4	23.4	38.1	67.0

a) Standard free energies of formation of gaseous sulfanes formed from gaseous H_2 and gaseous S_2 (after Hyne and Raymont, 1975);
b) All values are kcal/mole.

SOLUTIONS. It was long assumed that elemental sulfur was insoluble in liquid hydrogen sulfide. It was found that the solubility is 0.14 weight percent at -60°C, 0.49% at 0°C, and 1.3% at 80°C (Smith, 1970). Thus, the sulfur plugging observed in sour gas wells might be due to a combination of a change in solubility and the simultaneous chemical decomposition of polysulfides, as described below. As mentioned above, hydrogen sulfide also dissolves in liquid sulfur and slowly establishes an equilibrium with polysulfides (Muller, 1969; Hyne, 1966).

Due to its dipole moment of 0.88D, hydrogen sulfide can act as a solvent for both polar substances, such as hydrogen chloride, carbon disulfide, ether, ethanol, and alkali halides, as well as non-polar compounds such as pentane and hexane. Hydrogen sulfide reacts with bromine, chlorine, and fluorine yielding sulfanes; with iodine it forms a charge transfer complex which contains ions, as evidenced by the 10^4 fold increase in conductivity (Jander, 1949). Substituted ammonium halides also dissolve as ions, as do the heavy metal halides.

Liquid hydrogen sulfide forms mixed phases with many solvents and compounds, for example, the hydrogen halides. The phase equilibrium with carbon dioxide, with which it occurs in some gas wells, has been reported by Bierlein (1953). It is not yet established whether hydrogen sulfide and carbon dioxide form an azeotropic mixture or a complex. Carbon disulfide and hydrogen sulfide form a solid, hydrate $CS_2 \cdot 6H_2S$ with the same formula as the corresponding $CO_2 \cdot 6H_2O$. The former melts at -102°C. The formation of this compound is reversible. The adduct $CS_2 \cdot H_2S$ is not identical with thiocarbonic acid, H_2CS_3, even though it has the same molecular weight!

With 30 mole % ether, hydrogen sulfide forms an eutectic mixture which melts at -60°C. The two molecules also form a 1:1 adduct which melts at -148.6°C. Hydrogen sulfide and sulfur dioxide mix in all proportions; an eutectic containing 76 mole % hydrogen sulfide melts at -110°C. Pure liquid hydrogen sulfide and sulfur dioxide do not react. At higher temperature, in the presence of moisture or a catalyst, they form complex mixtures of elemental sulfur, polythionates, and thiosulfate, commonly called Wackenroder's liquid (1946).

CHEMICAL REACTIONS. With ketones and aldehydes, hydrogen sulfide forms adducts which slowly react and form thioaldehydes and water. Cyanamide, NH_2-CN, reacts with hydrogen sulfur forming thiourea in a reaction which is analogous to that with water. However, cyanogen forms $(NH_2)_2(C{=}S)_2$, and not HCNS, as one might expect. The reaction of acid chlorides with hydrogen sulfide is slow, and, unfortunately, the equilibrium is unfavorable for synthesizing dithioacids. More details about the chemistry in liquid hydrogen sulfide can be found in the series by Feher (1970) and Jander (1949). The latter conducted a large number of acid-base titrations using liquid hydrogen sulfide.

Fluorine, chlorine, and bromine react with hydrogen sulfide yielding disulfane and higher sulfanes. It is not clear whether the reaction involves free radicals. Elliot (1974) and Strausz (1966) showed that HS radicals are formed in all three phases of pure hydrogen sulfide and in matrices, by UV light at 2,200 A or shorter wavelength. Forys (1976) studied the radiolysis of hydrogen sulfide and found very high rate constants.

2. Polysulfides and Sulfanes

Polysulfides produced from liver of sulfur were discovered by Scheele (1711) and were extensively used and studied by alchemists and chemists for the last 200 years. An excellent 34 page paper by Berthollet, published in 1798, describes much of our present chemical knowledge. These compounds are mixtures of molecules of the type X-S_n-X in which n values range between 1 and 10. The sulfur chains are terminated by hydrogen, halogen, or other inorganic or organic terminal groups. Since these compounds are the sulfur analogues of alkanes, H-$(CH_2)_n$-H, their official IUPAC name is now sulfane.

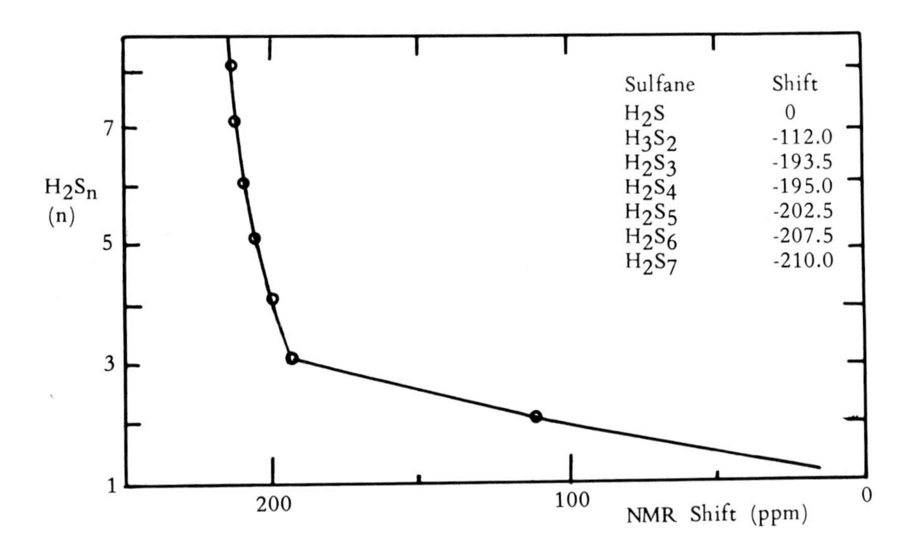

Fig. 3.11. NMR shifts of sulfanes in CS_2
(Schmidbaur, 1964; Muller, 1969; Jensen, 1971).

Polysulfides are not branched. Feher (1969) developed a series of elegant synthetic methods for selective synthesis of pure compounds having uniform chain length. Schmidbaur (1964) discovered that in the sulfanes–in contrast to the peroxides and ozonides–the proton exchange is slow enough to yield a characteristic NMR spectrum for each chain length value n. This effect has since been studied for pure sulfanes and sulfanes in various solvents by Muller (1967, 1969) and Morelle (1970). The NMR spectrum of sulfides in liquid sulfur was studied by Hyne (1966) and Wiewiorowski (1966, 1970), Fig. 3.11. Schmidt (1960-1973) promptly exploited this analytical tool to synthesize 12 new sulfur allotropes and many other novel compounds. Feher (1959) reported the UV spectra, and all physical properties of the sulfanes, chlorosulfanes, and some organic sulfanes. The electronic properties of the sulfur chains are of theoretical interest and have been studied by Muntzer (1969) and Meyer (1969, 1977). The structures of various polysulfides have been discussed by Winnewasser (1968), Wieser (1968), Rahman (1970), Banister (1969), and Wart (1976). Cardone (1972) described analytical procedures and also reviewed our present knowledge of organic sulfides.

Crude sulfane, a mixture of sulfanes with different chain lengths, is produced by fusion of 500 g $NaHS{\cdot}9H_2O$ and 250 g sulfur at 100°C:

$$Na_2S \quad + \quad nS \quad \rightarrow \quad Na_2S_{n+1}$$

The sodium sulfides are quenched in concentrated hydrogen chloride, and can be separated from the aqueous phase, because they are non-polar. The average chain length of the product is 5.5. The same product composition can be obtained by dissolving sulfur in ammonium sulfide (Piskorski, 1972).

Pure sulfanes with a given chain length can be obtained by condensation of chlorosulfanes and sulfanes, which are simultaneously but separately added to dry ether at -40°C (Schmidt, 1974; Feher, 1963):

$$2H_2S_n \quad + \quad Cl_2S_m \quad \rightarrow \quad H_2S_{m+2n} \quad + \quad 2HCl$$

The sulfanes are fairly stable in solution. The pure substances equilibrate (Leinecke, 1967; Morelle, 1970) and decompose (Muller, 1969) until a mixture is formed in which the resulting elemental sulfur is saturated with sulfane with an average chain length between 4 and 7. Apparently, the dissolved sulfur has a stabilizing effect, because the mixture decomposes quickly if elemental sulfur is extracted. The equilibrium composition of hot liquid sulfur saturated with 1 atm hydrogen sulfide was discussed by Wiewiorowski (1965) and Rubero (1964). The stability of sulfanes has been calculated (Raymont, 1975). Feher (1958) determined the heat of vaporization, the vapor pressure, the critical temperature, and Trouton's constants for all sulfanes known at that time.

3. Chlorosulfanes and Other Halosulfanes

Pure chlorosulfanes and bromosulfanes can be prepared by the method described above (Feher, 1958-1969), if excess chlorosulfane is used:

$$H_2S_n \quad + \quad 2Cl_2S_m \quad \rightarrow \quad Cl_2S_{2m+n} \quad + \quad 2HCl$$

The most stable chlorosulfane is the red, oily disulfur dichloride, S_2Cl_2, commonly called 'sulfur dichloride'. It forms whenever sulfur is reacted with chlorine gas, almost independent of the ratio of the reagents. A large excess of chlorine yields 'sulfur monochloride', which is monosulfurchloride, SCl_2. The stoichiometry of both compounds was already established by Dumas (1832). S_2Cl_2 is a very peculiar molecule (Bradley, 1967; Frankiss, 1968) in that it contains two very tightly bonded sulfur atoms at a distance of 1.89 A, while the chlorine atoms are unusually distant and weakly bonded, Table 3.8. Thus, upon photolysis S_2Cl_2 forms S_2 and not SCl (Morelle, 1971). The chlorosulfanes react with sulfur (Hammick, 1928) forming an equilibrium mixture of chlorosulfane. They also react with all substances which react with elemental chlorine or sulfur. With water S_2Cl_2 decomposes and forms polymeric and colloidal sulfur. In industry, S_2Cl_2 has been used since 1846 for vulcanizing rubber (Penati, 1975). It is also used as an antioxidant additive to cutting oil and drilling mud (Braithwaite, 1967); in many organic reactions for chlorination, and for other purposes. An excellent review of the chemistry and applications is given in a pamphlet by Stauffer Chemical (1975).

The bromosulfanes are in every respect similar to the chlorosulfanes, but the S-Br bond is yet weaker. Iodine-S bonds are not stable at room temperature (Minkwitz, 1975), despite sporadic earlier claims (Feher, 1963). The fluoro sulfanes have been prepared and reviewed by Seel (1964-1975). They form a fascinating family of compounds, but so far they have only academic interest. As in the case of Cl_2S_x and Br_2S_x, the halogen-S bond distance is unusually weak (Kuczkowski, 1964; Companion, 1972), since the sulfur has a positive charge, as opposed to the polysulfides in which it has its normal octet structure preserved by negative charges at the chain end. Reactions of H_2S and S_2F_2 have been studied by Meyer (1971). Fluorinated organic polymers formed from polysulfides have been described by Krespan (1962, 1968). SF_4 (Becher, 1974) and SF_6 are important chemicals. The latter is almost as inert as nitrogen and is used on a substantial scale, but it is not important in connection with this book.

4. Alkali Polysulfides

Alkali metal polysulfides can be obtained by direct reaction of the elements (Jellinek, 1968), by fusion *in vacuo*. The chain length of the product depends on the ratio of the reagents. Lithium sulfide and sodium sulfide have recently drawn substantial attention, because they are suitable for use in bulk power storage batteries. This subject is discussed in Chapter 14. The stoichiometry of sodium polysulfides has been studied by Rule (1914), by thermalanalysis (Oei, 1973; Rosen, 1972), X-ray and Raman spectroscopy (Janz, 1976), and the phase diagram is fairly well established. Tegman (Tegman, 1974), Fig. 3.12. The potassium sulfides are well established (Jellinek, 1968; Diot, 1972) and their heat of formation is known (LeToffe, 1974); the lithium polysulfides are under intensive

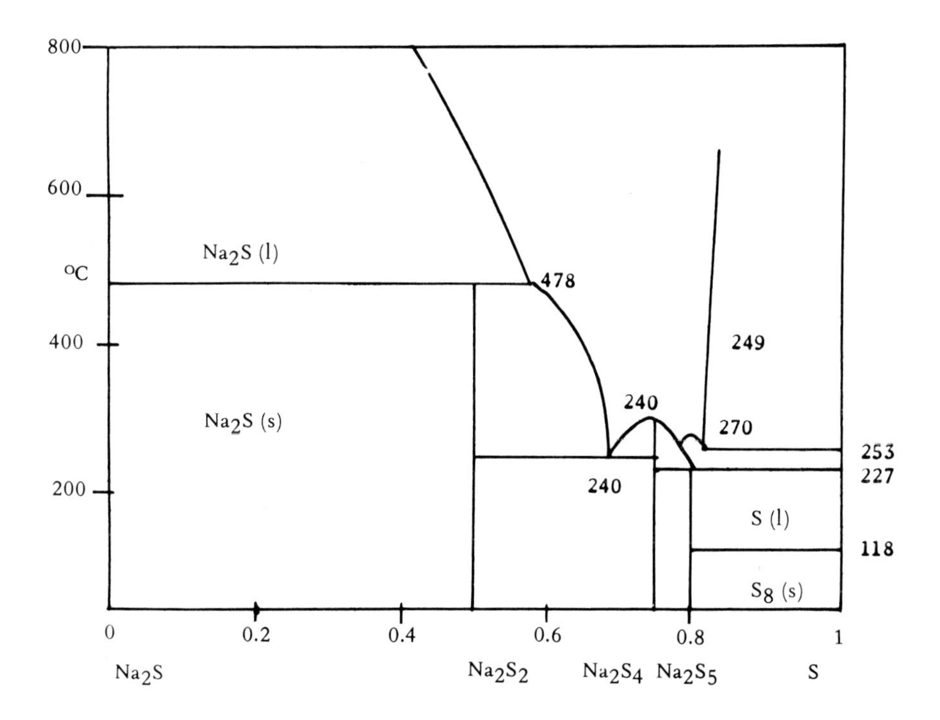

Fig. 3.12. Phase diagram of sodium-sulfur (after Tegman, 1974).

study. Hofmann (1903) prepared heavy metal tetra-, penta- and hexasulfides in which the heavy metal ion forms part of a sulfur ring (Schmidt, 1972). They have over-all formulae of the type $PtS_{15}(NH_4)_2$, and were originally widely believed to consist of undefined mixtures.

Sulfide and Polysulfide Solutions

Hydrogen sulfide and its base salt NaHS are commercially available. The latter is sold as a pale yellow aqueous 45% solution which crystallizes below 25°C (77°F). This solution attacks copper, zink, and aluminum alloys rapidly. It is used for metal refining, in the production of high quality papers (for example, cigarette paper), for applying dyes to leather or cotton, to regenerate lead sulfide in the gasoline sweetening process, and in organic chemistry. A review of uses, application, and analysis is available from Stauffer (1974).

Table 3.14
ACIDITY CONSTANTS OF SULFANES

Sulfane	K_1	K_2	Reference
H_2S	6.9	17.1	a
H_2S_2	5.0	9.7	b
H_2S_3	4.2	7.5	b
H_2S_4	3.8	6.3	b
H_2S_5	3.5	5.7	b
H_2S_6	3.2	5.2	c
H_2S_7	3	4.8	c
H_2S_8	2.9	4.4	c

a) After Giggenbach (1971); b) Schwarzenbach (1960; c) Meyer (1977).

The reactions of hydrogen sulfide, the sulfanes and the alkali polysulfides are determined by their acidity constants, Table 3.14, which were determined by Schwarzenbach (1970). Giggenbach (1971) showed that the second dissociation of hydrogen sulfide has a value of $pK_2 = 17.1$ plus or minus 0.2 at 25°C, and not 14, as was generally accepted. The equilibria between aqueous polysulfide ions at different pH and temperatures have been studied by Giggenbach (1972, 1974) and Teder (1971, 1969). Both used the UV spectrum. The measurements are difficult, because at high pH the sulfides oxidize readily, even with traces of dissolved air. Furthermore, they undergo temperature dependent autooxidation (Bowers, 1966). Giggenbach and Teder also measured the reaction rates. Hartler (1967) determined the rate at which sulfur dissolves in aqueous sodium sulfide, and carefully observed that the reaction has two stages, both of which require only little activation energy. The conversion of elemental sulfur to aqueous polysulfide was also studied by Hartler (1972), Teder (1971), and Chen (1973). Teder (1969) determined the effect of pH on the equilibrium. Giggenbach (1974) studied the kinetics of the decomposition of polysulfides to thiosulfate, up to 240°C. Morth (1966) studied the formation of sulfate. Teder (1970) studied the vapor pressure over liquid polysulfides. The oxidation of sulfide in sea water and air is discussed later. Ikeda (1972) developed a potentiometric titration method for sulfide and sulfur dissolved in polysulfide, for example, in kraft cooking and waste liquors. Non-aqueous solutions of polysulfide have been studied by Cleaver (1973), who determined cryoscopic effects of alkali polysulfides fused in thiocyanates.

Table 3.15
SOLUBILITIES OF HYDROGEN SULFIDE

Solvent	T (°C)	Solubility ml/g
Water	25	2.21
2.4% H_2SO_4	25	1.96
5% NaCl	25	1.8
5% $(NH_4)_2SO_4$	25	1.6
5% Na_2SO_4	25	1.4
Hexane	20	8.9
Benzene	20	16.6
10% Monoethanol amine	25	30
at 20 Torr	45	21
	60	15

The solubility of hydrogen sulfide in water is 2.21 ml gas per gram water (3.35 g/l) at 25°C. The temperature and pressure dependence under geological conditions has been reported by Selleck (1952). Other solubilities are summarized in Table 3.15.

C. SULFUR OXIDES AND OXYACIDS

This section deals with the chemical reactions between sulfur and oxygen in the presence or absence of water, bases, or other reagents. These reactions are important in the atmosphere, the oceans, in soil, as described in Chapter 6, and during combustion, described in Chapter 8.

The two best known sulfur oxides are the gaseous sulfur dioxide, SO_2, and the sulfate ion, SO_4^{2-}, which occurs in solids and solutions. The first is a common air pollutant, the second occurs in gypsum, copper sulfate, alum, and many minerals. It is not always recognized that these two compounds contain sulfur in different chemical forms:

Sulfur burns in air yielding sulfur dioxide

$$S_n \quad + \quad nO_2 \quad \rightarrow \quad nSO_2$$

At room temperature sulfur dioxide is a stable colorless gas which can be stored and shipped in pressure cylinders in liquid form (Hitchcock, 1931). However, inspection of the electron configuration shows that the sulfur atom in sulfur dioxide has not yet reached the highest oxidation state and can be further oxidized. In the presence of water and oxygen, sulfur dioxide can be converted to sulfuric acid:

$$H_2O + \frac{1}{2}O_2 + SO_2 \rightarrow H_2SO_4$$

The last equation describes two chemical steps, hydration and oxidation, which can take place in either order. In ambient air, SO_2 is first hydrated, for example by carbonized water, and then slowly oxidized:

$$SO_2 + HCO_3^- \rightarrow HSO_3^- + CO_2$$
$$HSO_3^- + \frac{1}{2}O_2 \rightarrow HSO_4^-$$

In a sulfuric acid plant, dry sulfur dioxide is oxidized first and then hydrated:

$$SO_2 + \frac{1}{2}O_2 \rightarrow SO_3$$
$$H_2O + SO_3 + H_2SO_4 \rightarrow 2H_2SO_4$$

The above reactions show that the oxidation rate of sulfur is not a simple process. As a matter of fact, today over 20 intermediate oxides and oxyacids of sulfur have been identified and synthesized. In some of them, for example thiosulfate, $S_2O_3^{2-}$, sulfur atoms in different oxidation states co-exist in the same molecule. The complexity of the sulfur oxide systems was recognized by alchemists around 1700. They knew, for example, the difference between the chemical behavior of volatile acid (sulfurous acid) (Homberg, 1710, Seehl, 1744), obtained by condensing the vapor of burned sulfur, and that of acid obtained by distillation of Glauber's salt (sulfuric acid). The oxyacids of sulfur were so intensely studied that they occupied more space in the first edition of Gmelin's Handbook of Inorganic Chemistry of 1848 than elemental sulfur and all other compounds of sulfur together. Curiously, interest in the subject slackened before the subject was fully elucidated, and now, 150 years later, many who work with sulfur are unaware of the earlier work or cannot obtain copies of the older articles, because they are neither referenced nor quoted in modern textbooks. This loss has caused substantial inconvenience to those who were rushed into building large scale sulfur dioxide plants in order to meet the impending air quality laws. Several oil companies, for example, discovered that a deep yellow, sticky, oily suspension accumulated in recovery tanks built to receive elemental sulfur from Claus tail gases. This liquid was correctly described by Wackenroder in three papers in 1846. The liquid contains a mixture of oxyacids of sulfur, among them thiosulfate and polythionates, which are stubbornly stable, even against boiling water. This discovery caught many

Table 3.16
OXYACIDS OF SULFUR

$H_2S_2O_4$	Dithionous acid
H_2SO_3	Sulfurous acid
$H_2S_2O_5$	Disulfurous acid
$H_2S_2O_6$	Dithionic acid
H_2SO_4	Sulfuric acid
$H_2S_2O_7$	Disulfuric acid
H_2SO_5	Peroxo monosulfonic acid
$H_2S_2O_8$	Peroxo disulfonic acid
$H_2S_2O_3$	Thiosulfuric acid
$H_2S_3O_6$	Trithionic acid
$H_2S_xO_3$	Sulfane monosulfonic acids
$H_2S_xO_6$	Sulfane disulfonic acids, polythionic acids

modern engineers and physical chemists by surprise, because they had relied on thermodynamic calculations which predicted only whether reaction conditions are theoretically favorable. Unfortunately, such calculations contain no information about the reaction mechanism or the kinetics. The latter both prevent the formation of elemental sulfur in the above reaction.

Table 3.16 lists some sulfur oxides and oxyacids. These compounds have been reviewed by Gmelin (1959), Lyons (1968), and Schmidt (1971). Only few of their properties will be mentioned here, and even these will be treated superficially.

1. SO_3, Sulfur Trioxide

For many years, commercial sulfur trioxide was stabilized by the addition of 2% $SOCl_2$, BCl, or $TiCl_4$ to keep it in the liquid phase. Bulk commercial sulfur trioxide is now sold 99.8%+ pure, no longer contains inhibitors, and is shipped at 100°F (40°C) as a liquid which melts at 16.8°C and boils at 44.8°C, Table 3.12. Details for handling SO_3 are described in a pamphlet by Stauffer (1975).

Oleum, fuming sulfuric acid, contains 25-65% sulfur trioxide in concentrated sulfuric acid. Sulfur trioxide is a planar molecule with an S-O bond length of 1.43 A, and a bond angel of 120°C. The gas phase IR and Raman spectrum has been discussed by Tang (1975), who assigned the frequencies: $v_1(a'_1) = 1065\ cm^{-1}$, $v_2(a''_2) = 480\ cm^{-1}$, $v_3(e_1) = 1380$, and $v_4(e_1) = 536\ cm^{-1}$. The photoelectron spectrum was measured by

Lloyd (1976) and Alderdice (1976). Three forms of sulfur trioxide are known: Gamma-SO_3 consists of trimers, alpha- and beta-SO_3 are polymers and are solids at room temperature. The polymerization is catalyzed by water and impurities.

In gamma-SO_3, three SO_3 form a S_3O_9 molecule in which alternating sulfur and oxygen atoms form a six membered, puckered ring in which each sulfur is surrounded by four oxygen atoms located at the corner of a tetrahedron. The bond angles are 99° and S-O-S is 121°. The substance forms orthorhombic crystals, with the space group Pbn. Each unit cell contains four molecules; the dimensions of the unit cell are a = 5.2, b = 10.8, and c = 12.4 A.

Beta-SO_3 consists of helical chains which crystallize as needles by spontaneous polymerization. Since the reaction is catalyzed by traces of water, the polymeric molecules can be considered to be polysulfuric acids: $H_2O + S_nO_{3n} = H_2S_nO_{3n+1}$. This asbestos-like substance forms monoclinic crystals of the space group $C^5{}_{2n}$-$P2_{1K}$, with a unit cell of a = 6.2, b = 4.06, and c = 9.31 A. The dihedral angle is 109°. Alpha-SO_3 melts at 62^oC. The solid has a layer structure.

The vapor pressure of liquid SO_3 is 265 Torr at 25^oC, the critical point is 217 ± 2^oC and 80.8 ± 0.3 atm. Sulfur trioxide is soluble in liquid sulfur dioxide and $SOCl_2$, sulfuryl chloride. In carbon disulfide and carbon tetrachloride, it slowly reacts with the solvent forming COS and $COCl_2$, respectively. With water, sulfuric acid is formed. Sulfur trioxide is very reactive. It acts as an oxidizing agent and as a Lewis acid. The equilibrium

$$SO_2 \quad + \quad \tfrac{1}{2}O_2 \quad \rightarrow \quad SO_3 \quad + \quad 22.85 \text{ kcal/mole}$$

is dicussed in the section on sulfur dioxide. The equilibrium is shown in Figs. 3.13 and 3.14 . Above 400^oC, sulfur trioxide decomposes without catalyst; however, the conversion remains inefficient up to 1000^oC. The reactions of sulfur trioxide are reviewed by Gmelin (1953-1963) and Schenk (1968). The commercial uses are listed in Chapter 12.

2. SO_2, Sulfur Dioxide

Sulfur dioxide was one of the first chemicals ever used. Its function as a bleaching agent and disinfectant was known at least 4000 years ago. Today, over 100,000 tons of sulfur dioxide are produced, and about 60 million tons are released as impurity, together with 4000 million tons of

Table 3.17
VAPOR PRESSURE OF H_2S, SO_2 AND SO_3

p (Torr)	H_2S	SO_2	SO_3 (alpha)
10^{-1}	119.6	161.6	215.4
1	138.26	177	234.3
10	156.70	195.8	256.7
100	180.78	225.26	283.9
760	212.97	263.08	324.3
5 atm	250.9	305.3	351.1
10 atm	272.8	328.7	367.9
40 atm	329.0	391.2	406.8
60 atm	349.5	414.9	419.8
T_K	373.5	430.4	492.40
p_K (atm)	88.9	77.7	83.4

carbon dioxide which enters the atmosphere as a result of coal and oil combustion.

The sulfur dioxide molecule has an S-O bond length of 1.432 A, and a bond angle of 119.5°. Some physical properties of sulfur dioxide are shown in Table 3.12. The colorless, poisonous gas condenses at -10.02°C and freezes at -75.48°C. The vapor pressure is shown in Table 3.17. The solubility of sulfur dioxide is important in air pollution abatement, and will be discussed in Section 8C. Sulfur dioxide acts as a Lewis acid; it reacts with water and forms a complex system containing HSO_3, SO_3^{2-}, $S_2O_5^{2-}$, but no H_2SO_3. These reactions will be discussed below.

The IR spectrum has been reported by Maillard (1975). The UV spectrum of sulfur dioxide has been described by Meyer (1971), and many others. Sulfur dioxide has a low lying triplet state at 3150 A, i.e. in the blue region of the spectrum (Greenough, 1961; Meyer, 1968, 1970; Chung, 1975). The rotational structure of this transition has been analyzed by Hamada (1975). This usually long-lived singlet state, with a half life of about 0.02 sec, can be activated by light, radiation (Lalo, 1974), or chemical reactions. It is quenched by flue gases (Okabe, 1974), and contributes to the reactivity (Morelle, 1971; Bottenheim, 1976), for example, the oxidation (Allen, 1972). However, the blue emission which is used in the flame photometric detection is not due to SO_2, but to the very efficient 2900 A emission of the S_2 molecule which is formed in the hot reducing zone of the detector flame:

$$2SO_2 \quad + \quad 4H_2 \quad \rightarrow \quad S_2^* \quad + \quad 4H_2O$$

The same emission is also observed in the emission of shock heated SO_2 (Levitt, 1967). The IR and Raman spectra indicate the following values for the fundamental frequencies: v_1 = 1151.38 cm^{-1}, v_2 = 517.69 cm^{-1}, and v_3 = 1361.76 cm^{-1}.

Sulfur Dioxide in Solutions

Sulfur dioxide dissolves in pure water without chemical reaction. It forms hydrates of the type $H_2O{\cdot}SO_2$. Sulfurous acid, H_2SO_3, has never been observed. It probably does not exist. In the presence of base, a complex series of reactions take place, and the bisulfite ion, HSO_3^-, disulfite, $S_2O_5^{2-}$, and sulfite, SO_3^{2-}, are formed. This system is discussed in the section on oxyacids.

As a Lewis acid, sulfur dioxide strongly interacts with all inorganic and organic bases, and forms adducts as well as ionic compounds. Alcoholic solutions have been studied by Maine (1957), who identified donor-acceptor effects in the UV absorption. The Raman spectrum was studied by Simon (1955). The solubility in methanol, ethanol, and n-propanol was measured by Tokunaga (1974). In water it decreases from 15.66 weight percent at 20°C to 10.9 at 20°C, 7.77 at 30°, and 5.76 at 40°C. In methanol it decreases from 131.6 weight percent at 10°C to 41 at 40°, and in ethanol it decreases from 71 wt % at 10° to 24 at 40°C.

Aqueous solutions will be discussed in the section on sulfurous acid and sulfite in Chapter 8.

Liquid Sulfur Dioxide

Sulfur dioxide was first liquefied by Northmore (1808) by compression in a glass cylinder. Walden (1899) was the first to recognize its value as a solvent. Liquid sulfur dioxide is a commercial chemical (Hitchcock, 1931) and an excellent chemical medium for a variety of reactions (Audrieth, 1975). It is a polar solvent, with a dielectric constant of 14.3 at 0°C; thus, it promotes ionic, especially nucleophilic, reactions. Sulfur dioxide acts as an electron acceptor, and also forms complexes with donor molecules.

In industry, sulfur dioxide is used to extract sulfur, i.e. desulfurize, kerosene during refining. Krause (1955) described the sweetening of heating oil; Arnold (1956), Haney (1954), Rylander (1956), and Seelig (1955) described the desulfurization of petroleum products in the presence of $AlCl_3$, BF_3, $SOCl_2$, and SO_2Cl_2. Sulfur dioxide also extracts excess sulfur from nitro compounds, cellulose acetate, and many other organic and inorganic materials (Audrieth, 1975). Technical uses are listed in Chapter 12. With sulfites, liquid sulfur dioxide forms pyrosulfites, i.e. disulfites of the formula MeS_2O_5, as described below. With ammonia, a vigorous reaction is observed which yields amidosulfinic acid, $NH_2SO_2NH_2$, or NH_2SO_2H (Becke-Goehring, 1958; Meyer, 1977). Methyl- and ethylamine form the corresponding esters of thionyl imide, i.e. CH_3NSO and C_2H_5NHS, and the corresponding amine pyrosulfites. The assumed autodissociation, proposed by Jander (1949):

$$SO_2 + SO_2 \rightarrow SO^{2+} + SO_3^{2-}$$

has not been generally confirmed, but the concept gives good nominal, theoretical prediction of many solute systems, such as:

$$PCl_5 + SO_2 \rightarrow POCl_3 + SOCl_2$$

and of solvation of LiI, $(NH_4)_2CNS$, and other substances which form ions and electrically conducting solutions.

Sulfur dioxide is a good solvent for sulfur trioxide, and, thus, can serve for sulfonation. reactions. Dithiols form mono- and disulfated products which form cyclic ethylene sulfate *in vacuo*. Sulfur dioxide is an excellent solvent for Friedel-Crafts reactions, because it dissolves both $AlCl_3$ and aromatic hydrocarbons. Likewise, bromination can be conveniently conducted in this solvent.

The extensive literature on redox, solvolysis, electrochemical, and other reactions has been reviewed by Waddington (1965), Jander (1949), Audrieth (1975), Schenk (1968), and Burow (1970). Technical applications are discussed in Chapter 12. Radiation induced redox reactions have been reported by Rothschild (1966).

Reactions of Sulfur Dioxide

Sulfur Dioxide and Ammonia. Ammonia and sulfur dioxide are both present in the atmosphere; their reaction has been invoked to explain the atmospheric behavior of sulfur dioxide, and is used in air pollution abatement systems. Some scientists, expecially physical chemists and atmospheric scientists, have claimed that their reaction is purely physical and reversible. Others, mainly inorganic chemists, have described chemical reactions and colorful reaction products. Apparently, many members of the two groups are not familiar with each other's work; the work of both contains correct observations.

In 1826, Döbereimer observed that dry sulfur dioxide and ammonia form a yellow-brown gas which condenses as a brown solid. The latter dissolves in water, forming a colorless solution. Forchhammer (1837) described the reaction of ammonia with sulfurous acid. In the following twenty years, Rose published several papers on the subject, and in 1890 Fork reported that diammonium pyrosulfite could be formed, which was prismatic, unlike $K_2S_2O_5$, which he described as being 'monosymmetric'. In 1900, Divers, Ogawa, and Schumann independently published thorough studies. The first reported that in dilute organic solution ammonium aminosulfite, $(NH_4)SO_2NH_2$, was formed, which decomposed and formed diammonium imido sulfite $[(NH_4)SO_2]_2NH$. The latter reported that initally a yellow $NH_3{\cdot}SO_2$ adduct was formed, and that the colorless aqueous solution contained sulfite, sulfate, thiosulfate, and di- and trithionate. A thorough study of the reaction was made by Goehring (1951, 1956), who established that below 10°C, in aqueous systems, 90% of the reagent converts to bisulfite and ammonium ion. Above 30°C, completely different products are observed. The reaction products depend on whether water is present and on many other conditions. Frequently, secondary reaction products are formed. The yellow products were assigned to $NH(SONH_2)_2$, and the red to $NH_4{\cdot}N(SO_2NH_4)_2$. Similar results were obtained by Badar-ud-Din (1953). In 1969, Scott summarized the status of atmospheric research; it indicates that 90% of the aerosol in the stratosphere, 20 km above earth, consists of ammonium sulfate, and that ammonium sulfite in the troposphere is formed by vaporization of water droplets. For details see Chapter 6. In 1970, Scott measured the vapor pressure of adducts of sulfur dioxide and ammonia at -70°C to -10°C and the heat of vaporization, and described a violent reaction at -10°C. Cleghorn (1970) reported that sodium pyrosulfite decomposes above 95°C to sulfite and sulfate, and above 218°C to sulfate. In 1971, Baggio established the structure of ammonium pyrosulfite. In 1971, Scargill measured the heats of dissociation of ammonia sulfite (ΔH = -65 kcal/mole) and disulfite (ΔH = -80 kcal/mole) in air, and reported that in the presence of air water favors sulfite formation, while water alone favors the formation of pyrosulfite. He reported that at 50°C the dissociation products are gaseous sulfur dioxide and ammonia. McLaren (1974) re-measured the heat of dissociation of products containing sulfur dioxide, ammonia, and water in different ratios. In 1974, Landreth reported the heat of dissociation of ammonia-sulfur dioxide adducts with an ammonia to sulfur dioxide ratio of 1-1, 2-1, and 1-2, and gave values of 9.5 kcal/mole (S = 15.3 cal/mole), 23 kcal/mole (S = 82 cal/mole), and -53 kcal/mole (S = 113 cal/mole) for the reaction $(NH_3)_n{\cdot}(SO_2)_m(s) \rightarrow NH_3(g) + SO_2(g)$.

In 1975, Hisatsune reported an IR study of trapped $NH_3{\cdot}SO_2$ supporting the earlier findings. Hata (1964) reported formation of ammonium disulfite. Hartley (1975) and Vance (1976) studied the formation of aerosols. The first considered several reaction products, while the latter tried to explain his results in terms of simple adducts. Our own work indicates that the low temperature 'adduct' consists of NH_2SO_3H, $NH_2S_2O_4H$, or ammonium salt, depending on temperature, ratio of reagents, and the concentration of the reagents in the reagent gas (or air) mixture. Above 0°C secondary reactions are almost always fast, and lead to irreversible conversion of the reagents. The secondary reactions can involve several steps, and thus can be very complex. Our IR and Raman studies indicate that eventually N_4S_4 (Lippincott, 1953), water, and other nitrogen-sulfur containing products (Hofmann, 1956) are formed. The oxidation of the sulfur to sulfite is apparently slow, except in the atmosphere where oxygen and light are important. The atmospheric reactions are described in Section 6C; the abatement reactions in Chapter Eight. For technical applications, oxidation in the solution, such as proposed by Wolfkowitsch (1933), should be considered. The technical oxidation of ammonia-sulfur dioxide solution to ammonium sulfate during abatement processes, for preparing ammonium sulfate fertilizer, or regeneration of the ammonia without oxidation are also discussed in Chapter 8.

The Raman spectra of $N(SO_3)_3^{3-}$, $NH(SO_3)^{2-}$, $NH_3\text{-}SO_3^-$, KNH_2SO_3, $(NH_2)SO_2$, and $(NSO_2)_3$ have been described by Hofmann (1956). The matrix spectrum of NH_3SO_3 was reported by Lucazeau (1975). Thionylimid, a gas with the formula NH-SO, with bond angles of 110° and 130°, was studied by Richert (1961). Ito (1975) discussed the solubility of various imidobisulfate salts.

With amines, sulfur dioxide originally forms adducts (Bateman, 1944) rather than ionic products as Jander (1949) proposed. Norris (1954), who carefully analyzed isotopic exchange of sulfur compounds in several systems, studied the system triethylamine-SO_2, and confirmed Bateman's work. Donbavand (1974) established free radicals in N,N diethylmethyl amine and tri-n-propylamine. Maylor (1972) reported the formation of sulfites and pyrosulfite or tetra-n-alkylammonium hydroxides of the type R_4NSO_3 or $R_4NS_2O_5$.

3. Other Sulfur Oxides

SO, Monosulfur Monoxide

This molecule exists at high temperature and low pressure. It also occurs as transient species during photolysis. It is well characterized, but not important in connection with our subject (Barrow, 1970).

S_2O, Disulfur Monoxide

This molecule forms during low pressure combustion of sulfur, especially in oxygen deficient gas mixtures. Pure disulfur monoxide can be prepared by the reaction of Ag_2S with $SOCl_2$ (Schenk, 1968). Upon condensation it forms a red solid which can be dissolved in ether. It is possible, but not established, that S_2O occurs as an intermediate in the oxidation of the sulfur molecule S_8. Disulfur monoxide has the same number of valence electrons as ozone, S_3, and SO_2. The structure was identified by Meschi (1959), Blukis (1965). The UV was reported by Phillips (1969); the IR spectrum by Hopkins (1975).

S_8O, Octasulfur Monoxide

Octasulfur monoxide, prepared by the reaction of H_2S_x with $SOCl_2$, is a laboratory curiosity. Its structure, a strained S_8 ring with an oxygen bonded to one ring member, is well established (Steudel, 1975) and proves that sulfur can partake in far stronger bonding arrangements than chemists normally would like to consider.

4. Oxyacids of Sulfur

H_2SO_4, Sulfuric Acid

Sulfuric acid is not only the most important sulfur compound, but also the most commonly used industrial chemical. Various preparation methods have been discussed in the modern scientific literature since at least Cox's (1674) papers in the Philosophical Transactions Volume 9. During 1975 over 110 million tons of sulfuric acid were manufactured (Texas Gulf, 1976). The technical importance of sulfuric acid is described in Chapter 12. Several excellent books and reviews describe the properties

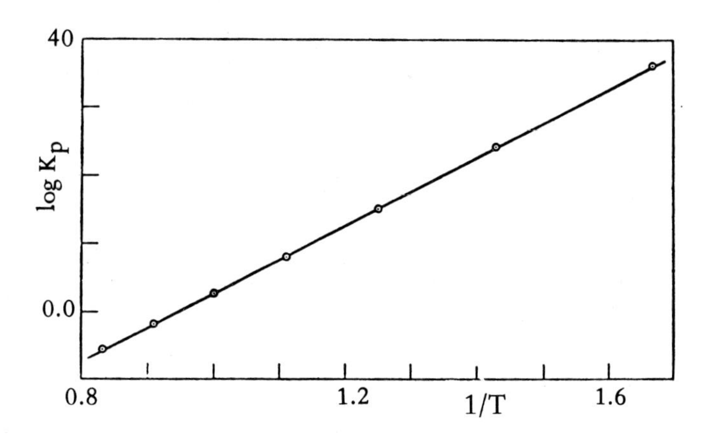

Fig. 3.13. Equilibrium constant $SO_2 + ½O_2 = SO_3$ from 350 to 1000°C.

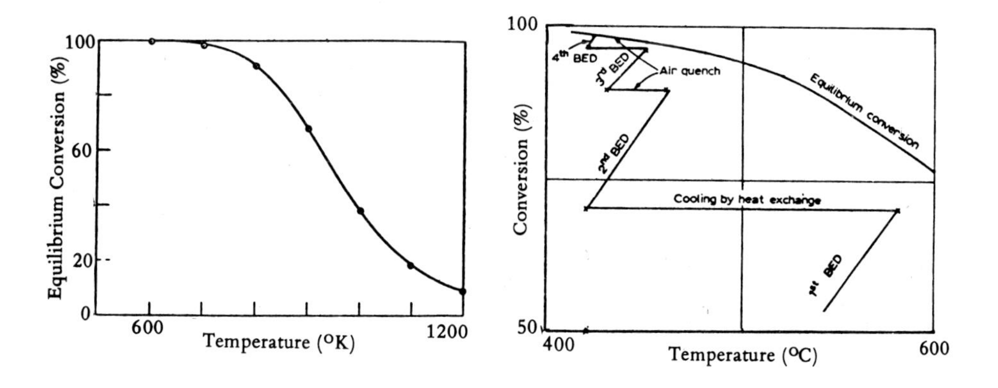

Fig. 3.14. Equilibrium conversion of SO_2 to SO_3 from 250 to 1000°C
a) conversion of 8% mixture, b) 4 step conversion.

of sulfuric acid: Duecker (1959), Pearce (1968), Ullmann (1964). In this section, despite the importance of the subject, only the most basic and superficial properties can be touched.

The acid is made from SO_2 by reaction with oxygen:

$$K_p\ (atm^{-1}) = \frac{PSO_3}{(PSO_2)(PO_2)^{1/2}}$$

$$\log K_p = \frac{5022}{T} - 4.765$$

Table 3.18
EQUILIBRIUM DATA FOR THE REACTION $SO_2 + ½O_2 = SO_3$

Temp. (^oK)	H_T (kcal/ mole)	F_T (kcal/ mole)	log K_p	K_p ($atm^{-½}$)
600	-23.42	- 9.94	3.621	4180.0
700	-23.27	- 7.72	2.410	237.0
800	-23.08	- 5.51	1.505	32.0
900	-22.87	- 3.34	0.811	6.47
1000	-22.61	- 1.18	0.258	1.81
1100	-22.35	+0.96	-0.191	0.645
1200	-22.06	+3.07	-0.559	0.276

The temperature dependence of the equilibrium constant is shown in Fig. 3.13; some select equilibria constants are listed in Table 3.18 and the equilibrium conversion is shown in Fig. 3.14a. At a temperature below 500°C (932°F), 90% conversion is possible in one step. Unfortunately, the kinetics are slow and catalysts have to be used. In a study of Vanadium catalysts, the following rate constant was established:

$$\frac{d(SO_3)}{dt} = \frac{k_1(SO_2)^{½}(O_2)^{½}}{(SO_3)} - \frac{k_2(SO_3)^{½}}{(SO_2)^{½}}$$

with k_1/k_2 = 13.4 at 575°C. The activation energy is 34 kcal/mole at 500°C. Mars and Krevelen (1954) described the reaction in three steps:

$$SO_2 + Cat\text{-}O \rightarrow SO_3 + Cat$$
$$Cat + O_2 \rightarrow Cat\text{-}O$$
$$SO_3 + Cat \rightarrow SO_2 - Cat\text{-}O$$

and derived an equation which fits much of the existing data:

$$k/V_s[(\bar{P}_{O_2})n/(P_{SO_2})] = a_{eq}[\ln (1/1\text{-}a^{\cdot}) - a^{\cdot}]$$

where V_s = space velocity; $\bar{P}_{O_2}$ = average partial pressure of O_2, P_{SO_2} = initial partial pressure of SO_2, and $a^{\cdot}$ = degree of conversion, i.e a/a_{eq} (Duecker, 1959).

On platinum, the observed kinetics are given by:

$$k(SO_2)/(SO_3)^{½}$$

and the activation energy is 10 kcal/mole. Vanadium oxide is commercially preferred. Promoted by potassium sulfate, it has a life of 7-15 years, has a reasonable reactivity, and is insensitive to almost everything except fluoride below 600°C.

Details of various conversion methods have been worked out and discussed by Gardy (1957). A typical conversion curve for an 8% sulfur dioxide gas flowing through a reactor is shown in Fig. 3.14b. The gas heats in each bed and must be cooled before entering subsequent beds. If yields above 98.5% SO_3 are to be reached, it is advisable to remove part of the SO_3 at an intermediate stage to prevent dissociation. Two stage plants are a necessity where new air quality laws demand more than 99.5% conversion. Schmidt (1968) discussed direct conversion of sulfur to SO_3.

The absorption of SO_3 by water yields theoretically an azeotropic mixture with 98.3% sulfuric acid; however, if sulfur trioxide is absorbed in dilute acid, the gaseous sulfur trioxide reacts with water vapor and forms an acid fog which must be scrubbed. The final extent to which sulfur trioxide can be scrubbed is limited by its vapor pressure above the acid.

For two hundred years, catalytic oxidation using nitric oxides was popular. In this 'contact process', sulfur dioxide first entered the Glover Tower:

$$SO_2 + 3H_2O + 2NO^+ \rightarrow 2NO + H_2SO_4 + H_3O^+$$

followed by the (lead) chamber reaction:

$$SO_2 + \frac{1}{2}O_2 + H_2O \rightarrow H_2SO_4$$

and finally the Gay-Lussac Tower reaction:

$$NO_2 + NO + 2H_3O^+ \rightarrow 2NO^+ + 4H_2O$$

The physical properties of sulfuric acid are shown in Table 3.12. It is a colorless, oily liquid. The Raman spectrum of the sulfate ion has been discussed by Hexter (1964). The vibrational spectrum was reported by Goypiron (1975). It forms well characterized hydrates, $H_2SO_4{\cdot}(H_2O)_n$ with n values of 1 to 4. Sulfuric acid exhibits strong hydrogen bonding. The heat of hydration is about 210 kcal/mole. The thermodynamic properties of dilute sulfuric acid have been discussed by Pitzer (1976). The dissociation constants for sulfuric acid are:

$$H_2SO_4 + H_2O \rightarrow H_3O^+ + HSO_4^-; \qquad k_1 = 10^3$$

$$HSO_4 + H_2O \rightarrow H_3O^+ + SO_4^{2-}; \qquad k_2 = 1.29 \times 10^{-2}$$

The first constant indicates full dissociation of a strong acid; the second step corresponds to a moderately strong acid. Thus, in a 1 normal solution, the apparent degree of dissociation is 51%, i.e. only the first proton has been donated; in a 0.1 normal solution, 59% apparent dissociation indicates that about 20% of the second dissociation step is completed.

Depending on the pH of solution, the sulfate ion can form either a hydrogen sulfate $MeHSO_4$ or a normal sulfate Me_2SO_4. If $MeHSO_4$ is heated and melted, a pyrosulfate is formed:

$$2NaHSO_4 \rightarrow Na_2S_2O_7 + H_2O$$

The solubilities of some common sulfates are listed in Chapter 9. Cold dilute sulfuric acid is a poor oxidant. Its action results entirely from its acid character, i.e. its strong proton donor tendency. In contrast, hot concentrated sulfuric acid is a strong oxidant. The latter reacts with phosphorus, yielding elemental sulfur. With many metals, carbon, and elemental sulfur it can form sulfur dioxide.

The following two facts show seemingly contradictory behavior of sulfuric acid in technically important systems: Concentrated sulfuric acid can be stored in iron tanks, but dilute acid attacks iron vigorously. This effect is due to passivity. In lead containers, the reverse is true. Dilute acid forms a protective lead sulfate coating, but concentrated acid severely corrodes lead. Methods for the purification of sulfuric acid have been reported in several patents (Nojima, Jpn P. 76 09,097; Knollock, U.S. P. 3,956,373). The exchange of oxygen between sulfuric acid and water has been measured by Hoering (1957), Mills (1940), and Hall (1940).

Sulfuric Acid as a Solvent. Sulfuric acid is one of the oldest non-aqueous solvents; after an intermission, interest has resurged with the discovery of polynuclear actions (Rheingold, 1977). The solubility of sulfur dioxide in sulfuric acid decreases from 15 wt % at 20°C to 1% at 120°C. The solubility decreases with increasing sulfuric acid concentration until it reaches a minimum at 85°C, Fig. 3.17. Dilute sulfuric acid, together with gypsum, has been used as a solvent for sulfur dioxide abatement in the 'Thoroughbred' 101 process (Tamaki, 1975), Chapter 8. The UV spectrum was studied by Gold (1950). Sulfuric acid autoprotolyzes according to:

$$2H_2SO_4 \rightarrow H_3SO_4^+ + HSO_4^-$$

Sulfur trioxide reacts with sulfuric acid and forms disulfuric acid, $H_2S_2O_7^-$. It protolyzes according to:

$$H_2S_2O_7 \quad + \quad H_2SO_4 \quad \rightarrow \quad H_3SO_4^+ \quad + \quad HS_2O_7^-$$

Disulfuric acid is also formed by self-dissociation:

$$2H_2SO_4 \quad \rightarrow \quad H_3O^+ \quad + \quad HS_2O_7^-$$

the dissociation constants are:

$$k = [H_3SO_4^+][HSO_4^-] \qquad k = 2.7 \times 10^{-4}$$
$$[H_3O^+][HS_2O_7] \qquad = 5.1 \times 10^{-5}$$
$$[H_3SO_4^+][HS_2O_7^-]/[H_2S_2O_2] \qquad = 1.4 \times 10^{-2}$$

ESR, NMR, and Raman spectroscopy have shown that radical cations can be formed if hydrocarbons are oxidized in sulfuric acid (Gillespie, 1968). Many solutions exhibit beautiful, deep color. Sulfur, selenium, and tellurium all form cyclic cations. The sulfur amines S_4^{2+}, S_8^{2+}, S_{16}^{2+}, and S_{24}^{2+} have all been proposed. the structure of S_8^{2+} is well established, Fig. 3.2 (Gillespie, 1968; Bali, 1975). Sulfuric acid is a good sulfonating agent. For more detailed information on sulfuric acid as a solvent, see Gmelin (1959), Stauffer (1975), and Gillespie (1968, 1970). The cryoscopic behavior has been described by Dacre (1967). Disulfuric acid can be considered a dimer of sulfuric acid and an intermediate of polysulfuric acid, $H_2S_xO_{3x+1}$. It has the same linked-tetrahedron structure as polymeric sulfur trioxide.

Dithionic Acid

Free dithionic acid, $H_2S_2O_6$, is not known. The dithionates, its salts, and ions can be obtained by oxidation of sulfite or sulfur dioxide (Bassett, 1935), for example:

$$2MnO_2 \quad + \quad 3MeHSO_3 \quad \rightarrow \quad MnSO_4 \quad + \quad MnS_2O_6 \quad + \quad 3MeOH$$

The formal oxidation state of sulfur is S^{5+}. The structure of dithionate was determined by Stanley (1956). The S-S bond is abnormally long, and the two SO_3 groups can rotate against each other, Table 3.19. The Raman spectrum (Gerding, 1950) indicates that solid dithionates have a different structure than the aqueous ion (Peter, 1977). Such solutions are stable up to the boiling point. The salts decompose above 200°C into sulfuric acid and sulfur dioxide:

Table 3.19
BOND PROPERTIES OF OXYACIDS OF SULFUR

Name	Formula	Symmetry	Bond distance (A)	Bond angles (deg)
Sulfate	SO_4^{2-}	T_d	S-O = 1.44	OSO = 109
Dithionate	$S_2O_6^{2-}$	D_{3d}	S-S = 2.15	
			S-O = 1.43	OSS = 104
Sulfite	SO_3^{2-}	C_{3v}	S-O = 1.40	OSO = 106
Bisulfite (hydrogen sulfite)	HSO_3^-	C_{3v}	-	-
Disulfite pyrosulfite	$S_2O_5^{2-}$	C_s	S_1-S_2 = 2.205	OS_1S_2 = 100
			S_1-O = 1.50	OS_2S_1 = 106
			S_2-O = 1.43	OS_1O = 110
			S-O = 1.47	OS_2O = 113
Dithionite	$S_2O_4^{2-}$	C_{2v}	S-S = 2.39	OSO = 99
			S-O = 1.50	OSO = 108
Thiosulfate	$S_2O_3^{2-}$	C_{3v}	S-S = 2.00	SSO = 109
			S-O = 1.46	
Polythionates	$S_n(SO_3)_2^{2-}$		S-S = 2.04	SSS = 107 to 113
			S-O = 1.41 to 1.46	OSO = 110
			S-SO_3 = 2.14	S-S-(SO_3) = 102
				d = 90-110

For references see Meyer (1977), Kao (1977).

$$H_2S_2O_6 \rightarrow H_2SO_4 + SO_2$$

Details about this molecule and its compounds are reviewed by Gmelin (1959), Schmidt (1973), and Lyon (1968).

Tri-, Tetra- and Polythionates. The nomenclature of these compounds is misleading, because they differ generically from dithionate. They contain sulfur atoms in the oxidation state 0 and are, in reality, sulfane-disulfonates. Thus, they will be discussed together with monosulfane-monosulfonate, i.e. thiosulfate.

Aqueous Sulphurous Acid

Contrary to the formula, 'H_2SO_3,' which can be found on the label of bottles in many chemistry labs, the free sulfurous acid, the oxyacid of sulfur dioxide, is not known. The 'volatile acid of sulfur' (Seehl, 1744; Homberg, 1710) consists of sulfur dioxide and water. The formula $(HO)_2SO$, currently used in the Substance Index of Chemical Abstracts to identify sulfurous acid, classification No. [7782-99-2], is ambiguous, because in all its presently known salts, the proton is bound to the sulfur and not the oxygen atom.

Species. The solution obtained by dissolving sulfur dioxide in water forms a complex system whose composition is not yet fully defined. Depending on the concentration, the system contains various amounts of SO_2-hydrate, $H_2O \cdot SO_2$, HSO_3^-, bisulfite or hydrogen sulfite, $S_2O_5^{2-}$, pyrosulfite or disulfite, and traces of sulfite, SO_3^{2-}. The best description of the system is still the summary in Gmelin (1963).

Below 12°C, $(H_2O)_n(SO_2)_m$ forms hydrates. Simon (1961) showed that the Raman spectrum of the solution is largely that of dissolved SO_2. Modern laser Raman spectra confirmed this finding (Herlinger, 1969; Meyer, 1977). The solubility of sulfur dioxide in water is 8 Mol %, i.e. 24 wt % at 120°C, and sulfur dioxide dissolves 1.34 wt % of water at the same temperature. Solutions of water in liquid sulfur dioxide and of sulfur dioxide in water have the same species, Fig. 3.15. The equilibrium between sulfur dioxide in the gas phase and unhydrolyzed sulfur dioxide in the liquid phase follows Henry's Law:

$$[SO_2] = H(pSO_2)$$

where $[SO_2]$ is the concentration of unhydrolyzed sulfur dioxide in water; (pSO_2) is the SO_2 partial pressure, and H is Henry's Constant, see Chapter Eight. A spectrum assignable to HO bonds has not been observed. In $H\text{-}SO_3^-$, the hydrogen is bonded to the sulfur atom, even though electronegativity consideration would predict the species SO_3H^-. The latter has not yet been identified. The $H\text{-}SO_3$ species, analogous to that of HPO_3^{3-}, is probably stabilized because of symmetry considerations (Meyer, 1977).

The UV spectra of sulfur dioxide species are broad and overlap (Flis, 1975). Schaefer (1918) made a most careful study of this system. His observations have been fully confirmed. The same is true for Dietzel (1925), Ley (1938), and Foerster's study of 1923. Ephraim (1923) studied the spectra shift in complexes. Ley (1938), Golding (1960), Bourne (1974), and Flis (1965, 1975) used the UV spectra to determine the equilibria between the different sulfur dioxide family members. Lüdemann (1958) extended UV work to pressures up to 6 kbar; Eriksen (1972) determined the extinction coefficient; Briquet (1974) used the IR spectrum to establish rotation in dissolved sulfur dioxide. The diffusion of sulfur dioxide in aqueous solutions was measured by Eriksen (1969).

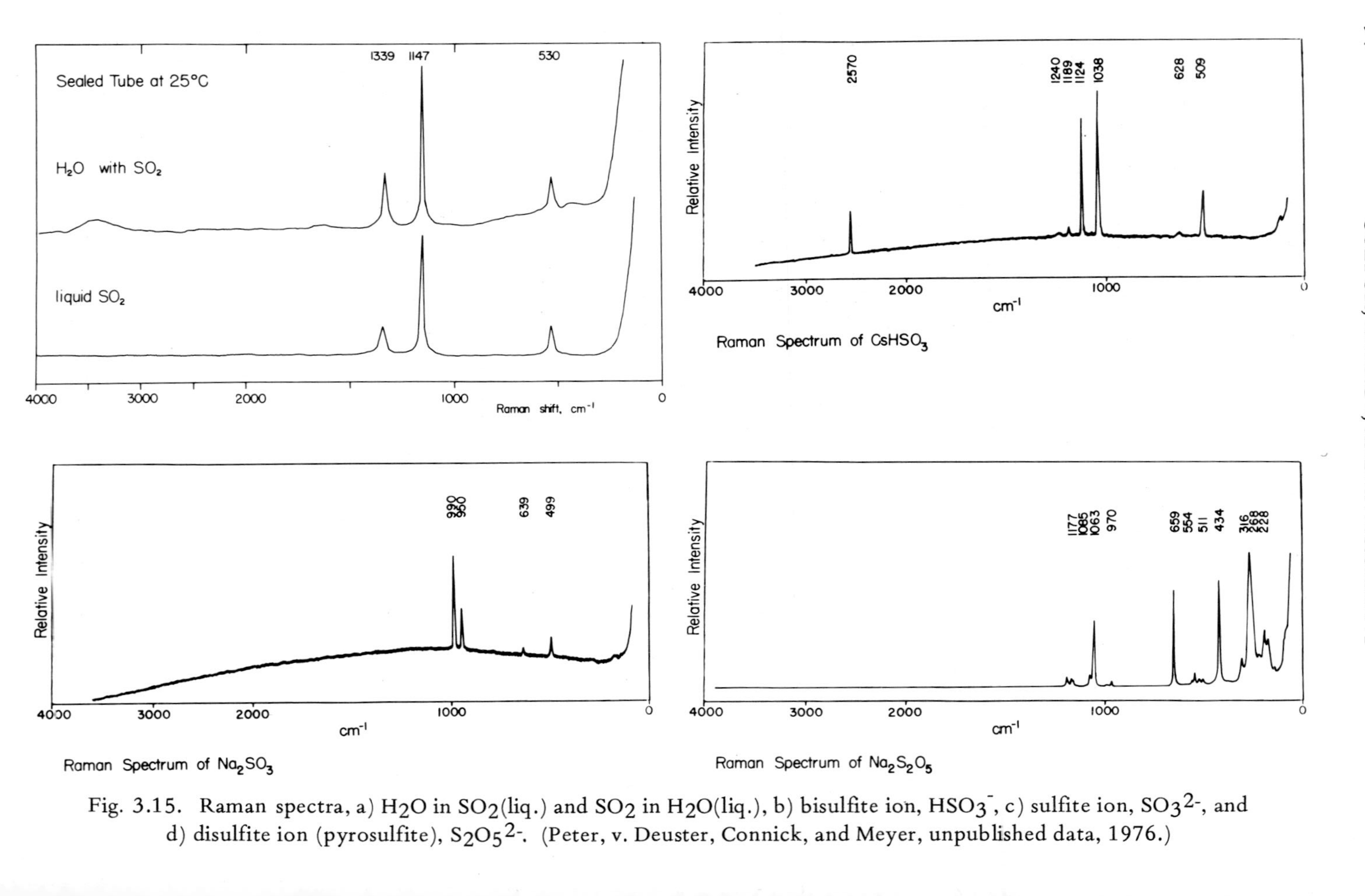

Fig. 3.15. Raman spectra, a) H_2O in SO_2(liq.) and SO_2 in H_2O(liq.), b) bisulfite ion, HSO_3^-, c) sulfite ion, SO_3^{2-}, and d) disulfite ion (pyrosulfite), $S_2O_5^{2-}$. (Peter, v. Deuster, Connick, and Meyer, unpublished data, 1976.)

The IR spectra have been measured by Antikainen (1958) and Goulden (1967). Brown (1972) reassigned the frequencies of sulfite. The solubility products of some alkali salts of $S_2O_5^{2-}$ and SO_3^{2-} are listed in Chapter 8. The solubility of the bisulfites, $MHSO_3$, is not known; they are all more soluble than $M_2S_2O_5$, except for Cs and Rb, which form both bisulfite and disulfites, depending on the preparation method (Simon, 1960). The Raman spectra are shown in Fig. 3.15. The species, $S_2O_5^{2-}$, is a disulfite with an S-S bond (Simon, 1956); the structure is well established (Zachariasen, 1932; Lindquist, 1957). The solid salts were already prepared by Rammelsberg (1846). They are often called pyrosulfite. Simon (1955) discovered that almost all bisulfite solutions convert to pyrosulfite during crystallization. Thus, the compounds $NaHSO_3$ and $KHSO_3$ exist in name only. In reality, they consist of pure $Na_2S_2O_5$ and $K_2S_2O_5$, Fig. 3.15, regardless of the labels attached to their commercial containers. Likewise, all literature referring to solid bisulfite in reality describes disulfites. This applies, for example, to literature describing the Wellman-Lord process. Since pyrosulfite is a disulfite, the stoichiometry and reactions undergo, for all practical purposes, only a semantic change. This fact is due to the relatively low solubility of the pyrosulfite. The thermal stability of the 'metabisulfite', $Na_2S_2O_5$, i.e. disulfite, was measured by Cleghorn (1970), Malanchuk (1971), and Foerster (1926). The pH dependence of the stability of $S_2O_5^{2-}$ was investigated by Heintze (1974).

Radiolysis of aqueous sulfur dioxide was studied by Eriksen (1974), who observed an absorption at 255 nm which he assigned to the SO_3^- radical, which dimerizes to $S_2O_6^{2-}$. At low pH, he also observed the SO_2^- radical, which also has been observed by Andrews (1976), Janzen (1972), Hayon (1972).

Solubility of Sulfur Dioxide. The solubility of sulfur dioxide depends on temperature and pressure, Fig. 3.16. The temperature dependence is of utmost importance in sulfur dioxide abatement chemistry. Many practically important systems have been measured by Johnstone (1935-1952), who established various relationships between variables, which have to be treated with upmost care, because he did not know about the existence of the disulfite ion; thus, his values are poor for extrapolating towards concentrated solutions (Schmidt, 1971). A very extensive and careful review of all sulfur dioxide work up to 1963 is contained in a

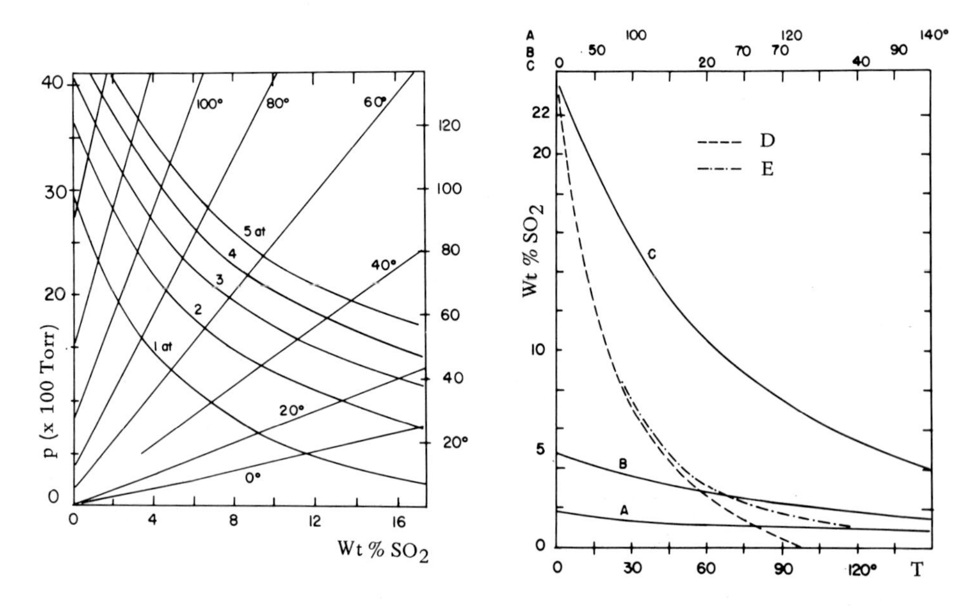

Fig. 3.16. Solubility of sulfur dioxide in water, a) as a function of pressure, b) as a function of temperature (curves A, B, C see top scale; curves D and E read bottom scale; D: $p(SO_2 + H_2O) = 1$ at., all other curves: $p(SO_2) = 1$ at.)

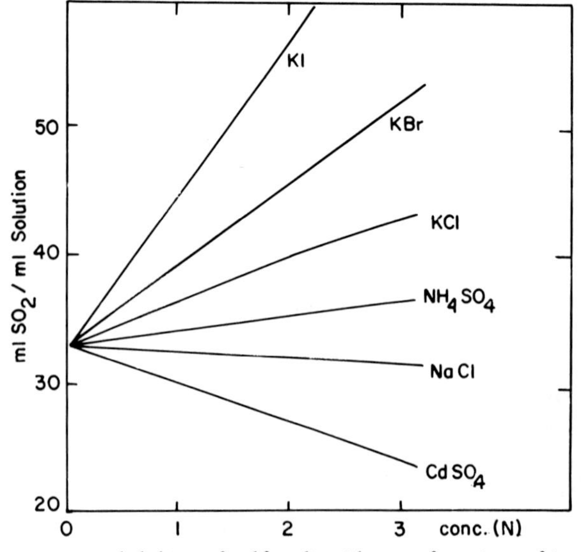

Fig. 3.17. Solubility of sulfur dioxide as a function of ionic strength.

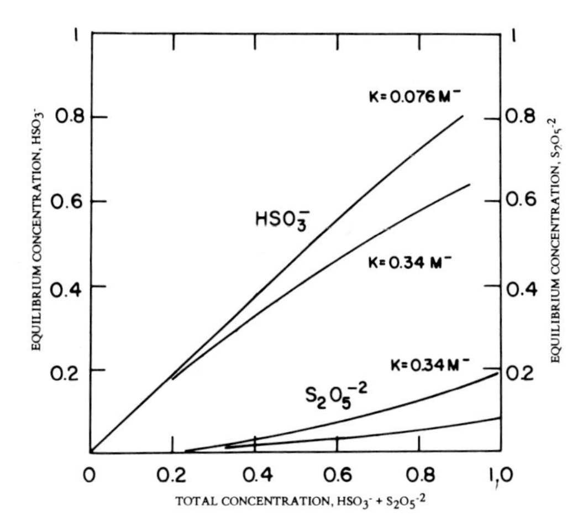

Fig. 3.18. Equilibrium concentrations of bisulfite (HSO_3^-) and disulfite ions ($S_2O_5^{2-}$) as a function of total concentration ($[SO_2] + [HSO_3^-] + [S_2O_5] + [SO_3^{2-}]$). After Bourne (1974).

special volume of Gmelin (1963) which contains large, fold-out graphs showing the relationship between all feasible variables.

The solubility of SO_2 is strongly affected by the ionic strength of the solution, Fig. 3.17, especially those which buffer the pH, such as the carbonate and Mg-salts present in sea water (Bromley, 1971), or in rain drops in urban atmospheres, which contain up to ten times more sulfur dioxide than pure water, because of their content of heavy metal ions (Barrie, 1976). Such solutions also promote oxidation.

The chemical equilibrium in solutions of bisulfite salts have been discussed by Golden (1960), Bourne (1974), Kerr (1974), Davis (1975), v. Deuster (1977), Gmelin (1966), and others. A saturated solution of '$NaHSO_3$' contains in reality in the solid phase $Na_2S_2O_5$ (and no $NaHSO_3$), and in the aqueous phase about 40 wt % dissolved sulfur dioxide in the form of HSO_3, $S_2O_5^{2-}$, and some SO_3^{2-}, as well as $H_2O{\cdot}SO_2$. Bourne (1974) measured the UV extinction of solutions with various ionic strengths between pH 3 and 5, calculated the equilibrium constant as a function of ionic strength, and, by extrapolation to zero ionic strength, determined the thermodynamic equilibrium constant:

$$k = [S_2O_5^{2-}]/[HSO_3^-] = 0.076 \pm 0.01 \text{ per mole at } 250^{\circ}C$$

Fig. 3.18 shows the relative concentration of $S_2O_5^{2-}$ and HSO_3^- as a function of total dissolved sulfur concentration for two different ionic

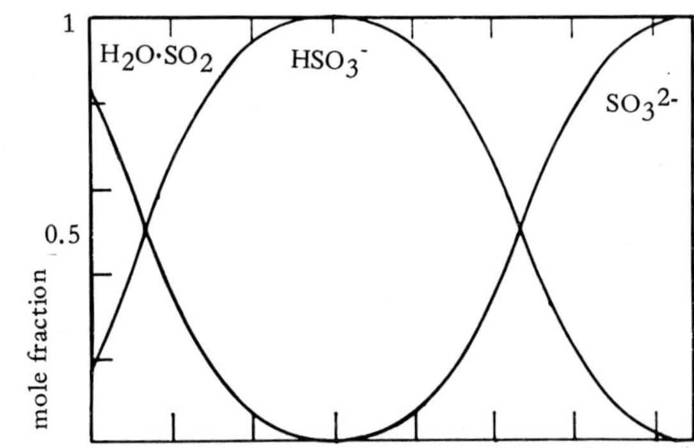

Fig. 3.19. Concentration of SO_2, HSO_3, and SO_3^{2-} as a function of pH.

strengths. Fig. 3.17 shows how the solubility of sulfur dioxide changes as a function of KCl, Na_2SO_4, and acetic acid content of the aqueous phase.

In practice, the solubility of sulfur dioxide is often not determined by the equilibrium constants, but by the rate of the reactions. This holds for the absorption of sulfur dioxide by rain drops in the atmosphere (Brimblecombe, 1974), as well as for the sulfur dioxide stripping in flue gases of industrial plants (Wen, 1975). Unfortunately, the absorption kinetics are not yet well understood, and published values differ from 2.5×10^9 per m·sec (Betts, 1970) to 10^6 (Jackson, 1965) and 10^3 Bänau, 1956), and by a factor of 10 million (Beilke, 1975). These discrepancies are very likely not due to incorrect observation, but due to different experimental techniques which might respond to different intermediate species. The exchange rate between the oxygen in sulfur dioxide and water has been measured by Norris (1965-1970). The kinetics in technical systems will be discussed in Chapter 8.

The dissociation constants (Salomaa, 1969) for SO_2 are:

$$H_2O{\cdot}SO_2 \rightarrow H_3O^+ + HSO_3^- \qquad pk_1 = 1.92$$
$$HSO_3^- \rightarrow H_3O^+ + SO_3^{2-} \qquad pk_2 = 7.20$$

The transition from sulfur dioxide gas to HSO_3^- is now commercially used to enrich the heavy sulfur isotope (Stachewski, 1975). Fig. 3.19 shows the range of existence for the various species. Obviously, the amount of sulfur dioxide gas dissolved in an aqueous system can be increased if the dissolved sulfur dioxide is chemically transformed into SO_3^{2-} or, by secondary reaction, to insoluble sulfites or oxidized to sulfates. Both remove it from the gas-liquid equilibrium. Thus, from a practitioner's viewpoint, such a system has 'increased solubility', and

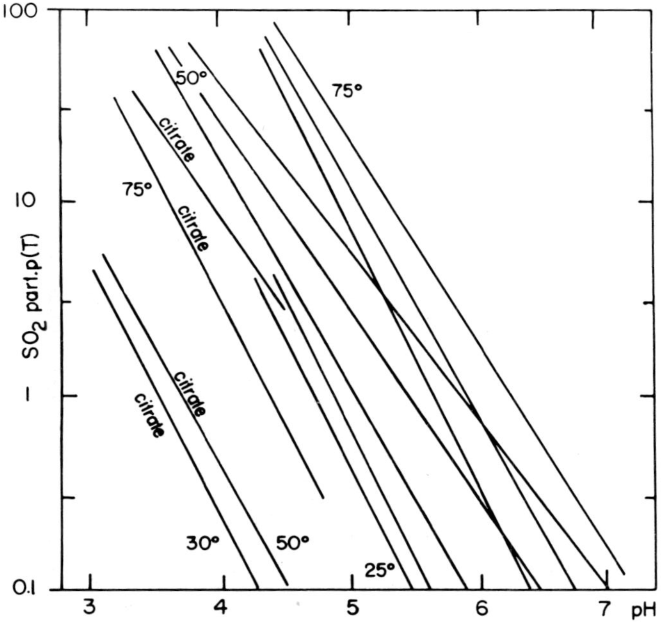

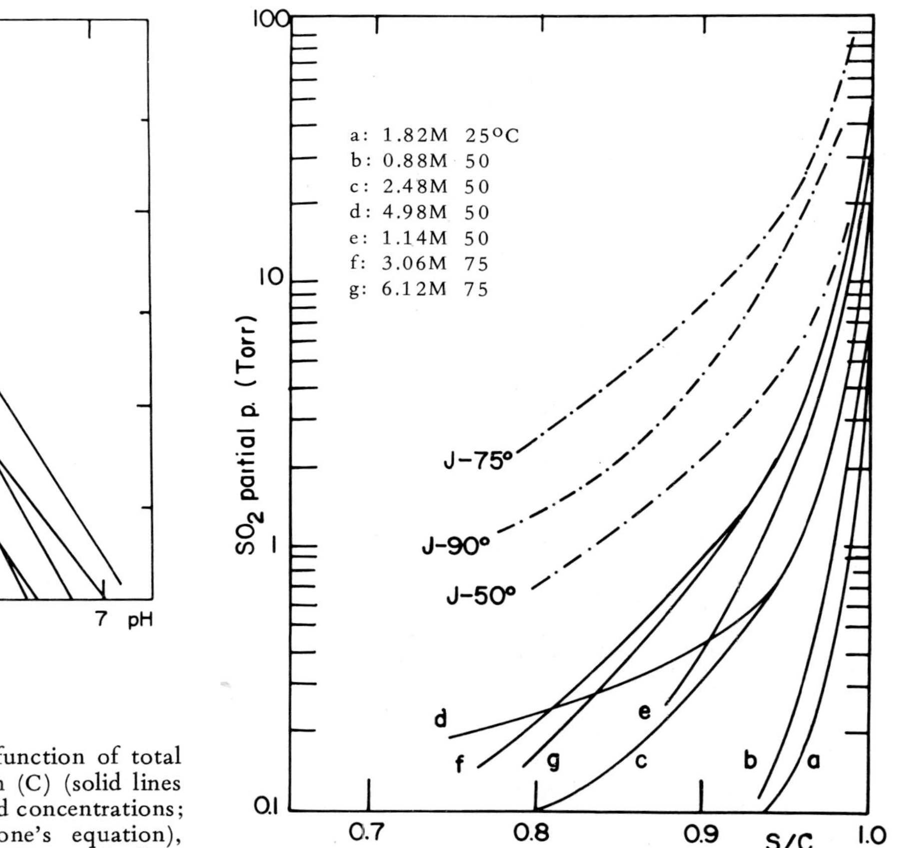

Fig. 3.20. a) SO_2 vapor pressure as a function of total sulfur (S) to total cation concentration (C) (solid lines measured at 7 different temperatures and concentrations; dotted lines computed from Johnstone's equation), b) SO_2 vapor pressure as a function of pH in two buffers at 3 temperatures (after Oestreich, 1976).

the addition of a base or buffer 'catalyzes the solubility of sulfur dioxide.' From a chemist's viewpoint, the equilibrium constants remain unchanged, the absorption of sulfur dioxide is due to formation of sulfite ion or precipitation of solid pyrosulfite to sulfite, the aqueous system is merely a medium for forming the final species, and the buffer or base is not a catalyst, but a reagent whose properties and concentration determines the ultimate quantity of sulfur dioxide absorbed. As this case indicates, the language barrier between these two groups of workers is often a more serious obstacle to progress than is the confusion about what is going on in thc system.

Sulfur Dioxide Removal. For removing sulfur dioxide from the gas into a liquid phase, the partial vapor pressure of sulfur dioxide is a decisive variable. Typical vapor pressure *vs.* pH curves for several typical scrubbing conditions are plotted in Fig. 3.20 (Oestreich, 1977). It is common among engineers to plot such data as a function of S/C, a term defined by Johnstone (1937). C stands for the total concentration of the cation belonging to the sulfur ions, and S stands for the total concentration of all dissolved sulfur. In the theory,

$$S_{th} = [H_2O{\cdot}SO_2] + [HSO_3^-] + [SO_3^{2-}]$$

were considered; the existence of S_2O_5 was not recognized by Johnstone. Fig. 3.20 shows the discrepancy between the measured values (Oestreich, 1977) and values obtained by extrapolation of Johnstone's work using his equation. Again, the discrepancy is due to the formation of $S_2O_5^{2-}$. Unfortunately, much practical work still relies on the old, and incorrect, data; the implication of such errors is discussed by Schmidt (1971) and Oestreich (1977). Oestreich carefully measured $p(SO_2)$ in the citrate, sodium, and potassium system, and developed the Vart'Hoff equations which should be used, instead of the Johnstone's equation which is incorrect:

$$p(SO_2) \neq (2S\text{-}C)_2/C\text{-}S \qquad \text{Johnstone's equation}$$

The actual efficiency of a chosen practical system depends on a great many factors. In addition to absorption kinetics which are determined by the as yet not accurately known kinetic constants, surface area, mixing, container geometry, and many other practical factors can be important. If a buffer or a base is used, the solubility of the reagent, the solubility of the product, and other factors which depend both on the intrinsic, theoretical suitability of the materials as well as the physical properties of the reagent at hand influence the observed efficiency. Thus, for a given system, a large number of different results may be observed. According to Chemical Abstracts, the literature is presently increasing by some 500 papers on aqueous absorption systems each year, and yet our basic knowledge about these systems remains virtually unchanged and only little systematic work (such as that reported by Borgwardt (1970), Slack (1968, 1975), Potter (1969), and Carlson (1969)) complements our incomplete basic knowledge.

In principle, any base can be used to remove sulfur dioxide from vapor in equilibrium phase (Chapter 8). The absorption of sulfur dioxide by sea water buffered by dissolved salts and carbonate has been studied by Spedding (1972) and Bromley (1971). Inspection of figures 3.19 and 8.7 shows, however, that bases which cause a pH greater than 7 will not only absorb sulfur dioxide, but also carbon dioxide, which reacts with water very much like sulfur dioxide, by forming a hydrate rather than the free carbonic acid, H_2CO_3.

$$H_2O \cdot CO_2 \rightarrow HCO_3^- \qquad k_1 = 6.52$$
$$HCO_3^- \rightarrow CO_3^{2-} \qquad k_2 = 10.33$$

This reaction must be evaded, because in the atmosphere, and especially in combustion gases, carbon dioxide is 50 to 100 times more abundant than sulfur dioxide, and if it would be absorbed, enormous quantities of absorber would be wasted. Thus, it is vital that the pH of the ‘liquor’ remains in the range between pH 4 and 6. This can be achieved by either of two ways: One can add any base with a pK greater than 5 up to any value, at a rate matching sulfur dioxide absorption in order to keep the pH in the proper range by what a chemist would call titration of sulfur dioxide, or one can add any desired excess of a base with a pK between 5 and 7, because such a base would intrinsically maintain a pH in the proper range until it is exhausted. Fig’s. 3.19 and 8.7 show some suitable reagents. Of these, the chemically most attractive are those that are two or triprotic in the proper range, because they can absorb a two or three-fold quantity of sulfur dioxide, respectively. These systems will be described in more detail in Chapter 8.

Oxidation of Aqueous Sulfur Dioxide Species. Aqueous sulfite-disulfite-sulfur dioxide solutions undergo a slow oxidation to sulfate at room temperature (Simon, 1956). They can also form dithionate

(Bassett, 1935). The speed of the reaction depends on concentration, ionic strength, temperature, oxygen concentration, pH, and the purity of the solution, i.e. the presence of catalysts. Knowledge about these systems has been thoroughly reviewed by Gmelin (1969). The influence of impurities, which are characteristic for flue gas systems, for example selenium, has been discussed by Schmidt (1975).

In flue gas abatement systems, which normally contain less than 5% oxygen, the oxidation rate is usually too small to quantitatively convert sulfite to sulfate (Bartlett, 1972), but it is large enough to hamper commercial regeneration of sulfur dioxide, unless precautions are taken. In atmospheric systems, the oxidation is substantially increased. The rate of the uncatalyzed system was studied by Beilke (1975). In urban atmosphere, particulates catalyze the oxidation (Barrie, 1976; Novakov, 1971-1976). The influence of photooxidants such as H_2O_2 has been measured by Hoffmann (1975), Junge (1958-1963), and Pitts (1975). The oxidation of sulfur dioxide with atom oxygen was studied by Kugel (1975) in rare gas matrices. Gas phase work will be discussed in Chapter 6; oxidation in abatement liquors will be discussed in Chapter 8.

Reaction of Aqueous Sulfur Dioxide and Hydrogen Sulfide. The reaction between sulfur dioxide and hydrogen sulfide can be nominally written as:

$$2H_2S \quad + \quad SO_2(aq.) \quad \rightarrow \quad 2H_2O \quad + \quad 3/8\ S_8$$

The reaction is exothermic, ΔH = -29 kcal/mole, and, thus, should proceed smoothly. In reality, the reaction yields a milky liquid which cannot be rendered clear by filtration. The suspended sulfur is soft, gummy, and contains only little S_8. If the liquid is diluted, a brown, transparent liquid is obtained; this liquid resists boiling, and slowly precipitates elemental sulfur over a period of many months. This liquid is named after Wackenroder. Thus,

$$2H_2S \ + \ SO_2(aq.) \xrightarrow{\text{instantly}} \text{Wackenroder's liquid} \xrightarrow{\text{months}} 2H_2O \ + \ 3/8\ S_8$$

This liquid was first observed by Berthollet (1798) and Dalton (1812). Wackenroder (1846) discovered and identified the existence of pentathionate, $HSO_3\text{-}S_3\text{-}SO_3H$, in this solution, and contributed much to our present knowledge, which, to a large extent, is still based on observations completed before 1850. Much about this liquid remains a mystery.

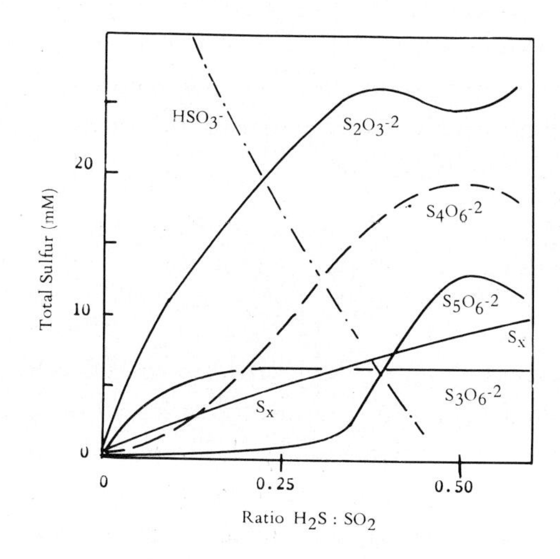

Fig. 3.21. Composition of Wackenroder's solution as a function of pH.

Debus (1888) and Steinle (1962) have carefully, and in great detail, described and analyzed the system which contains, among other species:

Wackenroder's liquid: HSO_3^-, bisulfite
$S_2O_3^{2-}$, thiosulfate
$^-SO_3$-S-SO_3^-, trithionate
$^-SO_3$-S-S-SO_3^-, tetrathionate
$^-SO_3$-S_x-SO_3^-, polythionates (disulfonic acids), and
$^+S_x^+$; S_8; sulfur

Fig. 3.21 shows the concentration of various species as a function of pH. The composition changes during aging (Stamm, 1941) and upon boiling, but most of the meta-stable species persist for months. Since the mixture is in a state of partial equilibrium, analytical methods aimed at separating components do not necessarily reflect the composition of the mixture. Often, the analytical results are not even reproducible. Steinle (1962) found that chromatography with a mixture of n-butanol, acetone, water, perchloric acid, barium perchlorate in the ratio 50:20:30:4:10 gives good separation and relatively little decomposition. The mechanism of formation of the liquid is little understood. During the last twenty years, several mechanisms for the formation of elemental sulfur from aqueous systems have been proposed. The most likely process is the reverse-reaction of the sulfur degradation by sulfite, which has been studied by Schmidt (1962, 1963, 1965, 1974) and others. Calculated charge distribution on sulfur atoms of the various intermediates (Meyer, 1977) supports a model which involves formation of thiosulfate, tetrathionate, and as yet unidentified intermediates:

$$HS^- + SO_3^- \rightarrow H\text{-}S\text{-}SO_3^-, \text{ followed by}$$
$$(HSO_3^-)_x + S\text{-}SO_3^- \rightarrow O_3S\text{-}S_x\text{-}SO_3^-, \text{ or}$$
$$HS\text{-}S_x\text{-}SO_3^-$$

Polythionates, of the type $^-SO_3\text{-}S_x\text{-}SO_3^-$ with up to 100 sulfur atoms, can be made (Weitz, 1928, 1952, 1958), and molecules with up to 20 atoms can be separated by paperchromatography (Pollard, 1955) or high voltage ionophoresis (Blasius, 1965, 1973). Schmidt (1956) prepared the free acids, which he properly calls disulfonic acids. He also prepared sulfanesulfonic acids, $H\text{-}S\text{-}S_x\text{-}SO_3H$. Steinle (1962) used radioactive 35-S in his attempt to establish the reaction mechanism, and found that hydrogen sulfide and sulfur dioxide both contribute equal amonts of sulfur to the product, and not 66% and 33%, as the nominal over-all stoichiometry for forming elemental sulfur would suggest. Barbieri (1962) studied the acidification of thiosulfate and supports the proposed mechanism. Earlier work is reviewed in Gmelin (1963).

During the last 15 years, several attempts have been made to use the Wackenroder reaction for recovering hydrogen sulfide or sulfur dioxide from tail gases of Claus or Kraft recovery boilers. The reaction is attractive, because the final concentration of both reagents can be kept very low. So far, large scale applications have been hampered by the capricious kinetics of the reaction, which is sensitive to minute perturbation. As is, undesirable concentrations of the liquor can accumulate, followed by precipitation of gummy sulfur wherever temperature or pressure gradients occur; for example, in corners, pumps and ducts. Teder (1975) studied the reaction kinetics after carefully determining the effect of pH, concentration of all components, temperature, and other parameters (Nordin-Fossum, 1974). The initial reaction is fastest at low pH, but the solubility of hydrogen sulfide in an aqueous solution is substantially higher at high pH, because of the formation of HS^-. The two factors are counterproductive and explain why the initial kinetics in industrial systems are complex. Teder studied the reaction from the viewpoint of hydrogen sulfide outlet concentrations, and found that the initial reaction rate, described by

$$k_x = 2.3(-d\log[H_2S]/dt)$$

is first order. $[H_2S]$ is the gas phase concentration. The second reaction step, the formation of polythionate and thiosulfate, has not yet been determined. However, Teder measured the ratio of the two reaction rates, and found that the second reaction is about ten times slower than the first, that it depends on mixing rates, and that it is autocatalyzed, as several workers have suggested before. This is supported by the fact that during the initial reaction a UV absorption at 2850 A builds up, but slowly decreases as the reaction proceeds.

Only at high concentrations is the reaction first order in regard to reagent concentration. At low concentration the kinetics are unclear. The apparent activation energy is 0.5 kcal/mole, and the hydrogen sulfide absorption rate changes only 10%, from 25° to 70°C.

More basic research on this reaction is needed before engineering procedures can make the reaction reliable. Such basic research is highly over-due, because Schmidt (1959) showed that the free polythionic acids can be synthesized in cold, dry ether. Thus, individual reactions and reaction steps can be isolated and analyzed.

Dithionites; Dithionous Acid

Free dithionous acid does not exist, and even dithionites decompose in solution by self disproportionation, according to:

$$2S_2O_4^{2-} + H_2O \rightarrow 2HSO_3^- + S_2O_3^{2-}$$

The solid salts are stable. They are prepared by reduction of bisulfite solutions:

$$H_2O \cdot SO_2 + 2HSO_3^- + Zn \rightarrow S_2O_4^{2-} + ZnSO_3 + 2H_2O$$

The kinetics of this reaction were studied by Suzuki (1966). The sodium salt is a dihydrate: $Na_2S_2O_4 \cdot 2H_2O$; it can be dehydrated *in vacuo*. The unit cell contains two dithionite ions in a monoclinic unit cell with the dimensions a = 6.404, b = 6.559, c = 6.586, and beta = 119°31 sec. Other preparation methods are reduction of bisulfite with sodium amalgamate (Rinker, 1969) and reduction of liquid sulfur dioxide with zinc powder. The structure of the molecule involves an S-S bond

(Simon, 1949) with a distance of 2.389 A (Dunitz, 1956), Table 3.19, the largest S-S bond of any oxygacid. The Raman spectrum of the solution differs from that of the solid (Peter, 1977), even if the latter can be recrystallized. Lynn (1964) found that the solution exhibits an ESR spectrum which indicates that 0.3% of the dithionite has disproportionated into the SO_2^- radical, which has also been observed in other systems by Janzen (1972), Hayon (1972), and Andrews (1976).

The decomposition of dithionite has been followed by Lynn (1964), Rinker (1965), and Burlamacci (1969) with ESR. Two reactions have been identified. The first is catalyzed by sulfite ions; the second, very fast reaction has an induction period. Lem (1970) determined $k_1 = 1.67 \times 10^{-1}$ l/mole·sec and $k_2 = 5.83 \times 10^3$ l^2/mole^2·sec at 23°C, using polarographic observation. He also discovered that the heavy cations Zn^{2+}, Cd^{2+}, and In^{3+} stabilize the the solution. Thus, ZnS_2O_4 is far more stable than $Na_2S_2O_4$. Wayman (1970) explained the reaction mechanism, while Burlamacchi (1971) and Wayman (1971) discussed each others' proposals. Cleghorn (1970) studied the IR spectra of the thermal decomposition products; the kinetics were also studied by De Poy (1974).

Peroxi Mono- and Disulfuric Acid; H_2SO_5; $H_2S_2O_8$

The peroxides of sulfuric and disulfuric acid are not important in connection with this book. They can be prepared by the following reaction:

$$ClSO_3H \quad + \quad H_2O_2 \quad \rightarrow \quad H_2SO_3\text{-}OOH \quad + \quad HCl$$

The free acid forms a colorless solid which melts at 45°C. This substance occurs in industry as an intermediate in the manufacture of H_2O_2 from peroxodisulfonic acid. The acid can spontaneously explode. The colorless solid, $HSO_3\text{-}O_2\text{-}SO_3H$, decomposes at 65°C. In aqueous solution it can thermally hydrolyze:

$$H_2S_2O_8 \quad + \quad H_2O \quad \rightarrow \quad H_2SO_5 \quad + \quad H_2SO_4$$
$$H_2O \quad + \quad H_2SO_5 \quad \rightarrow \quad H_2SO_4 \quad + \quad H_2O_2$$

The ammonium salt $(NH_4)_2S_2O_8$ is a strong oxidizing agent. A third peroxide, $K_2S_4O_{14}$, can be prepared from alkali metal peroxides and SO_3. In this molecule, the S atoms are linked together by a peroxide group:

$$(SO_2\text{-}O\text{-}O)(SO_2\text{-}O\text{-}O)$$

The peroxides of lower oxidation states of sulfur, for example, peroxo-sulphinic acid, HSO_2-O_2H, are not known. The latter molecule promptly rearranges and forms sulfuric acid.

Thiosulfate; Thiosulfurous Acid

Thiosulfate contains two different sulfur functions; an $-SO_3^-$ group in which sulfur has a nominal oxidation state of plus six, and an adjoining sulfide group in which sulfur has an oxidation state of minus two. The thiosulfate salts are amazingly stable and are excellent complexing agents for metal ions. Schmidt (1961) prepared the free acid in dry ether at -78°C. Upon warming, the dry free acid decomposes at 0°C according to:

$$H_2S_2O_3 \rightarrow H_2S + SO_3$$

The dry products do not react with each other. The decomposition is analogous to that of sulfuric acid at 300°C

$$H_2SO_4 \rightarrow H_2O + SO_3$$

The structure of the thiosulfate contains an S-S bond with an inter-atomic distance of 1.98 A. The Raman spectrum is well known. The UV spectrum was published by Eriksen (1972). Thiosulfate can be prepared from sulfute and elemental sulfur:

$$SO_3^{2-} + 1/8\, S_8 \rightarrow S_2O_3^{2-}$$

The reaction involves, first, ring scission of S_8, followed by stepwise degradation of the sulfur chain (Schmidt, 1965; Meyer, 1977). The reverse reaction, the condensation of thiosulfate to long sulfane-sulfonic acids, from which elemental sulfur can form, is well established. The oxygen exchange between thiosulfate and water has been measured by Betts (1971). He proposes that at an intermediate pH, sulfite appears as an intermediate species. Seibert (1975) showed that the sulfur atoms of thiosulfate can exchange positions in a thiocyanate melt at a high rate.

Technically, thiosulfates are prepared by oxidation of polysulfides (Teder, 1965; Gmelin, 1963) in air. The sodium salt, $Na_2S_2O_3 \cdot 5H_2O$ is very soluble. In photographic processing, thiosulfate is used to dissolve unreacted silverbromide, so that it can be washed out of the emulsion:

$$AgBr + 3S_2O_3^{2-} \rightarrow Ag(S_2O_3)_3^{5-} + Br^-$$

The oxidation of thiosulfate to tetrathionate:

$$2S_2O_3^{2-} + I_2 \rightarrow S_4O_6^{2-} + 2I^-$$

is used in analytical chemistry. The reaction has been carefully studied by Connick (1956). In this reaction, one sulfur atom is oxidized from S^{2-} to S^0, while the terminal $-SO_3^-$ groups remain unaffected. This redox reaction can also be achieved by bacteria in soil (Trudinger, 1967). Often, the oxidation of thiosulfate leads to sulfate; the intermediate sulfite is not stable in the presence of oxidation products such a tetrathionates, because of the chain degradation reaction explained above.

Thiosulfate is used in the bleaching industry as 'antichlorine' to remove excess chlorine:

$$5H_2O + S_2O_3^{2-} + 4Cl_2 \rightarrow 2HSO_4^- + 8HCl$$

In this reaction, the sulfide atom loses eight electrons, i.e. its entire valence shell.

The electrolytic oxidation has been studied by Yokosuka (1975), who observed a mixture of products, including sulfate and polythionates. He also oxidized it with peroxide and hypochlorite; at pH 4 only polythionates were obtained. At high pH, with an iron catalyst, he produced only sulfate. The thermal decomposition was studied by McAmish (1974). The analytical methods for separating thiosulfate, including iodometric work by Kurtenacker (1927), are described by Blasius (1968) and Karchmer (1971).

Sulfane-Sulfonic Acids

Schmidt (1957) prepared the free acids, $H-S_x-SO_3H$, sulfane-sulfonic acids, the thioanalogs of the peroxosulfuric acid. Homologue members of this series with n values between 1 and 7 can be prepared in dry ether at -78^oC from chlorosulfonic acid and sulfane:

$$ClSO_3H + H_2S_x \rightarrow H-S_x-SO_3H + HCl$$

The existence of these acids establishes an important link in the elucidation of the degradation of elemental sulfur by sulfite, which is now convincingly explained by Schmidt (1965). The first step is:

$$S_8 + SO_3^{2-} \rightarrow {}^-S_8 \cdot SO_3^-$$

The charge distribution in this and homologous chains, calculated with a semiempirical Hückel model, indicates that the sulfide end of the molecule can act as a nucleophile on other molecules (Meyer, 1977). In the sulfane-sulfonic acids, the intermediate chain members have a nominal oxidation state of zero. Oxidation of the acids can produce disulfonic acids:

$$2\,{}^{-}S_xSO_3^- \quad + \quad X_2 \quad \rightarrow \quad {}^{-}SO_3\text{-}S_{2x}\text{-}SO_3^- \quad + \quad 2X^-$$

as observed in the case of thiosulfate. The resulting species are discussed next.

Polythionates; Sulfane-Disulfonates

As discussed above, Wackenroder (1846) identified polythionates among the products of the aqueous reaction of hydrogen sulfide and sulfur dioxide. Schmidt (1960, 1964) prepared tetrasulfane dithionic acid from thiosulfate and sulfur chloride:

$$Na_2S_2O_3 \; + \; S_2Cl_2 \; + \; HCl \; \rightarrow \; HSO_3\text{-}S_4\text{-}SO_3H \; + \; NaCl$$

and demonstrated that it is identical with hexathionate. Thus, Oden's polythionate and sulfur colloids and Weitz' polythionates which contain up to 100 atoms, all consist of a long 'sulfane', i.e. an unbranched polysulfide chain in which the chain terminals are formed by the sulfonic groups. The Raman spectra of short polythionate confirm this conclusion (Gerding, 1950), which was first proposed by Dural and Stamm in 1941. Schmidt (1964) prepared the potassium salts of the pure tri-, tetra-, penta-, and hexathionates, and reported their solubility, dielectric constants, UV spectra, and polarographic titration. Their structure was determined by Foss (1956-1958). Mixtures of polythionates are difficult to separate, because they equilibrate. Ritter (1970) studied displacement reactions in trithionate and concluded that the divalent sulfur acts as a soft-acid center which responds primarily to the polarizability of the attacking nucleophile, and only secondarily to the basicity.

Pollard (1955) separated polythionates on paper, as did Steinle (1962); Seiler (1964) used thin layer chromatography, and Blasius (1965, 1968, 1973) developed a high voltage ionophoretic method which works by far the best. A detailed summary of the structure, bonding, and reactivity of the sulfane-disulfonates, i.e. polythionates, was given in a review by Schmidt (1973). The exchange reaction between sulfite and polythionates was studied by Wagner (1976).

The other inorganic compounds of sulfur are not discussed here. Sulfur nitrogen compounds are discussed in Section 13B, because S_xN_x forms polymers. Sulfur-phosphorus compounds and many other interesting and important compounds have been reviewed in many excellent chapters in books by Senning (1970-1972), Nickless (1968), Schmidt (1971), Meyer (1973), and Gmelin. The commercially important sulfur compounds, among them elemental sulfur, sulfuric acid, sulfur dioxide, carbon disulfide, carbonylsulfide, thiophosgene, hydrogen sulfide, the sodium sulfides, the sulfites and disulfite, thiosulfate, the polythionates, dithionites, and sulfur chlorides, are discussed in Chapter 12.

D. CORROSION

Sulfur compounds are blamed for damage ranging from the crumbling of old art treasures to early amortization of steam power plants. In ambient air the corrosive agent is almost always sulfuric acid, which acts as a proton donor and readily dissolves cement, stones, metals, and all carbonates. Gilette (1975) estimated that corrosion by excess sulfur dioxide in ambient air in the U.S. caused damage costing $900 million in 1968. Today, the damage is about $100 million per year. In acute pollution episodes, ambient sulfur dioxide acts as a redox agent which bleaches lawns and fabric dyes, and embrittles paper. In industry sulfur corrosion is also well known: In coal fired power plants, the sulfur dioxide and traces of sulfuric acid work synergistically, as explained below, and corrode steel walls and brick liners. In the reducing environment of the Claus plant, in coal gas plants, and in the presently tested future generation power plants, hydrogen sulfide attacks steel and causes serious sulfidization damage. In all these cases, lime and limestone are well established corrosion inhibitors (Clegg, B.P. 3,968, 1815; Philips, B.P. 4,142, 1817). Corrosion can also become a problem in the shipment of elemental sulfur if moisture is present, because nascent rust of the shipping vessel readily reacts with elemental sulfur and forms sulfides and sulfates. The first are pyrophoric.

Damage of cement, stone, paint, leather, paper, cloth, iron, steel, zinc, copper, and nickel by sulfur from power plants has been carefully studied for over 50 years in several countries. Schikorr measured corrosion of steel and cast iron outside Berlin–Dahlem in ambient air in 1935 and found that 36 mg SO_2/m^3 reduced the weight of steel plate by 420 g/m^2 per year, and that of cast iron by 640 g/m^2 per year. The values for eight mg/m^3 sulfur dioxide were 340 g for steel and 300 g for cast iron. English studies on steel galvanized with a 2 oz/ft^2 coating showed that base

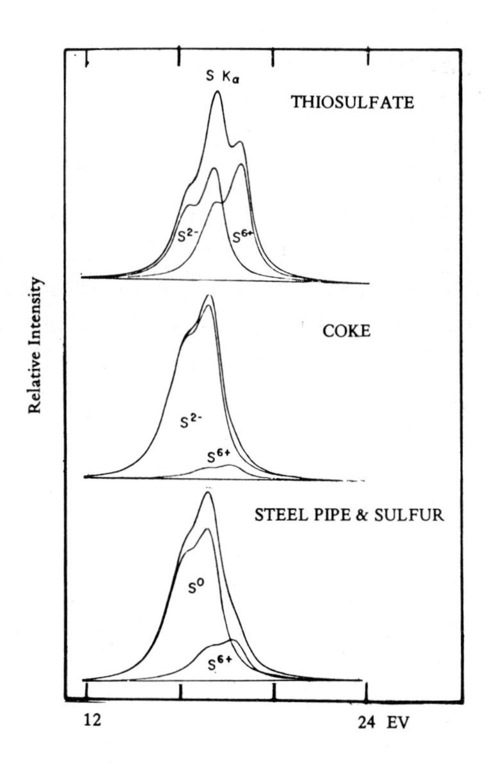

Fig. 3.22. High resolution X-ray spectrum of a) sodium thiosulfate, b) coke, and c) sulfur corroded steel pipe. The different oxidation states of sulfur can be resolved and are indicated in the figure.

metal oxide, due to corrosion of the protective layer, appeared in railroad tunnels after one year, while the same sheet metal lasted an average of fifteen years in rural areas of England, 70 years in Singapore, and 150 years in Khartoum. The blame was put on the combination of smoke, moisture, and sulfur dioxide. Tunnel smoke destroyed steel rails more than six times faster than sea air on the English coast.

The chemistry of corrosion has been of vital importance to the industry. Today, all large industries and universities have specialists who deal with sulfur corrosion chemistry. Enormous progress in this field has been made during the last fifteen years, because of progress in physical analytical instrumentation. The most prominent of these are based on spectroscopic techniques. Among them are low energy electron diffraction (LEEP), photoemission by X-ray excitation, called ESCA (electron spectroscopic chemical analysis), and Auger spectroscopy. Fig. 3.22

shows how high resolution spectroscopy makes it possible to recognize the oxidation state of sulfur *in situ*, in thiosulfate, and in iron coke. Such techniques make it possible to locally investigate both the chemical composition and structure in microscopic samples, on pure single crystal surfaces, and on surface defaults.

Corrosion takes different paths at different temperatures. At low temperature, corrosion preferentially takes place at the interface between the solid and a film of moisture. In this system, the rate of corrosion is determined by the ionic strength, the oxidation potential between ions and neutral metals, and the pH of the aqueous phase. Typical corrosion rates of metals in aqueous brines are of the order of 10^{-5} to 10^{-8} amperes per square centimeter. Dry surfaces corrode only very slowly. Corrosion at high temperature is due to phase changes described below. Sulfur compounds are far more corrosive than oxygen, because the latter form oxide films which yield passive, protective surfaces. In the case of aluminum, the surface films form readily and reproducibly and make it possible to use the metal under conditions where the pure, clean surface would rapidly corrode. The structural reasons for the difference between sulfide and oxide films are not yet well understood, but form the focus of the bulk of modern corrosion work.

1. Corrosion in Process Plants

An essential factor in the development of modern sulfur process plants was the control of corrosion. As mentioned in Chapter 7, several otherwise attractive, modern Claus absorber liquids are unsuitable because they slowly form decomposition products which corrode iron and, thus, force the manufacturer to use expensive specialty steels. In the Frasch industry, steam boiler corrosion was an important difficulty. Since 1500 gallons of water are needed to produce one ton of sulfur, it was necessary to develop sea water resistant steam plants before off shore mining became possible (Brogdon (U.S.P. 2,937,624, 1960; Axelrad, U.S.P. 2,947,690, 1960). The development of modern large scale coal gasification plants depends on materials which can resist the corrosive gas environment. The gas composition of a coal gasifier is shown in Table 3.20. The process conditions are far more stringent than in Claus or Frasch plants because of the high temperatures.

The construction materials must have good structural integrity and chemical resistance. They are exposed to corrosion as well as erosion.

Table 3.20
CONSTITUENTS OF HYDROCARBON SOURCES
(% weight)

	Crude petroleum	Petroleum residue	Tar sands bitumen	Shale oil	Dry bituminous coal
Carbon/hydrogen (ratio)	7.0	8.0	8.0	8.0	14-20
Sulfur	1.4	2–4	4.0	1.0	1-4
Nitrogen	0.2	0.5	0.4	2.0	1-2
Oxygen		1.2	1.0	1.5	5-10
Ash	0.1	0.2	0.6	0.1	10-13

These deterioration processes are presently under intensive study by industrial, academic, and government researchers world-wide. Corrosion includes several independent processes. Among the most important are: Oxidation, hot corrosion, sulfidation, and carbonization.

Oxidation has been well studied and is now quite well understood. In the initial process, oxides of all alloy component phases form in the approximate alloying proportions. As a surface film develops, the oxygen activity decreases at the oxide interface at a rate proportional to the dissociation pressure of the oxide, and the thermodynamically most stable oxide forms. At high temperature, the oxidation rate largely depends on the diffusion rate of the most reactive components of the alloy to the metal-oxide interface, and the oxygen diffusion rate into the alloy. The latter is limited by the dissociation pressure of the surface scale. Iron and nickel steels are protected by chromium addition. On such alloys, up to eighteen weight % $NiCr_2O_4$ and Cr_2O_3 can form. The alloy is formulated to contain the highest possible chromium content while maintaining the best possible structural strength. Aluminum provides similar protection by the formation of Al_2O_3. Simultaneous addition of aluminum and chromium works well in combination with iron and nickel steel.

Hot corrosion of alloys consists of oxidation induced by inorganic salts. Alkali metal chlorides and sulfur in the form of sulfides or sulfates are most corrosive. The effect is due to the fact that the salts react with the protective oxide coating and form an eutectic liquid phase which dissolves the solid coating, allowing the sulfur to form sulfide ions which penetrate into the metal. Furthermore, sulfur depletes chromium and, thus, prevents formation of a subsequent new oxide film. Hot corrosion can turn into a continuous reaction cycle which has catastrophic results. The effect is influenced by the total sulfur concentration and the sodium to potassium cation ratio. Hot corrosion can be suppressed by addition of magnesium or calcium oxide, first patented by Clegg (1815) and Philips (1817). The protection is partly due to neutralization, partly to the formation of the high-melting double sulfate, $K_2SO_4 \cdot 2(CaSO_4)$, which retains alkali which, in turn, tends to form the low melting complex salt, $Na_3Fe(SO_4)_3$. The corrosion rate of iron alloys has been experimentally correlated to the coal composition:

$$k(mg/cm^2) = 5.96 + 5.07(Na_2O/K_2O) - 0.42(CaO + MgO)$$

Alkali metal oxides are expressed in parts per million from acid extraction analysis. Alkaline earth oxides are determined on the basis of ash. The effect of silicates has been discussed by Natesan (1976). Vanadium has a negative effect. Fifty ppm suffice to penetrate 60 mil deep into normal type 304 steel at 900°C (1650°F) within 100 hrs. Fuel oils containing 200 ppm vanadium, 50 ppm sodium, and 3% sulfur destroy normal boiler alloys at 800°C within 10,000 hrs. High chromium Inconel 671, IN-687, and IN-589 have been developed for resisting such environments.

Sulfidation can be caused by sulfate, sulfur dioxide, elemental sulfur, or sulfide; it is due to the formation of sulfides. Fig. 3.23 shows the phase diagram of the nickel-sulfur-oxygen system; the operating condition of the presently tested coal gasification processes are indicated. Corrosion is due to the structure of the solid sulfides, which do not form protective films, and the low melting point of sulfides which causes them to form liquid phases under process conditions. These phases can cause severe mechanical problems in welding seams (Nakagawa, 1974). Chromium inhibits sulfidation by sulfur dioxide by oxide formation; it cannot protect alloys against sulfide at 600°C. Aluminum forms Al_2S_3 which provides partial protection. The isocorrosion curves, Fig. 3.24, show that austenitic steels are at least twenty times more resistant to sulfidation than low chromium steel. One hundred ppm hydrogen sulfide in hydrogen gives a corrosion rate of one mil per year at 600°C. The figures show that the stable iron and chromium sulfide phases form only at high hydrogen sulfide concentration. Corrosion is only influenced by the relative pressures of hydrogen sulfide; it is independent of the total gas pressure. In the proposed coal gasification processes, hydrogen sulfide to hydrogen ratios vary from 6×10^{-3} for the Hygas process to 10^{-2} for the Synthane process. Such ratios will cause serious problems with conventional steels. The Consol carbon dioxide acceptor process has a ratio of 10^{-4}, because calcined dolomite is present in the combustion bed.

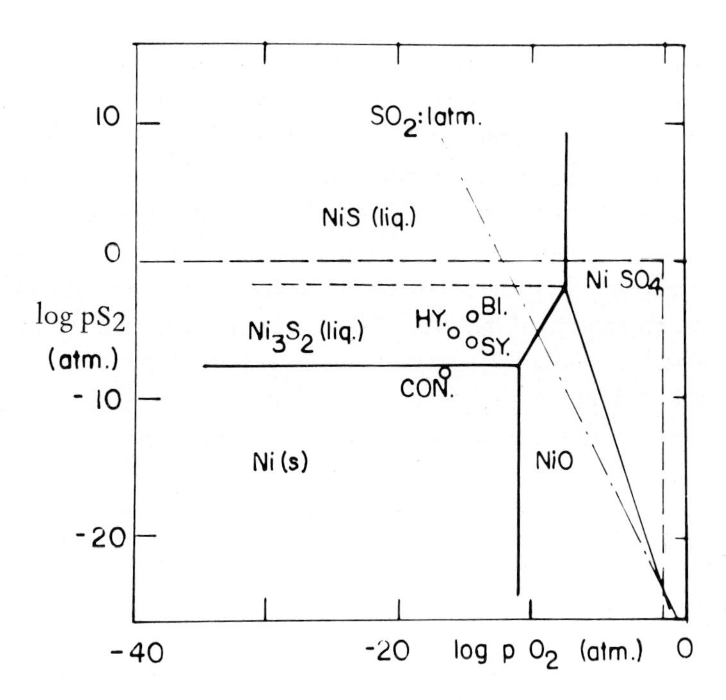

Fig. 3.23. Phase diagram of nickel sulfide; the operating conditions of the four presently contending coal gasification processes are indicated (Worrell, private communic., 1976).

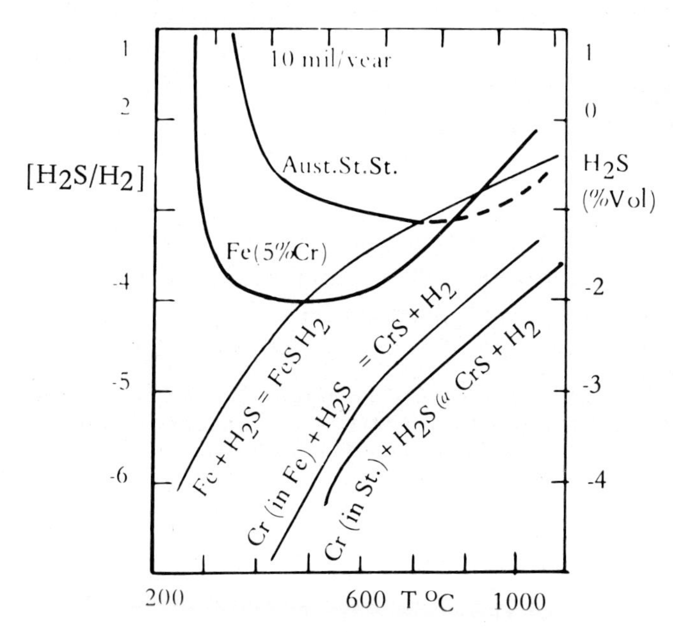

Fig. 3.24. Isocorrosion curves for austemitic steel and chromium steel under coal gasification conditions (after Natesan, 1976).

Carburization is caused by transfer of carbon from the surface to the bulk of the alloy. It depends on the temperature and gas composition. Carburization is not connected with sulfur and has been well studied. Often, it is accompanied by metal dusting which is caused by the formation of unstable metal carbides which can decompose to graphite and metal. This can lead to a continuous process of deterioration.

As mentioned above, corrosion leads to pitting, cracking, and, eventually, to structural failure. The latter is facilitated by mechanical erosion which is caused by particulates in the fluid medium of the combustion gas impact on the reactor walls, and inflicts cuts and frictional wear. In coal gasifiers, the gas velocity is relatively low, i.e. 10-30 m/sec. The structural damage cannot be measured by over-all loss in weight of the metal, because structural changes and localized damage are likely the limiting factors.

2. Corrosion During Shipment of Moist Sulfur

It is well known that moist sulfur rapidly corrodes iron vessels, for example, barges used for the trans-oceanic shipping of sulfur. The reaction products are known to ignite spontaneously, i.e. they are pyrophoric. The reason for the reactivity is easily understood if one considers reactions which can take place in the presence of air, which freely circulates through granular or slate sulfur. Nascent ruse, an iron oxide hydrate, can react with elemental sulfur as follows:

$$3Fe_2O_3 + 3H_2O + 9/8S_8 \rightarrow 6FeS + H_2SO_4$$

Obviously, the initial reaction products can vigorously attack iron and perpetuate the corrosion. The structures of the many iron sulfide modifications are complex (Section 6B3).

RECOMMENDED READING. Current references in this field are included in a monthly newsletter distributed by The Sulphur Institute, 1725 K St., N.W., Washington, D.C. A full set of references to Section 3A can be found in Meyer, Chem. Rev. 76 (1976) 367. The Bibliography of this book lists only references relevant to energy and the environment.

Chapter 4

Analytical Chemistry

It is comparatively easy to determine the presence of sulfur, but many excellent analytical chemists dread working with sulfur compounds, because accurate analytical determination of sulfur is tricky. It is hard to establish the identity of sulfur species, because they readily interconvert and often coexist in mixtures of homologues. Quantitative work is hampered because it is not easy to fully convert sulfur to the desired oxidation state and it is very difficult to separate sulfur from mixtures or other compounds. However, several excellent standard methods have now been established which give reproducible, if not always absolute, results. In this chapter basic methods for determining total sulfur are listed and some specially important sulfur systems are discussed. For a complete coverage of analytical sulfur chemistry, the reader should consult Karchmer (1971-1972), Blasius (1968), Gmelin (1959), Ashworth (1976), and the extensive literature.

A. QUANTITATIVE ANALYSIS OF TOTAL SULFUR

Table 4.1 lists standard extraction methods for establishing total sulfur in common materials. The basic separation methods include the lamp test in which, contrary to the procedure implied in the name, the sample is burned in a closed volume in a mixture of 30% oxygen and 70% carbon dioxide. The gas is oxidized with 3% hydrogen peroxide to yield sulfuric acid. In the high temperature combustion test the material is burned with pure oxygen in an oven kept above 1315°C, causing 97% conversion to sulfur dioxide. It is absorbed into an iodide-starch solution and titrated with iodate. This method is used in several variations. Another popular method involves soda-fusion. In the Eschka method, porous MgO and Na_2CO_3, in a 2:1 ratio, is mixed with the material and slowly heated to 1000°C, where the fusion is kept for at least an hour.

Table 4.1

STANDARD METHODS FOR THE EXTRACTION OF SULFUR FROM VARIOUS MATERIALS

Material	Suggested material	Finishing technique	ASTM standard
1. Petroleum	modified lamp	$BaCl_2$ titration	D1072-56
Fuel gases	gaschromatography	(tetrahydroxy-quinone)	
Low-boiling petroleum distillates	lamp	a. NaOH titration (methyl purple) b. turbidimetric	D1266-64T
Kerosine & mid-boiling distillates	lamp or x-ray	NaOH titration (methyl purple)	D1266-64T
Fuel & lub oils	HT combustion or x-ray	iodate titration	D1552-64
Residuals	HT combution	iodate titration	D1552-64
2. Solid fuels			
Coal	Eschka	gravimetric	D271
Oil shale	Eschka	gravimetric	
3. Agricultural materials			
Fertilizers	bromine-HNO_3 oxidation	gravimetric	
Plant material	carbonate or peroxide fusion		AOAC
Soil	carbonate or peroxide fusion	turbidimetric	
Insecticides & pesticides	acid oxidation	gravimetric	
4. Rubber			
Cured rubber	fusion, x-ray	gravimetric	D297-61T
Uncured rubber	Grote combustion x-ray	sodium hydroxide titration	
Polymers	x-ray		
5. Foodstuffs	peroxide fusion	gravimetric	
6. Drugs & medicines	flask combustion	barium per-chlorate titration	
7. Wood and pulp	wet oxidation	iodate titration	
8. Metals			
Stainless steels	combustion	iodate titration	E30-56
Carbon steels, low-alloy steels, cast iron	HT combustion	iodate titration	E30-56 E30-60T
High-Si steels, open-hearth iron, wrought iron	wet oxidation	gravimetric	E30-56
Alloy steels: High C, high S, molybdenum & others	evolution	iodate titration	E30-56
Iron alloys with Mn, V, Si, W, Mo	wet oxidation	gravimetric	E31-63
Chromium-nickel-iron alloys	wet oxidation	gravimetric	E30-56 E38-58
Brasses & bronzes	HT combustion	iodate titration	E54-66

Continued...

Table 4.1 (continued)

Material	Suggested material	Finishing technique	ASTM standard
9. Rocks			
Sedimentary materials, silicates, & insoluble sulfates	fusion, sodium carbonate, potassium nitrate	gravimetric	
10. Ores & minerals			
Zinc ore concentrate	wet oxidation	gravimetric	
V, Mo, W, Th	wet oxidation	gravimetric	
Alumina	modified combustion	iodate titration	E30-60T
Commercial sulfur	wet oxidation	gravimetric	O-I583C
11. Cement & related materials			
Portland cement	wet oxidation	gravimetric	E30-60T
Limestone	HT combustion	iodate titration	E30-60T
Sulfur mortar	extraction with CS_2	weighing	C287-62
12. Glass	HT combustion x-ray fluorescence	iodate titration	E30-60T
13. Carbon black	O_2 bomb combustion	gravimetric	D1619-60
14. Water	determine as $S^=$, SO_3, SO_4		
15. Air	determine as SO_2, SO_3, SO_4		

After Tuller (1970), Karchmer (1970).

Usually, the melt is then treated with hot water, bromine water, and acid (Chaudhari, 1973). Fertilizers are analyzed by a modified Carius method, i.e. nitric acid followed by bromine treatment. Many organic materials are best treated with a sodium peroxide, potassium chlorate or nitrate,, or sugar mixture in a bomb. The best and most reliable method for organic materials consists in burning sulfur in a closed flask. This method is simple and rapid: The material is weighed on a filter paper in which it is then wrapped. The paper is inserted into a platinum wire coil connected to the stopper of a 250 ml Erlenmeyer. In the flask, 10 ml sodium peroxide solution is prepared, the flask is flushed with oxygen gas, the filter paper is lit and inserted into the flask, which is kept closed until the combustion is completed. After 20 sec. shaking, the sulfate can be determined with barium chloride or any other desired method. Bataglia (1976) burned organic tissues in open flames. Wet oxidation includes Benedict's mixture, consisting of cupric nitrate and sodium chlorate. For metals nitric acid and chlorate are used, and also perchlorate. It should be noted that perchloric acid is very dangerous, as it can spontaneously decompose, i.e. explode, if traces of organic material or dust are present. In the evolution method, the total sulfur, which must be present as sulfide, is converted to hydrogen sulfide by acid treatment. In the extraction method, carbon disulfide-soluble sulfur is separated from all other material by solvent treatment.

The strengths and weaknesses of these methods have been discussed by Thompson (1970). The difficulties in comparing results are due to loss of volatile materials, incomplete conversion, adsorption of sulfur by solid residues, or chemical reaction between sulfur and other components. The determination of the sulfur content of impregnated wood is a good example demonstrating practical problems (Valentova, 1972; Surma-Slusarska, 1976). The determination of sulfur in polymers has been described by Majewska (1968). Qualitative and quantitative separation and analysis of sulfur in soils has been carefully reviewed by Beaton (1968), who also considered plant materials.

After conversion, sulfur can be determined by various methods. The most reliable is gravimetry, using barium chloride which forms insoluble barium sulfate; other methods are redox titration with potassium iodide-iodate, acid base titration with sodium hydroxide, turbidimetry or nephelometry, conductometry, and others described by Thompson (1970) and many inorganic analytical textbooks. In some cases hydrogen iodide, as described by Schmidt (1959, 1964), is a good reagent.

During the last 15 years, some modern, instrumental analytical methods have become available, which if properly adapted will make sulfur analysis simpler, faster, more accurate, and more sensitive. X-ray fluorescence, already widely accepted in industry can be applied *in situ* and yields several elements in one operation; its use is defined in standard ASTM D-2622. It can respond to as little as $10^{-4}\%$ sulfur. It is already available in connection with electron microscopes; thus, microscopic samples imbedded in heterogeneous samples can be analyzed without previous separation. Gamma rays are used to determine the sulfur content in drill core samples (Parus, 1976). The x-ray spectrum also yields structural information (Whitehead, 1973).

Among further separation methods, the various forms of liquid and gas chromatography, and masspectroscopy (Parfenova, 1975) must be mentioned. UV-spectra are not always specific, but give good quantitative data, while IR and Raman techniques give quick and reliable qualitative analysis, but are not yet useful for accurate, routine, quantitative work. In special cases, radioactivation analysis is useful. The most sensitive and accurate spectroscopic method is flamespectroscopy. It involves reduction of sulfur compounds in a hydrogen flame to elemental sulfur, which in the combustion gas efficiently emits the characteristic blue fluorescence of S_2. This method, combined with gas chromatography or alone, is now used extensively in air pollution monitoring. It is accepted by the U.S. Environmental Protection Agency as a standard equivalent method, as described below.

The most accurate determination of trace amounts of sulfur is probably the argentometric titration which yields 2% accuracy with 3 nanograms of sulfur (Blasius, 1973).

1. Sulfur in Soil

Beaton, Burns, Bixby, and Tisdale (Beaton, 1968) have compiled a thorough review of the common methods presently used to determine total sulfur in soil, and the optimum level of extractable sulfur. They list several hundred recent references and give a detailed discussion of the relative merits and problems. The same source also lists methods for establishing reduced inorganic and organic sulfur and discusses the separation of the organic fractions.

2. Sulfur in Water

A reliable and quick procedure for analyzing sulfates in water has been described by Tabatabai (1974). This method is based on barium chloride and uses a turbidimetric detection method. Tabatabai used it for quantitative work on snow containing as little as 1 ppm sulfate, for ground water with 15 ppm and for well water with 40 ppm. The determination of hydrogen sulfide in water is difficult, because the standard detection methods are far less sensitive than the human nose. Occasionally, this causes problems for city engineers responsible for domestic water supplies.

3. Sulfur in Air

The concentrations of air pollutants are usually small, often smaller than those of other impurities which sometimes interfere with the analytical study of sulfur. The sudden public interest in air pollution control caught analytical air chemists by surprise, because the legally tolerable levels of pollutants were below the level of accurate quantitative measurements. The demand for scientific, quantitative determination of malodorous and obnoxious substances caused some well-meaning over-eager people to release unfinished methods and sell half-tested equipment. For some industries, this led to unnecessary, economic hardship. For others, it created loopholes. For the regulating agencies, it caused confusion and embarrassment which will persist for some time, even though the problems have been largely solved. In January, 1977, in his last appearance before the U.S. Senate Subcommitte on Environmental Pollution, the outgoing EPA Administrator, Russell Train, summarized the situation as "distressing". He mentioned the "reoccurring need" to make urgent regulatory decisions on a scientific basis that can "charitably be described as barely adequate." However, contrary to many doubts (Farwell, 1976), total sulfur determinations are now fairly reliable, have well established limits of accuracy, and have never been as controversial as the nitric oxide measurements which are not yet dependable. The nitric oxide methods were in the past based on the Jacobs-Hochheisen technique which gave such exaggeratedly high readings that in 1972 a reevaluation of U.S. air quality control regions became necessary, during which it was found that 43 of the 47 areas classified at the critical level I, in reality belonged to the non-critical level III (Burchard, Bowen, 1976). In contrast, sulfur measurements, if in error, always tend to be low.

Sulfur occurs in air as gas, aerosol, and in particulates; in each it can exist in several different chemical forms, for example as ammonium sulfate, sulfur dioxide, sulfuric acid, sulfite, and as sulfate. Not all of the sulfur forms are soluble, and their relative importance in ambient air is not yet well established. This has made it difficult to develop a reliable air sampling technique (Noll, 1976; Butcher, 1972; Leithe, 1971; Balint, 1975; Dharmarajan, 1971; Treece, 1976; Roberts, 1976; Mamuro, 1976; Forrest, 1973). At the present time, the quality of analytical techniques changes so quickly that the U.S. EPA is forced by Federal Regulations (40 FR 7044, February 18.1975) to publish and update quarterly lists of acceptable detection modes, the necessary measuring procedures, and their equivalency as reference methods. Table 4.2 lists commercial equipment meeting U.S. EPA requirements as of October 1, 1976. A review of conventional analytical methods for determing sulfur as sulfur dioxide in air is given in Table 4.3. The fluorescence methods, Table 4.2, are the presently most widely used methods (Saltzman, 1975).

Table 4.2
EXCERPT FROM THE LIST OF DESIGNATED REFERENCE AND EQUIVALENCE METHODS IN ACCORDANCE WITH 40 CFR PART 53, PROMULGATED ON FEBRUARY 18, 1975, DATED OCTOBER, 1976

EPA No.	IDENTIFICATION
EQS-0775-001	"Pararosaniline Method for the Determination of Sulfur Dioxide in the Atmosphere-Technicon I Automated Analysis System."
EQS-0775-002	"Pararosaniline Method for the Determination of Sulfur Dioxide in the Atmosphere-Technicon II Automated Analysis System."
EQSA-1275-005	Lear Siegler model "SM1000 SO_2 Ambient Monitor," operated on the 0-0.5 ppm range, at a wavelength of 299.5 nm, with the "slow" (300 sec) response time.
EQSA-1275-006	"Meloy Model SA185-2A Sulfur Dioxide Analyzer," operated with a scale range of 0-0.5 ppm.
EQSA-0276-009	"Thermo Electron Model 43 Pulsed Fluorescent SO_2 Analyzer," operated on the 0-0.5 ppm range.
EQSA-0676-010	"Philips PW9755 SO_2 Analyzer" consisting of the following components. PW9755/02 SO_2 Monitor with: SO_2 Source, Filter Set SO_2, Electrolyte SO_2, Supply Unit/ Coulometric, either Sampler or Dust Filter; operated with a 0 to 0.5 ppm range and with a reference voltage setting of 760 mV.
EQSA-0876-011	"Philips PW9700 SO_2 Analyzer" consisting of the following components: Chemical Unit with: Electrolyte SO_2, Electrical Unit, Filter Set SO_2, Sampler Unit or vendor-approved alternate particulate filter source operated with a 0 to 0.5 ppm range and with a reference voltage of 760 mV.
EQSA-0876-013	"Monitor Labs Model 8450 Sulfur Monitor," operated with a 0 to 0.5 ppm range, a 5 sec time constant, a model 8740 hydrogen sulfide scrubber in the sample line.

Further information may be found in the Federal Register 40 FR 7044 or from the U.S. Environmental Monitoring and Support Laboratory, Department E (MD-76), U.S. EPA, Research Triangle Park, North Carolina, 27711.

It has been known for over a hundred years that even traces of sulfur compounds, regardless of their chemical nature, color flames blue. The emission is an extremely sensitive measure for sulfur. It is caused by fluorescence corresponding to the allowed electronic transition from the first excited triplet state to the ground triplet state of the S_2 molecule, which was discussed by Salet (1869), as described in Section 3A. The transition, $T_o = 29{,}100\ cm^{-1}$, with the corresponding wavelength of 280 nm, has been extensively studied (Barrow, 1955-1977). The excited state can undergo pre-dissociation. The reverse reaction, the combination of atoms, leads to energy-rich electronically excited S_2 in the B state; the fluorescence results by radiative decay of the state (Meyer, 1971).

Table 4.3
STANDARD METHODS FOR SULFUR IN AIR

Method	Description	Detection Limit	Interferences	Comments
Emission Flame Photometry	Sulfur containing gas ignited in H_2 flame S_2 peak at 394 nm	0.1 ppm	H_2S, mercaptans, SO_3 NO_2 & ozone (when SO_2)	Limited in this field due to instrument requirements
Acidimetry (Brit. Std. method)	$H_2O_2 + SO_2$ – H_2SO_4; H_2SO_3 titration with NaOH or Na_2CO_3	0.005 ppm (in 24 hr test)	Acidic or basic substances in the air	Slow & subject to interference
West-Gaeke Method (1956)	$SO_2 + Na_2HgCO_4$ – Na_2HgSO_3Cl or $HgSO_3^{=}$ + pararosaniline + formaldehyde – pararosaniline methyl sulfonic acid (red) det. colorimetrically at 560 nm	0.005 ppm	NO_2 greater than 2 ppm West & Ordovera; amidosulfonic acid addition; NO_2 greater than 10 ppm	The mercury complex is stable for at least 24 hr - field sample; Concentration can then be det. in lab
Continuous Colorimetry	West-Gaeke, on a continuous scale; Air (250 ml/min) & Acidic pararosaniline (3.3 ml/min) on cont. feed; color complete in 5 min	West-Gaeke	2 ppm NO_2 causes loss of 0.1 ppm SO_2	Well suited for industrial use; many variations
Conductometry	a) $SO_2 + H_2O_2$ – H_2SO_4 delta conductivity measured b) Direct thermocouple could be used, but only in constr. background situations	1 ppm (?)	H_2S, HCl, NO_2 NH_3, Cl_2	For plant use; modications used in industry
SO_2 by semi-continuous pH measurements	Air & phosphoric acid bath removes NH_3, HCl, SO_3; Air(SO_2) + H_2O_2 –H_2SO_4 (pH monitored) fresh H_2O_2 cell automatically changed.	0.003 ppm	Modified version not sensitive to interferences	Best suited for industrial use
Silica gel Reduction by Stratman	SO_2 adsorbed on silica gel then eluted by H_2; SO_2 reduced to H_2S by hot Pt H_2S trapped in urea - $(NH_4)_4Mo_7O_{24}$ det. as Molybdenum Blue or trapped in zinc acetate then Fe^{3+} & dimethyl-p-phenylene diamine added det. as Methylene Blue	0.3 micro grams	None	Suited for lab use
Polarographic Det. Blasius (1974)	SO_2 – H_2SO_2 E = -.050 V	0.008 ppm	NO_2 does not interfere	Limits: SO_3^{2-} 3.6, SO_4^{2-} .39, $S_2O_3^{2-}$ 5.5, $S_3O_6^{2-}$ 13, S^{2-} 5.5, S_x^{2-} 4.8 x10^{-4} mole/l Rec. by Am. Conf. Govt & Ind. Hygen.

Continued...

Table 4.3 (continued)

Method	Description	Detection Limit	Interferences	Comments
Iodometry	$SO_2 + I_2 - I^- + SO_3$ I_2 in a starch solution	1 micro gram	NO_2	Has been replaced by Gaeke-West
Ferric test	SO_2 reduces Fe^{3+} to Fe^{2+} in the presence of 1,10-phenanthroline – colorimetry at 510 nm	0.05 ppm	H_2S, ozone formaldehyde	Good in lab; has been adapted to automatic measurements
^{85}Kr	SO_2 + chlorite – ClO_2 ClO_2 + quinolclathrate of ^{85}Kr; ^{85}Kr is released & det. radiochemically	(?)	None	Presently a lab oddity
Impregnated paper	a) KOH traps SO_2; SO_2 is det. as in West-Gaeke b) Ammoniacal zinc nitroprusside + SO_2 – red compared with std colors	1 ppm	H_2S	Good for empirical observation
Bell Method	Lead peroxide powder + gum tragacanth paste on gauze, exposed 30 days in a cylinder $PbO_2 + SO_2 - PbO + SO_3$ $SO_3 + BaCl_2 - BaSO_4$, thoron titration	mg/day 30 day test		Still used to get integral exposure values
Thoron	$SO_2 + H_2O_2 - H_2SO_4$ $SO_4^{=} + Ba^{2+} - BaSO_4$ thoron indicator turns pink with Ba^{2+}		Heavy metals Na salts exceeding 0.14	

See W. Leithe, The Analysis of Air Pollutants, Ann Arbor Science Pub., 1971, pp 149-167.
S.S. Butcher & R. J. Charlson, An Introduction to Air Chemistry, Academic Press, N.Y., 1972, pp 107-112.

The transition energy of this state, 29,1000 cm^{-1}, is much larger than the thermal energy of the flame at 1000°C; thus, the flame is not reponsible for exciting S_2 emission. The function of the flame is to break up the sulfur reagent and produce reactive intermediates in the correct oxidation state which can form S_2. The conversion for sulfur to S_2 is amazingly efficient, but the correlation between the sulfur reagent and light emission is not always fully linear. However, the fluorescence method is now sufficiently understood to serve as a reliable detector for as little as 1 ppb sulfur dioxide.

Much active research is underway in this area. A very fast response method for medical studies was described by Devel (1976); it is based on conductometry, using oxidation of sulfur dioxide with hydrogen peroxide to sulfuric acid as source of protons which influence the conductivity. A method for determining the total sulfur, total sulfate, sulfur trioxide, and sulfuric acid collected on perimidylammonium bromide impregnated

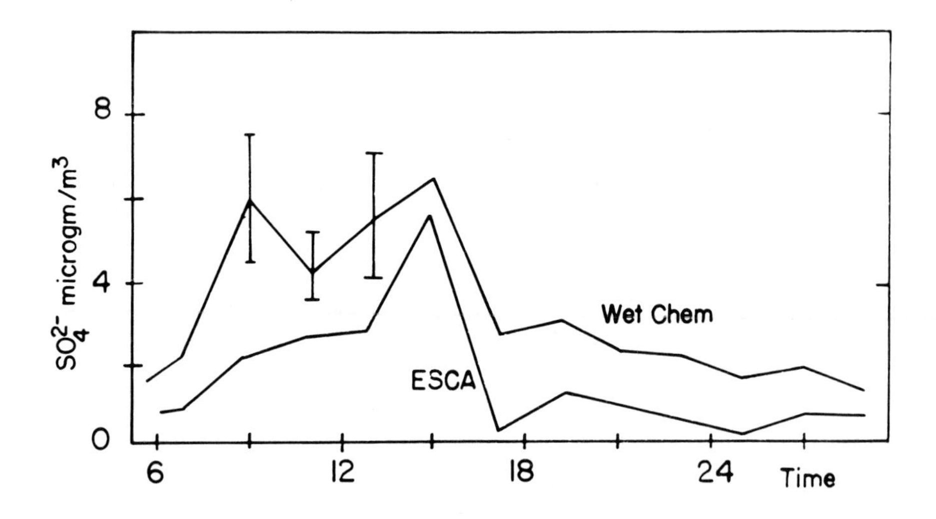

Fig. 4.1. Diurnal sulfur content of air in Pomona, California, 1975 (after Novakov, 1975).

filters has been published by Thomas (1976). The method is advantageous because the reagent reacts quickly with sulfuric acid before it can react with particulates. The working range extends from 1 to 50 micrograms, with a limit of 0.3 micrograms. Another method for processing samples collected by unskilled field workers has been described by Lorenzen (1975). Reactive sulfur is collected on silver nitrate filters, and sulfur dioxide adsorbs on sodium hydroxide. The filters are kept in cassettes which can be processed in a central lab by X-ray fluorescence and evaluated with a computer. Hydrogen sulfide from pulp mills can be measured by a method described by Natusch (1972); the sensitivity is 0.005 ppm, because fluoro-mercury-acetate is used to process the concentrate extracted with cyanide ion from silver nitrate impregnated filters. Stevens (1971) has used gas chromatography to determine ppb of odorous pulp gases. A split beam detector which can discriminate between total sulfur and reducible sulfur was described by Lang (1975). A very sensitive method for measuring atmospheric hydrogen sulfide was described by Natusch (1972). Holt (1976) used oxygen-18 to identify the chemical history of air-borne sulfate. Novakov (1976) made a careful comparison of wet-chemical methods with X-ray photoelectron spectroscopy, ESCA, using laboratory and field samples. Fig. 14.1 shows the results with an

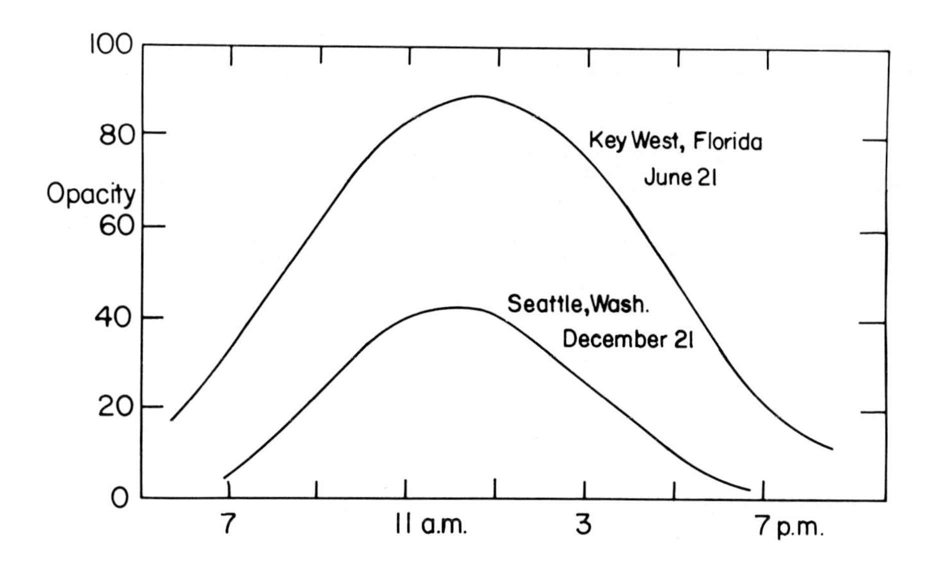

Fig. 4.2. Effect of geographic location and time of day on opacity (after Weir, 1976).

approximate range of accuracies for the diurnal pattern for sulfate in Pomona. The ESCA method can measure insoluble sulfate which escapes wet-chemical methods. IR and laser spectroscopy are also being used to analyze gaseous pollutants, but they serve mainly qualitative purposes (Novakov, 1975, 1976). The sulfate particulate can be determined by light scattering (Waggoner, 1976), but the opacity measurements are controversial, since they depend on the angle of observation, the brightness of the sky, climate, geographic location, and other factors (Weir, 1976). The effects are shown in Fig. 4.2. Among the factors are the stack diameter, mean particle size, stack gas temperature, water content, wind speed, air temperature, color of sky, the sun level, and the time of day and year, as well as the geographic location. Obviously, this convenient method is archaic.

B. QUALITATIVE ANALYSIS

1. Sulfur in Air

The state of sulfur in ambient air is not yet well established, because the concentrations are low, and because sulfur can occur in the gas phase, in the form of aerosols, as chemical components of particulates, and can be absorbed on particulates. Fig. 4.1 shows a practical example of "inconsistent" "total sulfur" values, obtained because the analytical methods respond to different phases. Conventional analysis checks only for soluble sulfur, while ESCA analysis can discriminate between sulfur dioxide, sulfur trioxide, and elemental sulfur, sulfide, sulfite, and sulfate in solid samples (Novakov, 1974). This type of analytical work has suggested that much of the sulfur in aerosol results from catalytic oxidation on particulates (Novakov, 1973-1977) rather than from homogenous gas phase oxidation, as has been assumed for over 30 years. The implications are discussed in Section 3B. Unfortunately, the ESCA, IR and Raman methods are not yet reliable as quantitative tools. Gas chromatography is widely used to study the components of effluent gases (Hasinski, 1975; Fujii, 1976; Moeckel, 1976). The most widely accepted methods are still based on wet chemistry, which has been perfected on micro scale (Sullivan, 1964).

2. Solid Sulfur Compounds

The identification of sulfur compounds in mixtures is an art. Lorenzen (1975) and Goshi (1975) used high resolution X-ray spectroscopy in the range from 10 to 30 eV to identify thiosulfate, sulfate, sulfide, and elemental sulfur in coke and corrosion products, and obtained better resolution than ESCA, which uses energies between 120 and 200 eV (Novakov, 1974). These physical *in situ* methods, together with IR and Raman spectroscopy, will become important techniques. Auger and ESCA spectra can only be recorded on expensive machines and will not become routine tools. The spectra of CS_2, COS, H_2S, SO_2, and SF_6 were recorded and compared by Asplund (1976).

Often, gaseous and liquid sulfur compounds can be separated by chromatography. Hiller (1976) successfully used methanol to separate aliphatic polysulfides by reverse-phase bonded-phase chromatography, and obtained a linear relationship between k-values and the sulfur rank, as well as the carbon rank. Inorganic sulfanes equilibrate on all known columns and cannot be quantitatively separated by chromatography alone.

The separation of mixtures of sulfur compounds in aqueous solution is difficult. Even a superficial discussion of this topic exceeds the scope of this book. Gmelin (1959) gives a good review of older work. More recent work has been reviewed by Karchmer (1970). Aqueous systems have been extensively studied by Blasius (1968, 1972), who developed a polarographic method for determining the components in mixtures of sulfate, sulfite, thiosulfate, trithionate, sulfide, and polysulfide in the eluants of industrial furnace ashes (Blasius, 1974). The method responds to concentrations of 10^{-4} mole/liter with an accuracy of 2%. The titration is very fast. Earlier, Blasius (1974) perfected a high voltage ionophoretic method for separating mixtures of polythionates, which he used to identify the product of the reaction of aqueous hydrogen sulfide and sulfur dioxide. Even earlier, he had used this method to show that thiosulfate and sulfate form tetrathionate (1965, 1973). Polythionates can also be analyzed by thin layer chromatography using a mixture of methanol/propanol. Paper chromatography (Pollard, 1955) is of limited value, because it is slow and the phases can equilibrate during dilution (Steinle, 1962).

C. SULFUR ISOTOPES

The chemical path of sulfur reagents can be far better traced if isotopes are used. Terrestrial sulfur normally contains 96% 32-S; 34-S accounts for 4%. Sulfur-35 is radioactive and has a half-life of about 3 weeks. Vennart (1976) studied the influence of nuclear power plants on the natural concentration of 35-S, and its effect on man. He derived that not more than 170 micro Curries of 35-S from emission from nuclear power plants would reach the food ingested by adults each year. Bain (1965) has used 35-S very successfully to establish the exchange rate of sulfur in aqueous and non-aqueous systems. Blasius (1965, 1974) used it to identify components of polythionate mixtures. Sulfur-35 was also used to determine the kinetics of sulfur dioxide absorption (Happel, 1973). Changes in the relative abundance of natural sulfur isotopes have aided Thorpe (1953) in unravelling the geochemical fate of sulfur, an accomplishment which found practical applications in the search for oil and sulfur. Artificially enriched 34-S has proven useful for identifying the structure of many sulfur compounds, for example that of pyrosulfite (Meyer, 1977), and for determining the reduction of sulfate (Harrison, 1957). Ninety percent plus sulfur-34 is now available at a very reasonable price because of recent industrial separation efforts which use the isotope exchange between sulfur dioxide and aqueous bisulfite (Stachewski, 1975). The separation of SF_6 isotopes by IR laser, mentioned above, might soon become economically advantageous.

The nuclear properties of sulfur have been described elsewhere (Texas Gulf, 1959; Vernotte, 1976).

D. IMPURITIES IN ELEMENTAL SULFUR

Frasch and Claus sulfur are often 99.5% pure, or more. Their production methods are described in Chapter 7. The "bright" sulfur grades contain traces of As, Se, and organic compounds, depending on their origin. The analytical methods have been described by Tuller (1954, 1970), and in a pamphlet by Texas Gulf (1959). Amorphous sulfur has been tested for impurities by Banerjee (1975).

The moisture content is commonly measured by weight loss after drying at 80°C for 24 hours. The ash content is determined by combustion. The carbon content is such an important factor that all large sulfur companies have developed their own methods. IR spectra give a quick and reliable measure of the carbon content and its chemical form (Matson, 1963). The common industrial procedure is based on dry combustion. Arsenic is determined by the bromine-carbon tetrachloride oxidation method (Tuller, 1970); arsenic occurs in volcanic sulfur and smelter tail gases. It can be separated by treatment with a lime solution containing calcium chloride (Nelen, 1975). Selenium is also determined by wet chemistry, as is tellurium.

Several methods for preparing ultra high purity elemental sulfur have been described in Section 3A. Pure sulfur isotopes are now separated cheaply; progress in this field has been such that these are now not much more expensive than normal high purity sulfur of comparable quality.

Liquid sulfur slowly reacts with hydrocarbon and organic impurities and releases hydrogen sulfide. As mentioned above, it dissolves in liquid sulfur and slowly forms polysulfides which can be detected with NMR (Hyne and Wiewiorowski, 1966). Novel methods for determining hydrogen sulfide in sulfur have been proposed by Hanson (1975) and Schwalm (1975). The gases released by the reaction can be analyzed by gas chromatography (Wiewiorowski, 1972).

Chapter 5

Occurrence and Sources of Sulfur

Knowledge of planetary chemistry is far from complete (Lewis, 1973). It is generally believed that the sulfur content of the planets is determined by their iron content and the condensation temperature. Sulfur is mainly retained as FeS. In the pressure range between 10^{-7} and 10 atm., this reaction occurs at about 400°C. On the sun, sulfur constitutes only 0.002% of the weight, iron 0.005%, and oxygen 0.05%. The bulk of the sun consists of 92% hydrogen and 7.5% helium. The sulfur content of Mercury is nil, because the planet has only 0.1% FeO, and thus can retain neither sulfur nor water. Venus is earthlike in composition, but its FeO content is ten times lower. Accordingly, sulfur and water are almost absent there. It has been suggested that the Venus clouds contain sulfur, together with mercury, bromine, silver and antimony. On Mars iron oxidation to FeO is higher than on the earth. It is believed that Mars has a core of pure FeS. The chemistry of Jupiter, which contains 70% of the mass of our planetary system, is not well known. Its density is only 1.3 g/cm^3, and it is generally believed that its sulfur content is similar to that of the sun, i.e. about 10^{-3}%. Khare (1975) has proposed that the clouds of Jupiter contain cyclo-S_8.

Oppenheimer (1974) studied the chemistry of sulfur in interstellar clouds, and developed a model for the equilibrium between hydrogen sulfide, carbonyl sulfide, CS, CH_2S, SO, NS, NS^+, and SO^+, all of which are present in concentrations of 10^{12} to 10^{15} molecules/ml, i.e. 10^{-8} to 10^{-10} moles. He postulates rate coefficients for describing their formation, on the basis that sulfur ionizes more easily than oxygen.

A. NATURAL DEPOSITS

The earth consists of three different layers; a metallic core, which contains about one third of all its mass, a massive mantle of silicates, which contains the bulk of all remaining mass, and a thin crust which makes up a mere 1% of the total mass. The core consists of a solid inner core, with a density of 5.2 g/cm^3 and a liquid outer core which is at 200 kbar and 3,500^{o}C. Fifteen percent of it is light components, mainly sulfur. The mantle contains about 10% FeO. Sulfur constitutes 0.0520% of the weight of the earth's crust.

Table 5.1

WORLD RESERVES OF SULFUR
(MILLION LT)

Area	Petroleum	Natural Gas	Native Ore	Dome	Sulfide Ores	Pyrite	Coal
United States	30	25	100	100	25	25	3,000
Canada	5	100	*	–	25	25	2,000
Mexico	5	*	*	50	10	5	*
Central and South America	30	*	100	–	50	50	*
Western Europe	*	25	5	–	5	25	800
Africa	5	*	*	–	20	20	2
Near East and South Asia	340	500	200	–	5	50	*
Far East and Southeast Asia	10	10	50	–	20	20	3,000
Oceania	–	–	–	*	5	5	400
Total	334	680	555	150	265	375	12,000

*Insignificant. After Horseman (1970).

Table 5.1 gives a summary of the minerable inventory of sulfur. For the purpose of this book only four types of sulfur containing materials are important: Volcanic deposits, pyrite deposites, elemental sulfur deposits and sulfur in fossile deposits. Sulfate deposits are very large, but are not presently commercially useful. Thus, they are not further discussed here.

I. Volcanic Sulfur Deposits

The largest presently known volcanic deposits are located in a 3,000 mile long zone in the Andes Mountains in South America. This area contains over 100 deposits with a total of over 100 mil. tons of elemental sulfur, of which only about 5% have been mined during the last 100 years, because of the remote location and the altitude of up to 6,000 m. In Japan, total volcanic reserves are estimated to be 50 mil. tons, distributed over 40 locations. Of this, about 5 mil. tons have been mined during the last hundred years. The other volcanic deposits, found around the Pacific Ocean, in Asia and in Europe are too small to have commercial potential.

2. Pyrites

Igneous, metamorphic, and sedimentary pyrite are abundant. The formation during the sedimentary marine cycle is discussed in Chapter 6. Large Pyrite reserves are found in over 30 countries. The largest U.S. reserves are found in Tennessee. U.S. reserves are estimated at 80 mil. tons; Spain and Portugal have over 500 mil. tons from which about 200 mil. tons of elemental sulfur seem recoverable. Total world reserves might be as high as 2,000 mil. tons. During the first half of this century, about 100 mil. tons of sulfur, i.e. about one half the world production, stemmed from pyrites. Most of it originated in Norway, Spain and Portugal where a large fraction was converted to sulfuric acid.

3. Elemental Sulfur Deposits

Elemental sulfur constitutes a metastable, intermediate state in the geological conversion of sulfate to sulfide or vice versa. By far the most detailed description and analysis of sedimentary sulfur deposits is contained in Ivanov's book (1964), in which he reviews the history of research, and the geological nature of most of the well known sulfur deposits. Already 80 years ago, scientists agreed that the elemental sulfur deposits were formed from hydrogen sulfide which resulted from the chemical reaction of sulfate with organic matter. However, there was disagreement between two schools whether the element was formed during or after formation of the enclosing rock. The microbial nature of the reaction was postulated around 1900. The geological bacterial sulfur cycle has been explored since the 1940's, when sulfate reducing bacteria were found in Indian and Sicilian deposits. Ivanov (1964) showed that oxidizing and reducing bacteria can occur side-by-side, and that the active stage cannot be determined simply by locating the bacteria.

Sulfur Deposits in Italy

In the area of Sicily, sulfur deposits are spread over an area of about 1000 square miles. The structure of these deposits is shown schematically in Figure 5.1. So far, some 50 mil. tons have been mined, most of it before 1900. If sulfur prices should increase, this area could again become a very productive source.

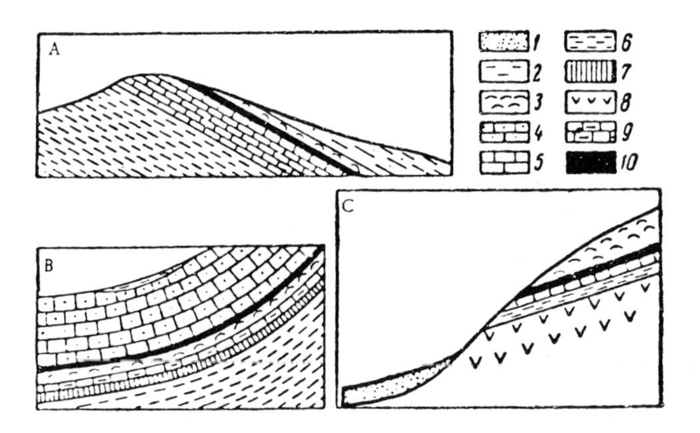

Fig. 5.1. Sulfur deposits in Sicily
l-alluvium; 2-marl, rich in foraminifers; 3-gypsum; 4-limestone; 5-flinty limestone; 6-marl; 7-tripolite shale; 8-rock salt; 9-concretionary sulfur in marl; 10-sulfur rock. (After Ivanov, 1964.)

The Gulf Coast Salt Domes

The structure of these deposits is shown in Figure 5.2. The dome structure makes it possible to contain superheated water to melt sulfur in situ, so that it can be collected and produced like oil. A major improvement of our understanding of the history of sulfur geochemistry is due to the work of Thode (1954) who developed sulfur isotope ratio measurements to the advanced degree which makes it useful for measuring movement and relationships between various strata. Jones (1957) and Feely (1957) used the isotope measurement to study the reaction rate of anhydrite reduction. From field sample measurements and laboratory experiments, it is concluded that sulfur in salt domes of the gulf area resulted from reduction of anhydrite of the cap rock of the salt dome to hydrogen, followed by oxidation to elemental sulfur. For this, a period of 150 mil. years, the age of the deposits, seems to be insufficient for petroleum to reduce anhydrite, but tests showed that bacterial reduction is sufficiently quick, as is the reaction between hydrogen sulfide and dissolved sulfate. The isotope work showed that sulfur from eleven domes all had the same ^{32}S to ^{34}S ratio of 21.85. Anhydrite cap rock has the same isotope composition as the anhydrite inclusions of the salt. The hydrogen sulfide is enriched in ^{32}S by several percent. Such, and even larger, enrichments can be achieved by a bacteria. Furthermore, it is found that the sulfate in the calcite rock has very uneven isotope distribution and is depleted by a comparable amount. The native sulfur has a lower 32/34 ratio than does

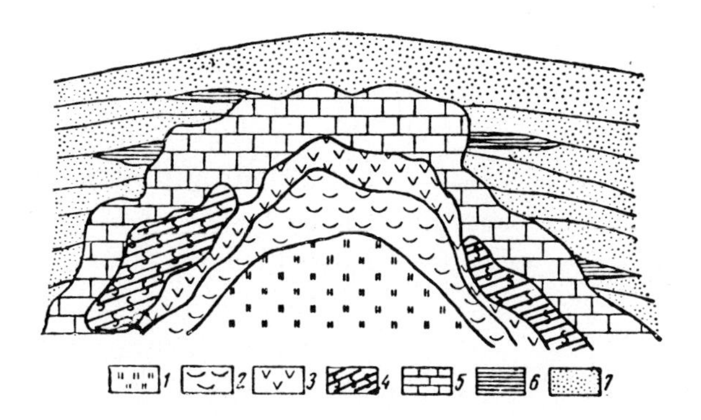

Figure 5.2. Section of a typical salt dome with a well developed cap rock
1-salt core; 2-anhydrite zone of cap rock; 3-intermediate gypsum zone of cap rock; 4-calcite sulfurous zone of cap rock; 5-calcite zone of cap rock; 6-oil; 7-sedimentary rock faulted and dislocated during formation of the saline dome (after Ivanov, 1964).

hydrogen sulfide of the dome, as is expected for the reaction products between hydrogen sulfide and calcium sulfate. Representative isotope ratios are shown in Chapter 4. Silver (1968) described a field trip through sulfur bearing deposits in Texas and compiled an extensive bibliography, which constitutes one of the most valuable sources.

The salt beds of the Gulf of Mexico are pre-Cretaceous and were buried below 10,000 meters. Even before the Pleistocene, differential stress caused dome formation and intrusions. The plugs advanced intermittantly up to 3,000 m below the surface where the anhydrite cap began forming. By the time the cap rock reached 2,000 m the temperature had dropped to 50-60°C, and bacterial action began as soon as petroleum had seeped into the cap rock. Most the the hydrogen sulfide escaped as it was formed. At 200 to 500 m below surface, the salt domes stabilized. Oil and gas, escaping from adjacent upturned sedimentary formation, filled the cap rock; trapped hydrogen sulfide underwent oxidation and formed elemental sulfur deposits. The amount of sulfur deposited depended on the motion of the salt, tightness of the cover, and the amount of petroleum in the cap. Fractured caps let hydrogen sulfide and polysulfides escape and seeping oxygenated water oxidized residual sulfur. The native elemental sulfur is metastable and has a comparatively short geological lifetime.

Highest sulfur concentrations are found in domes with a thick anhydrite and calcite rock cap and with large petroleum deposits in the flanking sediment, Fig. 5.2.

Orr (1975) reviewed the balance between hydrogen sulfide formation and removal. The most important three sources of hydrogen sulfide are microbial sulfate reduction, thermal decomposition of organic sulfur during maturation, and thermochemical sulfate reduction by reaction between hydrogen sulfide and dissolved sulfate. Hydrogen sulfide is removed by several processes. The most prominent are formation of pyrite and extraction by water. These systems are discussed below.

Polish and Volga Deposits

The formation of sulfur deposits in Poland and the Volga deposits have been explained by Ivanov (1964) and Osmolski (1973). Figure 5.3 shows the structure of the epigene Shor-Su deposits. Sulfur is always in the vicinity of gypsum, from which it was produced.

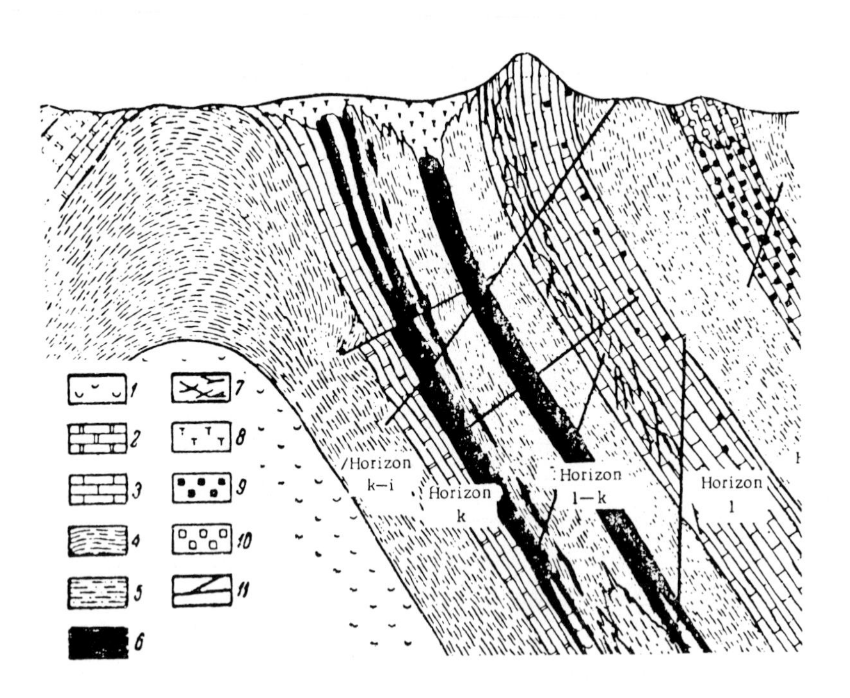

Fig. 5.3. Schematic section of the northern limb of the 2nd Shor-Su anticline 1-gypsums; 2-dolomitized limestone; 3-limestone; 4-green clays and marls; 5-red marls; 6-sulfur beds; 7-streaks of sulfur; 8-zone of oxidation of sulfur ores; 9-ozokerite beds; 10-zone of oxidized ozokerite; 11-ruptured tectonic dislocations (after Ivanov, 1964).

B. SECONDARY SOURCES

1. Sulfur in Petroleum and Gas

Hydrogen sulfide from natural gas presently constitutes the most important source of elemental sulfur. A large amount of sulfur is obtained by separating sulfur from petroleum. Oil and gas deposits are spread all over the world. From the viewpoint of sulfur, the sour gas deposits in France and Canada have been the most important sources. During the next 20 years, the Middle East will become a dominant source of petroleum sulfur. The estimates of world reserves are not yet fully known. During the last 50 years, estimates of "proved reserves" have doubled every ten years, together with production.

Crude oil contains from 0.05 to 14% sulfur. The great majority of oils contains 0.1 to 3%. A detailed analysis of sulfur compounds in oil was published by Rall (1972). Some 60 of the 300 representative compounds so far isolated are listed in Table 5.2.

During refining, sulfur compounds decompose and react with other petroleum components; thus, the composition of sulfur in tail residues is not necessarily representative of that in the original crude. The observation which British petroleum scientist Perkin made 60 years ago is as valid in 1977 as in 1917:

> To the refiner, the chemical nature of the sulphur-compound is a matter of indifference, provided he can remove it. On the other hand, if more were known as to the chemical structure of the sulphur compounds, the chemist might be able to assist the refiner much more than at present. It is therefore a matter of great importance that further extended research into the composition of the sulphur compounds in petroleum oils should be undertaken. This is much more important at the present than in the past, because, owing to the exhaustion of some oil supplies, and the much greater demand for oil fuels, many new sources of oil are being exploited, a number of which are very rich in sulphur.

Far more is known about low molecular weight oil fractions than about the residues, because they are easier to extract and analyze. However, most of the sulfur accumulates in the heavy, tail section, which is only slowly being explored. As in the case of coals, the composition varies greatly, and generalizations are impossible. The sulfur composition is determined during deposition, by post-genetic time-temperature cycles, and the enclosing rock formation. The initial formation involves thermal and catalytic bond cleavage in kerogen and bitumen in sedimentary rocks. At this stage, high sulfur content is favored by reducing environment, and

Table 5.2
SULFUR COMPOUNDS IDENTIFIED IN A WASSON, TEXAS, CRUDE OIL DISTILLATE BOILING FROM 111° TO 150°C

Cyclic Sulfides	Thiols	Chain Sulfides
C_4H_8S: Thiacyclopentane	C_4H_9SH: 1-Butanethiol 2-Butanethiol 2-Methyl-1-propanethiol	
$C_5H_{10}S$: 2-Methylthiacyclopentane 3-Methylthiacyclopentane Thiacyclohexane	$C_5H_{11}SH$: 1-Pentanethiol 2-Pentanethiol 3-Pentanethiol 2-Methyl-1-butanethiol 3-Methyl-1-butanethiol 2-Methyl-2-butanethiol 3-Methyl-2-butanethiol 2,2-Dimethyl-1-propanethiol C_5H_9SH: Cyclopentanethiol	$C_5H_{12}S$: 2-Thiahexane 3-Thiahexane 3-Methyl-2-thiapentane
$C_6H_{12}S$: 2,2-Dimethylthiacyclopentane 2,3-Dimethylthiacyclopentane cis- and/or trans-2,3-Dimethylthiacyclopentane cis-2,4-Dimethylthiacyclopentane trans-2,4-Dimethylthiacyclopentane cis- and/or trans-3,4-dimethylthiacyclopentane cis-2,5-Dimethylthiacyclopentane trans-2,5-Dimethylthiacyclopentane 2-Ethylthiacyclopentane 2-Methylthiacyclohexane 3-Methylthiacyclohexane 4-Methylthiacyclohexane	$C_6H_{13}SH$: 1-Hexanethiol 2-Hexanethiol 3-Hexanethiol 2-Methyl-1-pentanethiol 3-Methyl-1-pentanethiol 4-Methyl-1-pentanethiol 2-Methyl-2-pentanethiol 3-Methyl-2-pentanethiol 4-Methyl-2-pentanethiol 2-Methyl-3-pentanethiol 3-Methyl-3-pentanethiol 2,2-Dimethyl-1-butanethiol 2,3-Dimethyl-1-butanethiol 2,3-Dimethyl-2-butanethiol 3,3-Dimethyl-1-butanethiol 3,3-Dimethyl-2-butanethiol 2-Ethyl-1-butanethiol $C_6H_{11}SH$: Cyclohexanethiol 1-Methylcyclopentanethiol cis-2-Methylcyclopentanethiol trans-2-Methylcyclopentanethiol cis- and/or trans-3-methylcyclopentanethiol	$C_6H_{14}S$: 2-Thiaheptane 3-Thiaheptane 4-Thiaheptane 3-Methyl-2-thiahexane 2-Methyl-3-thiahexane 4-Methyl-3-thiahexane 5-Methyl-3-thiahexane 3,3-Dimethyl-2-thiapentane 3,4-Dimethyl-2-thiapentane 3-Ethyl-2-thiapentane 2,2-Dimethyl-3-thiapentane 2,4-Dimethyl-3-thiapentane
$C_7H_{14}S$: 2,2,5-Trimethylthiacyclopentane	$C_7H_{15}SH$: 2-Methyl-2-hexanethiol 3-Methyl-3-hexanethiol	$C_7H_{16}S$: 2-Methyl-4-thiaheptane 2-Methyl-3-thiaheptane and/or 3-methyl-4-thiaheptane 4-Methyl-3-thiaheptane and/or 4-ethyl-3-thiahexane 2,2-Dimethyl-3-thiahexane 2,4-Dimethyl-3-thiahexane 2,5-Dimethyl-3-thiahexane 4,4-Dimethyl-3-thiahexane 4,5-Dimethyl-3-thiahexane
$C_8H_{16}S$: 2,2,5,5-Tetramethylthiacyclopentane		

After Rall, 1972.

high sulfate content. During the maturing process, the sulfur content can undergo drastic changes by migration and accumulation. The difference in the isotope ratio between gas and oil deposits sometimes contains clues about the deposition mechanism. The use of isotopes is described in Chapter 4. In some cases, most of the sulfur seems to be deduced from hydrogen sulfide. But in others, hydrogen sulfide is clearly irrelevant as a source. In deep high-temperature reservoirs containing hydrogen sulfide, the stationary equilibrium between sulfurization and desulfurization can equilibrate isotopes and mask the history.

Ansimov (1971) reviewed the hydrogen sulfide content in sediment and natural gas. Its concentration is highly variable and not related to the sulfur concentration in associated crude oils. Many high sulfur crudes have no hydrogen sulfide in the gas, and vice versa. The hydrogen sulfide content of gas is always connected with the geological structure of the area, and is dependent on carbonate stratigraphic horizons. The reason for this is that carbonates generally contain little iron and, thus, cannot bind hydrogen sulfide, and also that anhydrite, $CaSO_4$, is commonly found in carbonates. The main sources of hydrogen sulfide are microbial sulfate reduction and thermal desulfurization of organic sulfur compounds in petroleum or kerogen. In deep high-temperature reservoirs high hydrogen sulfide can be caused by the reaction of sulfate with hydrogen sulfide, which yields intermediate elemental sulfur (Toland, 1960). The latter dehydrogenates organic matter. This reaction is favored at temperatures too high for microbiological reactions to occur, and has been invoked for explaining hydrogen sulfide in the Aquitaine basin where the deposit is at a temperature over 80°C. The evolution of liquid hydrocarbons, gaseous hydrocarbons, and hydrogen sulfide in the Aquitaine Basin (LeTran, 1974) is exlained in the temperature *vs.* depth profile of Figure 5.4.

Hydrogen sulfide in petroleum reservoirs is lost primarily by formation of pyrites, by extraction with water, and by oxidation to sulfate.

Sulfur in Athabaska tar sands has been described by Allen (1974), and analyzed by the Canadian Department of Energy (Tostevin, 1973). The rich tar contains some 400 mil. tons of sulfur. It is buried under 0-500 m of Muskeg. The mining is challenging and involves very large scale operations.

2. Sulfur in Coal

The world's coal reserves are well established and immense. The proven reserves of bituminous coal alone are at least 1,300,000 mil. tons. This corresponds to a sulfur reserve of over 20,000 mil. tons; this is ten times more than all other proven sulfur reserves together. Half of it is located in North America, a quarter in Asian Russia and fifteen percent in Europe. Coal contains between one and fourteen percent sulfur which is released during combustion. Presently, the largest coal producers are, in decreasing order of importance, East Germany, Russia, England, West Germany, Czechoslovakia, Poland, Hungary, Australia and North America.

Presently, sulfur released from coal combustion about equals all other forms of sulfur production. If coal combustion becomes more important, sulfur released from coal will make a major impact, regardless of whether it is recovered or released to the atmosphere.

Coal contains pyrite, sulfate and organic sulfur. The nature of organic sulfur compounds in coal is less established than that in oil or gas. Sulfur is introduced in the early stages of peat formation, together with nitrogen and oxygen. Additional sulfur is obtained by transfer from microbiological action. During the metamorphic stage some sulfur is lost as hydrogen sulfide, together with methane, water, ammonia and carbon dioxide. Proteins and amino acid from vegetable matter in peat swamps contribute small amounts of sulfur. In high sulfur coals, up to one percent or more sulfur stems from high sulfate marine waters, which under anaerobic conditions is reduced and incorporated as hydrogen sulfide. The ratio of pyrite to sulfate and organic sulfur varies as much as the total sulfur content. Sometimes appreciably different compositions are found within one coal seam. It is generally believed that the functional groups are largely the same as in crude oils. Thus, disulfides, thiols, and thiophene are present. However, coal contains more nitrogen and oxygen than oil. Accordingly, heterocyclic compounds are more common. The relative abundance of sulfur compounds depends on the coal rank. The thiophenes, which are thermally most stable, are most abundant in the high rank coals, the anthracite. Sulfur in coal tar is not necessarily attached to its parent molecule, because coal decomposition is not a purely physical process. Coal extraction might yield more valuable data to increase the presently generally meager information we hold about coal chemicals (Ozdemiv, 1971).

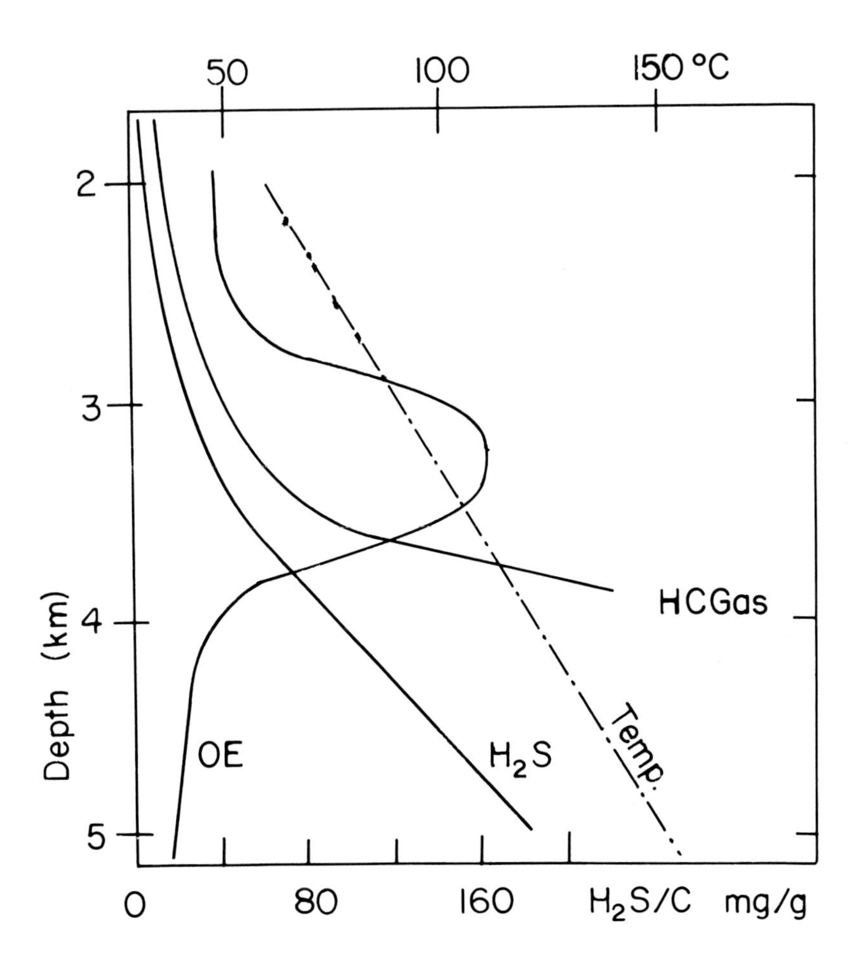

Fig. 5.4. Evolution of liquid hydrocarbons, gaseous hydrocarbons and hydrogen sulfide in the Aquitaine Basin, France. (After LeTran, K., Bull. Centre Rech. Pau. 8 (1974) 111; 'Diagenesis of Organic Matter and Occurrence of HC and H_2S in Southwestern France;' and Orr, 1975.)

The sulfur production trends and the history of sulfur production are discussed in Chapters 1 and 7. The production methods are described in the following chapters.

Chapter 6

The Sulfur Cycles

A schematic model of the sulfur cycle is shown in Fig. 6.1. Our knowledge about details of the sulfur transfer is still limited. Sulfur concentrations in the air are so small that new, analytical methods had to be developed before accurate concentration measurements and rate transfer measurements became possible. In the ocean sulfur concentrations are equally small, and methods have yet to be improved before changes in concentrations of intermediates can be tracked. In the sediment problems are caused by inhomogeneity of strata and deposits, which exhibit erratic changes in total sulfur content and in the ratio of the various sulfur forms. Furthermore, some steps of the sulfur cycle involve geological time periods, which are far beyond the meaningful extrapolation range of laboratory experiments. The residence time of sulfate in the ocean, for example, is believed to be about 40 million years; that of pyrite in sedimentary rocks is over 250 million years (Garrels, 1974). The sulfur cycle has undergone quite drastic changes and alterations during the last 200 million years, and the present cycle is not fully balanced. A further problem is that many of the most important chemical steps are only qualitatively and superficially known, because the data was measured long before the modern, instrumental techniques were developed. Thus, only the ingenuity of individual researchers has made it possible to gain at least a superficial glimpse of geochemical events. One of the most useful tools so far applied has proven to be the masspectrometer with which changes in isotopic composition of sulfur compounds can be determined and monitored. Thode (1954) pioneered this field and led research for almost two decades. The normal isotopic abundance of terrestrial sulfur is 96% sulfur-32 and 4% sulfur-34. In chemical reactions the two isotopes have a slightly different equilibrium constant; thus the products contain "fractionated" isotopes. The fractionization is expressed as:

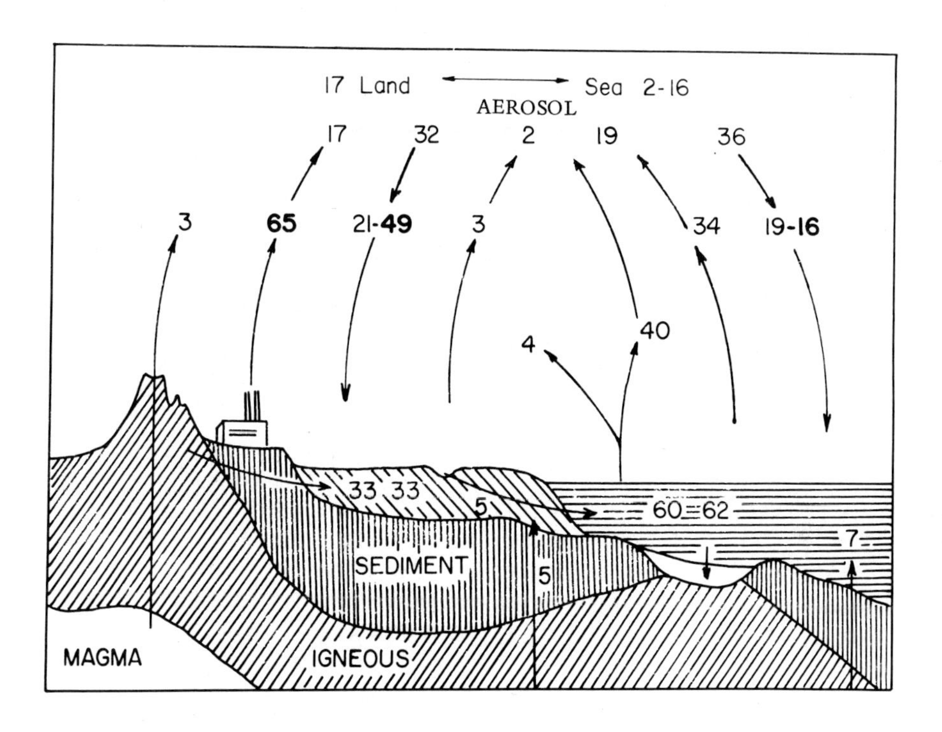

Fig. 6.1. Global sulfur cycle (after Bolin, 1976; Granat, 1976). (Bold numbers are human contributions.)

$$\Delta\ {}^{34}S\ \% = \frac{{}^{34}S/{}^{32}S\ \text{sample} - {}^{34}S/{}^{32}S\ \text{standard}}{{}^{34}S/{}^{32}S\ \text{standard}} \times 10^4$$

Leskovsek (1974) showed that it is advantageous to measure the sulfur as SF_6, rather than as sulfur dioxide. Harrison (1957) found that $^{34}SO_4$ is reduced to sulfide 2.2% faster than the heavier $^{34}SO_4$. Similar fractionization is observed in other reactions. Δ S values for various terrestrial materials are shown in Fig. 6.2 (Holser, 1966). Similar values were obtained by Jones (1957) in measurements for elemental sulfur. Δ S values below -1.5% indicate biological processes. Δ S values above 0.0% result from igneous or hydrothermal reaction (Orr, 1975).

Maritime sediment is markedly enriched in 32-S. The enrichment can be used to determine the age of sediment: Vinogradov (1975) reasons on the basis of ΔS = 0.08 % for the apatites and scapolites of the Aldan

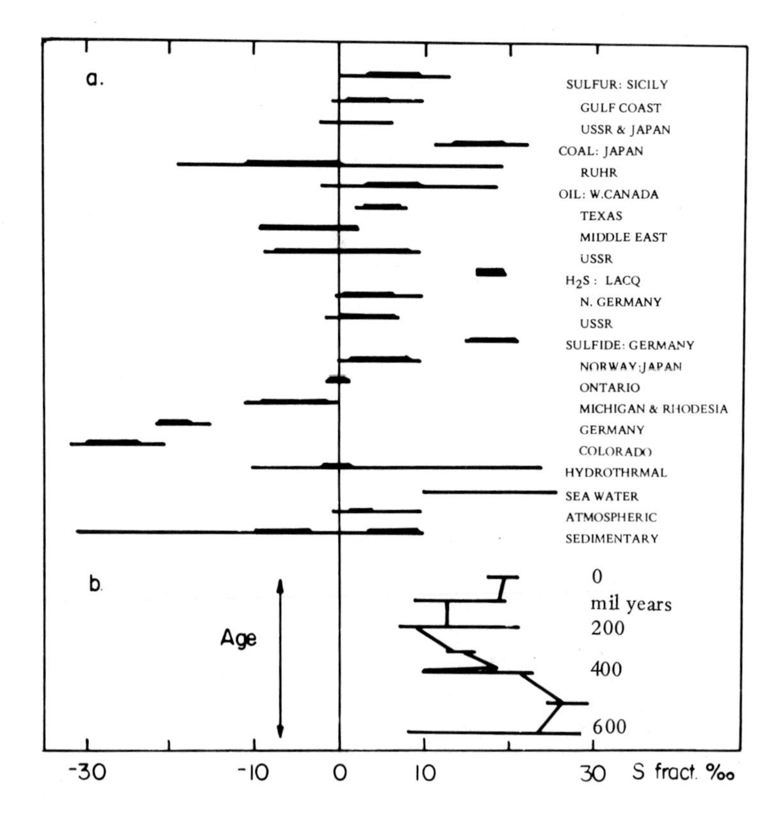

Fig. 6.2. Sulfur isotope fractionization a) for various materials, b) as a function of geological age.

Shield that the Earth's sulfur cycle commenced 3 billion years ago, and that the oxygen atmosphere was generated 2 to 3.5 billion years ago. Laton (1974) used sulfur isotopes to make a model for the evolution of oceans.

Garrels (1972) developed a quantitative model for the sedimentary rock cycle in which he balanced the average dissolved and suspended load of streams, the concentration in the ocean, in sedimentary rocks and in precipitation. The sulfate in ground water is determined by the pH and solubility. Limestone, sand, gravel, and metamorphic rocks all contain about 10-11% dissolved sulfate. Shale contains 10%, gravite 8%, sandstone 7%, and basaltic 5%. Streams contain 8% sulfate. The geochemical inventory of sulfur is shown in Table 6.1. The distribution and relative abundance of sulfur species in water are determined by their solubility, dissociation constant, and oxidation potential. All have been reviewed by

Table 6.1
GEOCHEMICAL INVENTORY OF SULFUR

Rock Type	Estimated Mass of Rock (10^{15} T)	Estimated Mean S Content (%)	Total Sulfur (10^{12} T)
Deep oceanic			
Sediments	300	0.025	75
Mafic rocks (to Moho)	4,400	0.053	2,300
Sedimentary			
Sandstone	280	0.090	250
Shale	750	0.27	2,000
Limestone	290	0.13	380
Evaporites	30	17	5,100
Volcanics	120	0.04	50
Connate water	140	0.019	27
Total evaporites and connate water	170		5,100
Total other sediments	1,440	0.19	2,700
Total all sediments	1,600	0.49	7,800
Fresh water	0.3	0.0011	0.003
Ice	35	0.00003	0.006
Atmosphere	5.3		($3.6 \cdot 10^6$)
Sea	1,420	0.090	1,280
Total in sedimentary rocks and sea			9,100
Continental igneous and metamorphic rocks (to Moho)			
Granitic	10,500	0.021	2,200
Mafic	8,700	9.053	4,600
Total	21,000	0.032	6,800
Total in the crust			18,200

After Holser (1966).

Garrels (1958). From this a residence time of 250 million years for sulfate in rocks and 22 million years in the ocean can be estimated. In geological terms these residence times are short. The sulfate has direct influence on the calcium and magnesium cycle. The total sulfur in rocks appears to be fairly stable with time, but the isotope ratio has gone through several variations (Holser, 1966), Fig. 6.2b. This is due to a change in the ratio of oxidized to reduced sulfur. Today, sulfate and sulfide are about equal in weight. It is estimated that at earlier times, the ratio varied from 4:1 to 1:4. This has a profound effect on the oxygen concentration in the atmosphere. During the last one million years, sulfur has been accumulating in the ocean in reduced form.

Table 6.2
SIX ESTIMATES OF THE ATMOSPHERIC FLUX OF SULFUR
(Tg)

	Eriksson 1960;63	Junge 1963;72	Robinson & Robbins 1968	Kellog 1972	Friend 1973	Granat Bolin & Charlson 1976
Sources						
Pollution	40	40	70	50	65	65
Biological decay sulfides						
from land	110	70	68		58	3
sea	170	160	30	89	48	28
Sea spray	44		44	43.3	44	44
Volcanism				.75	2	3
Sinks						
Oceanic absorption	100	70	25		25	16
Vegetative & soil absorption	100	70	26	15	15	18
Precipitation & dry deposition over oceans	100	60	71	72.3	71	56
Precipitation over land	65	70	70	86	86	49
Dry deposition over land			20	10	20	4
Total Atmospheric flux (input = output)	365	270	212	183	217	143

1 Tg = 1 mil tons/yr.

A. THE GLOBAL SULFUR CYCLE

The estimated values of the different steps of the global sulfur cycle vary considerably from author to author, Table 6.2. Geochemical data of this type is based on difficult-to-test assumptions about "global" processes, which are obtained by computing the sum of a variety of different regional and local sulfur cycles, comparing processes in different areas, weighing the relative importance of different models, and adding, i.e. extrapolating, data. Thus, only few steps in the sulfur cycle can be directly determined; the remaining links must be deduced by balancing the cycle.

Estimations of the yearly atmospheric sulfur flux vary between 90 and 365 Tg (1 Tg = 1 mil tons/yr). The values are usually based on the sulfur content in rain. The earliest estimates were the highest. They were based on observations in highly populated countries. The lower, more recent values are based on measurements in remote areas. It is believed that the difference between the values is due to local industrial activity, i.e. "anthropogenic" sources. The present results indicate that the sulfur flux in the unperturbed atmosphere is far smaller than had been assumed. The regional variations in the sulfur cycle prove that atmospheric sulfur flux of man-made emission is a regional problem, and, thus, that the residence time of sulfur is very short. Table 6.5 shows that 90% of sulfur emission settles within 24 hours and within 200-500 miles of the source. In the case of large scale volcanic emissions such as that of Fuego, Guatemala in 1971, up to 10,000 tons of sulfur dioxide are emitted per day (Stoiber, 1974). The sulfur dioxide is quickly thrust into very high levels and can remain there for weeks. There are reports that such emissions, which are accompanied by lava-dust, have discolored the sky, i.e. caused opacity, over large parts of the hemisphere.

Estimates of the global sulfur flux are based on the assumption that sources and sinks for all sulfur—except anthropogenic—are balanced in the pedosphere. The primary sources are weathering, volcanic sulfur, sea spray; they reach the earth as wet or dry depositions. Loss of sulfur from the pedosphere is due to river run-off and volative reduced sulfur reaching the atmosphere.

Sulfur weathering estimates are based on calculations of an annual denudation rate of 10,500 Tg. Of this about 9,000 Tg are assumed to be carried by rivers, 100 Tg by glaciers, and 6 Tg by winds. The river load consists of about 90% undissolved material, 9% dissolved, and 1% abrasive. The estimated river load seems unduly high (Granat, 1976), as it would correspond to 12 km^3 per year. At this rate, the continents would be reduced to sea level in 10 million years.

Sulfur weathering is obtained from the above total by considering the relative sulfur content of predominant sedimentary rocks: 80% of the sediment is shale with 0.22% sulfur; 10% limestone containing 0.13% sulfur; sandstone and graywacks contain 0.09% sulfur and make up 9% of the sediment. Evaporities contain 17% sulfur, but make up only 1% of the sediment. The average sulfur content of weathered rock is 0.33%. This yields a natural weathering rate of 33 Tg sulfur.

Estimates for volcanic sulfur, excepting major eruptions, vary from zero to 12 Tg. These figures are speculative. The best estimate seems to be that of Holser and Kaplan (1966). It is based on hot-spring flow rates, and suggests that a large fraction of aerial hydrogen sulfide originates from geysers. Sea sprays are another source of disputed importance. Evaporated reduced sulfur from land is clearly a major sulfur source. Grey (1972) deduced from isotopic data that biogenic sulfur is comparable to anthropogenic sulfur in the Salt Lake City area. The global importance is disputed.

Volatile sulfur from oceans and stagnant waters is about 85 Tg. Lovelock (1972, 1974) and Rasmussen (1972) showed that organic sulfur–mainly dimethylsulfide and carbon disulfide at concentrations of about 10^{-10} g/l–is an important source, because organic sulfur can survive the oxygenated layers of the surface sea, which quantitatively oxidizes hydrogen sulfur to non-volatile sulfate in the oceans.

B. HYDROSPHERE

The sulfur cycle in the hydrosphere involves a complex system of reactions which is only slowly being unraveled (Nriagu, 1976). Goldhaber and Kaplan (1974) have prepared a detailed review of the sulfur cycle in the sea. Important data is contained in a review by Orr (1974). Earlier work was reviewed by Berner (1972) and in a book by Ivanov (1968). Sulfur reaches the oceans by river run-off (60 tril tons/yr), by wet or dry deposits from the atmosphere (a total of about 20 tril tons/yr), and by volcanic eruptions, as indicated in Fig. 6.1. All sulfur is in the oxidized form, either as sulfate, or, initially, as sulfur dioxide from the air. Beilke (1974), Granat (1976), and Spedding (1972) believe that the ocean is a sink for up to 20 tril tons of sulfur per year. The latter measured the absorption rate in the laboratory. The most prominent sulfur species in the ocean is sulfate. Trapped or deep sea water looses its oxygen to dissolved and particulate matter faster than diffusion or convection can replenish it. Thus, anaerobic microorganisms can reduce sulfate to sulfide. At the present time, such reduction is only observed in the Black Sea, in fjords, lagoons, and estuaries which suffer from stagnation. It is feasible that in the past sulfate reduction in the water columns was of geological importance.

Anoxic conditions are frequently found at the interface between water and sediment, or around sulfate dissolved in porewater of the sediment. Bacterial action produces hydrogen sulfide which is toxic to respiratory organisms and produces metabolic products which influence metal sulfide precipitation, carbonate formation, pH, and enhances methane formation. This transfer of sulfate to reactive sulfide is an important link in maintaining a steady state sulfate concentration in the ocean.

1. Microbial Processes

The metabolism of sulfur compounds has been well studied. Work up to 1975 has been reviewed in fifteen chapters edited by Greenberg (1975). Siegel (1975) lists 473 references on recent biochemical research on the sulfur cycle.

All plants, animals, and bacteria in the sea metabolize dissolved or ingested sulfate for the synthesis of proteins. Decomposition of polysaccharide sulfate ester liberates sulfate, and in some cases methyl- or similar mercaptans. The sulfur content of some marine animals and algae was measured by Kaplan (1963). It varies between 0.4 and 3%. A large fraction is combustible. In algae the spread is equally large. The sulfur is distributed almost equally between ash, combustible, and soluble fractions.

Dissimilatory sulfate reduction is achieved by bacteria of the class of *Schizomycetes* in the order *Eubacteriales*. The genus *Desulfoibrio*,, a non-sporulating curved bacteria, is the most abundant of all sulfate reducing marine bacteria. It occurs over a large pH range and a large range of salt concentrations, often in sediments of saline or gypsum deposits, and in oil wells. The bacteria are normally heterotrophic, i.e. draw their carbon from organic matter. Energy is generated by use of lactic acid which yields acetic acid and hydrogen sulfide:

$$2CH_3{\cdot}CHOH{\cdot}COOH + SO_4{}^{2-} \rightarrow 2CH_3COOH + HS^- + CO_2 + HCO_3{}^-$$

or pyravic acid:

$$4CH_3COCOOH + SO_4{}^{2-} \rightarrow 3CH_3COOH + CH_3COO^- + HS^- + 4CO_2$$

The pathway of dissimilatory reduction of sulfate to sulfide is summarized in Fig. 6.3. Microbial oxidation of sulfur, for example by *Thiobacillus thiooxidans* is unimportant in the sea. It can be important in sediment, where it yields elemental sulfur or sulfate. A list of the most common *Thiobacilli* members is shown in Table 6.3. In the conversion of sulfur to sulfate, polythionic acids, Chapter 3, occur as intermediates. Very little is known about their metabolism, Fig. 6.4. The interrelationships between the various bacterial functions are shown in Fig. 6.5. This bacterial community was named the *sulfuretum* by Baas-Becking (1925). The reduction of sulfate to sulfur seems to be catalyzed by intermediates. The reduction rate depends on concentration of bacteria, i.e. on their organic carriers, sulfate concentration, and temperature (Murzaev, 1968;

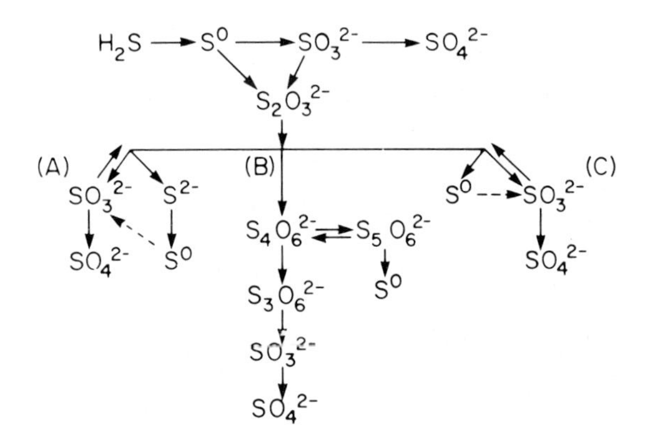

Fig. 6.3. Dissimilatory reduction of sulfate (after Goldhaber, 1974).

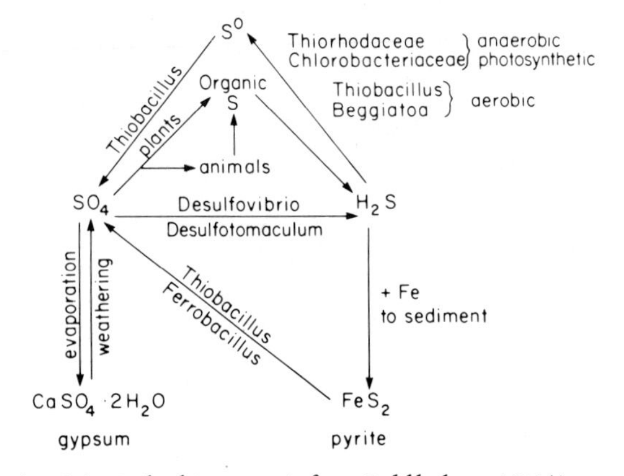

Fig. 6.4. Polythionates (after Goldhaber, 1974).

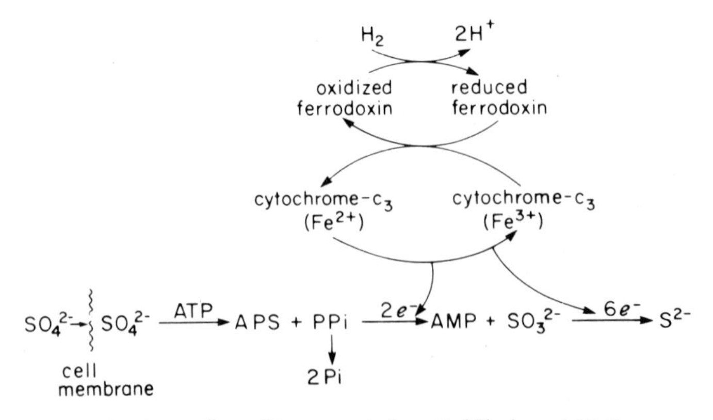

Fig. 6.5. The sulfuretum (after Goldhaber, 1974).

Table 6.3

THE *THIOBACILLI.* MEMBERS OF THIS GENUS ARE GRAM-NEGATIVE, NON-SPORULATING RODS MEASURING 0.5 BY 1-3 MICRON. MOST OXIDIZE SULFIDE, ELEMENTAL SULFUR AND THIOSULFATE TO SULFATE

Organism	Habitat	General Characteristics
T. thioparus	Canal water, mud, soil	Growth range pH 7.8-4.5; generally aerobic and motile. Some strains grow anaerobically in the presence of nitrate
T. neapolitanus (Thiobacillus X)	Seawater, corroding concrete structures	Strict aerobe with properties very similar to those of T. thioparus
T. dentrificans	Canal and river water, salt water, peat composts, mud	Optimum growth pH 7; oxidizes sulfur compounds anaerobically in the presence of nitrate.
T. thio-oxidans	Soil	Optimum growth pH 2; withstands 5% H_2SO_4; strict aerobe; motile
T. concretivorus	Corroding concrete structures	Very similar to T. thio-oxidans
T. ferro-oxidans	Acid mine and soil waters containing hydrogen sulfide	Strict aerobe; optimum growth pH 2.5-5.8; motile
T. novellus	Soils	Facultative autotroph; nonmotile; optimum growth near pH 7
T. intermedius	Fresh-water mud	Facultative autotroph; growth pH range 2.0-7.0; motile; autotrophic growth stimulated in the presence of organic matter
T. thiocyanoxidans	Gas works liquor; sewage effluent	Similar to T.thioparus; thiocyanate oxidation serves as an energy source; oxidizes formate
T. perometabolis	Soil	Motile; no growth in mineral salts without yeast extract or casein hydrolysate: reduced sulfur compounds are oxidized to sulfate; oxidation stimulates growth

After Roy and Trudinger (1970).

Tuttle, 1976). The isotope fractionization measured in laboratory reactions is shown in Fig. 6.2. The microorganisms are only active below 70°C. Their activities tie into the C, H, N, O, and P cycles by respiration, catabolism, photosynthesis, and other chemical reactions.

2. Sulfur Deposits in Hydrogen Sulfide Rich Water

In stagnant hydrogen sulfide rich, shallow water, sulfur can be produced in noticeable quantities if the temperature is around 30°C and the water contains sufficient salts. Lake Ain-ez-Zania, with a surface area of 8 km^2 produces about 200 tons of elemental sulfur yearly (Butlin, 1954). The bottom water contains 108 mg/l hydrogen sulfide; the surface contains 20 mg/l. The salt domes in the Gulf of Mexico are described in Section 5B.

High hydrogen sulfide containing water is found in sewage (Oswald, 1976). It attacks concrete piping by the same microbial mechanism as anhydrite. Remedies using elemental sulfur are described in Chapter 14. Artisian springs (Casati, 1975) can contain up to 3 mmoles/l hydrogen sulfide and 1 mg/l sulfate in the presence of 0.05 mg/l ammonium ion. Oxidation by air (Teder, 1969; Giggenbach, 1971; Chen, 1972; Avrahami, 1968) is slow because of lack of diffusion and mixing. In contrast, Almgren (1973) attributes part of the oxidation in the Baltic Sea, at pH 8, to air oxidation. Zrobin (1975) measured the kinetics as a function of pH and oxidation potential. Aizatullin (1975) computed the oxidation rates.

3. Pyrite Formation

During diagenesis, i.e. between the settling of the sediment and solidification, bacterially reduced sulfate reacts with iron and forms a complex family of iron sulfides: Amorphous iron sulfide, (Fe:S = 0.9), mackiniwite (Fe:S = 0.9), cubic iron sulfide (Fe:S = 1), hexagonal pyrrhotite (Fe:S = 1.1), greigite (Fe_3S_4), smythite (Fe_3S_4), orthorhombic marcasite (FeS_2), and cubic pyrite (FeS_2).

The iron-iron bonding in these minerals has been studied by Morice (1969) using Mössbauer spectroscopy. Only pyrite and pyrrhotite are thermodynamically stable, but the initial reaction product between sediment and hydrogen sulfide is black amorphous iron sulfide. The pyrite is formed by slow addition of sulfur to the sulfide. In the S_2^{2-} ion of pyrite sulfur has a nominal oxidation state of -1. Pyrite often contains excess sulfur, probably as polysulfide (Section 3B). Boulege (1973) demonstrated that under geochemical conditions, such as a pH between 7.5 and 8.3, a redox potential of 0.7 and a total sulfur ion concentration between 10^{-5} and 10^{-2} g/l, polysulfides can be formed. The solubility of pyrite in water is 10^{-28}. The complex reactions, which are partly non-stoichiometric, are not fully understood. Speculations about their geochemistry, and about their intermediates has been published by many authors.

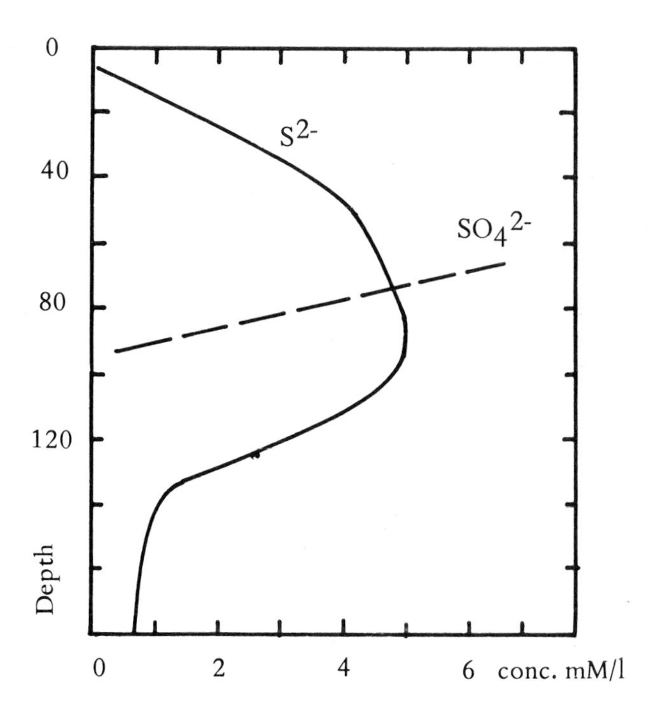

Fig. 6.6. Sulfate and sulfide concentration in surface sediment of Saamich Inlet.

During sedimentary pyrite formation, characteristic textured spheroids are formed, which geologists call "Framboids". They are due to the biogenic history of the material. In contrast, volcanic pyrite forms in distinct crystal clusters. It is believed that the Framboids reflect the metastable iron sulfide phase of their precursors.

Reduced sulfur normally is only found in stagnant, undrained swamps; eutrophic, high nutrient lakes; tidal lagoons, on deltas and estuaries; landlocked sea basins; and trenches in the sea bottom. The pyrite content of these sediments is directly correlated to the organic content, but is not related to depth, as is often assumed. Iron is very rarely a limiting factor in pyrite formation (Berner, 1970). Studies in the Saamich Inlet, British Columbia, show that sulfate can be completely reduced from porewater within a few hundred years, if sufficient organic material is present in a rapidly sedimenting area. The Saamich Inlet has a settling rate of 4 m/1000 years. The sulfate and sulfide concentration in a drill core of this area is shown in Fig. 6.6. At the surface, all sulfur occurs as sulfite. Between 80

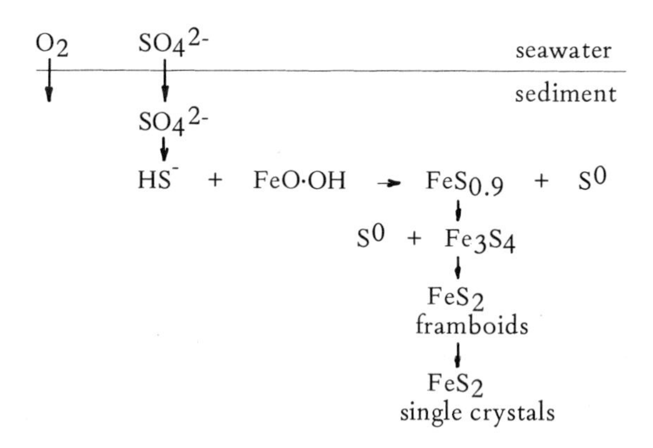

Fig. 6.7. Summary of pyrite framboid formation in the upper layers of marine sediments (after Orr, 1973, and Goldhaber, 1974).

and 100 cm below the surface, sulfate sharply drops, while sulfide goes through a maximum. At neutral pH, 50% of hydrogen sulfide exists as bisulfide, Section 3B; see also Goldhaber (1975).

Below 2 m, sulfide has fully reacted with the sediment. Calculations suggest that about half of the sulfate entering the sediment by diffusion from the sea will be spread over the sediment as sulfide. Sulfide cannot diffuse far because of its rapid reaction with the iron. Oxidation of sulfide which might occur in stagnant hydrogen sulfide rich water is negligible. Gardner (1973) developed a model for computing the pH of the aqueous phase. As mentioned above, isotope measurements indicate that the pyrite formation rate has changed over the last 600 million years. Kaplan (1963) measured the sulfur distribution and isotope factors in recent marine sediments in California.

The sulfur cycle in the ocean and sediment is summarized in Fig. 6.7. The oceans contain sulfate, introduced by river run-off, and pyrite which is formed *in situ* by bacterial reduction. Orr (1973) measured the reduction rate in the Pettaquamscutt River basin, where every few years seasonal changes cause overturns in an otherwise stagnant basin. Eighty percent of the hydrogen sulfide in the original mixing had been depleted within one month by oxygen rich surface waters.

Stanton (1972) reviewed the formation of strataform pyrites. Such strataforms are layers of base metal sulfides embedded in the sediment.

Table 6.4

THE DIMENSIONS OF THE EARTH AND THE MASS OF THE ATMOSPHERE

Surface Areas (km^2)	
Total earth	5.1×10^8
Northern Hemisphere land	1.03×10^8
Northern Hemisphere ocean	1.54×10^8
Southern Hemisphere land	0.46×10^8
Southern Hemisphere ocean	2.10×10^8
Mass of the Atmosphere (g)	
Total mass	5.2×10^{21}
Mass of the troposphere (to 11 km)	4.0×10^{21}
Number of moles in the atmosphere	1.8×10^{20}
Sulfur flux in the atmosphere	2.0×10^{14}

After Butcher (1973).

Table 6.5

ESTIMATED HEMISPHERIC SULFUR EMISSION

(mil tons sulfur)

Source	Total	Hemisphere Northern	Southern
Biological H_2S			
Land	68	49	19
Marine	30	13	17
Sea Spray	44	19	25
Coal	51	49	2
Petroleum	14.3	13.6	0.7
Smelting			
Copper	6.4	4.3	2.2
Lead	0.7	0.6	0.7
Zinc	0.6	0.6	0.05
Total	215	149	66
% of anthropogenic		93%	7%
% of total		69%	31%

After Robinson (1970).

The mechanism of formation is limited by availability of metals, because the oceans contain sufficient sulfate and organic matter (Orr, 1973) to deposit pure iron, copper, zinc, or lead sulfide, if the latter metals are available. The mechanism by which the metals are enriched is not yet known. In the Red Sea, where such a strataform is presently forming, it is believed that a submarine basalt or a Miocene shale is eluded, and that the brines are transported by some geothermal gradient.

Table 6.6
ATMOSPHERIC CONCENTRATION OF SULFUR DIOXIDE AND SULFATE AEROSOLS
(micro gr/m^3)

Location	SO_2	SO_4^-	SO_2/SO_4^-
Antarctic (60° S)	0.13	1.57	0.08
Subantarctic (40° S - 60° S)	0.18	1.68	0.11
South Pacific Ocean (20° S - 40° S)	0.12	1.15	0.10
North Atlantic Ocean (50°N - 8°N)		2.33	
Mediterranean Sea	2.27	8.43	0.27
Boulder, Colorado			
Ground	2.1	1.7	1.3
5.2 km (above ground)	0.4	0.1	4.0
Germany			
Ground to 1600 m	4-25	3-7	5-3
at approx. 2800 m	1	4	0.25
Sweden			
400-2800 m; unpolluted area	1-7	0.1-1.6	10-4

After Nguyen (1974).

C. ATMOSPHERIC SULFUR BUDGET

Table 6.4 shows the relationship between the size of the atmosphere and the surface of the continents and oceans. The residence time of sulfur in the atmosphere lies between a few hours for sulfur dioxide and several weeks for sulfate aerosols. Table 6.5 shows that the estimated atmospheric sulfur budget of the Southern Hemisphere is about 66 Tg, i.e. one third that of the Northern Hemisphere (Robinson, 1970). The difference is attributed to anthropogenic sources. The total global atmospheric sulfur flux of 200 Tg accounts for an average of about 10^{-7} of the total mass of the atmosphere; it corresponds to an average of about 400 mg/m^2 · year sulfur on the earth's entire surface. In reality, the local concentrations vary extremely. In remote areas, the average deposition lies between a value of 100 mg/m^2 in Australia, Kiyuhu (Africa), Colorado (U.S.), Leningrad, and Greenland (Granat, 1976), an average value of 1 g/m^2 in Europe in 1965, and values of up to 500 g/m^2 in Tacoma, Washington (U.S.) and other cities close to incompletely controlled metallurgical smelters. Aerial concentrations vary between 0.05 (in the South Pacific) and 1 micro g/m^3 in the Carribean. Sulfate ranges between 0.006 in the Indian Ocean and 1 micro g/m^3 in the stratosphere of the Northern Hemisphere. Authors disagree on values because of differing measuring methods and because of differing assumptions. The values of Nguyen (1974, 1975) are shown in Table 6.6.

The atmospheric flux has been extensively discussed by Eriksson (1960-1963), Junge (1963-1972), Robinson (1961), Kellog (1972), Friend (1973), Bolin (1965-1976), Granat (1976), Nguyen (1973-1975), and Jost (1974).

In the stratosphere, sulfur occurs mainly as aerosol. Junge found in 1961 that aerosol particles, probably consisting of sulfate, form a stratospheric layer which is now called the Junge layer. He and Friend (1973) proposed mechanisms to account for the presence of sulfur. Three main reactions are assumed: 1) Formation of 'acid embryos' containing a water to sulfur dioxide ratio of about 10, through slow oxidation of sulfur dioxide with oxygen atoms or ozone, depending on weight. The ozone concentration depends on altitude, latitude, and on local effects and industrialization; it can vary between 20 and 50 ppb, and in urban areas often reaches a maximum at about 5000 ft altitude; 2) neutralization of the embryos by ammonia; and 3) rapid conversion of remaining sulfur dioxide to sulfate. Castleman (1974) used sulfur isotopes to confirm the mechanism, and identified volcanic sulfur as a major stratospheric sulfur source, as did Nielsen (1973). This model explains why sulfate forms in the upper atmosphere while sulfur dioxide is washed out promptly in the lower atmosphere. The presence of these particles can contribute to opacity (Bolin, 1976) and possibly to the 'greenhouse effect' of the atmosphere. The total contribution of sulfur to this effect has been estimated to be about 0.03% (Wang, 1976). It is negligible, no matter what present distribution model is chosen, compared to the effect of carbon dioxide (0.8%), nitric oxides (0.7%), or the fluorocarbons (0.5%).

The removal of sulfur dioxide and sulfate from the atmosphere is shown in Fig. 6.1. Dry to wet removal ratios vary regionally between 4:1 and 1:4 depending on climate, season, and on whether total aerial sulfur is locally large or not. It is presently believed that oceans and land areas absorb about equal amounts of both sulfate and sulfur dioxide. The latter must be chemically removed. In water, the buffer capacity of carbonate and dissolved minerals helps retain sulfur dioxide (Brimblecombe, 1972; Spedding, 1972; Beilke, 1974). Over land, sulfur dioxide is very efficiently absorbed by plants (Garland, 1973; Creed, 1973; Castleman, 1973; Whelpdale, 1974; Hill, 1976). If sulfur dioxide concentrations are over two mg/m^3, the plants suffer stomatal damage (Unsworth, 1972; Naegele, 1973), Chapter 11. Smith (1973) determined that soils absorb 1 to 15 mg of sulfur dioxide per gram soil, 15-65 mg/g hydrogen sulfide, and 2-32 mg/g

Table 6.7
FIVE REACTIONS OF SULFUR DIOXIDE, AND THEIR RATE[a]

Reactions	Expressions	Constants
$SO_2(g) + H_2O(l) \overset{H_S}{\rightleftharpoons} SO_2 \cdot H_2O$	$H_S = [SO_2 \cdot H_2O]/P_{SO_2}$	1.23 mole l^{-1} atm^{-1}
$SO_2 \cdot H_2O \underset{k_{-1}}{\overset{k_1}{\rightleftharpoons}} HSO_3^- + H^+$	$R_1 = k_1[SO_2 \cdot H_2O] - k_{-1}[H^+][HSO_3^-]$	$k_1 = (3.3)10^{-2} s^{-1}$ [b] $k_{-1} = (2.7)$ l $mole^{-1}$ s^{-1} [b] $k_1 = (3.4)$ 10^6 s^{-1} [c] $k_{-1} = (2.0)10^8$ l $mole^{-1} s^{-1}$ [c]
	$K_{1S} = [H^+][HSO_3^-]/SO_2 \cdot H_2O]$	$K_{1S} = (1.7)10^{-2}$ mole l^{-1} [d]
$SO_2 \cdot H_2O + OH^- \underset{k_{-10}}{\overset{k_{10}}{\rightleftharpoons}} HSO_3^- + H_2O$	$R_{10} = k_{10}[SO_2 \cdot H_2O][OH^-] - k_{-10}[HSO_3^-]$	(?)
$HSO_3^- \overset{K_{2S}}{\rightleftharpoons} SO_3^= + H^+$	$K_{2S} = [H^+][SO_3^=]/[HSO_3^-]$	$(6.24) \cdot 10^{-8}$ mole l^{-1}
$SO_3^- + ½O_2 \overset{k_o}{\rightarrow} SO_4^=$	$R_o = k_o[SO_3^=]$	(?)

a) After Beilke (1975); b) Wang (1964); c) Eigen (1961); d) Beilke (1974).

hydrogen sulfide, and 2 to 32 mg/g methylmercaptan. The absorption is apparently almost independent of soil pH or organic and clay content.

1. Chemistry of Atmospheric Sulfur

Sulfur dioxide is chemically unstable. Thus, atmospheric sulfur can enter a variety of reactions. The most widely studied are oxidation of sulfur dioxide to sulfate, reaction of sulfur dioxide with ammonia, and reaction of sulfur dioxide with air pollutants.

In all systems, gaseous sulfur dioxide converts to non-volatile sulfate which forms aerosols. The aerosols can affect the opacity of the air, (Chapter 4), they can participate in reaction with other pollutants or catalyze reactions in polluted atmosphere, and they are known to aggravate pre-existing health problems (Chapter 10). Furthermore, unreacted sulfate is a strong acid, and excessive acid can inflict damage to land, vegetation, buildings, and corrode marble and steel. There is substantial controversy about the nature of the conversion of sulfur dioxide to sulfate; the reaction mechanisms in the atmosphere are not known. Some reaction rates have been studied in the laboratory, Table 6.7. Laboratory experiments at the low concentrations found in the atmosphere are difficult, and thus most data is deduced by extrapolation of laboratory experiments on concentrated samples. A serious hazard in applying such data is that the atmosphere contains other trace gases and particulates which are normally ignored in laboratory experiments. Novakov (1971-1976) has shown that the most effective path for oxidation of sulfur dioxide in a real atmosphere is on the surface of such particulates, which are ubiquitous wherever sulfur dioxide occurs. McCain (1973) recently measured the size distribution of particulates from power plants. They extend over the entire reactive range. Thus, an extensive discussion of the various merits of the far better studied and better known relative reaction rates of homogenous gas phase reactions seems irrelevant for the purpose of this book. Novakov (Novakov, 1971-1976; Chang, 1975; Harker 1975; Craig, 1974) used ESCA, IR, Raman, and other instrumental methods and studied heterogeneous reactions on soot, manganese oxide, and other particles as a function of particle size and gaseous impurities, including nitric oxide. Barrie (1976) studied similar systems and estimated that sulfur dioxide in an urban atmosphere can be reduced at a rate of up to 2% per hour. Junge (1961-1976) studied aerosols over the entire range of the atmosphere from ground level to the stratosphere. He also tried to determine their acid content. The influence of light on the oxidation of sulfur dioxide (Becker, 1975; Cox, 1971; Allen, 1972) is very important, because it is well known that sulfur dioxide can absorb visible sun light, and transform it into electronic excitation in the form of its triplet state (Meyer, 1971; Schmidkunz, 1963), which can enter a variety of new chemical reactions (Finlayson, 1976) or form sulfuric acid (Penkett, 1974; Castleman, 1975). These reactions are most important in urban areas where smog, or at least nitric oxides, are present (Roberts, 1976; E. Meyer, 1969). Their interaction can lead to particle formation (Cox, 1974; Kasahara, 1976).

Studies of the atmospheric sulfur chemistry depend on analytical tools. Such tools are still in an early developmental stage (EPA, 1976). In addition to flame spectrometry, UV spectroscopy, X-ray fluorescence, gas chromatography, and other analytical methods described in Chapter Four and many references (EPA, 1976; Appel, 1976; California, 1976), sulfur isotopes (Castleman, 1974; Nielsen, 1974) and oxygen isotopes (Holt, 1976) and Radon (Nguyen, 1973-1974) were used. New field methods and sampling methods are described weekly (Thomas, 1976; Stevens, 1971).

2. Sulfur Dioxide Absorption by Water

The first step in all reactions of sulfur dioxide is the absorption by water. It is discussed in detail in Section 3C. Atmospheric absorption processes were discused by Brimblecombe (1972), Spedding (1972), and Beilke (1974). The absorption reaction involves physical absorption, without chemical reaction:

$$H_2O \quad + \quad SO_2 \quad \rightarrow \quad H_2O{\cdot}SO_2$$

Estimates of the rate vary by a factor of 10^6, Section 3C. The present knowledge of the equilibrium between gaseous and absorbed sulfur dioxide goes back to the work of Schaefer (1918). The absorption is strongly dependent on pH, because sulfur dioxide reacts with base:

$$SO_2 \quad + \quad OH^- \quad \rightarrow \quad HSO_3^-$$

The HSO_3^- ion (in which the proton is attached to the sulfur, and not to the oxygen as is still widely assumed) is rapidly oxidized at low pH (Borgwardt, 1971-1976; Chen, 1972; Beilke, 1968). The oxidation has been reviewed by Junge (1958), Bufalini (1971), and in Section 3C.

Studies are further complicated by the formation of disulfite:

$$H_2O \quad + \quad HSO_3^- \quad + \quad SO_2 \quad \rightarrow \quad S_2O_5^{2-} \quad + \quad H_3O^+$$

The latter, commonly called pyrosulfide, has been known since Rammelsberg's work of 1846; its structure has been known since the X-ray work of Zachariasen (1934); but its existence is still not fully appreciated. A summary of the pertinent reactions is given in Chapter 3.

3. Ammonia and Sulfur

Ammonia is a well-known constituent of air, even though its concentration is not yet clearly established and its origin is not yet fully elucidated. Again, the basic reactions have been known since the early days of modern chemistry (Bergrather, 1826; Forchhammer, 1837), but it is still not known how much the physical and chemical reactions compete in binding sulfur dioxide and ammonia. In air, sufficient moisture is normally present to induce the formation of $(NH_4)HSO_3$, which readily oxidizes to sulfate. This reaction has been described and studied by Schumann (1900), Divers (1900), Gold (1950), Becke-Goehring (1954), Baggio (1971), Din (1953), Hata (1964), Scott (1969-1970), Scargill (1971),

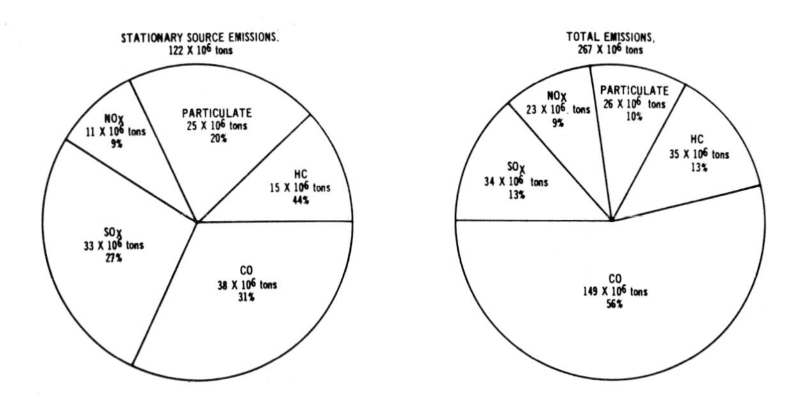

Fig. 6.8. U.S. emission factors, 1970; a) stationary emission, b) total emission.

Kling (1973), Charlson (1974), Arrowsmith (1973), Beilke (1974), McLaren (1974), Rehme (1975), Hartley (1975), Vance (1976), Landreth (1974), Hisatsune (1975).

D. THE ANTHROPOGENIC SULFUR CYCLE

Fig. 6.8 shows the emission factors for eight pollutants in four segments of the U.S. economy. Transportation contributes 86 mil tons per year, industry 23, power 20, space heating 8, and refuse burning 5. The power industry contributes 14.09% of all pollutants. Forty-six percent of this, i.e. 10 mil tons, is sulfur dioxide.

1. Nature of Activities

Modern activities have influenced the sulfur flux in two basically different manners: 1) Sedimentary sulfur is mined or recovered from oil and gas and used in industry and agriculture, and 2) sulfur from coal, minerals, and oil is inadvertantly released during combustion.

In both cycles, a total of about 80 mil tons per year are presently involved, but both sources are local or regional in nature; otherwise, they have little similarity or connection. In the first case, sulfur is deliberately produced for two different purposes: Agricultural sulfur is applied and spread for the purpose of manipulating the sulfur soil cycle and the entire soil cycle, including that of nitrogen and phosphorus. The goal is to increase productivity of the soil. Ideally, a large fraction of this sulfur enters the biological sulfur cycle, and a fraction of it enters protein. The rest remains in the soil and eventually reenters the sediment. Industrial sulfur is almost totally converted to sulfate, is used as an acid, and promptly returns to the sediment as sulfate. So far, only comparatively little sulfur enters industrial products. It is possible that in the future a much larger fraction of sulfur will be incorporated into industrial products. If this becomes the case, it will be mainly in elemental form or in the form of sulfide, both of which readily biodegrade to sulfate, which reenters the sedimentary cycle. We need not dwell here on the use of gypsum, which is mixed and used as such, because in it sulfur does not change oxidation state. In all the above applications, sulfur is deliberately produced for a specific use; if it is inadvertantly lost during production, it is lost to its intended purpose, and the process becomes inefficient, i.e. uneconomical.

Sulfur released from coal constitutes an entirely different situation, because it is inadvertantly produced as by-product by industries which traditionally consider all their combustion products as wastes, and have no intrinsic appreciation of the use and handling of chemicals. Almost all of their by-product sulfur enters the atmosphere as sulfur dioxide. Only very little remains in the ashes as sulfate. Since sulfur dioxide is heavier than air, it does not mix well, and about 90% returns to the soil within a day and usually within a radius of less than 50 miles. Depending on local geography and climate, this emission is noticed or ignored. In rare cases, it contributes to the fertility of the soil; more commonly, it damages forests and lawns, and is considered a nuisance or, even, a menace. In the past, complaints have been handled by building taller stacks and spreading the sulfur dioxide thinner and wider. It is now recognized that this method only compounds problems. The present trend is to capture sulfur dioxide with lime (Chapter 8), and dump it in ponds or sinks. While this shifts the burden further from the source, it cannot and is not considered a permanent solution. A viable solution depends on integrating sulfur dioxide emission in a useful cycle in which the sulfur has a positive purpose and use. Such a cycle can only be found if combustion products are considered as raw materials, rather than as waste. Such a change of attitude is slowly developing, but psychological changes take more time than technical innovation. The smelter industry is far ahead of the coal industry in this respect. COMINCO has made a successful business of sulfur dioxide since 1935, and almost all smelter companies now explore uses for their sulfur.

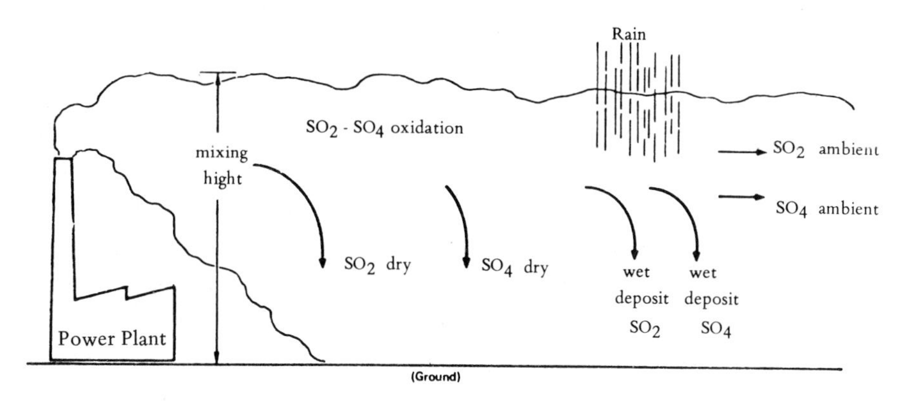

Fig. 6.9. Fate of sulfur dioxide from power plants.

2. Fate of Flue Gases

The coal fired electric power plants now constitute by far the largest over-all source. Their flue contains, for each pound of sulfur, 100 lbs of carbon dioxide, 1 lb of nitric oxide, and several grams of particulates and water. Smelters, in contrast, emit such highly concentrated sulfur dioxide, up to 5%, that unabated emission causes acute local damage. In power plant plumes, the impact is rarely acute, but the delayed release of sulfur dioxide, and its reaction products, has an equally serious impact, which is even more insidious and annoying than the short range impact of smelter plumes. The best known form of delayed and remote sulfur dioxide release is the so-called 'acid rain', described below. The fate of sulfur dioxide in dilute flues depends on a variety of factors. Geography, climate, and season all play a role. Fig. 6.9 shows a typical sulfur profile. Meyer, (1969) discussed the effect of microclimatic effects, for example the so-called 'inversion' layers which can serve to accumulate emission over a period of hours or even days; sometimes this enriched air is released and deposited within a period of hours, and it can cause severe, acute damage to health (Chapter 10) and vegetation (Chapter 11) (Robinson, 1971). It has been customary to use tall stacks to overcome inversion, and to better mix the flue with air. It is now widely felt that the process of diluting wastes before dumping is useless, and asocial, as is implied in the word 'pollution'. It is now widely recognized that the often well-intended dilution causes severe secondary problems, because sulfur dioxide released by this method is no longer confined to a local microclimate which can be reasonably well predicted, but enters a regional zone with complex, uncontrollable, and often unforeseeable behavior. Only recently have detailed studies of regional sulfur flux drawn detailed attention. Garland (1976) studied the mixing height and its effect on the range of the sulfur dioxide transport over Great Britain. Davis (1974) measured trace amounts of ozone, nitric oxide, and sulfur oxides from an aircraft over Washington, D.C.; White (1976) measured secondary pollutants, aerosols, and ozone over a 200 mile area NE of St. Louis. Opacity measurements in St. Louis were made by Charlson (1974) and others. In all cases, the inevitable was observed: If large amounts of sulfur dioxide are released, they will return downwind to earth, and, eventually, will be noticed wherever they reach the ground. Thus, if the local damage is reduced by dilution, the exposed area is increased, and the damage is merely spread. While dilution might be a suitable method in remote areas, it is not suitable in populated areas, because dilution increases the affected population. Since it has now been established that even low concentration can cause substantial damage that can be documented, the disposal via large stacks is obsolete. A further consequence is that plumes with high buoyancy might release sulfur dioxide into higher air levels, where the residence time is known to be far longer. This might lead to a steady state increase in sulfate concentration in the upper atmosphere. This now widely discussed effect has been proven to constitute presently negligible damage compared to other anthropogenic atmospheric additives such as nitric oxide, hydrocarbons, etc. The other effect, acid rain, constitutes a more tangible nuisance.

3. Acid Rain

The scientific and popular literature contains a series of alarming articles dealing with 'acid rain'. It has been long known and well established that not all terrestrial water has a pH of 7. The pH of stagnant water is often severely unbalanced, even though such waters are naturally buffered. Rain water is very pure and thus not protected with a buffer. Therefore, a given concentration of acid or base has a larger effect on the pH of rain than on the pH of lake or seawater, which contain dissolved salts. Even though the effect is larger and easier to measure, the amount of acid, i.e. the power to react, is not increased.

The pH of rain was measured for the first time in 1939, in Maine. Regular measurements have been made for less than two decades,

Table 6.8
SO_2 FROM TCM METHOD AND $SO_4^=$ BOTH IN RAINWATER MEASURED IMMEDIATELY AFTER PRECIPITATION

SO_2 Air (mg/m^3)	SO_2 Rain (mg/l)	$SO_4^=$ Rain (mg/l)	SO_2:$SO_4^=$
0.055	0.32	3.4	0.1
0.027	1.70	7.2	0.2
0.040	0.30	2.6	0.1

After Jost (1974).

starting in Scandinavia (Eriksson, 1950-1970), and, later, in the U.S. (Junge, 1956). It is believed that the 'normal' pH of rain water is above 5. Table 6.8 shows the correlation between sulfur dioxide in air and sulfur dioxide in rain (Jost, 1974). It has been claimed (Likens, 1976) that indirect calculations seem to indicate that the pH of rain in New York was "not acid prior to about 1930." This statement does not seem tenable. Wherever large sulfur dioxide point sources exist, some sulfur dioxide is oxidized and precipitation must contain measurable amounts of acid. Depending on the climate and climatic fluctuations, the 'rain-out' will be irrelevant or not noticeable. Under special conditions, when large amounts of sulfur dioxide accumulate in higher oxidation layers, some rains are bound to have a low pH. Larson (1975) very carefully measured the sulfate and pH content of a single storm in the Northwest U.S. The sulfur dioxide source was a smelter in Tacoma. The results, Fig. 6.10, show that the spreading of sulfur dioxide measurably affects an area of about 500 square kilometers. A very similar situation exists in the Eastern U.S. (Murphy, 1976; Schaug, 1976; Huebert, 1976; Wright, 1976; Galloway, 1976); where sulfate concentrations are approximately 2 to 5 mg/l, nitrate is 3 to 5 mg/l. The pH varies between 3.6 and 7; in an extreme case, a pH of 2.1 was observed (Likens, 1975). In the Eastern U.S. and in England, the sulfur dioxide sources are not normally well defined point sources. Thus, the sulfur dioxide and sulfate gradients are not as well defined and not as easily perceived as in the case of point sources. However, it is now well documented that the pH of rain in Scandinavia is caused by sulfur dioxide pollution from outside its national border (Rodhe, 1972). Unfortunately, some climatic conditions can perpetuate conditions by which sulfur dioxide from a point or a surface source accumulates in an

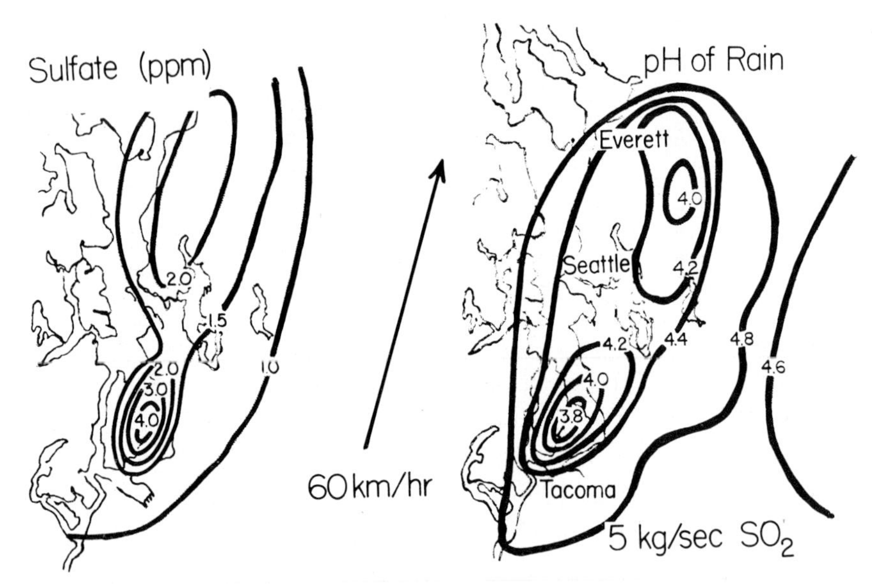

Fig. 6.10. Tacoma acid rain pH and sulfate content (after Larson, 1976).

atmosphere layer and is preferentially precipitated in a climatically predistined region. The result of such a continued condition could affect not only vegetation, which is far more sensitive to sulfur dioxide than is man, but also fish and other participants in the ecosystem. It is indeed well known that at a pH below 5.5, the fish population decreases rapidly. At a pH below 4.4, fish life in lakes is 80% barren. This situation is clearly undesirable, but it is not as acute as recent scientific reports try to paint it, because England and Germany, the sources of Scandinavian acid rain, burned far more coal during the period from 1920 to 1940 than now, and the effect cannot have started only ten years ago when the first measurements were made and the danger realized. Furthermore, the damage is quickly remedied, since the sulfur cycle is short lived; it is at least 200 times shorter than the ground water cycle into which over-eager environmentalists dump incompletely reacted lime and sulfite; and it is millions of times shorter than the desert cycle, where some urban governments propose to dump the solid waste.

4. The Pedosphere

The sulfur cycle in the pedosphere is described in Chapter 11. It involves bacterial oxidation of organic and inorganic sulfides, and elemental sulfur yielding sulfate. Sulfate, in turn, can be absorbed by plants, or it can be reduced in an anaerobic location to sulfide or elemental sulfur. Thus, the pedosphere is recipient and source of both oxidized sulfur and reduced sulfur. The pedospheric cycle varies greatly in different locations, and, except for emission from swamps and sulfate leading into rivers, the cycle is often quite local.

5. Industrial Impact

We have seen that the atmospheric sulfur cycle establishes locally a stationary state. Imbalances in the sulfur cycle are regionally absorbed. Thus, the sulfur cycle differs from the carbon dioxide cycle, the fluorocarbon cycle, and from radioactive fallout, which affect the entire hemisphere and thus have global effects. Most of the intentionally produced sulfur goes into sulfuric acid, and almost all of the 110 million tons of sulfuric acid produced each year is locally and quickly consumed, because the storage and transport of acid is expensive. Most of the product is fertilizer, which is produced with the intent to modify the natural growth cycle and to improve the yield of soil. Thus, much of the sulfur is quickly recycled and restored to nature. The net effect is a redistribution of sulfur. The manufacture of acid has very little secondary effect, and sulfur depletion due to mining is not a problem. Whether the redistribution is beneficial or not is an ethical question.

Inadvertent production of sulfur constitutes a different problem. In the case of the petroleum and gas producing industries, nobody questioned whether removal was necessary or not. The removal of sulfur is a prerequisite for making a saleable product. Both industries are modern and accepted standards of good neighbors, and, since the value of their primary product is sufficient, they have developed efficient and economic methods for removing sulfur. Furthermore, the oil industry even found it useful to convert sulfur to acid; and in the gas industry, elemental sulfur is considered a potential resource for the future which is carefully stockpiled. The smelter industry works with chemistry and procedures steeped in exactly 500 years of valuable tradition. Vested techniques and thinking had confined the industry to manufacturing metals, and all by-products were considered a nuisance. The case of COMINCO, Chapter 9, shows that an aggressive, foreward looking approach can open new markets for by-products, and that use of the sulfur in ore can be profitable, if a company is willing—or forced—to diversify into professional activities outside its traditional train of skill and thought. It is only a matter of time before all sulfur from smelters will be recycled in an acceptable way.

Electric power generation is a different case: Industry, government, and large segments of the public now push for nuclear power in the full knowledge that dangerous waste will accumulate for which we have no disposal method (Rochlin, 1977). If no serious effort is made to find some responsible methods for recycling the radioactive wastes from future power generation methods, one cannot be seriously astonished if the coal industry, which suffers from a well-documented history of social and economic problems, is not eager to abate sulfur dioxide which, in most locations, has little direct impact on man, animals, or vegetation, and which does not inflict the lasting or genetic damage of radioactive waste.

Chapter 7

Sulfur Production

During the last two thousand years sulfur has been produced from almost every mineral which contains the element. Elemental sulfur is found in volcanic regions, and, in some areas, in salt domes, and in sulfides, either bound to inorganic elements in "pyrite" and other metallic ores, or bound to hydrogen and carbon in petroleum and natural gas. Coal contains both inorganic and organic sulfides.

The abundance of the different sulfur forms varies regionally. Therefore, sulfur production practice has varied considerably in different regions and countries. During the Middle Ages, alchemists employed untreated elemental sulfur crystals, which Homberg (1710-1724) recognized as possessing inconsistent purity. Sulfuric acid was produced by distillation of vitriol, or by burning sulfur. When synthetic fertilizers were introduced on a large scale, the need for acid increased, and the sulfur mines in Southern France and on Sicily quickly prospered and dominated the market. With advancing industrialization, roasting of pyrites and other sulfur rich metallic ores became more important. This method was threatened by the Frasch method, first commercialized in 1896 in Texas, but pyrite production has persisted in Spain, Scandinavia, and other countries. Today, the ore recovery technique has regained importance because of environmental pressure for recovery of sulfur dioxide as a by-product from copper and other smelters. Fig. 1.4 gives an indication of the relative importance of some sulfur production methods used since 1750, but this figure indicates only general trends. For a history of the sulfur industry, other sources should be consulted, such as Gmelin (1953), Ullmann (1969), and historical statistics of Europe, the U.S. (1975), and other countries.

Involuntary sulfur evolved jointly with smoke and fumes as a by-product of combustion ever since coal and oil were used as fuels. By-product emission from Agricola's smelting operations (1615) must have been breathtaking, but such emission was accepted as inevitable until quite recently. The first need to remove sulfur and other by-product gases arose when city gas was manufactured and used to illuminate Murdock's home in Cornwall in 1792, houses in Paris in 1791, and, eventually, in December of 1813, Westminister bridge and public streets. The need to remove sulfur from the gas became immediately evident, but the original purpose of Clegg's and Philips' patents of 1815 and 1817 was only to use lime and limestone to reduce corrosion, and not to recover the sulfur values. In 1852 Laming patented a method for the simultaneous removal of ammonia and hydrogen sulfide, and indicated that ammonium sulfate fertilizer might be prepared via the sulfite. It was 1855 before oils and paraffins were com mercially produced from bituminous coal. Claus (1893) was the first to develop a viable method of recovering elemental sulfur. Thirty years later when the need arose to desulfurize oil and natural gas on a large scale, his method was quickly accepted and established itself in its modified forms as the dominant method for producing sulfur. The incredible increase of oil and natural gas consumption made recovered sulfur increasingly important. As stated in Chapters 1 and 15, it is reasonable to predict that we are at the beginning of a 30 to 50 year period during which involuntary sulfur recovery as a by-product from energy production will be a dominant factor. This situation will prevail until nuclear power production replaces fossile fuels.

The use of sulfur dioxide recovered from smelter gases did not commence until early in this century; commercial processes for producing sulfuric acid and making fertilizer were developed only around 1920. The pioneering experience of COMINCO in Canada is history and is described in Chapter 9. Recovery of smelter tail gases is now only a question of local economy. Unfortunately, the recovery of sulfur from coal combustion gases is nowhere nearly economical. It has not yet been decided whether coal desulfurization will be achieved by recovering sulfur dioxide from flue gases or by removing hydrogen sulfide before combustion. Thus, a general, broad discussion of the various recovery methods is still necessary.

All sulfur production methods can be fitted into a chemical system of the type shown in Fig. 7.1. It is immediately obvious that elemental sulfur is the most desirable product and source, because it can be easily handled, is not toxic, and contains more sulfur per weight than any other form. It will be shown later that Frasch's process is specially elegant, because it achieves extraction and purification in situ and in one step. In most other modern processes chemically bound sulfur is first converted to

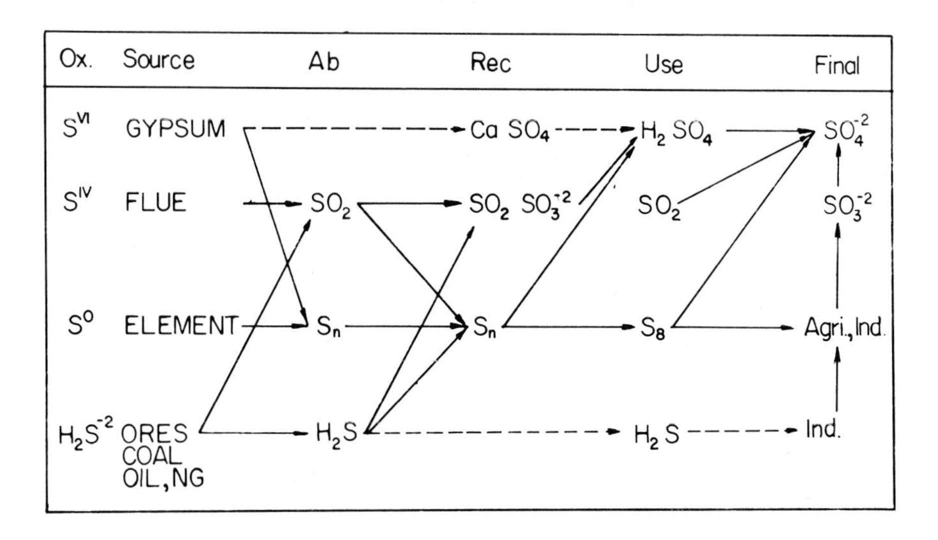

Fig. 7.1. Chemical conversions during sulfur production and use.

hydrogen sulfide or sulfur dioxide which can be extracted, concentrated, and purified. In the second step, elemental sulfur is always the preferred product, because it is non-corrosive, is a solid, and constitutes the most economic form for transportation. If sulfur dioxide or acid is produced, it is almost always consumed promptly and locally. In the oxidative environment of the pedosphere or atmosphere, the final sulfur form is invariably the sulfate which can be recycled or disposed. In the next section we will discuss the production of sulfur, sulfur dioxide, and hydrogen sulfide. In the following section, the extraction and concentration of these three sulfur forms will be discussed.

A. PRODUCTION OF ELEMENTAL SULFUR

Traditionally, the industry has distinguished three sources of sulfur: Brimstone (Frasch, Claus, and native), pyrite (iron or cupreous pyrites and pyrrhotites), and sulfur in other forms (gypsum, sulfur dioxide from smelters, spent pulping oxide, raw, unbeneficiated ores). This grouping ignores the chemical history of the material, and addresses itself merely to the chemical form available for marketing.

Table 7.1

WORLD BRIMSTONE PRODUCTION - MAJOR SUPPLY SOURCES BY COUNTRY

(Million tons S)

	1970	1980
TOTAL PRODUCTION	22.78	37.85
Frasch Sulfur	9.70	15.90
United States	7.00	8.90
Mexico	1.30	2.70
Poland	1.40	2.70
USSR	-	1.00
Iraq	-	0.60
Recovered Sulfur	10.43	18.95
Western Canada	4.70	8.00
United States	1.60	3.10
France (Lacq)	1.70	1.80
Germany, Federal Republic of	0.20	0.50
Near East	0.53	1.60
USSR	0.60	1.50
Others	1.10	2.45
Native Refined Sulfur	2.65	3.00
Western Europe	0.48	0.52
Poland	0.80	1.20
USSR	1.20	0.80
China	0.17	0.48

After Horseman (1970).

1. Brimstone

Until five years ago, the largest producers of brimstone were the Frasch mines in the United States, Mexico, and Poland. Some of the larger producers are listed in the Appendix. Mining of native sulfur, once the producers are listed by Horseman (1970). Mining of native sulfur, once the China. Smaller mines are operated in Sicily, Chile, Bolivia, and Argentina. As mentioned above, the prime source for brimstone is presently Claus sulfur. The main production centers are Western Canada, where production is expected to peak in 1985, Southern France, where the production will decrease within the next ten years, and the U.S. where production is quite steady. The Middle East is rapidly becoming an important source of recovered sulfur, and will be an important source for the next thirty years. If the U.S. or Russian governments decide to sponsor coal gasification research, recovered sulfur could well become a very important source of brimstone in countries with large coal deposits within about 50 years. The production figures for the various brimstone groups are shown in Table 7.1. In the future sulfur recovered from oil shale and from the

Athabasca Tar Sands in northern Alberta could become large brimstone sources if the posted world prices would warrant production.

Elemental sulfur can be produced by mechanical and physical refining of ores containing elemental sulfur, or by chemical conversion of sulfates, sulfur dioxide, or sulfides. The mining of elemental sulfur is the easiest and most obvious method and will be discussed first.

Sulfur Mining

The three classic methods for producing elemental sulfur have already been described by Agricola (1616). They consist of: Refining and flotation, melting, and distillation or sublimation. In all these methods the sulfur remains chemically unchanged in its elemental form. Sulfur can be separated easily from ground ore by flotation. With pine oil and other similar sulfactants, up to 90% of the sulfur can be separated. It can be recovered by settling, drying, and melting. Modern sophisticated flotation methods are used in many of the very productive mines in Poland.

In Sicily, the favored production method was melting of sulfur in "Calcarelli," i.e. open furnaces in which sulfur was ignited. Part of the sulfur burned and its heat of combustion served to heat the residue, and melt the residual sulfur. The latter was collected in urns. A substantially improved yield was achieved when the ovens were closed in order to contain the heat. These kilns, called "calcarone," contained up to a hundred cubic yards of ore; from which about half of the sulfur could be recovered. After 1890 Sicilian sulfur was refined in Gill's furnaces. These contained two chambers, in one sulfur was burned as fuel, in the other it was melted for recovery. In Chile and Japan sulfur has been traditionally sublimed in "Yakutori" retorts. About half a ton of coal is necessary to recover a ton of sulfur from ovens which contain up to 100 modern stainless steel retorts, but corrosion is a problem in this method which is not economically competitive.

Between 1899 and 1903 Hermann Frasch developed and perfected a melting method which is suitable for producing the elemental deposits in salt domes of the Gulf of Mexico region in the U.S. and Mexico. Three coaxial tubes, Fig. 7.2, are mounted in a 25 cm diameter shaft. Steam at 165°C and 16 atm is pressed through the outer pipe to heat the deposit and melt the elemental sulfur, which is produced by foaming it with compressed air which is pressed at 20 atm through the center pipe. In this manner some 300 tons of sulfur of 99 to 99.8% purity can be produced

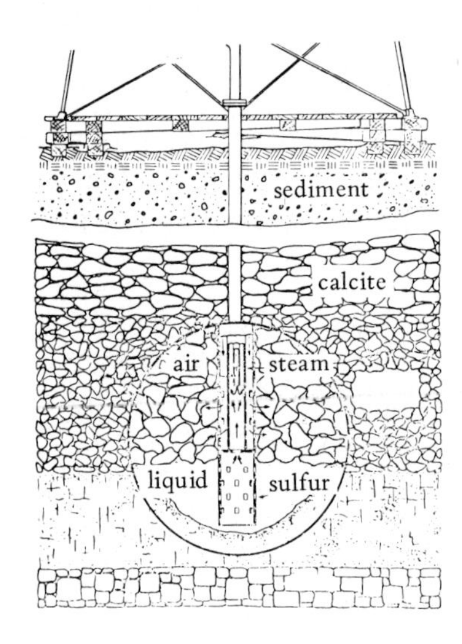

Fig. 7.2. Frasch sulfur production method. (After Gmelin, 1953).

daily. Frasch's method works only in geologically favorable locations. It is not suitable for producing the slanting deposits of Sicily, Fig. 5.3, which it made obsolete. These huge Sicilian deposits of pure sulfur now wait for a new, economical production method. Frasch's method started a golden age for North American sulfur, which quickly became so competitive that it could be shipped world-wide and replaced Sicilian sulfur, even in Europe. The development of the method, as well as the subsequent marketing strategy were vividly reported by Frasch himself (1912), who through his earlier work on oil desulfurization had already deeply influenced the growth of the young oil industry (Haynes, 1959).

2. Sulfur from Pyrite and Other Ores

Russia, Japan, and Spain are presently the largest pyrite producers. China, Italy, and Cyprus, as well as Canada, also produce large amounts of it, as is seen in Table 7.2. Producers distinguish between primary pyrite, mined and sold for its sulfur content, and concentrate, which is a by-product from ores primarily mined for lead, zinc, or copper content. Primarily, sulfur pyrite is produced in Italy, Portugal, and Spain, while Cyprus, Sweden, the Federal Republic of Germany, North America, and South Africa produce by-product pyrite.

Table 7.2
WORLD PYRITE PRODUCTION
(Thousand tons S content)

	1969
TOTAL PRODUCTION	11,403
Economic Commission for Europe (ECE) area	7,560
Cyprus	485
Germany, Federal Republic of	265
Italy	600
Norway	400
Portugal	275
Spain	1,300
Sweden	230
Romania	340
USSR	2,800
Economic Commission for Africa (ECA) area	432
Morocco	140
South Africa	250
Economic Commission for Asia and the Far East (ECAFE) area	2,548
Japan	1,475
China	590
Korea, Democratic People's Republic of	200
North America	830
Canada	470
United States	360
Economic Commission for Latin America (ECLA) area	33

After Horseman (1970).

The following production and recovery methods have been developed: Distillation of sulfur from ore by heating without air, roasting or melting of ore under reducing conditions, treatment of sulfide ore with sulfur dioxide in stoichiometric ratio , and leaching of ore with acid.

In all the above production methods, sulfur exists originally in the oxidation states 0 and -2, and the element is produced by conversion to the oxidation state 0 of elemental sulfur. In all these processes, some of the sulfur is first oxidized, and subsequently reduced. All present methods suffer from serious drawbacks, and only a few have been proved in practice.

In the Outokumpu process, developed in Finland, pyrite ores are heated to 1200°C. Liquid FeS results; sulfur vapor is cooled to 320°C, and condensed by washing with liquid sulfur. Arsenic is removed by washing with lime solution under pressure. FeS is prilled, granulated, and roasted. Sulfur dioxide is recovered for the manufacture of sulfuric acid. Heat from the process is used for steam generation. The other well established process, the Orkla process, was developed in Norway. In this process copper ores are roasted with coal, and the sulfur is condensed at 130°C. The gas is reheated and treated for further sulfur recovery by washing with liquid sulfur. The reaction heat is used to produce steam, and the main product is copper metal. Iron is lost in the slag.

Several processes have been proposed to extract sulfur from pyrites by leaching with hydrogen sulfide or sulfur dioxide. None of these methods have found wide acceptance. An excellent summary of the various proposed processes was given by Thieler (1936).

Sulfates as a Source of Sulfur

Whenever sulfur is in short supply, a variety of proposals surface for recovering the element from gypsum, anhydrite, or magnesium sulfate which are abundant. The possible pathways include: Reaction with ammonia and carbon dioxide to yield $(NH_4)_2SO_4$; catalysed, thermal decomposition yielding cement and sulfur dioxide; thermal reduction with coal to elemental sulfur; thermal reduction to calcium sulfide; and bacterial reduction to elemental sulfur. All these schemes have failed so far, because the sulfur price has always fluctuated, and cheaper sulfur became available before the technical development was completed.

3. Sulfur in Other Forms

It is estimated that about 50% of all sulfuric acid never reaches the open market and enters few statistics, because it is consumed internally by the producer. The prime sources are smelter gases, gypsum, anhydrite, oil refinery acid sludge, and hydrogen sulfide. The best smelter gases are obtained from zinc blende, while lead is least suitable. Copper smelter gas is an important source in Japan, Finland, Zambia, Canada, and recently in the U.S. In the latter case, air pollution regulations, described in Chapter Nine, have enhanced the use of formerly discarded sulfur values. A hundred years ago oxidation of hydrogen sulfide from coal gas was a significant source. This situation could reoccur if coal gasification is re-developed. The impact of sulfuric acid from fossil fuel buying electric power plants was thought to contribute to sulfur in other forms in the U.S. in the early 1970's. Since the U.S. Government will sponsor mainly throw-away sulfur dioxide recovery projects until 1985, the result of this source is unlikely to become relevant before 1995. In contrast, in Japan the production of gypsum from recovered SO_2 is already an important market factor (Ando, 1976). Ammonium sulfate is sold as a fertilizer, and liquid sulfur dioxide finds limited use for ore leaching and other uses described in Chapter 12.

A detailed analysis of the world sulfur production between 1960 and 1970 was made by Horseman (1970) for the United Nations Industrial Development Organization. The same report also analyzes the relationship between supply and demand, and includes a forecast up to 1980. A computer aided forecast of sulfur supply and demand was produced by Shell (1972) for the U.S. Environmental Protection Agency. Both are discussed in Chapter 15.

B. BY-PRODUCT SULFUR

Sulfur occurs as a by-product of several industries and activities. In this book only those sources are considered from which substantial quantities of sulfur value are now being recovered. Thus, the pulp and paper industry is not included. The two most important by-product forms of sulfur are sulfur dioxide and hydrogen sulfide (fig. 7.1). The first occurs in combustion products, especially the flue gases from electric power plants and smelters; the latter occurs whenever sulfur is removed before combustion.

1. From Sulfur Dioxide

Sulfur dioxide is not a natural staple, but it can be easily produced by burning sulfide or elemental sulfur contained in ores or minerals. Sulfur dioxide is often an involuntary by-product of roasting. For conversion to its elemental form, sulfur must be reduced from the oxidation state four to zero.

Dry methods. Sulfur dioxide can be reduced with coal, charcoal, hydrocarbons, hydrogen or hydrogen sulfide. The basic reaction steps are:

Coal gas: $SO_2 + 2CO \rightarrow 1/8\,S_8 + 2CO_2 + 70$ kcal
Hydrogen reduction: $SO_2 + 2H_2 \rightarrow 1/8\,S_8 + 2H_2O + 60$ kcal
Natural gas or methane: $SO_2 + 1/2\,CH_4 \rightarrow 1/8\,S_8 + 1/2\,CO_2 + H_2O + 24$ kcal

In reality, the main product is not S_8, but a mixture of S_2, S_3, S_4, S_5, S_6, S_7, etc. The equilibrium between these species depends on temperature, as shown in Fig. 3.10. Coal gas is the most economic source for reduction. In this case, the process is conducted in two steps: In the initial reaction sulfur dioxide is usually brought into contact with a coal bed at 600-900°C:

$$2SO_2 + 2C \rightarrow S_2 + 2CO_2 \quad \text{(fast)}$$
$$CO_2 + C \rightarrow 2CO \quad \text{(slow)}$$
$$S_2 + C \rightarrow CS_2$$
$$2SO_2 + 4CO \rightarrow S_2 + 4CO_2$$
$$2CO + S_2 \rightarrow 2COS$$
$$2COS \rightarrow CO_2 + CS_2$$

The second reaction step, at a temperature between 1000° and 350°C, has the purpose of converting the intermediates S_2, COS, CS_2, and residual S_2 to elemental sulfur vapor. Various catalysts are used to enhance the conversion:

$$2SO_2 + 4CO \rightarrow S_2 + 4CO_2$$
$$2SO_2 + 4CS_2 \rightarrow 3S_2 + 2CO_2$$
$$2SO_2 + 4COS \rightarrow 3S_2 + 4CO_2$$

The basic chemistry was already described by Berthollet (1883). A very important factor in practical processes is the dew point of sulfur, which, depending on the system, lies between 160°C and 325°C. It is highest in the presence of carbon reduction gas, and lowest if hydrogen sulfide is present, which can form sulfanes of the formula H_2S_n. The equilibrium between CO_2, COS, CO, SO_2, and S_2 is shown in Fig. 7.3. The conversion of CS_2, COS, and SO_2 to carbon dioxide and sulfur is shown in Fig. 7.4. Processes using carbon monoxide as a reagent or as intermediates are most favorable.

Normally, the reaction between hydrogen and sulfur dioxide must be conducted above 500°C. With catalysts, the reaction can be induced at 300°C; it yields sulfur dioxide as an intermediate. The thermodynamics of the equilibrium between sulfur and hydrogen sulfide have been studied by Hyne and Raymont (1974-1975).

Commercial reduction of sulfur dioxide is complicated by impurities which shift equilibria and might chemically interfere with the reaction or

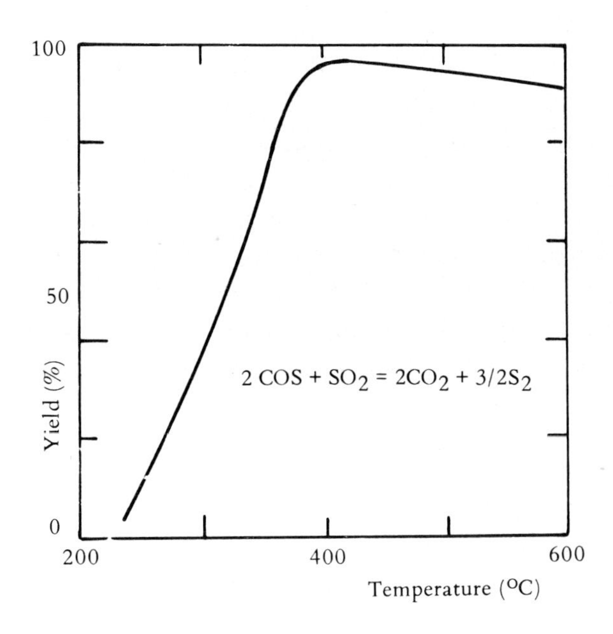

Fig. 7.3. Gas phase equilibrium between CO, COS, SO_2, and S_2.

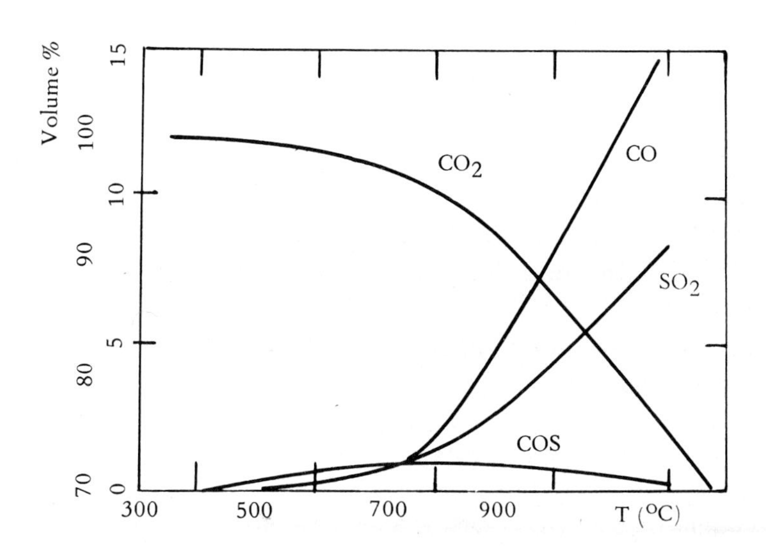

Fig. 7.4. Gas phase equilibrium between CO_2, COS, and SO_2.

the catalyst. Natural gas, which contains hydrogen and methane, is normally used. Averbukh (1969), who studied the equilibria and kinetics of this system, found that the reaction path is different and the process more efficient, if hydrogen sulfide occurs as an intermediate. The first process was developed by ASARCO (1942-1947). Bauxite and aluminum oxide are the standard catalysts, but iron oxide, sodium sulfide, and alkali carbonates have also been used. A summary of the various catalysts used in the Claus process is given by Gmelin (1959). Fly ash also works well. An excellent reducing gas is reformed natural gas, because it contains CO and H_2, both of which are efficient reagents. A commercially successful process using coal was developed by Bolidens Gruv and operated from 1935 to 1946.

Wet methods. Wet reduction of sulfur dioxide to sulfur is kinetically not favorable. The best studied reduction is that with hydrogen sulfide, the "Thylox" process, discussed below, which yields not only sulfur, but also polythionates, which are described in Section 3.3. They are difficult to handle and difficult to convert. Allied Chemical and Texasgulf have developed processes for low temperature conversion to sulfur. Wiewiorowski (U.S. P. 3,447,903, 1969) developed a process in which sulfur dioxide is absorbed in liquid sulfur, and directly converted to elemental sulfur by hydrogen sulfide which is added in proper proportions.

2. From Hydrogen Sulfide

The basic chemical methods for extracting hydrogen sulfide from natural gas and oil refineries have been adapted from those developed earlier for sweetening coal gases. Typical sulfur concentrations in natural gas vary between 0.1 and 40%, Table 7.3 (Grotewold, 1972; Hyne, 1965; Beddome, 1969; Cunliffe, 1970). If more than 12% hydrogen sulfide is present, elemental sulfur precipitates and tends to plug the tubing and the bottom of the well.

Fig. 7.5 shows the fate of sulfur during oil refining (Hoffman, 1976). The stream is adjusted to produce the desired products in the desired quantities. During the last decades about 30% of all sulfur was removed by amines treatment, and of this about 28% was recovered in Claus plants. About 9% was emitted, about 22% was concentrated in the residual asphalt and coke, while about 40% remained in various petroleum fractions. Trends in this distribution are discussed by Mol (1976).

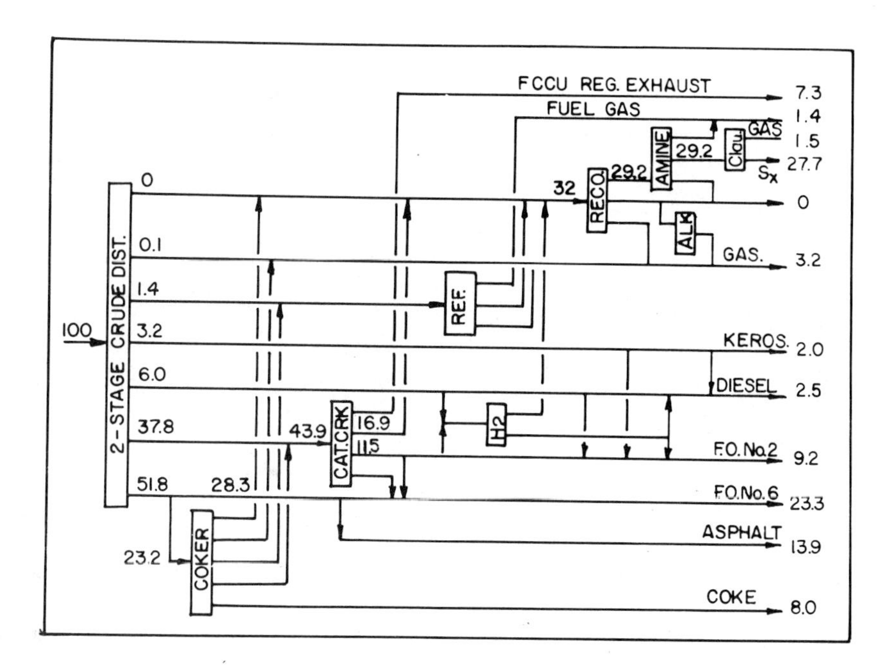

Fig. 7.5. Distribution of sulfur in the oil refinery stream (after Hoffman, 1976).

The present average recovery yield is about 98%. Thus, 500,000 tons of sulfur are wasted each year. Research is under way to reduce this loss (George, 1975). One approach is to add a (fourth) recovery step; another consists of searching for better catalysts (Dalla Lana, 1972). Today Claus tail gases are increasingly treated for further reduction of emission, as described in Section 8C. It is estimated that the U.S. oil industry will invest about $10 billion during the coming ten years to reduce emission by about 80%. Operating costs will increase by about 17% (Kittrell, 1976).

Gas Phase Oxidation; the Claus Process

The reaction of hydrogen sulfide with oxygen is exothermic.

$$H_2S + 1/2\ O_2 \rightarrow 1/8\ S_8 + H_2O + 50\ \text{kcal/mole}$$

Commercially, the reaction has been conducted either in one step, as indicated in the above equation, or in two steps, in a boiler followed by a converter.

$$\text{boiler: } H_2 + 3/2\ O_2 \rightarrow SO_2 + H_2 + 130\ \text{kcal/mole}$$
$$\text{converter: } 2H_2 + SO_2 \rightarrow 2H_2O + 3/8\ S_8 + 30\ \text{kcal/mole}$$

Table 7.3

TYPICAL FEED GAS COMPOSITION (VOL %)

Gas	Solvent	Gas Processing (Amine)		Refining and Tar Sands	
	Rich Feed	Rich Feed	Lean Feed	Amine	Water
H_2S	73.0	75.0	18.0	89.5	90.0
CO_2	20.0	21.0	78.7	5.0	
COS	-	-	-	0.7	-
NH_3	-	-	-	0.2	9.0
H_2O	3.8	3.8	3.0	4.0	1.0
C_{1+}	3.2	0.2	0.3	0.6	-

After Berlie (1973).

In the original process Claus (1882, 1883) used air to burn hydrogen sulfide, recovered from coal gas, in a furnace filled with bauxite or other catalysts. The air and hydrogen sulfide mixture streamed downwards. Claus (1883) and Chance (1888) adapted the process for recovering hydrogen sulfide from the Le-Blanc process. In 1918 the reaction was divided into two steps so that the yield could be increased by equilibrating the vapor at lower temperature. In the IG-Claus process (Bähr, 1932; Ger. P (686,520, 1932) the hydrogen sulfide is burned in a boiler and the reaction is completed by combination of hydrogen sulfide and sulfur dioxide at 300°C. In this way, the large heat of reaction can be dissipated without causing damage to the catalyst, as had occurred in the earlier versions. In the modified IG-Claus process (K. Braus, 1936; Ger. P. 666,572, 1936; B.P. 481,355, 1938), the reaction gases are again combined. Seventy percent of the sulfur is formed in the first steam boiler, and the rest is recovered in a second reactor.

An extensive list of catalysts is given in Gmelin (1959). Claus used iron oxide, manganese oxide, aluminum hydroxide, zinc oxide, limestone or calcined lime, and bauxite. Later workers used charcoal, porous coal, and fly ash (I.G.Farber, 1931), all transition metal oxides, and hydroxides, and, finally, zeolites (SULCO, 1937). In the second, low temperature, converter bauxite, porocel (bauxite containing iron oxide), aluminum oxide, silica gel, silicates, zeolites, sodium oxides, pyrite, and calcined bauxite were proposed. Older work has been well reviewed by Sawyer (1940).

A review of the Claus process was given by Chandler (1976). Earlier work was discussed by Garrison and Elkins (1953), who analyzed the

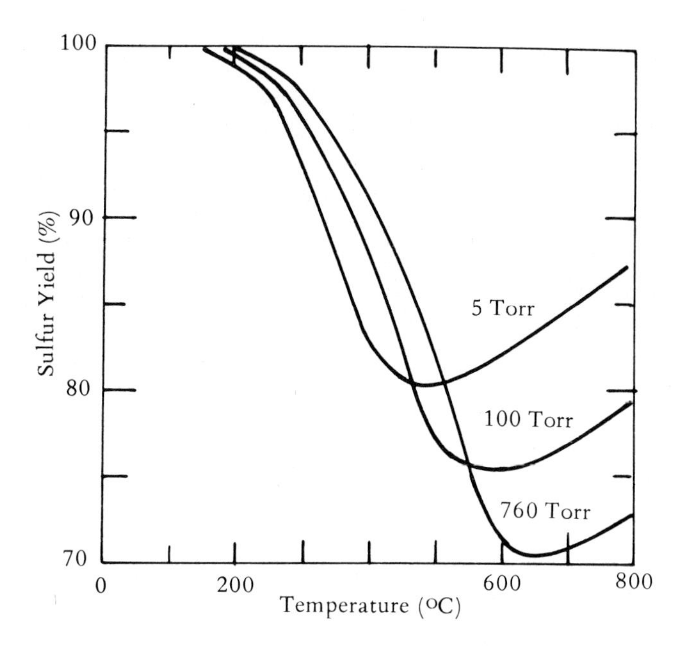

Fig. 7.6. Equilibria in the Claus process. (After Elkins, 1953 and Peter, 1969)

kinetics and thermodynamics of the entire process and calculated the equilibria, Fig. 7.6, and the ultimate theoretical yield. Using Detry's 1965 thermodynamic data, Peter and Woy (1969) recalculated process conditions and determined that 97% yield is theoretically possible in two contact steps. The correlation between Claus yield and modern air pollution requirements have been explored by Barry (1972), Fig. 7.7.

The combustion process of hydrogen sulfide and other sulfur compounds is not well understood. Among the intermediates in an oxygen rich flame are probably the following steps (Cullis, 1972):

$$H_2S + O_2 \rightarrow O_2H + HS \quad - 42 \text{ kcal/mole}$$
$$H_2S + M \rightarrow H + HS + M \quad - 89 \text{ kcal/mole}$$
$$HS + O_2 \rightarrow OH + SO \quad + 21 \text{ kcal/mole}$$
$$H_2S + OH \rightarrow H_2O + HS \quad + 30 \text{ kcal/mole}$$
$$SO + O_2 \rightarrow SO_2 + O \quad + 13 \text{ kcal/mole}$$
$$H_2S + O \rightarrow H_2 + SO \quad + 53 \text{ kcal/mole}$$
$$H_2S + O \rightarrow OH + HS \quad + 13 \text{ kcal/mole}$$

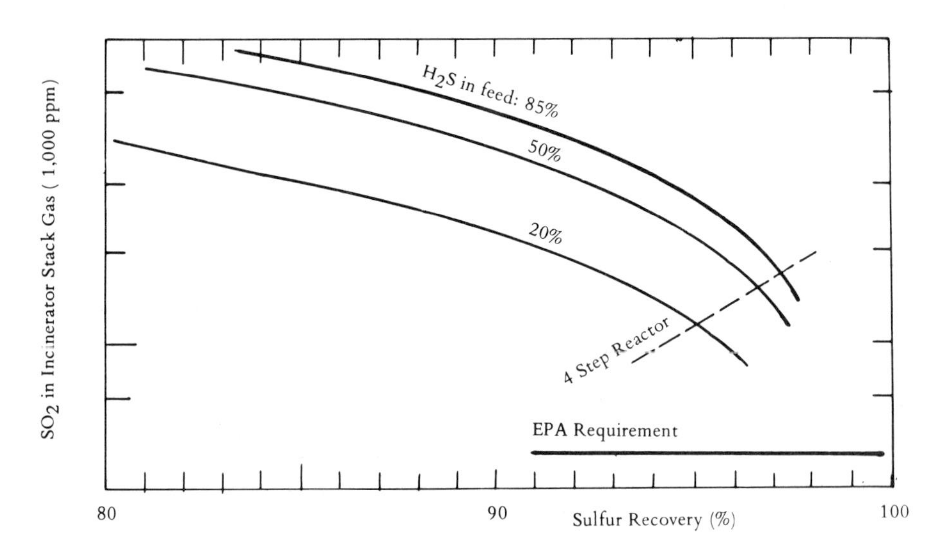

Fig. 7.7. Sulfur recovery yield in a four stage Claus reactor *vs* present U.S.EPA air pollution emission goals. (After Barry, 1972).

In the second step, energy transfer by an inert molecule M is important. In the hydrogen sulfide rich flames, disulfur monoxide might become an important intermediate. The composition of the final Claus gas depends on the temperature, pressure, and concentration of all gases, including inert gases. The various sulfur species which are in equilibrium are shown in Fig. 3.10.

Almost all modern hydrogen sulfide conversion methods, including those by Lurgi, Foster-Wheeler, Hancock, Koppers, Shell, Kellogg, Parsons, Pan America, etc. are based on the modified IG-Claus process. In modern Claus furnaces care is taken to keep the hydrocarbon concentration low to prevent the formation of soot, which contaminates sulfur. Furthermore, the process is electronically regulated to maintain stoichiometry of all reagents when the hydrogen sulfide concentration of the feedstock varies. Claus off-gases are now generally further processed. A well established process is that by Shell (Swaim, 1975).

Catalytic decomposition of hydrogen sulfide

Hyne and Raymont (1972-1975) have shown that hydrogen can occur as intermediates in the claus furnace. The data demonstrates that up to 20% hydrogen could be produced in the thermal decomposition of

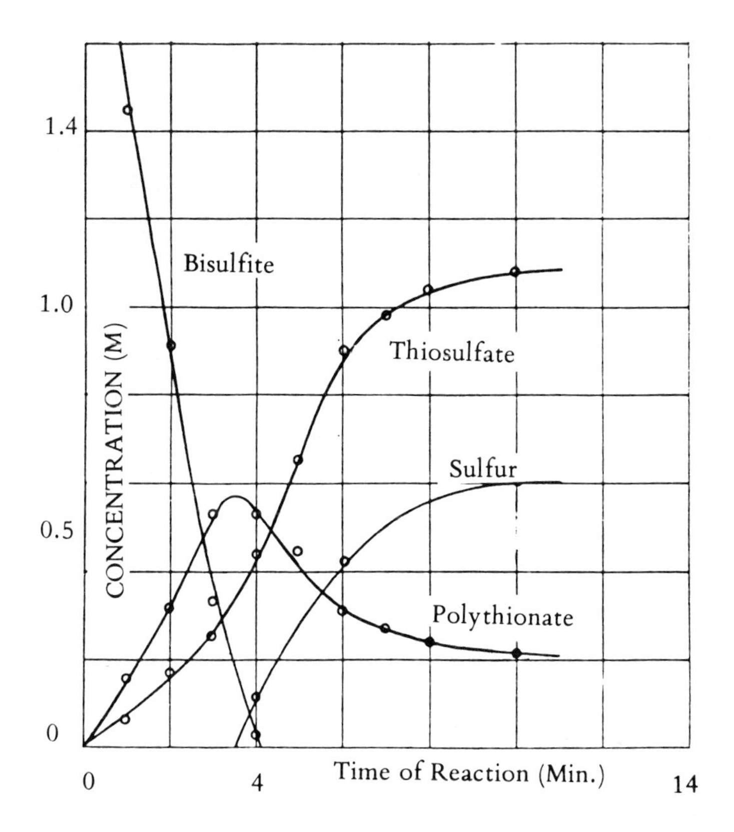

Fig. 7.8. Products of aqueous reaction of hydrogen sulfide with sulfur dioxide as a function of time. (After Korosy, 1975).

hydrogen sulfide at 1000°C. If the proper catalysts and permeable membranes were available, hydrogen could be withdrawn from the equilibrium. Suitable membrane materials must be inert to sulfur, or must form sulfides which do not form protective coatings on metal surfaces, because the latter would poison both the catalysts and plug the membrane. Both reaction products of the endorthermic reaction

$$H_2S \rightarrow H_2 + 1/n\ S_n - 20\ \text{kcal/mole}$$

would be useful. Further work in this field might well lead to new processes in which the resulting hydrogen could be fed back into the natural gas, or directly to petrochemical process plants.

2. Aqueous Reaction of Hydrogen Sulfide

The aqueous reaction

$$2H_2S + SO_2 \rightarrow 3/n\ S_n + 2H_2O$$

has been described in Section 3.3. The reaction proceeds via several intermediate steps, during which, depending on pH, concentration and temperature, a mixture of thiosulfate, polythionates, and even sulfates is obtained, as indicated in Fig. 7.8. The resulting sulfur is a mixture of long polythionates and colloidal sulfur. Wackenroder studied this system extensively in 1846-1848. It is amazingly stable and survives even boiling. Attempts to use this reaction for treating sulfur dioxide and hydrogen sulfide date back to 1840, but they all failed. In recent years, the process has been revived to remove sulfur dioxide and hydrogen sulfide traces from Claus tail gases, which have to be recovered—regardless of economy—to fulfill modern air pollution laws. Schaffner (1878) tried to produce crystalline sulfur by using aqueous solution of high ionic strength, containing $BaCl_2$, $MgCl_2$, and rock salt. The polythionate process can be enhanced by adding dilute sulfuric acid. Gmelin (1959) discussed 22 patents for this process. In the Girdler (1935) process, one-third excess hydrogen sulfide is used and recovered for combustion to sulfur dioxide. I.G.Farber (1940-1942) and Stauffer (1945) developed a process which yields colloidal sulfur for agricultural use.

The future of the Wackenroder process depends on a better understanding of the basic chemical reactions involved. If a catalyst can be found which guides the reaction in such a way that elemental sulfur is formed, this process would have superior promise. The proper catalyst would have to enhance the formation of ionic sulfur chains, such as sulfanemonosulfonates, but would also have to enhance the ring closure of the latter molecules, yielding S_8, rather than the long, polymeric polythionates which constitute Oden's (1911) mixtures.

Chapter 8

Recovery from Combustion Gases

A. COAL COMBUSTION CHEMISTRY

Combustion chemistry looks deceptively simple: Everybody knows that fire consumes fuel and releases heat. For many practical purposes, the burning of oil or gas can be satisfactorily described by equation (8.1). However, this equation violates the law of mass preservation. The alchemists recognized the need to identify combustion products and to find a substance intrinsic to all fire. They called this substance 'phlogiston,' eq. (8.2), and searched for it in vain for several hundred years. Lavoisier established that phlogiston does not exist and that matter gains–rather than loses– weight during combustion by extracting oxygen from vital air, eq. (8.4). His discovery was startling, because it is in conflict with the common observation that burning solids and liquid fuel reserves such as coal, candles, oil, and gasoline disappear and measurably lose weight, but his reasoning gained acceptance among scientists and it put the phlogiston theory to sleep. Despite Lavoisier, the common notion that fuel simply converts to energy has prevailed in the consumer's mind, including that of many industries and private car owners. However, among a large fraction of the present generation of students the instinctive urge to find a substance which is intrinsically connected with the emotions released by flames, combustion, energy, and power has reawakened. Many believe that they have found their phlogiston, which they call 'pollution,' eq. (8.3).

$$\text{Fuel} \rightarrow \text{(Combustion)} \rightarrow \text{Heat} \qquad (8.1)$$

$$\text{Fuel} \rightarrow \text{(Combustion)} \rightarrow \text{Heat} + \text{phlogiston} \qquad (8.2)$$

$$\text{Fuel} \rightarrow \text{(Combustion)} \rightarrow \text{Heat} + \text{pollution} \qquad (8.3)$$

$$C + O_2 \rightarrow \text{(Combustion)} \rightarrow CO_2 + 94\ \text{kcal/mole} \qquad (8.4)$$

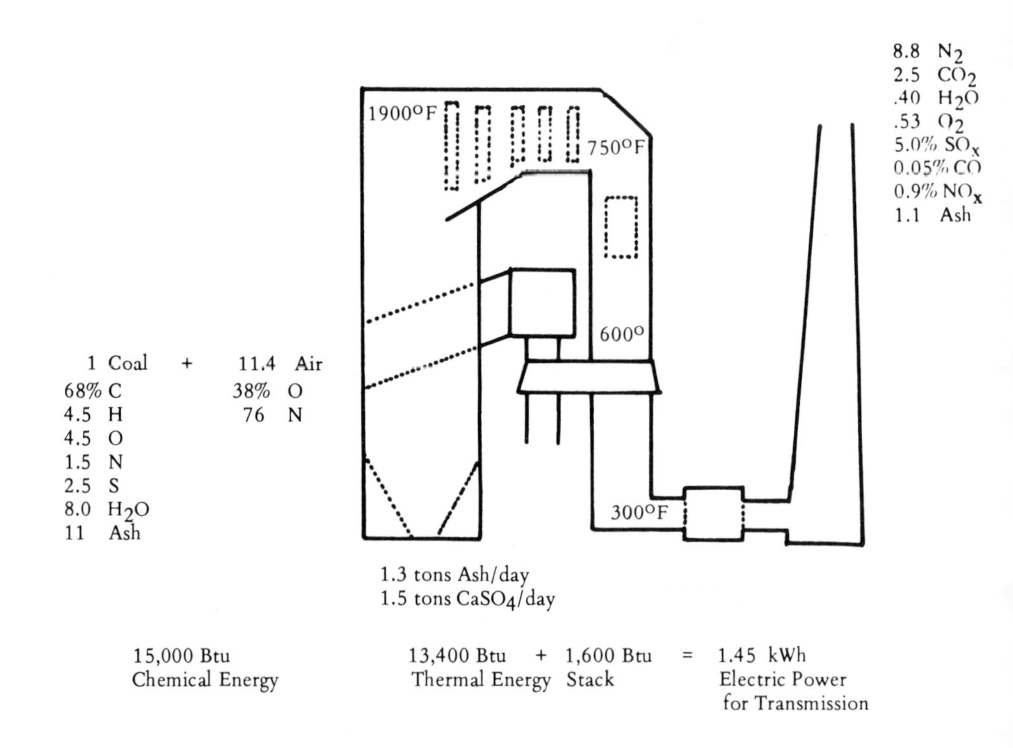

Fig. 8.1. Chemical balance in a coal fired power plant. A 1000 MW plant uses 35 tons of coal and 400 tons of air per hour. It rejects 6×10^9 Btu/hr (1.5×10^9 kcal/hr) as waste heat.

The chemical reactions during combustion are very complex. Every scientist who begins to study combustion and flame chemistry has to overcome the instinctive hurdles connected with the study of a phenomenon which is so deeply tied to emotions. Fig. 8.1 shows the weight ratio of some of the chemical components of a coal fired electric power plant. The chemistry involved in the combustion is discussed in Section 8B3. An excellent review of the subject is given in several books by Gaydon (1970).

B. ABATEMENT METHODS

Sulfur can be removed at any of four process stages by cleaning of coal, extraction from fuel gases, extraction during combustion, and extraction from combustion gases. All methods have been tried; not all are equally successful.

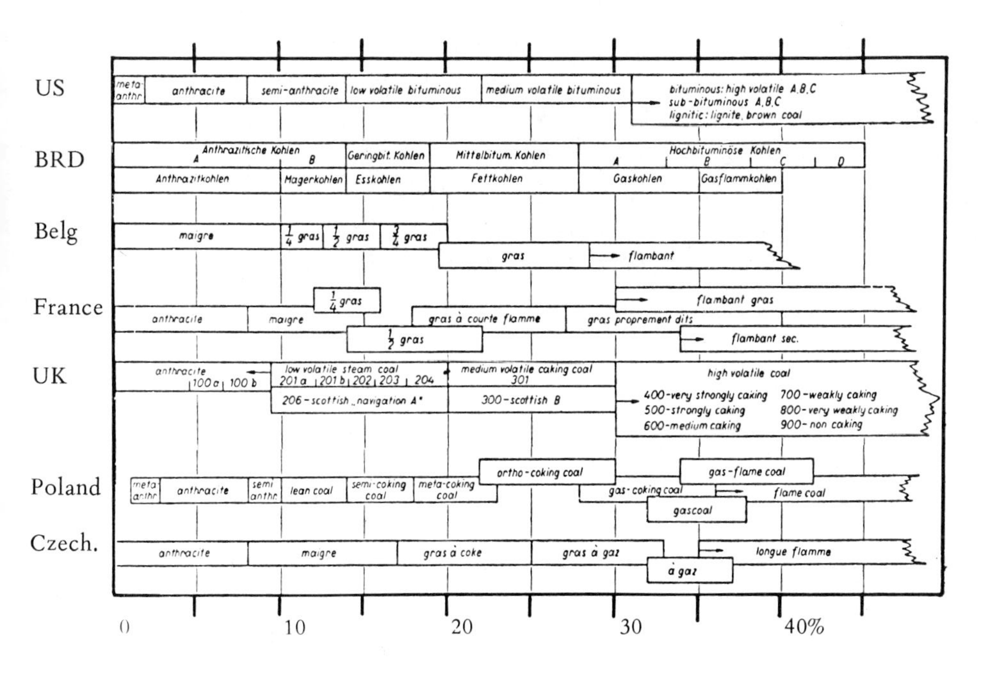

Fig. 8.2. Coal grading methods of eight countries, according to volatile content (after Ullmann, 1953).

1. Coal Cleaning

Coal is not a homogenous material, nor is 'coal' a well defined term. Coals constitute a continuous range of materials, because the geological coal formation involves a continuing series of changes during which peats are step-wise degraded to lignites, bituminous coal, and finally to anthracites (Van Krevelen, 1961). Coal can be graded according to heat content, chemical composition, geological age, or many other criteria. Fig. 8.2 shows the grading as a function of volatile, non-carbonaceous material. A comparison of the various terms used in eight countries shows that there is little agreement for grading among producing countries. Even the grading by size is confusing and can lead to international misunderstanding. A short review of process techniques was prepared by Gorin (1973).

Sulfur is ubiquitous in coals: It occurs in the pyritic fraction, as sulfate sulfur, and in the organic fraction which is part of the matrix. The concentration of all sulfur forms in coal can vary greatly, even with a single coal seam. About half of all sulfur is bound in organic form and forms part of the coal matrix. It cannot be removed by physical methods. Coal sulfate is normally soluble, and can be washed out. It usually constitutes less than 0.05% of total sulfur present. Pyritic sulfur occurs as microscopic particles with a specific gravity of 5.0. Since coal has a gravity of 1.8, the pyrite can be washed if it can be freed from the coal matrix. For this purpose, coal must be crushed. In the process heating value is lost. In a very heterogeneous coal, separation is easy and the loss of heating value is small. In a highly organic coal the loss is large without much tangible improvement.

The U.S. Environmental Protection Agency has sponsored a thorough inventory of the sulfur content of various coals (Cavallaro, 1976). The Bituminous Coal Research Inc. (1969-1976) analyzed a very large number of samples. On the average Alabama coal contains 0.7% pyritic sulfur, and 1.33% total sulfur. Thirty percent of this coal has such a low sulfur content that it does not need to be cleaned; 30% contains sulfur in a form that cannot be mechanically removed, and in the remaining 40% up to half of the heating value is lost if sulfur is mechanically removed. In contrast, in the Appalachian region, where the average analysis is 2% pyritic and 3% total sulfur, only 4% of all 227 coal samples tested were clean enough as is; however, 30% could be cleaned by crushing to 14 mesh. Overall, in this region 80% of the sulfur can be removed with a loss of 30% heating value. In 44 samples of Western coals 70% could be used without treatment and 94% could be used by crushing to 1½ in. size. However, the Western coals are subbituminous, contain much moisture, and have a low heat content. Thus, even moderate cleaning of coal leaves little heating value. These examples show that even seemingly simple mechanical processes have complex effects on the practical value of different coals. Of all 455 U.S. coal samples analyzed only 14% yield less than 1.2 lbs of sulfur dioxide per billion Btu without separation of pyrite. If crushed to 14 mesh, 32% could be cleaned by purely mechanical means, if a waste of up to 50% of the heating value would be tolerable. The loss of heating value is, of course, in conflict with the economic goals of coal utilization, and it is also in conflict with the increasing demand and requirement that all, and not just the most profitable fraction, of coal be used. In

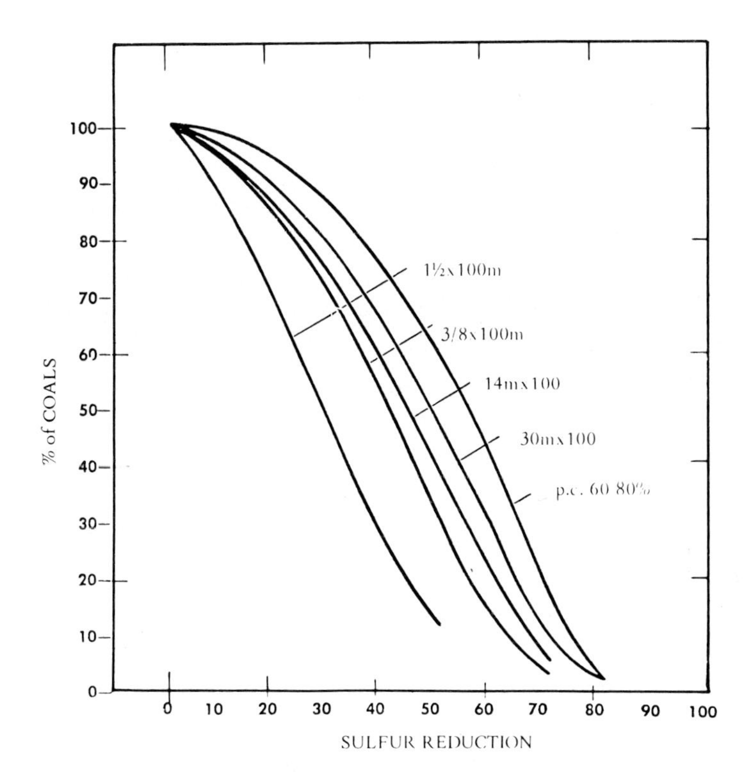

Fig. 8.3. Mechanical sulfur reduction as a function of sizing for 90 U.S. coals (Bit. Coal Res., Inc. for EPA, 1970); 'pc grind' is pulverized as fired, minus 200 mesh.

anticipation of inefficient use of low Btu, several states have prepared legislation requiring full utilization of coal seams. Montana has already passed a law to that extent to prevent superficial strip mining. Fig. 8.3 shows the reduction of sulfur as a function of the crush size, and indicates the percentage of samples which can be successfully treated. The standard grades of coals are reduced to 30 m x 0 at the power plants, and pc grind corresponds to minus 200 mesh to which coal is pulverized when fired.

Chemical coal treatment has been discussed for many decades, but no economically viable method has yet been demonstrated. Recently a process was proposed by Battelle (Hammond, 1975) in which sulfur is extracted as sodium sulfide and converted to elemental sulfur. Caustic is regenerated and recycled. Ever so often biochemical desulfurization is proposed (Davis, 1976). It is known that thiobacillus oxidans can convert organic sulfur to sulfate, but the practical problems have never been faced and would be staggering.

2. Extraction From Fuel Gases

Thermal conversion of coal to gaseous and liquid fuels was widely practiced during the 19th Century. This sophisticated art was virtually lost when natural gas and petroleum with high heating values became available. In 1900 there were over 50 companies capable of building coal gasification plants in the U.S. Today only 3 companies world-wide have the necessary know-how. Coal gasification survived in South Africa and England, and during World War II again reached a high technological stage in Germany. A crash effort is now underway to revive the art and translate it into the scale of modern power plants.

The first patent for producing coke and coal tar was awarded J. J. Becher (B.P. 214, 1681). In Holland in 1748 J. p. Minkelers was the first to produce gas on a continuous basis. The first U.S. gas plant was built by Henfrey in 1802 in Baltimore. Samuel Clegg, born in 1781 in Manchester, worked with Murdock for Boulton and Watt, Ltd. in England. They were the first to introduce gas lights in Soho, Birmingham in 1798. Clegg was the first to patent the use of lime for recovery of sulfur, but it was only much later recognized that the hydrogen sulfide content of gas for use as a lighting material had to be kept below 20 mg/m^3.

Gasification of coal has two independent origins. One is based on the technology of making steel, which dates back to the 15th Century. Faber (1837) succeeded in recovering metallurgical gas for outside use, and in 1840 invented generator gas. Water gas resulted from laboratory experiments. It was discovered by Fontana (1780); his research was confirmed by Lavoisier (1793), and practical application was commenced by Vere and Crane in 1823. Increasing demand for illuminating oil and waxes, and economic pressures prompted motivated efforts for processing bituminous coal. The first commercial plant was built in 1855 in Germany. Rolle developed a series of improved procedures, but his technology was only used in Europe, because of the cheap gas and oil which was discovered in North America. In Germany production of bituminous coke increased from 400,000 tons in 1910 to 4,600,000 tons in 1940. Tar production grew steadily from 75,000 tons in 1910 to 200,000 tons in 1930 and 1,600,000 tons in 1940. During the same time, the export oriented English coal production suffered despite increased domestic use. The coal production in England and Germany during that time is shown in Fig. 8.4.

During the last three decades not one single coal gasification process was in operation in the U.S. Thus, the present coal utilization program

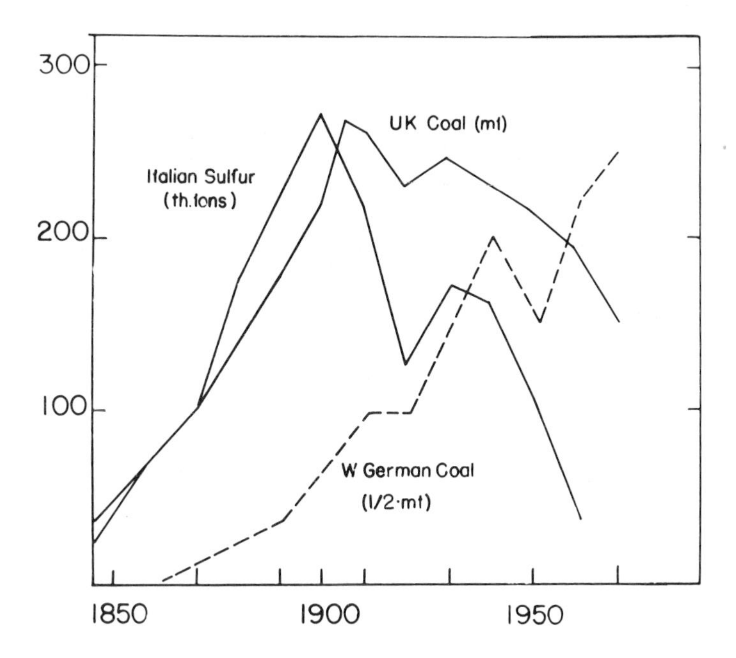

Fig. 8.4. Coal production in England (in million tons) and Germany (half scale), and sulfur production in Italy (in thousand tons) from 1850 to 1975 (after Mitchell, 1971).

had to start from scratch. The status and progress of this effort is regularly reviewed by several governmental agencies, industry, and the scientific community (see for example Science 193 (1976) 665 and 750; 194 (1976) 172, etc.). The four key processes which are under development are the carbon dioxide acceptor, Hygas and Bi-gas processes of the U.S. Office of Coal Research and the Synthane process developed by the U.S. Bureau of Mines. The main difference between modern coal gasification processes and their predecessors is that the former must yield synthetic gas with a heat content of 10^3 Btu/cft. so that the gas can be substituted for natural gas. Traditional coal gas has only half of this heat content. The process concept remains based on the Lurgi gasifier which produces 'town gas' and on the Koppers-Totzek process which produces 'synthesis gas' (CO and H_2). Reviews of modern efforts are given by Cochran (1976), Osborn (1974), and in many books, for example, Schora (1967).

Oil production by coal hydrogenation–with or without catalysis–is not ready for large scale production (Land, 1974; Squires, 1976;

Schlupp, 1976), even though the basic idea for this process is old (Lubin, 1923; Reid, 1973). Solvent refining constitutes another approach. Coal gasification has been long promoted by Squires (1971, 1974) and others. The redevelopment of a coal-chemical industry has been discussed by Franck (1964), who reviewed natural coal chemicals. Synthetic fuels and synthetic natural gas programs have been reviewed by Land (1974), Perry (1974), Schlupp (1976), Zahradnik (1976). The large scale use of methanol as fuel was discussed by Reed (1973).

Among other reevaluated concepts are bold proposals to gasify coal *in situ* and extract the coal gas directly from the coal mine (Stephens, 1975). For this it is necessary to develop vertical cracks to maintain continued gasification. Modern explosives and nuclear force are important in this concept.

In the U.S. the future development will depend to a significant extent on the new energy agency which President Carter is presently shaping from the Bureau of Mines, the Energy Research and Development Agency, the Environmental Protection Agency, and other government branches which have led the development of new technology.

All the above processes have one problem in common: They must cope with sulfur. The technology for desulfurization developed during the 19th Century parallel to that of coal gasification. It was early and successfully adapted by the natural gas and petroleum refining industry (Section 7B). The fate of sulfur in these processes depends on its original chemical form and on processing conditions, but in the final products, sulfur occurs in all fractions. During coke manufacture, about half of all sulfur is in the gas, 5% in the tar, and 40% is found in the coke. During coal hydration, most of the sulfur is produced in the form of gaseous hydrogen sulfide. Normal coal gas contains about 4 to 8 grams hydrogen sulfur per cubic meter. It must be removed to prevent poisoning of catalysts, corrosion, and to prevent toxic effects during use of the gas.

Dry Treatment

Originally dry treatment of gases was preferred. With increasing plant size, the wet methods became more economical. Often it was found advantageous to use several purification steps in sequence. Twenty-one basic sulfur removal processes are listed in Table 8.1.

Palme (1809) was the first to use iron oxide which quickly became the most popular scrubbing material, eq. (8.5). The residue can be

Table 8.1

REAGENTS AND PROCESSES FOR THE RECOVERY OF HYDROGEN SULFIDE FROM COAL GAS IN HISTORIC PERSPECTIVE

Reagent	Name	Product	Inventor or Developer	Reference
Dry Processes				
Limestone	-	CaS, $CaSO_4$	Clegg	BP 3,968 (1815)
			Philips	BP 4,142 (1817)
$Fe(OH)_3$	Dry cleaning	H_2SO_2, or	Laming	(1846)
		S extraction	Palmer	(1809)
		Bauxit-wet or dry	Lux	Ger. 16,456 (1881)
Coal	Charcoal		Behrens	Ger. 303,862 (1921)
Animal coal			Luogo	BP 1,410 (1879)
Wet processes				
$Fe(OH)_3$	Ferrox	S	Koppers Co.	US 1,578,650 (1926)
			Rambush	BP 153,665 (1920)
	Ges.Kohlen-technik	S + $(NH_4)_2SO_4$	Gluud	Ger. 415,587 (1927)
$K_3[Fe(CN)_6]_3$		S	Müller	(1931)
$Fe_4[Fe(CN)_6]_3$		S	Imp.Chem. Ind.	BP 585,381 (1947)
Arsenate sol.	Thylox	S	Koppers Co.	Ger. 619,847 (1935)
Metal salt (Mn,Cu,Ni,Zn)	Several	S or H_2SO_4	Several	
Fe^{3+}-NH_3 suspension	Burkheiser	$(NH_4)_2SO_3$,	Burkheiser	Ger. 212,209 (1909)
			Siegler	Ger. 300,383 (1917)
	Ges.Kohlen-technik	$(NH_4)_2S_2O_3$	Gluud	
Ammonium polythionate sol.	Feld, Polythionate	S + $(NH_4)_2SO_4$	Feld	Ger. 237,607 (1909)
			Overdick	Ger. 506,043 (1930)
Fe-Ammonium-polythionate sol.	C.A.S.	$(NH_4)_2SO_4$	Hansen	Ger. 583,938 (1933)
$H_2S - SO_2$ cat. then NH_3 sol.	Katasulf	S, $(NH_4)_2SO_4$ $(NH_4)_2S_2O_3$	Bähr	Fr. 609,931 (1926)
$H_2S - S_x$	Stretford	V^{5+}	Stretford	US 2,997,439 (1961)
			Nicklin	
NH_3 sol.	IFP	$(NH_4)_2S$-H_2S conc.	-	
Na_2SO_3 sol.	Seabord	Conc. H_2S	Koppers Co.	BP 391,833 (1933)
K_2CO_3 sol.	Petit	Claus products	Petit	Ger. 396,353 (1924)
	Potasch		Koppers	Ger. 712, 026 (1941)
Alkaliphosphate solution	-	-	Shell	US 1,945,163 (1934)
			Bergfeld	Ger. 270,204 (1914)
Alkaliborate sol.	-	-		
Phenolate	-	-	Crawford	US 1,724,909 (1929)
Organic base triethanolmine	Girdler		Bottoms	Ger. 549,556 (1932)
	Girbotol			
	Alkazid NaO		Bähr	Bel. 451,442 (1944)

regenerated by moist combustion, eq. (8.6), or as 'Preussian blue,' eq. (8.7):

$$2Fe(OH)_3 + 3H_2S \rightarrow Fe_2S_3 + 6H_2O \quad (8.5)$$
$$Fe_2S_3 + 3/2\,O_2 + 3H_2O \rightarrow 2Fe(OH)_3 + 3/8\,S_8 \quad (8.6)$$
$$3Fe(OH)_3 + 6HCN \rightarrow Fe_2[Fe(CN)_6] + 6H_2O \quad (8.7)$$

The iron oxide was usually obtained as by-product from the bauxite melt of aluminum plants. Details of the development of the classic absorber materials, developed by Lauda and Lux, are provided by Ullmann (1963) and other technical books. Pressure and temperature gradients were exploited to improve the yield. One cubic meter absorber could clean only a hundred thousand cubic meters of gas. Thus, the method involved large volumes of material and much manual work, because attempts to circulate the solid were unsuccessful. Instead, scrubbers were used parallel. The difficulties of the process development soon overshadowed the intrinsic chemical assets of this method which consists in excellent conversion yields and outstanding reliability.

Wet Processes

Ullmann (1964) categorically states: "No field of combustion technology has received as much attention as the wet scrubbing of hydrogen sulfide." Only highlights of the field are listed here.

Iron Oxide Slurry. The use of iron hydroxide, developed by Koppers (1926), was further developed in England by the addition of carbonates. Hydrogen sulfide can be reduced to less than 1 mg/m^3, but only 79% sulfur is produced. Thiosulfate, iron sulfate and CNS are by-products.

Oxidation with Soluble Catalysts in Aqueous Phase. In the Stretford process, hydrogen sulfide is washed with a solution of anthraquinone and oxidized to anthraquinone sulfonate, from which hydrogen sulfide produces elemental sulfur which can be separated. Ninety-five percent sulfur is obtained if the pH is kept around 9. 99.9% Sulfur can be achieved by further purification. Carbonate interferes with the pH regulation and the reaction is slow.

Absorption by Bases. The borate system has never been developed to a commercial level. Phosphate has been successfully introduced by Bergfelt (1914) and is part of the Shell process (1932). Dilute solutions are used to prevent precipitation of carbonate.

A mixture of sodium hydroxide and sodium phenolate of 1:3 constitutes an excellent absorber, but it is difficult to regenerate. Furthermore, phenol tends to precipitate, but the absorber is insensitive to HCN, carbon disulfide, and mercaptan impurities.

Amino Acids have been widely used. Diethyl and dimethylglycine work well in the presence of carbon dioxide. Such bases are commercially sold under the tradenames Alkazid, Sulfasolvan, Sulfursid, and Sulfazil.

Ammonia and Amines. Aqueous ethanolamine was used in the original Girdler process (1932). Monoethylamine (MEA), diethylamine (DEA), methyldiethanolamine (MDEA), and triethanolamine (TEA) are all commercially proven. MEA is vulnerable to reaction with carbonyl sulfide and carbon disulfide; DEA is more stable but less efficient. The Girbotol process works vcry well for refinery gases. In natural gas plants, corrosion of the steel is a problem, because the base must be heated to regenerate it if carbon dioxide is present. The Shell ADIP process uses diisopropanolamine which works well and exhibits a good pressure dependence.

Hot Potash Process, first patented by Behrens (Ger.P. 162,655, 1904) was further developed by the U.S. Bureau of Mines. The best operating conditions are 100°C and a pressure of 16 atm. Carbonyl sulfide and carbon dioxide react and are thus removed.

Rectisol-Recovery, using lime with methanol at -20°C yields extremely pure gases.

Fluorsol-Process. This process employs propylene carbonate, glycerine triacetate, butoxy-diethyleneglycol acetate, or methoxy-triethyleneglycol acetate as absorber; they absorb hydrogen sulfide selectively in the presence of carbon dioxide.

Dimethyl Formamide has been proposed by Woodhouse (1941). Absorption of hydrogen sulfide is excellent, but corrosion is severe.

Purisol Process. In this process water soluble heterocyclic carbonyls with a boiling point of 200-250°C, for example, N-methylpyrrolidone, are used. The process was developed by Lurgi.

Sulfinol Process, developed by Shell, uses alkane amines dissolved in tetrahydrothiophene dioxide for washing hydrogen sulfide from gases. This method combines high pressure solubility with the advantage of amine reactions. Thus, higher loading is achieved and less absorber is needed.

The EFCo Process uses zeolite sieves for absorption in three or four stages. This method (Dillman, U.S.P. 3,085,380, 1959) produces extremely pure gas from very dilute gases and has much potential for fulfilling air quality programs.

Sulfite-Bisulfite Recovery. Burkheiser (1909) proposed the absorption of hydrogen sulfide on iron oxide, followed by oxidation to sulfur dioxide which he proposed to absorb in ammonium sulfite for further oxidation to sulfate. Burkheiser's efforts received wide attention at the time, but his method failed because of the incomplete oxidation of sulfite to sulfate.

Polythionate Recovery, proposed by Feld (1909), is based on partial oxidation of hydrogen sulfide. He used the resulting polythionate solution to scrub hydrogen sulfide and sulfur dioxide; the liquid is regenerated by reaction with thiosulfate which yields ammonium sulfate.

A combined ammonia-hydrogen sulfide removal process for producing ammonia was developed by Gluud (1931). This process was tested on large scale for three years, and was said to remove 99% hydrogen sulfide in one stage. It was discussed by Ullmann (1963).

Katasulf Process. In this process (Bähr, 1926), hydrogen sulfide is catalytically oxidized to sulfur dioxide and subsequently absorbed in ammonia and reacted with thiosulfate. Ammonia sulfite and sulfate is obtained.

A cost comparison for various processes is given by Ullmann (1963) and by Beddome (1969). The latter also discusses many technical problems encountered in extraction of hydrogen sulfide from natural gas. Grotewold (1972) discusses three industrial cases of hydrogen sulfide recovery from sour gas in Germany, based on the Alkazid, Purisol, and Selexol processes.

General information about gas purification is contained in the valuable books by Nonhebel (1972), Buonicore (1975), Gmelin (1963), and Ullmann (1963). More details can be found in the general references listed below for sulfur dioxide recovery.

3. Sulfur Recovery During or After Combustion

The study of combustion and flames constitutes an independent, separate field of physical and chemical science to which entire journal series are devoted. The over-all chemical reactions of coal combustion were discussed above at the beginning of this chapter. The reaction mechanism and kinetics of the individual steps during the combustion process are not yet fully understood. During combustion in the boiler, the coal undergoes almost explosive changes during which physical and chemical changes compete. The chemistry depends on many factors, for example, the original composition of the fuel. As indicated in Chapter 5, the functional groups in coals are similar to those of crude oils. The flammability limit, the flame velocity, the gravity, and boiling points of some select fuel components are summarized in many handbooks. The heats of combustion are shown in Table 1.6.

Among the first to describe the combustion process was Michael Faraday, whose 'Candle Lecture' of 1860 is still interesting and a pleasure to read. However, we now know much more about combustion reactions (Minkoff, 1962; Gaydon, 1970): Flames contain both reducing and oxidizing zones. Carbon polymers and small carbon molecules such as C_n, especially C_3 and C_2, are abundant and account for much of the light emission. Hot flames also contain CN. Furthermore, positive radical ions such as H_3O^+, CHO_3, C_3HO_3, $C_3H_3^+$, and negative ions such as C_2^-, $CO_2H_2O^-$, $CO_2(H_2O)_2^-$, OH^-, and others occur as intermediates in the plasma. The concentration of these species is now quite well established for several fuel systems (Gaydon, 1970) by optical spectroscopy and mass spectroscopy.

Industrial coal fired steam boilers are much better defined stoichiometric systems than basic researchers tend to think. If too much oxygen is present, valuable heat is wasted. If oxygen is lacking, the fuel is not fully used. In order to extract optimum yield a gas composition of about 10.5% volume carbon dioxide, 3.4% oxygen, 9.9% water, and 75.9% nitrogen must be achieved. The sulfur concentration in the gas varies from 0.05 to 1%. The flue contains about 90% of the sulfur as sulfur dioxide, the rest as sulfur trioxide. The formation of sulfur and nitric oxides has been reviewed by Palmer (1973). During coal evaporation, organic sulfur is originally converted to hydrogen sulfide, carbon disulfide, and carbonyl sulfide. Some organic compounds, for example thiophenes, can vaporize before they decompose. If coal is liquefied or gasified before oxygen is admitted, hydrogen sulfide can be easily separated (Schora, 1974) with methods well tested for oil and gas desulfurization, as described in the preceding section. In the presently used coal boilers, a 4% excess of oxygen is necessary to make maximum use of all carbon and prevent emission of unreacted carbon monoxide. Thus, all intermediate sulfur forms are converted to sulfur dioxide and sulfur trioxide. Among the many intermediate reactions are many free radical reactions involving small high temperatures molecules such as CN, NH, CH, OH, CS, SH, NO, NS, SO, S_2, C_2, SOH, NSH, NOH, C_3, C_2S, C_2O, C_2N, C_2H, CS_2, COS, N_2C, CSN, CNO, CH_2, CSH, COH, and even corresponding negative and positive ions (Benson, 1976).

During coal combustion nitric oxides can form both from organic coal nitrogen and from reaction of nitrogen in air with the hot plasma. In a well managed power plant, the combustion from the latter reaction is negligible, but the first is inevitable. Until very recently, the nitric oxide emission from power plants was acknowledged but not taken seriously, even though the total emission from power plants in the U.S. is about 10 million tons per year. It was thought that its existence could be ignored. The situation is rapidly changing. In Japan legislation already limits both nitrogen and sulfur emission, in combination.

The chemistry of the SO_x-NO_x system was studied by Armitage (1971) and Haynes (1974). Benson (1976) has reviewed the kinetics and thermodynamics of a large number of the possible intermediate reactions. In the real-life system, reactions are yet more complex, because of turbulent flow, temperature gradients, and because of fly-ash. Fly-ash contains several transition elements such as manganese, nickel, iron, cobalt, and vanadium which are known to be excellent catalysts. The structure of the fly-ash surface is very complex, because of uneven heating (Fisher, 1976, Gutenmann, 1976). Thus, gas-solid reactions must be considered an important factor.

For the removal of sulfur, additives must be chosen which absorb or react selectively with sulfur dioxide and sulfur trioxide. For this, Lewis bases are suitable, if their carbonates are less stable than their sulfites and sulfates. The most suitable of the naturally available substances is limestone. Fortunately, it is abundant, occurs in very high purity, and can be readily mined.

Fig. 8.1 shows the masstransfer in a 1000 MW power plant. The temperature profile is shown in Fig. 8.5. For technical reasons four points are preferred for treatment of flue gases with chemical additives: Injection, together with coal, into the flame; injection into the main combustion chamber, above the water wall; injection into the super heater, or injection before the air preheater.

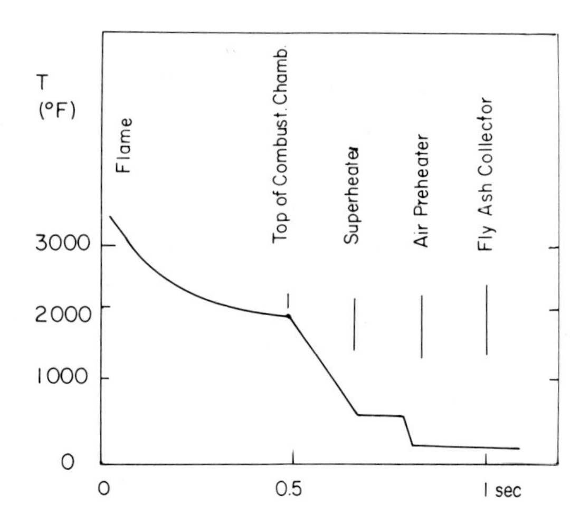

Fig. 8.5. Temperature profile in a 1000 MW power plant (Fig. 8.1).

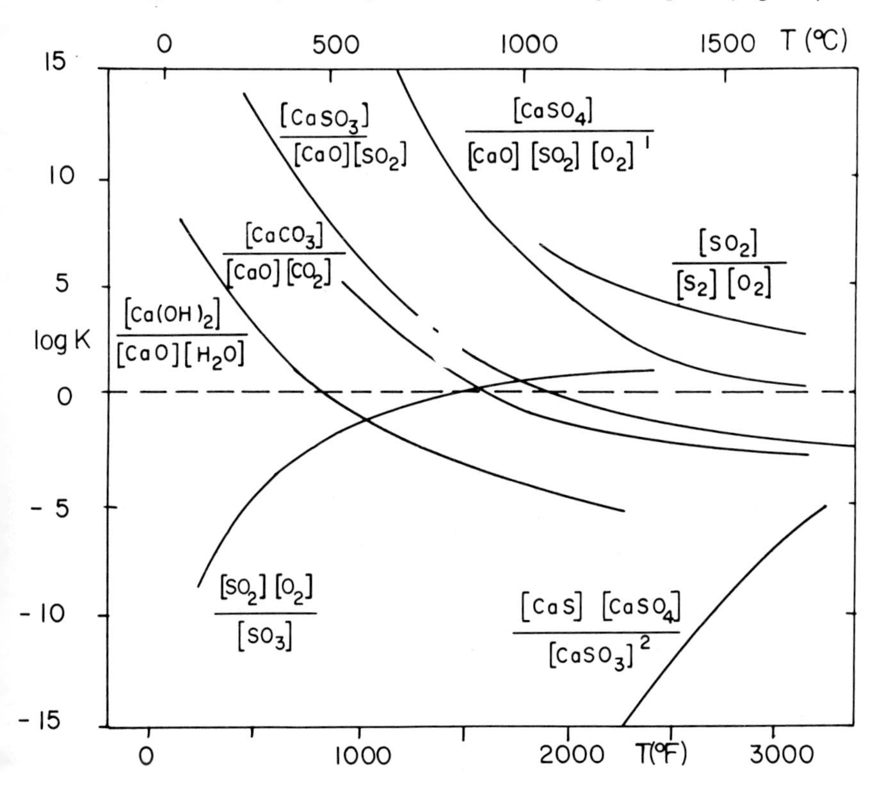

Fig. 8.6. Chemical equilibria relevant to gas phase abatement reactions in coal fired power plants.

Injection into the Flame

Lime or limestone has been added to coal for over 130 years to reduce boiler corrosion. The use of this system for full flue desulfurization was carefully studied in the 1920's and again after the last World War (Wickert, 1963, Kettner, 1965). Injection into the flame zone is convenient because the chemical can be premixed in the correct ratio before combustion. This is not useful in present boiler configurations because the flame temperature is 3000-4000°F (1700-2200°C); thus, the equilibrium, Fig. 8.6, is shifted in the wrong direction, i.e. the sulfates decompose and sulfur dioxide is released. In addition, lime is calcined at this temperature and the residual carbonate forms an eutectic mixture with CaO, the material melts, granulates, and is 'burned dead' because it loses surface area which is crucial for gas-solid reactions. In the fluidized burners of the 1980's, the coal combustion will take place at lower temperature and it is expected that almost all nascent sulfur dioxide can be recovered with limestone in the combustion chamber.

The products of the reaction of sulfur dioxide with lime and limestone have been studied by Grekhov (1975), who found that the aluminum silicates of the slag react with the calcium sulfate above 1050°C yielding free sulfur dioxide, clinker aluminates, and calcium aluminum.

Injection into the Main Chamber

In the present generation boilers, the preferred point for injection of carbonates is above the water wall at 1800-2400°F (1000-1300°C). The temperature is low enough for the equilibrium to favor sulfite and sulfate, but it is still high enough for the reaction rate to be sufficient to complete reaction with the short fraction of a second that remains before the gas reaches the cool part of the furnace.

A large amount of engineering and full scale research and development has been performed in the U.S. during the last ten years with natural limestones which are very pure carbonates. Coutant (1969) determined the reaction rates for various particle sizes and found that the best temperature for sulfite formation was between 1800-2100°F. Below 1500°F the reaction took over 1.5 sec., i.e. longer than the transit time of the particles in a normal boiler. Borgwardt (1970) measured the reaction kinetics in the temperature range between 540°C and 1100°C in a laboratory reactor and found values of 2.5 to 5 x 10^3 per second with an activation energy of 8 to 10 kcal, depending on the stone. He found that for particles with a diameter below 0.5 mm, the rate is almost independent of the size. Reaction occurs initially by diffusion in the particles, but the rate reduces quickly after about 20% of the particle has been converted. It is now accepted that this premature deactivation is due to sulfate which forms on the particle surface and covers it. This clogs up pores and prevents diffusion. Young (1975) studied the kinetics of half calcined dolomite and found an activation energy of 7 kcal/mole. O'Neill (1976) used a thermogravimetric technique, working at pressures between 1 and 10 atmospheres from 750 to 1050°C. The reaction rate corresponded to that found by Borgwardt. The dominant influence on the sulfation rate was found to be the carbon dioxide decomposition pressure during calcination.

Meyer and Carlson (1970) showed that hydrolysis of carbonate and the calcined oxide increases the kinetics. This is due to the fact that sulfur dioxide, a Lewis acid, reacts with the base function of the oxide surface in an ionic reaction:

$$H_2O + CaO \rightarrow Ca(OH)_2 \qquad (8.8)$$

$$Ca(OH)_2 \rightarrow CaSO_3 + H_2O \qquad (8.9)$$

It is believed that oxides react only very slowly and that the reaction rate of oxides is determined by the concentration of hydroxide 'impurity' sites. The above equations show that water is regenerated and acts as an acid-base catalyst. Young (1975) confirmed that the steam concentration affects the reaction rate.

Some spectacular 'more than stoichiometric' rates were achieved with some limestone samples which—as it turned out later—contained some alkali. The high reactivity of these samples confirms that the absorption involves an ionic reaction mechanism. The addition of alkali to activate limestone has been patented by Shell. However, despite tireless efforts on all scales and at all stages, the conversion of limestone that could be achieved in practice has been disappointingly far below the chemically predicted rate. Thus, limestone injection has been abandoned in England and is now deemphasized in the U.S.

Injection into the Superheater

Injection of lime or limestone into the superheater at about 700°F would lead to a thermodynamically favored yield of calcium sulfite, but the process is far too slow for the available flue gas transit time.

Injection into the Air Preheater

Injection before the air heater would be thermodynamically ideal, but is kinetically hopeless. In practice it is preferred to treat the gases after the air heater.

4. Sulfur from Stack Gases

The sulfur in combustion gases has annoyed people for over four hundred years. Chemical abatement efforts are at least a hundred years old. Gmelin summarizes these efforts with a quote from an article published in 1938: "Economical recovery of sulfur dioxide from flue gases is not possible. At best, the gases can be rendered innocuous." This statement still holds. The reason for this state of affairs is not connected with the chemical nature of sulfur dioxide. It is due to the low concentration of sulfur which is mixed with enormous volumes of combustion gases. It certainly would seem easier to remove sulfur from fuel where it is eleven times more concentrated than in combustion gases. The question follows why the major sulfur abatement efforts during the last ten years were concentrated on waste recovery rather than on emission prevention. The answer has to do with the financial and time scales involved in the design, the testing, and construction of a new type of power plant. It takes 15 to 20 years to design and test a prototype, and once a prototype has proven viable, at least another five years of work have to be invested (about 1.5 million man hours of technical labor and 10 million man hours of craft and manual labor worth over 1 billion dollars) before a new process is fully operational. During the last hundred years, the time scale for such a development has consistently exceeded the stamina of those who finance it, partly because the driving force for abatement is not industrial efficiency but social. Thus, all efforts have been abandoned before completion. It remains to be seen whether the present efforts will obtain the continued support they need, and it is, therefore, politically necessary and socially desirable to develop as a stop-gap measure flue gas desulfurization which can be retrofitted to existing plants.

Already a hundred years ago limestone was found to be the most suitable reagent for neutralizing acid formed during combustion in steam boilers. However, only enough was added to bind sulfur trioxide and sulfuric acid, while all sulfur dioxide was emitted unabated. When emission gases were found to be objectionable, a taller stack was built or a stack heater was added to give the flue gases sufficient buoyancy to penetrate above the inversion layer and to diffuse sufficiently to spread over a large area. This deliberate spreading of wastes is what came to be called 'pollution,' as discussed in Chapter 9.

The first process to extract and concentrate sulfur dioxide from flue gases was developed by Hämish and Schröder (1884). In 1906 England introduced the Alkali Act which restricted sulfur emission to 4 grains/cft before admixture of air, smoke, or other diluting gases. The first sulfur dioxide removal process tested on a full scale power plant was at the Battersea Station in London in 1927. The River Thames was used for providing 35 tons of water for each ton of coal burned. The gas was scrubbed in five banks of steel channel scrubbers, followed by a final wash with water containing suspended chalk. The process worked well, but it was expensive and on cold days it was found that the cool gases lingering over the river were still objectionable, even without sulfur. The process was discontinued during World War II. In 1931 seawater was first used by the Virginia Smelting Co. (This plant has been closed, but the process is periodically reconsidered and developed; Bromley, 1971; Ikenaga, 1974). Within the following ten years, over thirty sulfur dioxide removal processes were developed and tested. The war interrupted activities until the 1950's when the present cycle of interest began. During the last ten years, research and development have reached such a pace that it is almost impossible to follow the current events: Since 1973 the U.S. Patent Office has allowed about one patent related to this field per day. The research and technical literature is equally confusing: Chemical Abstracts excerpts some 15 abatement related articles per day. A thorough, critical review shows that much of the recent activity has been devoted to solving engineering problems. It is often difficult to recognize the novelty of inventions, partly because the basic variation of the about 30 basic processes combined with some 15 different regeneration reactions (applied in different sequences with various engineering devices, scrubber methods, and other equipment techniques) yields more than a thousand combinations of abatement methods. Obviously, it is impossible to review, or even mention, them all in this book. Therefore, we will only review some of the basic chemical systems and ignore the engineering considerations despite their obvious importance.

Table 8.2

PROCESS CLASSES FOR THE RECOVERY OF SULFUR DIOXIDE FROM STACK GAS

RECOVERY PROCESSES	Absorption by Liquids
Catalytic Reduction	Alkaline Earth Absorbents
Reduction by Carbon	Calcium Compounds
Reduction by CO	Magnesium Compounds
Reduction by Hydrogen Sulfide	Alkali Absorbents
Catalytic Oxidation	Molten Carbonates
Wet Systems	Potassium Compounds
Dry Systems	Sodium Compounds
High Temperature	Ammonia Scrubbing
Low Temperature	Organic Liquids
Adsorption	**THROWAWAY PROCESSES**
Regeneration by Reduction	Wet Systems
Regeneration by Washing	Alkali Lime/Limestone
Regeneration by Heating	Lime or Limestone
Sorption by Solids	Scrubber Injection
Organic Solids	Boiler Injection
Metal Oxides	Dry Systems
Sodium Compounds	Nahcolite Injection
	Limestone Injection

C. ABATEMENT CHEMISTRY

The following section deals with methods to extract sulfur dioxide. Processes for utilizing or converting sulfur dioxide are described in Section 8B. The extraction procedures can be fitted into the scheme shown in Table 8.2 above. Basic reviews of earlier work have been provided by Katz (1950), Rickles (1968), Cortelyou (1969), Chilton (1970), Yulish (1971), Harima (1974), Squires (1967), Semrau (1975), NAS Report to U.S.Senate Subcommittee on Labor (1973), U.N. Economic Commission (1970), Ando (1973, 1976), Elder (1972), Slack (1971), Rosenberg (1975), Hyne (1970, 1972), Nonhebel (1972), Pfeiffer (1975), Ponder (1974-present), Raben (1973). Process chemistry has been discussed by Wickert (1963), Bienstock (1958), Erdös (1962), Kettner (1965), Spengler (1965), Battelle (1966), Mita (1975), Gmelin (1959), Ullmann (1964), Kirk-Othmer (1969), and Haas (1973). Many organizations have prepared internal reports which are available upon request. Battelle Institute and the Electric Power Research Institute, among others, maintain complete libraries; computer searches of patent and literature are available from Chemical Abstracts, and PEDCo prepares very valuable quarterly, updated status reports of the future, present, and past full scale testing activities. The U.S. Bureau of Mines and Energy Research and Development Agency

Table 8.3
1976 U.S. EPA FORECAST OF FLUE GAS DESULFURIZATION (FGD) ATERNATIVES OF THE PERIOD 1977-1987

SO_2 Control Method	Date of Commercial Availability	Potential Applicability by 1985, % of Total Power Plant Fuel	Costs $/kwh	Average Total Cost of Power Mills/kwh
Conventional Coal-fired Boiler	–	–	500-700	31.5
Low-Sulfur Coal	Current	less than 50	20	39.6
Physical Coal Cleaning	Current	less than 14	12	33.0
Chemical Coal Cleaning	1980	30	75	36.0
Flue Gas Desulfurization				
Limestone	Current	85	54.7	34.7
Lime	Current	85	48.9	34.8
Magnesium Oxide	Current	85	57.4	34.1
Sodium Sulfite	Current	85	67.8	35.5
Aqueous Carbonate	1981	–	65-95	34.3
Double Alkali	1981	–	65-95	34.3
Citrate	1981	–	65-95	34.3
Coal Gasification	1980-85	less than 10	339-495	38.7
Coal Liquefaction	1980-85	0	350-500	–
Fluidized Bed Combustion				
Atmospheric	1984-86	0	632	30.0
Pressurized	mid-1980's	0	723	35.5

The bases for these predictions are explained in a detailed report by Ponder (1976).

(Formerly AEC) produce invaluable information. Most of all, the Environmental Protection Agency (formerly NAPCA and, yet earlier, NIH) through its Control Systems Laboratory produces a myriad of first class reports on original research and development on full scale work. It also conducts regular review meetings, of which proceedings are quickly published (1973, 1976) and which also contain the latest reports from Japan.

Table 8.3 constitutes the 1976 forecast of the U.S. EPA (Ponder, 1976) of the most likely practical availability of sulfur dioxide control alternatives. We will discuss these first. The presently installed commercial capacity for the various processes is shown in Table 8.4 (PEDCo, 1976). All present flue gas desulfurization methods are based on aqueous chemisty. The chemically elegant molten carbonate process is not yet economical with presently available materials.

Table 8.4

STATUS OF U.S. FLUE GAS DESULFURIZATION SYSTEMS BY CHEMICAL METHOD

PROCESS	NEW OR RETROFIT	OPERATIONAL		CONSTRUCTION		PLANNED: CONTRACT AWARDED		PLANNED: LETTER OF INTENT		PLANNED: REQUESTING/ EVAL. BIDS		PLANNED: CONSIDERING FGD SYSTEM		TOTAL NO. OF PLANTS	
		NO.	MW	NO.	MW	NO.	MW	NO.	MW	NO.	MW	NO.	MW	NO.	MW
LIME SCRUBBING	N	1	835	7	3575	6	3150	0	0	0	0	0	0	14	7560
	R	7	1407	1	183	0	0	0	0	0	0	2	660	10	2250
LIME/ALKALINE FLYASH SCRUBBING (PROCESS NOT SELECTED)	N	2	720	1	450	2	1400	0	0	0	0	0	0	5	2570
	R	0	0	0	0	0	0	0	0	0	0	0	0	0	0
LIME/LIMESTONE SCRUBBING (PROCESS NOT SELECTED)	N	0	0	0	0	0	0	0	0	1	527	2	900	3	1427
	R	2	20	0	0	0	0	0	0	0	0	2	1410	4	1430
LIMESTONE SCRUBBING	N	5	2067	17	7146	8	3970	0	0	0	0	3	1280	33	14463
	R	7	860	1	550	1	425	0	0	0	0	0	0	9	1835
SUBTOTAL - LIME/LIMESTONE	N	8.	3622.	25.	11171.	16.	8520.	0.	0.	1.	527.	5.	2180.	55.	26020.
	R	16.	2287.	2.	733.	1.	425.	0.	0.	0.	0.	4.	2070.	23.	5515.
AQUEOUS CARBONATE SCRUBBING	N	0	0	0	0	0	0	0	0	0	0	0	0	0	0
	R	0	0	0	0	0	0	0	0	0	0	1	100	1	100
DOUBLE ALKALI SCRUBBING	N	0	0	1	575	1	250	0	0	0	0	0	0	2	825
	R	1	32	0	0	1	277	0	0	0	0	0	0	2	309
MAGNESIUM OXIDE SCRUBBING	N	0	0	0	0	0	0	0	0	0	0	0	0	0	0
	R	1	120	0	0	0	0	1	240	0	0	2	486	4	846
NOT SELECTED	N	0	0	0	0	0	0	0	0	3	1800	16	7920	19	9720
	R	0	0	0	0	0	0	0	0	0	0	6	2320	6	2320
REGENERABLE NOT SELECTED	N	0	0	0	0	0	0	0	0	0	0	2	1000	2	1000
	R	0	0	0	0	0	0	0	0	0	0	1	650	1	650
SODIUM CARBONATE SCRUBBING	N	1	125	0	0	1	509	1	125	0	0	0	0	3	759
	R	2	250	0	0	0	0	0	0	0	0	0	0	2	250
THOROUGHBRED 101	N	0	0	0	0	0	0	0	0	0	0	0	0	0	0
	R	1	23	0	0	0	0	0	0	0	0	0	0	1	23
WELLMAN LORD/ALLIED CHEMICAL	N	0	0	1	375	0	0	0	0	0	0	0	0	1	375
	R	0	0	2	455	0	0	0	0	0	0	0	0	2	455
TOTALS	N	9.	3747.	27.	12121.	18.	9279.	1.	125.	4.	2327.	23.	11100.	82.	38699.
	R	21.	2712.	4.	1188.	2.	702.	1.	240.	0.	0.	14.	5626.	42.	10468.

PEDCo Quarterly Report, March 1977.

Table 8.5
DISSOCIATION CONSTANTS OF AQUEOUS BUFFER SYSTEMS

Material	Dissociation Constants		
	pK_1	pK_2	pK_3
SO_2	1.92	7.2	
CO_2	6.52	10.33	
Citrate	3.128	4.761	6.395
Arsenate	2.2	6.9	11.5
Phosphate	2.0	7.21	12.4
Formate	3.75		
Acetate	4.75		
$AlCl_3$	4.85	7	8
H_2S	6.9	16.1	
Borate	9.24		
Ammonia	9.24		
Phenolate	9.98		
$Ca(OH)_2$	12.5	-14	
H_2SO_4	- 3	2	

Basic Aqueous Chemistry

In the first absorption steps, described in Chapter 3.3, sulfur dioxide acts as an acid; the absorber as a base. For example:

$$Ca(OH)_2 \quad + \quad H_2O \quad \rightarrow \quad Ca^+ \quad + \quad 2OH^-$$
$$SO_2 \quad + \quad OH^- \quad \rightarrow \quad HSO_3^-$$

The dissociation constants are listed in Table 8.5. The constitution of the solution is shown in Fig. 8.7.

The extensive engineering efforts (Wen, 1976) cannot influence the basic equilibria, Section 3C, but they determine how much of the theoretically feasible value is approached. If the pH is too high, the base also absorbs carbon dioxide which is far more abundant than sulfur dioxide:

$$CO_2 \quad + \quad OH^- \quad \rightarrow \quad HCO_3^-$$

To prevent this, the pH must be kept in the range between 3.0 and 7.0, Fig. 8.7. This can be achieved by gradual addition of absorber, or by use of a base which by itself establishes a pH in this range, for example citrate. If excess HSO_3^- or calcium is present, i.e. if the solubility product of $CaSO_3$ is exceeded, a solid forms. The solubilities of various species are shown in Table 8.6. In some processes a collector is provided where this reaction is desired; in others it occasionally occurs unintentionally along the transfer lines, pumps, or in the heat exchanger; it is then called 'scaling.'

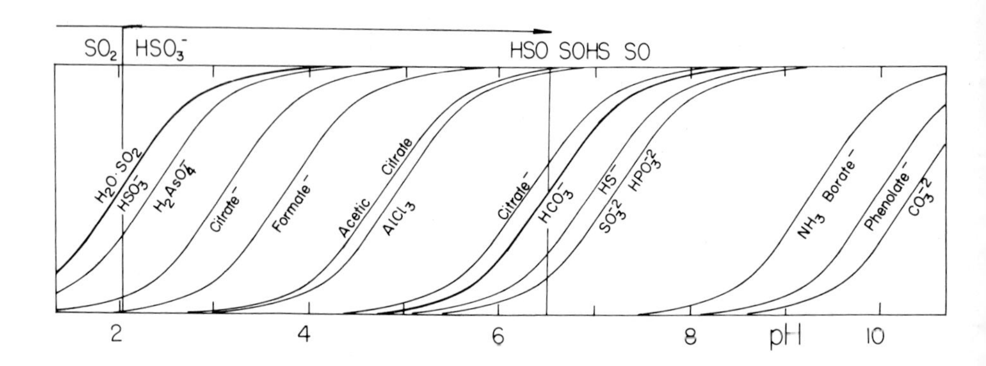

Fig. 8.7. Acid-base equilibria for 16 abatement 'catalysts' (buffers). The corresponding pK values (-log K) are listed in Table 8.5.

Table 8.6

SOLUBILITIES OF CARBONATES, SULFATES, AND SULFITES

	Na	K	Ca	Mg	NH_4
SO_4^{2-}	40[a]		0.2[a]	34[b]	43
SO_3^{2-}	11.7[a]	11.5[a]	0.135[a]	1	40
$S_2O_5^{2-}$	38	24			
CO_3^{2-}			0.014	7.46	

a) Solubility is not strongly temperature dependent in the range between 20°C and 100°C; b) solubility increases.

For desorbing sulfur dioxide the equilibrium is shifted backward by adding acid, by heating, pressure change, or other methods. If oxygen is present in the system, part of the sulfite is oxidized:

$$HSO_3 \quad + \quad \frac{1}{2} O_2 \quad \rightarrow \quad HSO_4^-$$

This process is not easily reversible. The sulfate cannot be desorbed, because the equilibrium

$$H_2SO_4 \rightarrow H_2O + SO_3$$

is shifted to the left. In reality, an entire series of complications can occur, Chapter 3: For example, bisulfite dimerizes and forms disulfite:

$$SO_2 + HSO_3^- + H_2O \rightarrow S_2O_5^{2-} + H_3O^+$$

Disulfite has lower solubility than bisulfite; thus CaS_2O_5 or other solid disulfites, also called 'pyrosulfites,' can form. The equilibrium between bisulfite and disulfite (fig. 3.18)influences the solubility of sulfur dioxide in aqueous solution which is not described by Johnstone's equation, as is still often assumed, but behaves in the manner reported by Oestreich (1976) (fig. 3.20). This reaction is neither quick enough to give full conversion, nor slow enough to be insignificant. Thus, a very substantial effort is aimed at finding process conditions in which the oxidation can be controlled (Borgwardt, 1970-1976). It also depends strongly on ionic strength and chlorine ion concentration.

Limestone Process

The most widely used reagent is limestone. It is readily available and very pure. Its chemistry has been reviewed by Boynton (1966). The aqueous chemistry is described in several advanced inorganic textbooks.

Composition of a Limestone Solution. The solubility of limestone in pure water is described by the solubility product, which is a characteristic constant of each chemical. For limestone it is:

$$[Ca^{2+}]\,[CO_3^{2-}] = K_{sol} = 10^{-8.5}$$

The concentration of $[Ca^{2+}]$ and $[CO_3^{2-}]$ cannot be directly determined in this case, because the carbonate ion interacts with water or acid of the solvent (Kerr, 1974):

$$H_3O^+ + CO_3^{2-} \rightarrow H_2O + HCO_3^-, \text{ and}$$
$$H_3O^+ + HCO_3^- \rightarrow H_2O + H_2O{\cdot}CO_2$$

The ratio of the acid base partners in these equations is determined by the dissociation constants:

$$[H_3O^+][CO_3^{2-}]/[HCO_3^-] = k_2 = 10^{-10.3}$$
$$[H_3O^+][HCO_3^-]/[HCO_3^-] = k_1 = 10^{-6.2}$$

The water self-dissociation constant is $[OH^-][H_3O^+] = 10^{-14}$. In a full evaluation, the sum of all positive and negative ions must be equal and balance the total charge in the solution. The following species must be taken into consideration:

$$[Ca^{2+}] + [H_3O^+] = [CO_3^{2-}] + [HCO_3^-] + [OH^-]$$

If only $CaCO_3$ and H_2O are originally present, each calcium ion must have some carbonate partner species in solution:

$$[Ca^{2+}] = [CO_3^{2-}] + [HCO_3^-] + [H_2O{\cdot}CO_2] = S$$

A full evaluation of all six unknowns is cumbersome, if not impossible. In reality, in all cases some species can be neglected: At high pH, $[H_3O^+]$ and $[H_2O{\cdot}CO_2]$ are very small, and at low pH $[OH^-]$ and $[CO_3^{2-}]$ will become negligible.

By substitution, the total amount of dissolved carbonate can be expressed as:

$$S = \frac{k_{sol}}{[Ca^{2+}]} + \frac{[H_3O^+][CO_3^{2-}]}{k_2} + \frac{[H_3O^+][CO_3^{2-}]}{k_1k_2} \quad \text{and}$$

$$S^2 = k_{sol}(1 + ([H_3O^+]/k_2) + ([H_3O^+]^2/k_1k_2))$$

It can be proven that $S^2 = k_{sol}/\text{alpha}[CO_3^{2-}]$, where

$$\text{alpha}[CO_3^{2-}] = [CO_3^{2-}]/\text{total carbonate}$$

S, the total dissolved carbonate, and S^2 can be plotted *vs.* pH on a logarithmic scale. In our case, S is about 10^{-4}. A yet easier approximation consists of first considering the pH of the solution, by treating the substance as a base:

$$\begin{array}{llll} CO_3^{2-} + H_2O & \rightarrow & HCO_3^- & + \; OH^- \\ 10^{-4}\text{-}x & & x & \quad x \end{array}$$

$$\frac{[HCO_3^-][OH^-]}{[CO_3^{2-}]} = \frac{x{\cdot}x}{10^{-4}\text{-}x} = \frac{k_w}{k_2} = \frac{10^{-14}}{10^{-10.3}}$$

$[OH^-] = x - 5.4 \times 10^{-5}$, and pH = 9.65. At any given pH, the ratio of all carbonate ions is governed by their dissociation constants:

$$[H_3O^+] = k_2([HCO_3^-]/[CO_3^{2-}])$$
$$[H_3O^+] = k_1([H_2O{\cdot}CO_2]/HCO_3^-])$$

Thus, in one case:

$$[HCO_3^-]/[CO_3^{2-}] = 10^{-9.65}/10^{-10.3} = 4.4; \text{ and}$$
$$[H_2O{\cdot}CO_2]/HCO_3^-] = 10^{-9.65}/10^{-6.5} = 7 \times 10^{-4}, \text{ i.e.}$$
$$\text{pH } 9.65\text{: } [H_2O{\cdot}CO_2] : [HCO_3^-] : [CO_3^{2-}] = 0.003:4.4:1$$

Since $[Ca^+]$ must be equal to the sum of all carbonate and the solubility product must be obeyed, we can now compute all concentrations:

$$[Ca][CO_3^{2-}] = 10^{-8.5} = [5.403x][1 \text{ alpha}]$$
$$[CO_3^{2-}] = 2.4 \times 10^{-5}$$
$$[HCO_3^-] = 10.6 \times 10^{-5}$$
$$[H_2O{\cdot}CO_2] = 7.25 \times 10^{-8}$$

Limestone and Sulfur Dioxide. If the above solution reacts with sulfur dioxide, it will act as a diprotic base neutralizing adiprotic acid. In a solution containing 1 mole sulfur dioxide and 1 mole $CaCO_3$, the latter would totally dissolve; the neutralization can be considered as a step-wise process. Again, the quickest procedure is to determine the approximate pH of the final solution, and then to proceed to calculate the ratio of all species from pH, their dissociation constants, and their total concentration. Looking at Fig. 8.7 and considering that the final pH of the solution must be somewhere between that of the original reagents, $[H_2O{\cdot}SO_2]$ and $[CO_3^{2-}]$, one concludes that the first neutralization step is:

$$H_2O{\cdot}SO_2 + CO_3^{2-} \rightarrow HSO_3^- + HCO_3^-$$

Thus, the resulting solution is a mixture containing equal amounts of bisulfite and bicarbonate ions, and the pH is given by

$$k_2(SO_2)k_1(CO_2) = [H_3O^+]$$
$$\text{pH} = (7.2 + 6.5)/2 = 6.8$$

At this pH, the ratio $[SO_3^{2-}]/[HSO_3^-] = k_2/\text{pH} = 2.5$, i.e. the solution contains 28% $[SO_3^{2-}] = [H_2O{\cdot}CO_2]$, 72% $[HSO_3^{2-}] = [HCO_3^-]$ and about 9.95% CO_3^{2-}, 0.001% $SO_2{\cdot}H_2O$, and, finally, from the k value in fig. 3.18, very approximately 5% $S_2O_5^{2-}$.

Now secondary processes must be considered: If $H_2O{\cdot}CO_2$ exceeds the solubility in water, carbon dioxide gas will be evolved. If the concentration of any combination of cations and ions exceeds the solubility product, a solid will precipitate. The solubility of important abatement chemicals is shown in Table 8.6. In our system, the least soluble salt is $CaSO_3$. Thus, if sulfur dioxide and limestone are continuously added, the overall products are calcium, pyrosulfate, and carbon dioxide.

Earlier work on abatement chemistry involving limestone has been thoroughly reviewed and analyzed by Goldschmidt (1968) and Coutant (1969). During the last two years, up to 95% utilization has been achieved (Borgwardt, 1975) and a large number of other studies on other parameters has been conducted, for example, by Lowell (1971) and TVA (Hatfield, 1975; Vivian, 1973). Potter (1969) studied the chemistry of 86 limestone samples.

Large Scale Testing and Applications. The original use of aqueous limestone by Clegg, patented 1815, was to recover hydrogen sulfide. Pilot scale work on sulfur dioxide goes back to 1920. During the last 15 years the U.S. EPA and its predecessor contributed very important research and development, and initiated full scale testing (Borwardt, 1965-1977). Large scale commercial applications were pioneered in the early 1970's in Japan. Substantial commercial experience is expected in the U.S. within the next few years.

A commercial process using limestone was described by Atkins (1975). The mechanism of sulfur dioxide absorption by limestone was studied by Carlson (1971) and Tanaka (1974). Suitable scrubbing conditions have been found for working at low pH (Gerstle, 1974, EPA, 1976) to evade carbon dioxide absorption. At low pH, the oxidation is far faster (Borgwardt, 1976). This is a very important factor in the 'once through' or 'throwaway' process, in which $CaSO_3$ and $CaSO_4$ are produced. Incompletely oxidized sulfite is such a problem that special 'sulfur oxide throwaway sludge evaluation panels' (Princiotta, 1976) had to be assembled to evaluate options. In Japan conversion to $CaSO_4$, sold as gypsum, has been successful on full range scale (Ando, 1976). A process based on lime is operated by Chemico-Mitsui. In the U.S. the economics are adverse (Ponder, 1976). ICI operated this process successfully at the Fullham Power Station, but the process proved expensive.

Disposal of Sulfur Dioxide Sludge

Gmelin (1959) states: "Unfortunately, the disposal of wastes has not yet been solved." The main problems are that sulfite does not fully oxidize to

Table 8.7
SO_2 ABATEMENT PRODUCTS AND THEIR ECONOMIC AND MARKETING POTENTIAL[a] IN THE U.S. (PRINCIOTTA, 1975)

Product	Production of 1000 Mw plant (metric t/yr.)	U.S. consumption (metric t/yr.)	Price ($/ metric t)	Maximum Credit[b] (mills/Kwh)	Maximum U.S. Market share	Disposal method
Sulfur	65,300 -47,000 m^3 volume	10,000,000	30-50	0.22	Fair: 5% of market, possibly 10-30%	Store/least bulky non-toxic, not soluble, possibly odor problem from impurities, susceptible to slow biological degredation
Sulfuric Acid	200,000	28,200,000	15-30	0.34	Good: 10-30% of market, possibly 75%	Neutralize & dispose as gypsum (ponds, piles, landfills)
Gypsum	351,000 =210,000 m^3	18,000,000	3-5	0.16	Questionable: (Not demonstrated that wall board grade gypsum can be made. Would have to be pelletized to ¼in.)	Store or dispose (ponds, piles, landfills)
Sodium Sulfate	290,000 (produced by Wellman-Power Gas process)	1,450,000	15-30	0.72	Limited (Eastern market: already by-product. Western market =output from one 1000 Mw plant)	Store, or neutralize & dispose as gypsum (ponds, piles, landfills)
Ammonium Sulfate	270,000	2,400,000	25-35	1.14	Poor (60% of market already by-roduct, will increase to 100% by 1980)	Store, or neutralize & dispose as gypsum
Liquid SO_2	130,000	86,740	N/A	N/A	Poor (Production from one 1000 Mw plant is greater than total US market consumption)	Convert to S or gypsum & dispose
Sludge	710,000 (784,000 short tons) (50% moisture) =480,000 m^3 (d=(0.7m^3/t))	—	—	—	Possible penetration of road bed & synthetic aggregate market	Untreated disposal[c] (ponds) Treated disposal (landfill) to reduce solubility, permeability, and leachability.

a) Assumptions: 6400 hr/yr operation, 3% S, 0.4 kg coal/Kwh, 85% SO_2 removal efficiency; b) Assuming 100% sale of product at lowest market price; c) There are potential ground and surface water pollution and land use/reclamation problems with scrubber sludge.

Table 8.8
TRACE METALS IN FLUE GAS ABATEMENT SLUDGES

Element	Concentration, ppm						
	Coal		Ash West		Ash Pond Liquor	Scrubber Liquor	PHS Water stand.
	East	West	Fly	Bottom			
Lead	30	4	30	20	(0.01)	0.01	0.05
Antimony	(0.05)	0.17	2.1	(0.01)	–	0.01	–
Barium	1800	400	5000	1500	(0.05)	0.07	1
Manganese	350	15	150	150	1.6	0.075	0.05
Mercury	(0.01)	0.05	0.01	0.01	–	(0.001)	–
Beryllium	(0.01)	–	3	(2)	–	0.002	–
Boron	46	15	300	70	11	0.5	–
Nickel	–	25	70	15	0.05	0.015	–
Cadmium	–	(0.5)	(0.5)	(0.5)	–	0.01	0.01
Selenium	–	1.6	18	1	–	0.035	0.01
Zinc	180	0.6	70	25	–	0.03	5
Arsenic	–	3	15	3	–	0.01	0.05
Vanadium	180	9	150	70	–	–	–

(After Princiotta, 1976); PHS = U.S. Pub. Health Service.

sulfate, that the sludge does not easily settle, and that the liquid phase, called 'leachate,' contains heavy metal ions which are potentially poisonous.

In current experiments in Japan (EPA, 1973), the sludge settling ponds still tend to form a floating crust which prevents evaporation and removal. It is anticipated that concentration of sludge will remain a very slow process and that improvement will be slow as well. Thus, Pennsylvania Power at Shippingport built an earth-filled dam across the Little Blue River Valley, and during the coming 25 years a 5 mile long and 150 meter deep pond will slowly fill with sludge (for a photograph of the site, see Hammond, 1976), because a plant with a 1000 MW generating capacity produces each year over 500,000 m^3 of a sludge with a density of 0.7 containing 50% moisture. Even if the moisture could be removed and solid gypsum could be produced, the volume would still be 200,000 m^3. The problems of finding markets suitable for absorbing 350,000 tons of gypsum are staggering. The equivalent quantities of various chemical products which could be conceivably produced from power flue wastes are show in Table 8.7

A major problem with present liquid sludge disposal is that trace elements from the coal which accumulate in flyash can leach from settling ponds. Table 8.8 shows the composition of trace constituents in coals, in ash, fly ash, and in the leaching ponds. A comparison with current U.S. drinking water standards shows that manganese and selenium exceed legal limits, and that cadmium, arsenic, and lead are close to the limits. Thus, long leaching periods could significantly affect the drinking water standard in the area (Princiotta, 1976). The selenium in the flyash has also drawn considerable attention among chemists (Fisher, 1976; Gutenmann, 1976; Schmidt, 1975), because selenium might interfere with the sulfur dioxide abatement chemistry in the slurry.

Other Processes

Presently almost all commercial processes recover sulfur dioxide as calcium sulfite, regardless of the original absorber reagent. Several excellent processes have been developed in Japan:

Thoroughbred 101 Process. In this process (Tamaki, 1976) iron sulfite and dilute sulfuric acid are used in addition to lime or limestone to stabilize the pH and catalyze oxidation. This process can be regarded as a direct descendent of the original Clegg process as modified by Laming (1946). Instead of using a slurry, Laming mixed his absorber with moist saw dust. The use of this process is popular and successful in Japan (Ando, 1976). The effect of iron upon the oxidation rate was studied by Freiberg (1975).

The Dolomite Process, or calcined dolomite process, is chemically identical with the limestone slurry process (Ludwig, 1968; Bratchikov, 1974; Yang, 1975) except that the magnesium ion replaces calcium. Therefore, the solubility product of the sulfite is higher. The original patent (Ger.P. 639,306, 1936) describes the problems of 'slow settling' all too well! The process has been used in Russia since 1956. Chemico has developed a Japanese version (Taylor, 1976).

The Carbonate Process is chemically similar to the above described limestone process, except that the reagent is far better soluble, but more expensive. It was first patented in 1890 (Ger.P. 81,773). Johnstone (1938) measured the sulfur dioxide vapor pressure. Gaither prepared elemental sulfur (U.S.P. 2,163,554, 1939) and the process was used in France (F.P. 861,191, 1941). In the modern version a lime precipitation is often added (General Atomics, Bolts, 1974). The Ca^{2+} ion concentration can be reduced before recirculation by precipitation as carbonate, i.e. by saturating the solution with carbon dioxide (Gehri, 1976; Bolts, 1975). Peabody Coal used this process to recover sodium sulfite for sale to the paper industry.

The Double Alkali Process. The U.S. EPA has come to believe that there is no cheap reagent in sight which can absorb sulfur dioxide from the stack and release it in a separator in a simple, one step process. Thus, it is now common to use two separate process chemicals: One that absorbs sulfur dioxide from the gas phase and keeps it in solution, and a second which is used to precipitate and separate the sulfite. This procedure is now generally called a 'double alkali' process. The preferred second base is presently limestone. It will be feasible in the future to substitute other methods to recover or separate the sulfur dioxide. A large amount of excellent basic and pilot plant work was originally performed by EPA internally. Unfortunately, funding for such work has been severely limited by Congress for political reasons. Double alkali processes are vigorously applied in Japan by Kureha-Kawasaki, Showa-Oenko, Kurabo, and others.

Popular absorbers are caustic, soda, sulfite, or ammonia. In the present EPA version, Draemel (1973, 1976) worked out a well functioning system in which the sulfite is precipitated with lime or limestone (LaMantia, 1976; Kaplan, 1973-1976). FMC (1976) and General Motors (Gall, 1975) developed other variations.

Sulfite Process. The sodium sulfite process was pioneered by Johnstone (1938), patented by Commonwealth Edison in 1939 (U.S.P. 2,161,056), and modernized by Stone-Webster. The potassium sulfite process is now commonly called the 'Wellman-Lord' process (Earl, 1975; Osborne, 1975; Pedroso, 1976; Davy-Powergas, 1976; Potter, 1973; Bailey, 1975). In this process disulfite is formed (Emerson, 1974):

$$HSO_3 \quad + \quad SO_3^{2-} \quad + \quad SO_2 \quad \rightarrow \quad S_2O_5^{2-} \quad + \quad HSO_3^-$$

which can be recovered as a solid and converted to gypsum (Mortita, 1975) or regenerated. The regeneration is aided by butyric acid and similar additives. Weyerhaeuser (U.S.P. 2,351,780, 1944) used magnesium sulfite to abate pulp plant emissions.

The Citrate Process. The citrate process has been developed by the U.S. Bureau of Mines. It constitutes an excellent chemical system, because the dissociation constants of citrate, fig. 8.7, all fall into the pH range between that of SO_2 ($pK_1 = 1.92$) and carbonate ($pK_1 = 6.52$). Thus, one mole citrate ($pK_1 = 3.13$; $pK_2 = 4.7$; $pK_3 = 6.4$) can absorb three moles of sulfur dioxide. Commercial forms of the process have been developed by Pfitzer (1975) and Stauffer (1974). In this process polythionate and thiosulfate can occur as intermediates (Vasan, 1975; Korosy, 1975), fig. 8.7.

Ammonia. The scrubbing of sulfur dioxide with ammonia has been studied for over 130 years, and a process was patented by Ramsey (B.P. 1,427, 1883). Johnstone studied the vapor pressure of ammonia solutions carefully (U.S.P. 2,134,481, 1938); by that time COMINCO had already extensive experience and held several patents (U.S.P. 2,021,558, 1935; B.P. 489,745, 1938). The process has been widely used in Russia and modern versions have been used by ASARCO in combination with aromatic amines to produce very pure sulfur dioxide (B.P. 564,734, 1944). The solution can be regenerated by various methods, or ammonia sulfate can be made, see COMINCO (U.S.P. 2,862,789, 1958). As an intermediate ammonia disulfite, i.e. pyrosulfite, is formed:

$$H_2O + 2NH_3 + HSO_3 \rightarrow (NH_4)_2S_2O_5 + OH^-$$

The pH dependence and temperature dependence of the solubility and stripping rates of sulfur dioxide were carefully investigated by Boone (1973). An entire family of processes can be derived from this system if it is combined with other reagents, such as amine, or oxides, and is treated to produce different products. If ammonium sulfate is produced, a stable market must exist or must be developed, as COMINCO did in its imaginative approach in the 1930's. In the Guggenheim process and COMINCO version, coke reduction can be used to produce elemental sulfur. ASARCO used this process in Garfield, and now uses natural gas for low temperature catalytic reduction in El Paso. Coke reduction is used in Spain. In the Fulham-Simon-Carves process, sulfuric acid is added and the mixture is autoclaved to produce sulfate and elemental sulfur. Mitsubishi-Showa Denko uses a relatively weak solution and oxidizes the bisulfite with air. Ozaki (1976) describes how this process can be used to treat coke oven gas. Aerojet Co. proposed hydrazine as an additive. Shell uses CuO which yields $CuSO_4$ which can be reduced to sulfur dioxide and nitrogen.

The ammonia processes might become important for the removal of nitric oxide. Dry catalytic reaction of NO_2 at 400°C yields:

$$2NO_2 + 4MeSO_3 \rightarrow N_2 + 4MeSO_4$$

Nitric oxide must be removed by oxidation of NO_2 with ClO_2, or by

$$2NO + 5SO_2 + 8NH_3 + 8H_2O \rightarrow 5(NH_4)_2SO_4$$

Water and Seawater. The original process of Hämisch (Ger.P. 36,721, 1886) is inefficient, because normal water is unbuffered (Johnstone, 1938). In the 125 MW Battersea Plant 35 tons of water were necessary to wash one ton of coal combustion gas. Seawater, used by Virginia Smelting Co. in 1931, contains carbonate and some magnesium and calcium salts which give it some buffer quality (Bromley, 1971, Ikenaga, 1974), but large volumes of water would be necessary. Special conditions, for example, in the vicinity of seawater desalination plants, might make this process practical.

ZnO, Zinc Oxide. Johnstone, Olin, and Wellman developed a dilute carbonate process in which ZnO was used for binding sulfur dioxide. COMINCO had a patent for this process in 1934. F. E. Townsend (U.S. P. 2,100,792, 1937) added oil to prevent oxidation. The process has been reconsidered by Kosev (1974).

Basic Aluminum Sulfate. ICI used originally $AlCl_3$ (B.P. 378,464, 1932) and then aluminum sulfate (B.P. 445,711, 1936). The first commercial application was at Imatra, Finland, by Outukumpy at a smelter (1936). A full scale plant has also been in operation in Manchester since 1958. This process is now used in Japan by Dowa Mining Co.

Borate Process. R. F. Bacon developed a process in 1939 using boric acid (U.S.P. 2,180,495). Fig. 8.7 shows that borate would be eminently suitable for an effective absorption process.

Phosphate Process. I.G. Farben (Ger.P. 553,910, 1932) and ICI (F.P. 755,255, 1933) developed a phosphate process which is chemically most attractive. A mixture of Na_2HPO_3 and Na_2HPO_3 in the ratio 3:1 and a 1.8 molar solution of $(NH_4)_2HPO_4$ were used. The process has been further developed as 'Power Claus' by Stauffer (Sheehan, 1974).

Acetate Process. In this process, which has been tested in pilot plant by Kureha in Japan, sodium acetate is used as a carrier and buffer and limestone is used to precipitate sulfite-sulfate.

Manganese Dioxide. Mitsubishi and Aerojet have developed a manganese process which uses an attractive chemical system, but the main procedure is ammonium sulfate which is already abundant. A description of recent work is given by Ludwig (1968). The kinetics have been studied by Bogun (1974).

Formate. Potassium formate is used as an absorber; at 100°C a complex is formed which reduces in the presence of excess formic acid to potassium hydrosulfide, KHS, which can be stripped with steam. Several organic absorption processes have been proposed. Many of the processes used for hydrogen sulfide are suitable starting points for developing such processes.

Xylidine. The toluidine-xylidine process was first patented by CIBA in 1931 (B.P. 339,926). The process is excellent in applications where the feed concentration is inconsistent. CIBA also patented an aniline process (B.P. 339,926, 1931). Several similar processes were patented after World War II in Sweden. ASARCO and Lurgi use a dimethylaniline-xylidine solution. The cost of the amine is substantial; thus, shrinkage must be prevented.

Cyclohexanol and tetrahydro naphthalene were tested in Germany in the early 1930's. Many other similar processes have been proposed, and many are being tested. This is not the place for an exhaustive review of all other possible processes. References are listed in Gmelin (1963), by Haas (1973), and especially by Bienstock (1958). Most of these processes are presently scientific curiosities. Among them is the bold molten carbonate process of Rockwell which might work on large scale if sufficient funds would be available to bring it to a full, technical stage. A very elegant system has been demonstrated by Wiewiorowski (1965), who reduced sulfur dioxide with hydrogen sulfide in liquid sulfur. Another potentially interesting proposal is the use of Western ashes:

Alkaline Flyash. Western ashes are intrinsically alkaline, and can be used as bases to absorb sulfur dioxide. Ness (1976) reported the absorber capacity of a large number of ashes from the Western U.S. The process, sponsored by the U.S. Energy Research and Development Agency, involves a complex sequence of chemical processes which take place on solid-liquid interfaces. It is practically attractive because the ash occurs on site and must be disposed of in any case.

Dry Processes

Dry processes are inefficient because of the large masses of absorber which are necessary to establish contact with large gas volumes. Furthermore, the solids are vulnerable to absorber poisons. The most famous processes are:

Charcoal Absorption. Charcoal absorption has been used in combination with both dry and wet processes. Boliden has successfully absorbed sulfur since 1936. ICI and COMINCO designed similar processes. Lurgi prepares dilute sulfuric acid from dilute sulfur dioxide in the

'Sulfamid process.' Reinluft developed three full scale plants for VW at Wolfsburg. During desorption with air—necessary to oxidize sulfur dioxide to sulfur trioxide—shrinkage of the absorber due to formation of carbon dioxide is severe. Hitachi's process also produces dilute acid, for which there is an insufficient market. Bergbauforschung and Foster-Wheeler, Bischoff (1975) and Steiner (1975) use vertically moving peletized charcoal from which sulfur dioxide is desorbed at 150°C with steam. Westvaco reacts the absorbed sulfur dioxide *in situ* with hydrogen sulfide and produces elemental sulfur. The latter processes are all in preliminary stages. The basic chemical reactions have been discussed by Kertamus (1974) and studied by Komiyama (1975) and Novakov (1972); basic related chemistry has been explored by Wiewiorowski (1962), Klabunde (1971), and others.

Zeolite. Union Carbide and others explored the use of zeolites which can absorb substantial quantities of gas. Steijns (1976) compared the activity of the zeolites with that of charcoal in the range from 150° to 350°C.

Alkalized Alumina was thoroughly investigated by the U.S. Bureau of Mines and the Central Electricity Generating Board in England. The process produces elemental sulfur, but suffers from excessive shrinkage. Its implementation has been abandoned, because the addition of trace metals, transition metals, etc., failed to improve it significantly.

Chapter 9

Environmental Control and Legislation

This chapter deals with public efforts to abate pollution. In this long, slow, and inefficient struggle, society has rarely used its best abilities, but has instead shown great tolerance for waste. Obviously, abatement is not only a chemical and technical problem, but involves human psychology. During the past fifteen years, rapid technological transition has caused the importance of psychological questions to steadily increase, as it has caused confusion as to which process and products were still viable, and which formerly valuable products were becoming obsolete and useless, and constituted unwanted waste. Thus, it is impossible to deal with the scientific problems of environmental control legislation without first considering some important psychological factors. No other raw material has experienced greater fluctuations in economic status than coal, which used to be considered "king," but slowly became undesirable compared to oil and gas. Then, during the "energy crisis" of 1973, it suddenly again became a public favorite.

A. WASTE, DISPOSAL, AND EDUCATION

The disposal of combustion wastes is a problem as old as community living. The handling of wastes is connected with pride and status. In all societies, the dumping of wastes and clean-up work onto others constitutes a natural and basic method of establishing one's status. Thus, wastes are removed from the sight of the powerful, and handed down to those in lesser positions. The total quantity of waste produced is correlated with gross national product. When resources are scarce, waste is re-used and re-circulated; but in a prospering country, everybody's social status tends to increase to the point that nobody is left responsible for other people's garbage. If education does not succeed in creating a sense of community responsibility, rich countries will become decadent and die in their own waste and stench, as did old Rome 2000 years ago.

In modern industrial society, the economy depends heavily on the consumer, who, according to Webster, is "one who uses goods, and so diminishes or destroys their utilities." It is now recognized that energy consumption creates wastes. A popular practice of the 19th Century consisted of diluting wastes and decreasing the burden on individuals by diffusion; by using tall stacks, for example. The population, onto whom the burden was shifted, quickly sensed the violation of their status, and named the procedure pollution. Few modern scientists realize that this seemingly technical term, according to Webster, involves an "action which makes or renders unclean; defiles, desecrates, or makes profane." The recent uproar about pollution has partly resulted from the rapid transition from a largely independent rural life style to an urban consumer society striving for a high GNP and high use of energy in densely populated centers. A large fraction of the populations of leading nations has, during this transition of the last forty years, quickly advanced beyond the state in which individuals work toward increasing their own wealth, actions being based on the premise that opportunity and happiness are limited by personal resources. The living standard in cities is no longer clearly related to personal effort, and further efficiency is often not only unnecessary, but yields a sense of over-consumption and saturation, and provides a negative incentive. At this stage, progress depends on education and acquisition of new values. Unfortunately, the human mind learns only slowly, and "modern" education still consists of teaching our children to deal with the future by responding as if to the parents' childhood. For example, city children are still taught to waste energy as if its source were free firewood collected on the family farm.

The waste problem is not alleviated by those who believe that the health of the economy depends on increasing consumption by, for example, building a new house every five years rather than every ten. The way to tackle waste is just the opposite: to hold the quantitative use of materials constant, but to increase their value by improving the efficiency and style of use. This cannot be achieved by economic or other stimulation, but only by exposure to different lifestyles, i.e. by education.

Many believe that modern education has already resulted in changed attitudes. We would like to believe that scientists and engineers, who have the use of computers, would never design a wasteful process, i.e., that a process designed by scientists is, by definition, efficient. Nuclear energy affords a test for this: It is a scientific and technological product of the 1950's. Its conception was not impeded by social, economic or technological traditions and it profited both from an enlightened public and from gov(nment support, free of pressure to establish profits. Despite this, no successful effort has been made to evade waste accumulation, and we now produce nuclear energy and plan to leave the digestion of radioactive wastes to the next generation (Rochlin, 1977). Such behavior will continue as long as exposure to wasteful practices begins in early childhood, and indeed such practices seem almost to be approved of. Many children observe at school that a good lunch consists of food which is only partly eaten and leaves wastes which can be flung at one's cohorts. This encourages them later, as adult scientists and engineers, to consider wastes a social weapon. This attitude has nothing to do with science and cannot be overcome by scientific education. It can only be changed by a different kind of education, which involves more time than the technological period with which this book deals.

1. Waste and Industry

According to a reliable weekly, the president of Dow Chemical Company has stated: "I cringe every time I hear a company say how much it's costing to clean up pollution. The opposite is true. We expect to make a profit at it." According to historic experience, the chances for success are excellent. During times of technological change such as the present, waste has always occurred inevitably, whenever new products which are cheaper or more desirable have suddenly replaced an established product. Furthermore, this waste has always caused economic instability and social friction, but once the social friction has abated, scientists and engineers have reliably been able to find efficient means for coping with waste, transferring it into useful resources. One of many examples from the field of pollution is the emission of HCl, which triggered the original Alkali Act of 1865, and caused the beginning of a new industry. The current problem with pollution is difficult, because of the economic and time scales involved in designing and testing new energy production processes. This discourages experimentation, and has caused several leading energy companies to abstain from bidding for government research on gasification of coal until, as one company spokesman stated it, the "unstable social and economic situation has been solved."

2. Sulfur Pollution

The Frasch sulfur industry has helped clean up the swamps of Texas and Louisiana, and convert them into clean land. The oil and gas industries have reduced sulfur emissions for a hundred years, and remove wastes before they sell their products. The smelter industry has learned how to reduce sulfur emission, and is in the process of abating SO_2 emission. In contrast, the coal burning industry, burdened by problems discussed in Chapter 1, is still at the beginning of industrial abatement efforts. The variation in the technical abatement skills of these industries is directly correlated to the value which they can gain from sulfur: for Frasch producers, sulfur constitutes the sole product; for gas and oil producers, sulfur constitutes a potentially valuable by-product; for smelters, sulfur now constitutes a tangible value, even if it is negligible to the value of the refined metal; to the coal industry, sulfur constitutes nothing but a nuisance. In view of the wide availability of cleaner and cheaper SO_2 from other sources, the electric industry cannot reasonably hope that SO_2 can easily be recovered for a profit unless drastic and fundamental progress in abatement chemistry is achieved. However, a comparison of common practice in the gas, oil, and coal industries shows some differences in traditional attitudes. The relatively young natural gas industry never questioned whether sulfur should be removed from the raw material. It was simply not feasible to ship gas containing sulfur compounds, because the market value of the material would have suffered. This was the case with oil, until 1888 when Herman Frasch succeeded in desulfurizing Ohio crudes, thereby increasing the value of oil from $.14 to $1.00. This, in turn, increased the value of shares of stock in Standard Oil Company from 168 to 820, and its dividends from 7% to 40% (Whitaker, 1912). Today, the oil industry is a chemical one and uses oil for better values than its latent heat of combustion; the recovered sulfur is transformed into profitable specialty chemicals which can be used to prepare high quality, high performance polymers, as described in Chapter 13. The heavy residual oils, from which sulfur is not easily removed, are used as asphalts and tars which profit from the sulfur content, as explained in Chapter 14. In contrast, the coal industry has not yet become enthusiastic about sulfur, because, despite more than a century of intensive research, no practical use for coal wastes has been found. Efficient sulfur recovery was invented by Claus in 1882, at a time when interest in coal research had already been overshadowed by that in oil and gas, and the coal industry was forced into hybernation until very recently. Accordingly, the chemical education of mining and civil engineers stops with freshman chemistry, and coal and electricity companies do not employ chemists. Thus, what chemists might have recognized as challenging research is not always recognized as such by the practitioners in the trade. A still more serious intrinsic handicap to the coal using industry is that coal is a solid, and yields solid wastes. The chemistry of solids yielding solids is a great deal more complicated than liquid chemistry, and involves complex, heterogeneous surface reactions which necessitate large and very expensive reactors. Therefore, a conventional coal combustion plant costs in the order of $200 million, and takes several years of planning, building and amortization. This scale exceeds the courage of even the most enthusiastic entrepreneur; and if it does not, the long time-lapse before the break-even point is likely to break his financial strength. A further problem is that future optimum working conditions cannot be predicted and provided for by a private entrepreneur. This is particularly true when the development of a new plant prototype takes at least ten times as long as the political term of the federal administrator who regulates the industry. It took the shipping industry 2000 years to develop modern ships, and the aircraft industry could not have developed within 50 years to its present state as a private enterprise. All communist countries, and most Western countries, including England, have nationalized their power industries, because they felt that the coal burning industry could not renew itself, especially when in competition with the nuclear industry, which is based on government research. Accumulated direct expenditures for nuclear research in the U.S. alone come to many billion dollars. In addition, nuclear research is bolstered by the work of many Nobel laureate scientists and the good will of a society which has judged it to be essential to military survival. Coal combustion research, on the other hand, is not considered glamorous, and suffers from the vested interests of industries which have to protect their private investments from threatening premature obsolescence. Thus, Federal leadership in research and development on the order of that for nuclear research is necessary. The Clean Air Act of 1970 provides for this, but still doesn't approach the order necessary for success. Another factor delaying progress is connected with society's changing attitudes. Calling to mind the fabled time of oil barons, and the fact that nuclear inventors have been rewarded with Nobel prizes, peer recognition, influence in government affairs, and large government research grants rather than with financial independence, people press for changes in patent laws which would prevent private inventors from receiving recognition for their inventions. Progress is still further hampered by the presently prevailing group approach, in which each member of a widely divergent group of regulators has a veto right, and each must measure novel ideas against standards of his own professional tradition. This alone, by definition, eliminates innovation. Unfortunately, it is only bold innovation which can help today's intrinsically inefficient and archaic coal combustion methods to bridge the gap to modern technology. Such innovation depends upon good will toward creative individuals and inventors. Such progress cannot be achieved in times of doubt, distrust or resentment. However, human inventiveness cannot be totally surpressed, only delayed. For the time being, the incentive for progress stems from those who suffer from pollution. This explains why the present scientific and engineering initiative is primarily taken by the Environmental Protection Agency, and only slowly shifts back to the Federal Energy Administration. It also explains why the present work proceeds via legal and political channels.

B. AIR POLLUTION LEGISLATION

In 1306 Edward I issued an apparently ineffective decree "compelling all but smiths to eschew the obnoxious material (coal) and return to the fuel they used of old." The struggle has continued in England for almost 700 years, and there has always been disagreement as to whether pollution constituted a nuisance or a hazard and a danger. Ramazzini, in 1713, describes an incident in Modena which—by the way—was not an underdeveloped or remote area, but was in the center of wealth and commerce, and at the cradle of Renaissance culture:

A few years ago a violent dispute arose between a citizen of Finale, a town in the dominion of Modena, and a certain business man, a Modenese, who owned a huge laboratory at Finale where he manufactured sublimate. The citizen of Finale brought a lawsuit against this manufacturer and demanded that he should move his workshop outside the town or to some other place, on the ground that he poisoned the whole neighborhood whenever his workmen roasted vitriol in the furnace to make sublimate. To prove the truth of his accusation he produced the sworn testimony of the doctor of Finale and also the parish register of deaths, from which it appeared that many more persons died annually in that quarter and in the immediate neighborhood of the laboratory than in other localities. Moreover, the doctor gave evidence that the residents of that neighborhood usually died of wasting disease and diseases of the chest; this he ascribed to the fumes given off by the vitriol, which so tainted the air near by that it was rendered unhealthy and dangerous for the lungs. Dr. Bernardino Corradi, the commissioner of ordinance in the Duchy of Este, defended the manufacturer, while Dr. Casina Stabe, then the town-physician, spoke for the plaintiff. Various cleverly worded documents were published by both sides, and this dispute which was literally "about the shadow of smoke," as the saying is, was hotly argued. In the end the jury sustained the manufacturer, and vitriol was found not guilty. Whether in this case the legal expert gave a correct verdict, I leave to the decision of those who are experts in natural science.

Scientists still battle over these questions. In 1925 the Royal Commission on the Coal Industry estimated that in the city of Manchester, coal flue gases increased the yearly household laundry bill of citizens by over .5 million pounds sterling during that year. It was estimated that 3 million tons of coal, at a value of $15,000,000, were wasted as soot; in 1927 it was claimed that some communities lost 80% of their sunlight, and the "loss caused by pulmonary and cardiac diseases which increase in direct proportion to an increase in the intensity of smoke fogs is beyond measurement." The situation was equally bad in some areas of Belgium and Germany, but in most locations damage occurred only during unfavorable weather conditions. In most parts of the world a smoke-belching stack was considered a normal part of daily life and a comforting sign of a healthy economy. This was reflected in the official seal of the U.S. Small Business Administration, Fig. 1.7, until 1968 when the seal was altered by removing the plume. Today, epidemiologists argue as indicated in section 10.G.2 that sulfur dioxide lowers the threshhold of resistance to poor health (Battigelli, 1976). Industry argues that standards are vague, contradictory and unworkable (Mallat, 1971; Megonel , 1975; Schimmel, 1975). The political leaders who try to establish practical standards have not only to cope with the social impact of control legislation (Johnson, 1976), the divergent goals of our pluralistic society, and with finding a democratic solution avoiding inequities caused by regional, economic and scale factors, but have to tackle divergent scientific views about equitable strategies (Robinson, 1970; SCEP Report, 1970; Larsen, 1970; EPA, 1970-76; Ross, 1972; Lemmon, 1974; Train, 1975; North, 1975; Stern, 1976) and with ill-defined ambient monitoring practices (Forrest, 1973). The problems of finding satisfactory scientific methods for identifying and measuring pollutant concentration, described in Chapter 4, "can be charitably described as basely adequate," according to the outgoing EPA administrator. And yet, substantial progress has been made ever since the British Alkali Act restricted the emission of HCl, which occurs during the Solvay Soda process in the molar ratio of 1:1 with the main product. Sulfur dioxide never constitutes more than a few percent of the reaction processes, and it is merely a byproduct of processes in which immense quantities of oxygen are extracted from air, and enormous quantities of invisible carbon dioxide are released (Fig. 8.1). Nonetheless, its control became necessary when smelter operations increased their batch size.

1. The Case of COMINCO

One of the earliest and best documented modern cases is that of the Consolidated Mining Company. Their copper smelter at Trail, British Columbia, Canada was put into operation in 1894. In 1901 lead smelting was commenced, and in 1910 a zinc facility was added. A lead-chamber sulfuric acid plant was built in 1916. In 1917 it reached a capacity of 30 tons per day. In 1931 three contact acid plants produced 110 tons of acid per day. It was quite easy to produce the acid from the roasting of gas containing 6% SO_2, but the closest market was 1500 miles away. COMINCO therefore decided at that time to build an ammonium plant, an ammonium sulfate plant, and an ammonium phosphate plant. The latter was the first full-scale plant in North America; in addition, a phosphate plant was developed in Montana. By entering into these enterprises COMINCO not only started in a business which was new to them, but pioneered two fields. First, it started a business for which a market had yet to be developed: prior to that time the farms in the Canadian prairies did not use fertilizer. Second, it was forced to become a leader in building SO_2 recovery plants. In 1926, to reduce the residual flue gas SO_2 concentration, COMINCO built two 400-foot stacks, as was standard at the time. However, Trail is located in a narrow valley, 11 miles north of the U.S. border. The atmospheric conditions in this micro-climate caused the frequent formation of an inversion layer at night, trapping all the stack gas in the valley. In the morning, the sun warmed the west side of the valley and caused the entire air volume to form a slow vortex along the valley's axis. This frequently inflicted extensive damage on vegetation, and led farmers in the U.S. Columbia River Valley to sue. Since COMINCO could not legally purchase land in the state of Washington, the case eventually (1929) reached the International Joint Commission set up under the Boundary Waters Treaty. In 1931 the Commission made a unanimous report, assessing accumulated damage at $350,000 up to that time. The U.S. Government, however, rejected this report, and the dispute was given to the International Tribunal for mediation. The Tribunal proceeded with due caution, as precipitous action might have led Canada to retaliate against U.S. companies along the eastern border in Detroit, Buffalo and Niagara Falls. In 1937, after an exhaustive study, the Tribunal proposed a temporary control regime, calling for continuous monitoring of sulfur dioxide at the international border, with limits set at 0.3 ppm for 40 minutes in summer, and 0.5 ppm for 60 minutes in winter. In 1940 the Tribunal made the control regime permanent as its final decision in the case. In the meantime, COMINCO had stepped up production to 600 tons per day in 1938; and in 1975 it reached 2,200 tons per day. One condition of the tribunal stipulated that emission had to be reduced by two tons per hour during rains. Today, COMINCO can no longer fulfill this condition, as its entire emission is less than one ton per hour, and the SO_2 concentration at the U.S. border is less than 10 parts per billion, the current monitor sensitivity. Despite greatly increased activity, COMINCO has not been assessed a fine since 1937.

This example (Robinson, 1972) illustrates that a forward-looking company can successfully solve its emission problems. However, COMINCO's fate was aided by simultaneous agricultural expansion in the prairies, by the advent of high-yield farming techniques, and by political conditions which encouraged consumption and export of farm products. The company claims that, despite these advantages, it earned only a 7% return on its capital over the period 1927 to 1975, and that were it starting over again under the present market conditions, the enterprise would be doomed to failure.

2. U.S. Clean Air Act

Since pollution is connected with industrial production, and, thus, with gross national product, the U.S. has accumulated substantial experience in the abatement field. Industrial leaders from all over the world observe trends in U.S. legislation, even though it is still very much in flux, and the final form of the law is still difficult to predict. Knowledge of the numerical listing of air quality standards is meaningless if it is not clear whether the law constitutes a presently enforceable regulation, a statement of a realistic future goal, or merely a programmatic expression of a theoretically desirable quality of life. Those who are engaged in corporation planning, and in transfer of the U.S. experience to other countries, must be aware of the basic cultural context in which the law is applied; particularly the legal climate of American common law. They must recognize the present trends as expressed in court findings, the mood of Congress as expressed in the size and conditions of funding appropriations, and the action of the President as evidenced by withholding or supplementation of budgets previously approved by Congress.

The Clean Air Amendment of 1970, the presently valid law, is based on two centuries of experience. Common law doctrines are plainly visible in the law in such concepts as "best available technology," land use controls, and control at the source. It employs administrative remedies based on common law court techniques, such as control efforts extending over time and monitoring and preparation of reports on emission data. The Amendment of 1970 integrates experience from earlier air pollution control laws, and from water pollution control. The latter is reflected in the "no significant deterioration" policy, which is at present hotly debated in Congress, in legal circles, and in public.

An excellent summary of the Clean Air Act of 1970 and its implications has been prepared by Rodgers (1977). The first step toward the present amendment was the Air Pollution Control Act of 1955, which confines the Federal role to research, and delegates control to the states. This delegation has been significantly diminished with the most recent three revisions. Specific and local problems were to be investigated by the Surgeon General, but only upon request of state or local government. The Clean Air Act of 1963 expanded the research and development scope, and established that, in cases involving interstate pollution, Federal intervention was no longer dependent upon requests from states. In these cases, a conference would be called to deal with the problem.

In the following 8 years, nine conferences were called. Of these, only one case survived, and it was settled out of court. The 1963 Act also established air quality criteria, based on scientific knowledge of pollution's effects. Its declared goal was "to achieve the prevention and control of air pollution," but legally trained readers noted that neither the scope nor the extent of the problem were spelled out anywhere in the text. The Air Quality Act of 1967 established atmospheric areas and air quality control regions. The contiguous U.S. was divided into 8 atmospheric areas; Alaska and Hawaii each formed an additional area. Since then, some 247 air quality control regions have been defined within these areas, their boundaries being determined by climate, meteorology, and topography. This Act also established air quality criteria, advocated the development of control techniques and reports, and urged states to adopt ambient air quality standards within the control regions. The word "ambient" is of significance. For scientists this term is well defined. It refers to open, freely moving air which can be sampled and quantitatively tested. For lawyers, it refers to the atmosphere above a community in general, without reference to origin, boundary, types or number of emission sources. Thus, "ambient air" does not constitute a meaningful base for legal expression or enforcement. For Rodgers (1977), such standards "rank with an official statement of a desire to reduce annual inflation or the unemployment rate below a given value." They are valueless as legal mechanisms.

The Act of 1967 also called for states to develop plans to implement ambient air quality standards. This provision developed into an implementation and enforcement plan. The Clean Air Act of 1970 sanctioned the Federal role in implementation and enforcement, and provided a comprehensive programmatic and regulatory system by providing technical and financial assistance for research programs, and by developing standards and implementation and enforcement channels. The Act contains three titles: Title I deals with stationary sources of air pollution; Title II deals with mobile sources of air pollution; and Title III deals with administrative and judicial reviews. Only some of the most important sections are listed here. Section 104 emphasizes research and development of new and improved methods for the prevention and control of air pollution resulting from fuel combustion, which would have industry-wide application. This includes laboratory and pilot plant testing, up to and including construction and operation of full-scale demonstration plants.

This provision has made it possible for the EPA and other agencies to establish scientific leadership and a high basic level of competence. Many reports quoted here demonstrate the impact of this provision. Section 105 establishes funding programs for air pollution control agencies on the state, interstate, city, county and local levels, with the goal being to produce information. This information constitutes a treasure trove for scientists, but it is not readily available to the public, because the reports must be individually purchased from the U.S. Technical Information Center, a process which requires knowledge of order number, author's name and detailed titles.

Amendment Section 107 deals with air quality regions, Section 108 with air quality criteria and control techniques, Section 109 with ambient air quality standards, and Section 110 with implementation plans. Section 111 deals with new sources performance standards, and Section 112 with hazardous air pollution standards. These two sections provide that state standards may not be lower than Federal standards. Section 115 preserves the old conference procedure, but only one conference was continuing in session during 1970. As to enforcement, Section 304 provides for citizens to sue violators, and Section 309 requires "Environmental Impact Statements," a procedure which is supplemental to the National Environmental Policy Act (NEPA), which became law on January 1, 1970. This act has been called by some "one of the most significant pieces of domestic legislation to be enacted into law this century." By others it has been charged that it forces the EPA to tackle too many problems with too little leverage to make its views felt. This act requires for the first time that all Federal agencies consider the environment, along with traditional and economic factors, before embarking on major projects. It also creates a "Council on Environmental Quality," (CEQ). The Environmental Impact Statement is required according to Section 102(2)(C), and is discussed below. This section may contain the seed for other, long-range developments for which the climate is not yet favorable. Among these are procurement disqualification for offenders and mandatory patent licensing provisions. The latter have not yet been implemented by any government agency. The present patent policies of the various branches of the Federal government are inconsistent. In the case of Federally sponsored research conducted in private laboratories, they range from full patent ownership by the government to full ownership by the grantees or contractees.

Section 312 of the Clean Air Act of 1970 requires annual reports by EPA estimating the cost of carrying out the provisions of the act. These reports are most valuable and are quoted elsewhere in this book. The 1970 Clean Air Act establishes a detailed process for setting and enforcing ambient air standards. It involves (a) the designation of air quality control regions, as discussed above; (b) the issuing of air quality criteria and information on air pollution control techniques; (c) establishing national ambient air quality standards for major pollutants, (d) preparation of an implementation plan by states; (e) administrative review and revision of such plans, and, finally (f) development of enforcement plans.

This Act also provides for the Commission to issue and regularly review lists which include potentially toxic materials that might occur in the air in critical concentrations. This list presently includes 22 air pollutants. All are high volume organic chemicals; sulfur dioxide is not among them. The Act takes several crucial steps toward providing ameliorating features. For example, EPA is required to accept State plans, if they are properly prepared. This has on at least one occasion led to what a federal judge called 'a paradigm of confession and evasion.' The Act also allows for exemptions, such as temporary suspensions with the possiblity of extensions, authorized as part of the Energy Supply and Environment Coordination Act of 1974, which was created to lessen U.S. dependence on foreign oil imports.

The enforcement provisions authorize EPA to ask industry to establish and maintain records, make reports, install monitoring equipment, and certify monitoring methods. As a result of the latter, EPA releases quarterly reports listing air pollution detection instruments which fulfill EPA equivalency standards if used according to the instructions provided (Chapter 4). This service was at least partly a result of complaints about the commonly used and well accepted opacity standards which are intrinsically insufficient to serve as absolute standards (Weir, 1976), because opacity readings depend as much on geography, time of day, weather, and other outside factors as on the plume effluents.

In addition to the above, the EPA commissioner can solicit other information as 'he may reasonably require,' and EPA can sue in U.S. court 'any person contributing to an air pollution problem presenting an imminent and substantial endangerment to the health of persons,' provided that 'appropriate state and local governments have not acted to abate sources.' Only one such case has occurred to date: A lawsuit filed in

Birmingham, Alabama, in November 1971 which was settled out of court. Possible further actions include administrative orders and supervision of federal facilities. Federal facilities 'have been notoriously laggard in abating pollution,' according to the Senate Committee Report on Public Works of 1970. Furthermore, the President can exempt federal facilities and the U. S. Supreme Court recently decided that they need not secure local or state permits, or comply with State laws. TVA and other federal organizations have regularly and successfully pleaded for lenient regulations because of their tight budgets. The future will determine if the Clean Air Act of 1970 will take and retain an acceptable form, whether, as its proponents fear, it will turn into a hollow policy statement, or whether, as its opponents claim, it will grow into a tool transforming the U.S. into a non-growth society with rigid land use laws and government interference. Part of this question will depend on developments involving the function of the environmental impact statement.

The Environmental Impact Statement

The Environmental Impact Statement (EIS) was established as Section 102(2)(C) of the National Environmental Policy Act (NEPA) which was unanimously supported by both the legislative and executive branches and became law on January 1, 1970. As one of its goals, the Act states that

> Congress, recognizing the profound impact of man's activity on the interrelations of all components of the natural environment–declares that it is the continuing policy of the Federal Government...to use all practicable means and measures...to create and maintain conditions under which man and nature can exist in productive harmony, and fulfill the social, economic and other requirements of present and future generations of Americans.

When signing the Act into Law, Former President Nixon stated: "The 70's absolutely must be the years when America pays its debt to the past by reclaiming the purity of its air... It is literally now or never." To give the policy credibility, a Council on Environmental Quality was formed, and the law requires that the environment, along with traditional economic and technical factors, be considered before major projects are funded or licensed and before permits and leases are signed. Thus, every recommendation or proposal for federal action 'significantly affecting the quality of the human environment' must carry a statement by responsible officials detailing (a) the environmental impact of the proposed action,

(b) any unavoidable adverse effects, (c) alternatives, (d) the relationship between local short-term advantage and the enhancement of long-term productivity, and (e) irreversible and irretrievable commitments of resources.

The law also calls for compulsory consultation with all federal agencies which have legal expertise in air pollution. This, as Rodgers (1977) states: "Makes EPA a national busybody and gossip on environment." However, the law in no way confers upon any federal agency or anyone else the power to veto any project. Thus, it is possible for projects to be declared environmentally damaging and yet gain approval. This was the case with both the Trans Alaska Pipeline and the Strip Mining bills.

Environmental Impact Statements must be signed by federal officials, but are based on information supplied by private sources. This has resulted in the sprouting of a minor industry of EIS consultant firms which can present project proposals from an environmental viewpoint. If Impact statements are found to be insufficient, court action may be brought; during the first three years of the law, some 200 suits were filed and court action delayed some projects. For example, U.S. AEC construction of nuclear power plants was postponed until an adequate review could be presented (Calvert Cliffs Coordinating Committee vs. AEC), and the Department of Interior was censored by the court for neglecting to consider in depth natural gas pricing policy and nuclear energy before issuing off-shore oil leases in the Gulf of Mexico (Natural Resources Defense Council vs. Morton). The law was successfully invoked in peripheral matters, for example, to stop construction of a federal prison in New York in order to prevent pollution of the human environment by drug addicts (Hanly vs. Kleindienst).

The main function of the Council on Environmental Quality is to help the system work, not to participate in judgment. It issues guidelines for the preparation of the Environmental Impact Statement and receives copies of all statements; a list of these is published in the federal register.

3. Air Quality Standards

Table 9.1 lists the present National Primary and Secondary Ambient Air Quality Standards. It was explained earlier that the word 'ambient' is interpreted differently by lawyers and scientists. Measurement of the ambient air concentration is difficult; thus, an entire family of monitoring instruments has had to be developed. These instruments are as yet not

Table 9.1

U.S. NATIONAL AMBIENT AIR QUALITY STANDARDS
(CLEAN AIR ACT SECTION 109, PROMULGATED APRIL 30, 1971)
(36 FR 8186; 40 CFR 50)

Pollutant	Primary Standards	Secondary Standards
Sulfur Oxides (Sulfur Dioxide)	(a) 80 micro g/m^3 (0.03 ppm) annual arithmetic mean (b) 365 micro g/m^3 (0.14 ppm) maximum 24-hour concentration not to be exceeded more than once a year	(a) 1,300 micro g/m^3 (0.5 ppm) maximum 3-hour concentration not to be exceeded more than once a year
Particulate Matter	(a) 75 micro g/m^3 annual geometric mean (b) 260 micro g/m^3 maximum concentration not to be exceeded more than once per year	(a) 60 micro g/m^3 (b) 150 micro g/m^3 maximum concentration not to be exceeded more than once per year
Carbon Monoxide	(a) 10 mg/m^3 (9 ppm) maximum 8-hour concentration not to be exceeded more than once per year (b) 40 mg/m^3 (35 ppm) maximum 1-hour concentration not to be exceeded more than once per year	
Photochemical Oxidants	160 micro g/m^3 (0.08 ppm) maximum l-hour concentration not to be exceeded more than once per year	
Nonmethane Hydrocarbons	160 micro g/m^3 (0.24 ppm) maximum 3-hour concentration (6 to 9 a.m.) not to be exceeded more than once per year	
Nitrogen Dioxide	100 micro g/m^3 (0.05 ppm) annual arithmetic mean	

Note: Congress is in process of preparing revision.

perfected and will in the future have to be continually checked for 'equivalency' in order to be effective. The choice of ambient standards rather than source emission standards has, for some time, encouraged polluters to employ dispersion tools, such as tall stacks, to achieve compliance without reduction in emission. Source standards prevent such evasion. A combination of source and ambient control had already been formulated in the international court decision of 1937 affecting COMINCO which provided that under certain specified weather conditions COMINCO should reduce sulfur emission by two tons per hour. This regulation is still in

Table 9.2
U.S. NEW SOURCE STANDARDS

Pollutant	Standard
Fossil Fuel Steam Generators	
Particulate Matter	0.10 lb. per million Btu maximum two hour average 20 percent opacity (40 percent opacity permissible for two min. in any hour
Sulfur Dioxide	0.80 lb. per million Btu (1.4 g/kcal) maximum two hour average when oil is burned 1.2 lbs. per million Btu (2.2 g/kcal) maximum two hour average when coal is burned
Nitrogen Oxides	0.20 lb. per million Btu maximum two hour average, expressed as NO_2, when gaseous fossil fuel is burned 0.30 lb. per million Btu maximum two hour average, expressed as NO_2, when liquid fossil fuel is burned 0.70 lb. per million Btu maximum two hour average, expressed as NO_2, when solid fossil fuel (except lignite) is burned
Sulfuric Acid Plants	
Sulfur Oxide	2 kg/metric ton produced 15 lbs./ton of acid
Acid Mist	10% opacity visible emission
Petroleum Refining	
Sulfur Oxide	0.10 gr H_2S/stand. cubic ft.
Kraft Pulp Mills	
Sulfur Recovery	230 mg/m^3

Source: EPA Regulations 40 CFR 60.42 to 40 CFR 60.44.

effect, even though COMINCO's total emission is now below this reduction requirement (Robinson, 1972).

Table 9.2 shows the new source standards for fossil fuel-fired steam generators. These are based on the present chemical and engineering art and serve mainly as guidelines for chemical engineers designing new plants and processes. Since the development of a new type of combustion system takes over ten years and can cost more than a billion dollars, EPA is in a difficult situation. If requirements for new source standards are too modest, abatement will be delayed for several decades; if the anticipated technology does not materialize in time for full-scale application, industry and, eventually, the public will face great financial loss. In the case of

Table 9.3
SULFUR DIOXIDE EMISSION STRATEGIES, BY U.S. STATES

State	Strategy	State	Strategy
Alabama	3	Nebraska	3
Arizona	4	Nevada	1,2,3,or 5
Arkansas	8	New Hampshire	2 or 6
California	2 or 7	New Jersey	2 or 4
Colorado	9	New Mexico	3
Connecticut	1 or 3	New York	2 or 4
District of Columbia	2	North Carolina	3
Delaware	2 or 8	North Dakota	3
Florida	4	Ohio	10
Georgia	2	Oklahoma	8 or 4
Hawaii	1	Oregon	7 or 2
Idaho	2	Pennsylvania	2 or 3
Illinois	4	Puerto Rico	1
Indiana	10	Rhode Island	5 or 3
Iowa	4	South Carolina	3
Kansas	9 or 5	South Dakota	3
Kentucky	4	Tennessee	3 or 4
Louisiana	7	Texas	8
Maine	1	Utah	2
Maryland	2	Vermont	1
Massachusetts	6	Virginia	3 or 7
Michigan	2	Washington	3 or 7
Minnesota	2	West Virginia	3
Mississippi	3	Wisconsin	8 or 4
Missouri	1 or 3	Wyoming	8
Montana	2		

Strategy	Description
1	% S for all fuels combined
2	% S for an individual fuel type
3	lb SO_2/million Btu for all fuels combined
4	lb SO_2/million Btu for an individual fuel type
5	lb S/million Btu for all fuels combined
6	lb S/million Btu for an individual fuel type
7	ppm SO_2 emmission regulation
8	ppm SO_2 ambient air quality standard
9	no State implementation plans
10	State implementation plan not enforceable (because of court action or other reason)

automobile emission, these difficulties have resulted in the well-known awkward situation of the present time. In the case of stationary sources, the new source standards seem realistic, tenable, and adequate.

Eventually, these new source standards will probably replace all older standards. However, air pollution is a regional problem and the standards will have to remain flexible so that emission from closely bunched sources can be kept at a tolerable level and unreasonable restrictions avoided in the case of isolated plants in regions environmentally not worthy of protection. The lively struggle between different interests will no doubt continue and discussion of the social (Johnson, 1976), medical, and economic impact and the value of clean areas will be a part of the transition period of the next twenty years. It is likely that a set of new political and social values and rules for society will be formulated and established during this time.

Table 9.3 lists the state air quality standards which are presently in effect. Many states have not yet acted and some state laws are already being challenged in court. The table also shows the different abatement strategies. Much of the pioneering work has been done by California which has one of the oldest state laws. It established a state operated public health laboratory with a special air pollution laboratory which for over thirty years has been a world leader in dealing with emission from mobile sources. The California Air Resources Board, established by Resolution Chapter 170 as recorded in the California Statutes and Amendments to the Code of 1968, has been among the leaders in initiating novel legislation. Most other U. S. states are not equipped with the scientific expertise for establishing or enforcing their own standards. Thus, most state standards are formulated by comparing the action of federal agencies and that of other states against local political and economical realities.

Chapter 10

Medical Use and Health Effects

From time immemorial until the middle of the Nineteenth Century sulfur was one of the basic staples of the household medicine cabinet. According to an old adage, all diseases can be assigned to one of three categories: "Those which can be healed with sulfur, those which respond to mercury, and the rest, which the devil cannot cure." In 1612, in his dictionary of alchemy, Martin Rulando listed 16 different types of sulfur from which dozens of different formulations could be made. Many of these are still used in dermatology and veterinary medicine.

A. ELEMENTAL SULFUR

1. Toxicity

Pure elemental sulfur is odorless, tasteless and not toxic. In order to become active it must change oxidation state. Sulfur reacts with proteins and alkalies to form sulfide, especially hydrogen sulfide, and possibly polythionates. In an aqueous environment, it can be oxidized to sulfate, reduced to sulfide, or both, and also can form thiosulfate or other compounds, as discussed in Chapter 3. If doses of over 10 grams of finely ground sulfur are ingested, headache, vomiting and intestinal pain can be experienced. The toxic effects are caused by hydrogen sulfide which is formed in the intestine; it can be noticed on the breath. Toxic effects can also be caused by impurities in the sulfur, such as arsenic, selenium, or heavy metal ions.

Dinegative sulfur, in the form of sulfide, is the most active sulfur form. The oxidation state of +4, as found in sulfite or sulfurous acid, is capable of participating both in reduction and in oxidation reactions, and sulfur in the oxidation state +6, as found in sulfate, is normally not active.

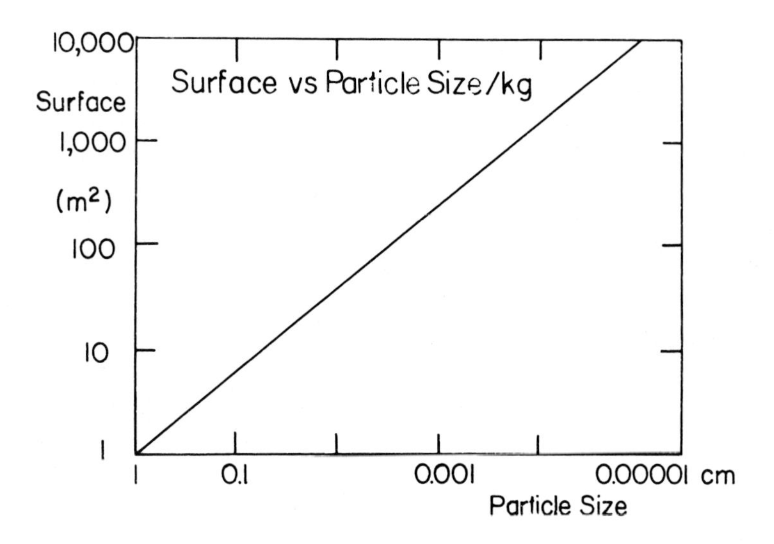

Fig. 10.1. Surface area of 1 kg sulfur as a function of particle size, assuming uniform, cubic single crystals.

The characteristic effects of elemental sulfur stem from the gradual and protracted manner in which, in the biological environment, it chemically converts into active intermediates. Since sulfur is insoluble in water, and slightly soluble in oils, the chemical reactions take place on the solid-liquid interface. Under such conditions, since the reaction products are readily soluble, the rate of reaction is proportionate to the surface area of sulfur particles. Figure 10.1 shows the surface area, and hence the trend in relative reactivities of one mole sulfur, as a function of particle size, for an hypothetical, idealized mixture in which all sulfur forms cubic single crystals of identical size. In a real system, the surface area is much larger, owing to cracks, faults and deformations. As Fig. 10.1 shows, small particles react much more effectively than large crystals. Curiously, the value of colloidal sulfur in medicine was recognized very late, in fact much later (Miller, 1935) than its value in plant pathology. In a mixture of particle sizes, the small particles react first. In the beginning, the large particles react slowly, but as they are consumed, their size decreases, and their reaction rate increases. Thus, by choosing the range of particle sizes, one can determine the rate and length of the release of active products.

Pharmaceutical sulfur is now sold in three common forms (Stecher, 1960): Precipitated sulfur (milk of sulfur), obtained by boiling sulfur with lime and precipitation with hydrochloric acid; sublimed sulfur (flowers of sulfur); and washed sulfur, made by treating sublimed sulfur with ammonia to dissolve arsenic and other impurities. All forms are 99.5% pure or better. For cosmetic and dermatological use, various other forms are prepared by proprietary methods; among these are, for example, sulfoderm, a powder containing 1% colloidal sulfur (Triebmann, 1931; Kloeppel, 1931; Zippert, 1929); and sulfidal, a solution containing 1% soluble colloidal sulfur (Heyden, 1937).

2. External Use

The external application of sulfur is still very much the same as it was several hundred years ago (Bulkley, 1880; Bouchardat, 1851). Some formulations are listed in the above references. On normal, healthy skin with pH 5.5, sulfur remains inert. In broken or damaged skin, as in the case of acne (Borelli, 1955), sulfur reacts with the serum and forms sulfides. In small concentrations, sulfides are believed to build up skin, as keratoplastics. A 5% ointment, for example, is applied to hasten keratinization of ulcers (Sollman, 1957), and 2-4% solutions are used to treat alopecia. Sulfur does not directly affect the growth of hair, but purportedly initiates early growth (Chase, 1954) by irritation of the scalp. The advantage of sulfur over other irritants, such as resorcin, salicylic acid, etc., is its low toxicity and the slow, continued release of the active reagent. Today, almost every large cosmetic firm markets some shampoo which contains sulfur, usually in 1-5% concentration. The shampoo formulation in the Appendix shows that the surfactant plays an important role. Typical medicated hair lotions have been described in numerous patents; for example, Maughan, Ger. P. 2,145,204 (1973), Armstrong-Prior, Austr. P. 440,609 (1973), and Delors, Fr. P. 2,112,085 (1971). In concentrations above 10%, sulfides soften the skin. This keratolytic action is used in cosmetics (Sarria, 1945) such as depilators (hair removal agents). Sulfur can be resorbed percutaneously (Rieth, 1966), where it can probably react with glutagen or cystein-cystine to form sulfide.

Almost every lotion and cream for treating acne contains elemental sulfur (Borelli, 1955; Knotts, 1970; Bernard, Fr. P. 2,128,187, 1972; Ljunggren, Ger. P. 1,947,179). In seborrhoea lotions (Gloor, 1973), its effect seems to be similar to that of bathing in hydrogen sulfide-containing thermal springs. One of the most effective dandruff shampoos for the treatment of seborrheic dermatitis is based on "selenium-sulfide" (Selsun,

U.S. P. 2,694,669, 1955), which, because of its mode of preparation contains polythionates. Sulfur is effective against many types of primary and secondary mycoses, i.e. fungus infections such as are often found on hands and feet (Rieth, 1966); it is mainly used in cases which do not respond to other pharmaceuticals. Sulfur is also still popular as an effective agent against parasites, such as scabies.

3. Internal Use

Sulfur passes unreacted through the stomach. In the large intestines at high pH up to 10% is converted to hydrogen sulfide, which is eventually oxidized and secreted as sulfate in the urine (Denis, 1927; Greengard, 1940). The mildly laxative action caused by very low concentrations of sulfide is well known and can be achieved by consuming the element as powder in form of pills or in mixture with molasses, as it has been used for several centuries as a "spring tonic." It is occasionally prescribed for patients suffering from hemorrhoids. A review of various other internal applications was given by Wild (1911). Contrary to occasional claims, sulfur has not been found effective in treating organic diseases, such as diabetes (Echtman, 1930).

In animals, the absorption of sulfur has been traced with the radioactive element S-35. In sheep, sulfur appears in the wool within two weeks. In their rumen, sheep can also convert sulfate, sulfite and thiosulfate to sulfide. The human body contains about 6 mg of sulfur per kg of body weight. Sulfur therapies, based on elemental sulfur are not longer popular. In many internal applications, sulfur has been replaced by more effective pharmaceuticals; among these obsolete uses are rheumatism, scalgia, diphtheritic, croup, atonic gout, and asthma, for which it was still listed in the 1889 version of the U.S. Dispensatory, and by Pereira (1889).

4. Intravenous Use

Intravenous application of colloidal sulfur was believed to prevent and cure carbon monoxide poisoning (Vita, 1932) and aid arthritic patients. It causes a rise in blood pressure and respiratory stimulations, often followed by anaphylactic shock (Koppanyi, 1942). The sulfur is excreted within 24 hours as sulfate in the urine. Different animals respond and recover at different rates (Chistioni, 1932).

5. Intramuscular Use

Injection of colloidal sulfur in oil or aqueous base causes very painful neurosis, leucocytosis, and fever. The latter sets in seven hours after injection, and can last two days. During the 1920's injections were used to immobilize advanced cases of dementia praecox (Power, 1932; McCowan, 1932), but the sulfur-shock treatment has not established itself.

B. HYDROGEN SULFIDE

In contrast to elemental sulfur, hydrogen sulfide reacts vigorously with living tissues. The biochemistry of sulfides has been well reviewed (Denis, 1927; Dziewiatowski, 1962). The colorless gas which has a smell characteristic of rotten eggs is very poisonous, except in very small concentrations, such as found in aqueous solutions in hot mineral springs. The smell is noticeable at a concentration of 0.02 ppm. Unfortunately, the strength of the odor does not increase with concentration, and often anesthesia sets in before any other effects are noted. Thus, many victims who recover from acute poisoning claim they were unaware that they were exposed to harmful concentrations. This makes it a most dangerous industrial gas, especially for those who regularly deal with it and become accustomed to working around the odor.

Hydrogen sulfide is widely generated in nature, for example in swamps, in decaying organic matter, and in discharges from volcanoes. Accidental release from industrial sources, as occurred in Poze Rico, Mexico, is extremely rare. An estimated dose-response curve for hydrogen sulfide as assembled from literature data is shown in Figure 10.2. This figure shows that hydrogen sulfide and ozone odors are readily detected at a concentration a thousand times lower than that causing injury. Table 10.1 shows that the span between the threshhold of the first effects and toxic action is far larger in hydrogen sulfide than in many common drugs, such as the widely used barbiturates. The present industrial threshhold level for continuous exposure is 5 ppm.

Hydrogen sulfide is readily absorbed by water, Chapter 3. At a pH above 6.9 the solubility increases because the HS^- ion forms. Hydrogen sulfide is readily absorbed in the lung and by the alimentary canal. It is rapidly oxidized to nontoxic sulfate. Five ppm hydrogen sulfide causes irritation of the conjunctiva and bronchial mucus. One hundred ppm

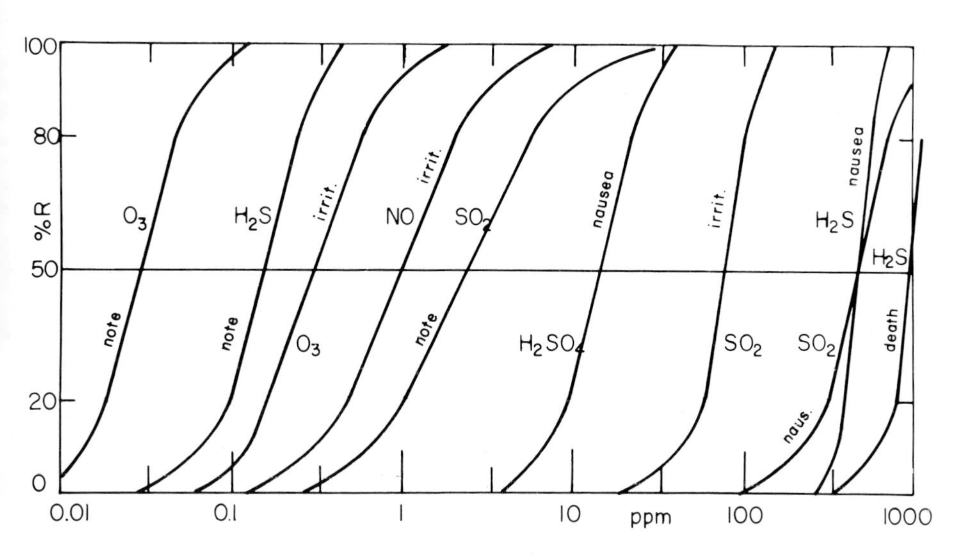

Fig. 10.2. Dose-response curve for ozone, hydrogen sulfide, nitric oxide, sulfur dioxide and sulfuric acid. The data is estimated from divergent values in many sources.

causes asphyxia, nausea, cerebral excitement, narcosis, and after an hour, death; 100 ppm leads to immediate paralysis and death (Haggard, 1925). In case of survival, pulmonary edema and pneumonia are liable to follow, and in serious cases, meningitis. While acute poisoning normally affects chiefly respiration, chronic exposure to low concentrations can also affect the skin and eyes. Internally, hydrogen sulfide causes diuresis, diaphoresis, irritation of the urinary passage, intestinal irritation and muscle pain. In very low concentration, 2-200 ppm, such as found in mineral hot springs, hydrogen sulfide can act as a keratolytic (Evers, 1958), antiprurite, antimicrobial and hyperuric agent. It has the same healing effect as sulfur lotions (Gloor, 1973). Internally, such waters have therapeutic value as laxatives, diuretics, and diapheretics. Toxic effects from overconsumption of hydrogen sulfide containing mineral waters are rare, probably because of the unpleasant odor which discourages overindulgence.

Workers who are continuously exposed to hydrogen sulfide vapors can suffer serious eye damage, due to keratitis. Such exposures have been observed on people who work in dye factories, mineral hot springs, city canalizations and sugar refineries. During the early days of liquid sulfur

Table 10.1
DOSE-RESPONSE LEVELS OF COMMON DRUGS AND POLLUTANTS

Drug	Therapeutic Level mg/100 ml	Toxic Level mg/100 ml	Lethal Level mg/100 ml
Barbiturates			
short acting	0.1	0.7	1
intermediate acting	0.1-0.5	1-3	3
phenobarbital	ca. 1.0	4-6	8-15
barbital	ca. 1.0	6-8	10
Bromide	5.0	50	200
Chlorpromazine	0.05	0.1-0.2	0.3-1.2
Ethanol	0.05-0.2	0.15	0.35
Ethyl ether	90-100	—	140-180
Lithium	0.6-1.2	2.0	2.0-5.0
Methaqualone	0.5	1-3	3
Nicotine	—	1	0.5-5.2
Paraldehyde	ca. 5.0	20-40	50
Aspirin	2-10	15-30	50
Pollutant	**Noticeable ppm**	**Illness ppm**	**Death ppm**
H_2S	0.2	500	1000
H_2SO_4	?	10	30
NO_x	1	1	?
O_3	0.05	0.5	?
SO_2, acute	3	80	500
acute[a]	0.1	0.12-0.27	0.5-1
chronic[a]	0.006	0.009	0.025
Sulfate[b]	0.007	0.01	0.025

a) epidemic threshhold; b) mg/m^3.

shipping, hydrogen sulfide caused health hazards to the workers who unloaded railcars and barges, because hydrogen sulfide accumulated in the holds. It was formed by the slow reaction of organic impurities with the liquid sulfur.

Since our senses cannot readily distinguish different concentrations of hydrogen sulfide, the only protection against it consists in constant electronic monitoring of hydrogen sulfide concentration in the air. Some industrial users, for example, the French sour gas industry, insist that workers carry gas masks in potentially hazardous locations.

C. SULFIDES

Metal sulfides, polysulfides and organic sulfides can be very toxic, both because of the sulfide group, and because of the metal or organic rest. Calcium sulfides and dithiocarbonic acid are used as agricultural fungicides, in some hair removing formulations, and to soften skin. If not properly used, it attacks the skin. However, it does not form the well known biologically important sulfhemoglobin, which has a dirty green-brown color and is responsible for the discoloration of cadavers. The latter is never formed *in vivo* and is not responsible for the asphyxial symptoms.

D. THIOSULFATE

Thiosulfate is not strongly toxic. It occasionally serves as an antidote to cyanide poisoning by forming thiocyanate. Injected intravenously, it can cause convulsions. It is also occasionally used as an antidote to arsenic. About 70% is excreted unchanged in the urine, the remainder is oxidized to sulfate before excretion. Internally, up to 12 gr can be administered per day. It acts as a cathartic. Externally, it is used as a 10% solution to treat parasitic skin disorders, such as ringworm.

E. POLYTHIONATES

Polythionates belonged to the chemically best studied sulfur compounds, but much of the work was performed during the early 19th Century. At that time much of the therapeutic effect of elemental sulfur was credited to tetra- and penta-thionates. During the last fifty years, biological action has been exclusively credited to sulfides. However, the polythionates deserve careful re-evaluation, because the thionates contain sulfur in the oxidation state of sulfide, in the presence of oxidized sulfur. Furthermore, a wealth of old literature gives a careful account of chemically convincing arguments. The effectiveness of "selenium sulfide" in treating seborrhoea is just one practical example which points to thionates, because this reagent is formed (Selsun, U.S. P. 2,694,669, 1955) by reaction of selenium dioxide with hydrogen sulfide. This reaction is known to produce a mixture of polythionates and selenates.

F. SULFITE

Sodium sulfite is extensively used for preserving foods. Even in gram quantities, it is not harmful (Sollmann, 1957). A 1% solution has been used as a mouthwash against aphthae, against ringworm and other parasitic skin disorders. However, intravenous application causes immediate toxic reaction, probably by interference with the oxidation mechanism. It also lowers the blood pressure, and depresses the central nervous system. *In vitro*, sulfite acts as a food preservative by destroying thiamine. Doses of more than 500 mg per kg body weight are toxic to rabbits and cats. The legally set limits in food (Great Britain) are shown in Table 11.5.

Bisulfite and Disulfite

Bisulfite and disulfite are used similary to sulfite and as "canning powder." Their toxicity is similar to that of sulfite.

G. SULFUR DIOXIDE

Sulfur dioxide obtained by combustion of sulfur has been used since antiquity for bleaching, as a preservative and for fumigation. It is not a reliable disinfectant against bacteria or parasites. The toxicity of sulfur dioxide depends on the form of application. Ingestion of sulfur dioxide and sulfites is harmless, hence their use as food stabilizers. Inhalation of sulfur dioxide is dangerous; chronic inhalation seems to have a cumulative effect. Fig. 10.2 shows the dose-response curve for acute sulfur dioxide exposure deduced from literature data. A detailed analysis of long range, chronic sulfur dioxide exposure is given in Table 10.2 (Higgins, 1974; Katz, 1975; NAS, page 604). Response *vs.* duration of exposure curves for different sulfur dioxide levels are plotted in Fig. 10.3a. Fig. 10.3b gives the corresponding curves for plants and vegetation which are far more sensitive to sulfur dioxide than man. Sulfur dioxide can be noticed at concentrations above 0.2 ppm. Irritation of the nose and throat occurs above 6-12 ppm. Twenty ppm causes eye irritation, and 1% causes skin irritation. The range between detection and injury is large, and acute sulfur dioxide poisoning is extremely rare, as the vapors become unbearably irritating to the eye and upper lung before serious injury is inflicted. The industrial threshold limit value (TLV) is set at 5 ppm. The mechanism of acute poisoning is not well understood. Sulfur dioxide reacts with water as a weak acid, and part of its effect is possibly due to the hydronium ion

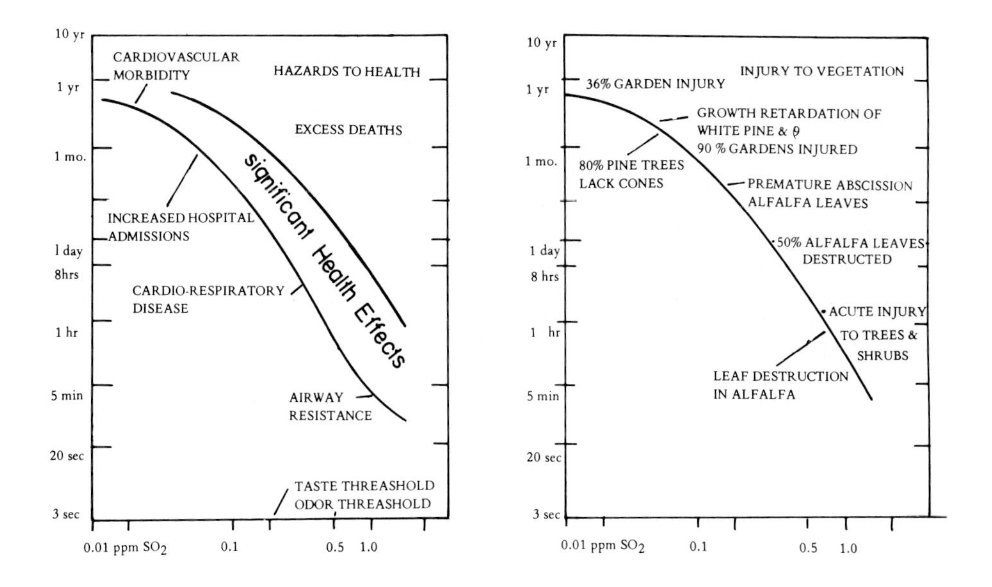

Fig. 10.3. Effects of SO_2 as a function of concentration and exposure: a) human health; b) vegetation.

Effect of Particles (2.0-50 micron) in Major Bronchi

1. Paralysis of Cilia
2. Hypersecretion
3. Mucus Gland Hypertrophy and Extension
4. Susceptibility to Infection Chronic Productive Cough

Effect of Particles (0.01-1.0 micron) in Terminal Bronchioles

1. Loss of Normal Defences
2. Effect on Surfactant
3. Goblet Celi Metaplasia
4. Inflammation and Obliteration
5. Premature Closure
6. Collateral Ventilation
7. Stress of Lung
8. Release of Proteolytic Enzymes

Effect of Particles (0.01-0.5 micron) in Alveoli

1. Increase of Cells: Causes Infection
2. Release of Proteolytic Enzyme: Causes Emphysema with Alveolar Destruction

Fig. 10.4. Paths of synergism of SO_2, NO_2, and particulates (after Bates, 1972)

Table 10.2
DOSE/RESPONSE RELATIONSHIP FOR PARTICLES AND SULFUR DIOXIDE[a]

Averaging time	Place	Concentration (Mg/m^3) Particles	SO_2	Effect
24 Hour	London	2.000	1.144	Mortality
24 Hour	London	0.750	0.710	Mortality
24 Hour	London	0.250	0.500	Deterioration of patients
Weekly mean	London	0.200	0.400	Prevalence or incidence of respiratory illnesses
24 Hour	New York	6	1.500	Mortality
24 Hour	Chicago	N.A.	0.700	Exacerbations of bronchitis in pts.
Winter mean	Britain	0.100-200	0.100-200	Incapacity for work from bronchitis
Annual	Britain	0.070	0.00	Lower respiratory infections in children
	Britain	0.100	0.100	Upper and lower respiratory infections in children
	Britain	0.100	0.100	Bronchitis prevalence
	Britain	0.100	0.100	Prevalence of symptoms
	Buffalo	0.100	0.0003	Respiratory mortality
	Berlin, N.H.	0.180	0.00073	Increased respiratory symptoms Decreased pulmonary infection

a) After Higgins (1971); b) 2.8 mg/m^3 SO_2 = 1 ppm.

formation in the lung, but most of the effects are probably caused by oxidation of sulfur dioxide. Possible biochemical effects have been reviewed by Goldstein (1975). The well known adverse epidermic effects of sulfur dioxide are not due to acute poisoning by the pure gas, but to the very large volume of gas which contains traces of sulfur dioxide and other wastes, which are released into the urban environment; an insidious process which is called pollution. All established injury by urban sulfur dioxide is due to chronic exposure in conjunction with moisture, particulate and other gases. These synergistic mixtures are found in ambient air, especially in the vicinity of fossile fuel flues. At high concentration they can form acute irritants; the threshhold concentrations are not yet established.

1. Clinical Exposure

Exposure of animals to sulfur dioxide was already reported by Wasserberg (1790), who studied its action on the skin and lungs of dogs by exposing the animals in elaborate containers fitted around their neck, so that either the air around their head or around their body could be enriched with sulfur dioxide. Many animal species, including man, respond to sulfur dioxide by bronchiorestriction (Lawther, 1975), i.e. by contraction of the upper respiratory tract. During normal breathing, 95% of sulfur dioxide is removed in the upper passage (Amdur, 1973) and forced, deep breathing is necessary for the study of the lower respiratory tract. In animals, after chronic exposure, one can observe a thickening of the mucous layer. In guinea pigs, the basic pulmonary-cardiac functions remain unaffected after one year's exposure at 5 ppm (Alarie, 1970). The dose-response curve, Fig. 10.1, shows that not all people respond at the same concentration. While many people can adjust and can continuously work with 50 ppm sulfur dioxide in the air, others might react to that concentration with bronchospasm, especially those with latent disposition to asthma and similar bronchial complaints. Thus, sulfur dioxide must be classified as a secondary environmental stressor. Table 10.3 is a partial list of preconditions which, according to a U.S. National Academy of Sciences study (Carnow, 1973), are to be considered specially vulnerable to the effects of sulfur dioxide. Fig. 10.4 shows possible paths of synergism in the action of sulfur dioxide on the major bronchi, terminal bronchioles, and the alreoli, as proposed by Bates (1972). As this figure shows, particulates and other gas components are important contributors in the scheme. As Amdur, in a careful review of all literature available up to 1969, states: "If sulfur dioxide present as an air pollutant remained unaltered until removed by dilution, there would be no evidence in the toxicological literature suggesting that it would be likely to have any effects on man on prevailing levels." Unfortunately, sulfur dioxide in ambient air emerges from sources jointly with particulates, other noxious gases, and moisture. Amdur (1969), Higgins (1971), Ferris (1973) and others have demonstrated that particulates with 1 micron diameter and smaller are very active. Charlson (1975) and others have shown that aerosols are equally effective. Hazucha (1975) showed the synergistic action of ozone and sulfur dioxide, both at 0.37 ppm. The data in Fig. 10.5 corresponds to a change in mid-expiratory flow rate from 6.49 plus or minus 1.72 to 5.60

Table 10.3
POPULATIONS AT HIGH RISK, DUE TO SO_2 AND PARTICULATE

Because of Pulmonary Deficiencies	Because of Cardiac Deficiencies
Genetically Deficient	Genetically Deficient
Asthma	Hypercholesterolemia
Cystic fibrosis	Diabetes with atherosclerosis
Cystic disease of lung	Congenital Disease
Chronic Disease	Tetrology of Fallot
Chronic bronchitis	Atrial or ventricular wall defects
Emphysema	Valvular defects
Kyphoscoliosis	Chronic Disease
Advanced tuberculosis	Cor Pulmonale
Bronchiectasis	Hypertension with left ventricular disease
Other Environmental Factors	Coronary insufficiency
Cigarette smoking	Rheumatic heart disease
Workers exposed to dusts, fumes, etc.	Congestive heart failure
Others	Other Environmental Factors
Prematures	Cigarette smoking
Newborns	Heavy work exposure
The obese	

After Carnow (1973).

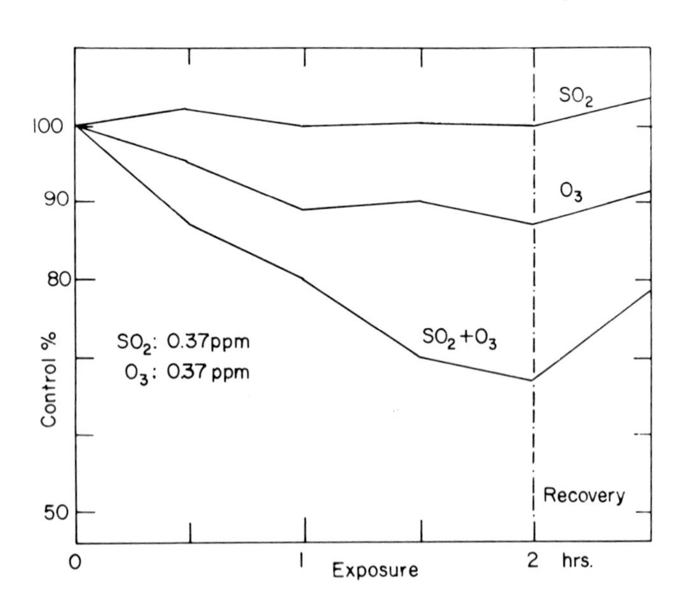

Fig. 10.5. Influence of sulfur dioxide on human mid-expiratory flow rate.

plus or minus 1.63; the forced expiratory volume fell to 78%, the mid-expiratory flow rate at 50% vital capacity fell to 54% of the initial value, and the static, unforced peak expiratory flow rate dropped to 94%. As in all such tests, some adaptive effects were also observed. Novakov (1974) has shown that moist surfaces of particulates are such effective catalysts for the oxidation of SO_2 to SO_4^{2-} that ambient concentrations are sufficient to account for most of the air pollution effects which formerly have been explained by gas phase mechanisms. This is not astonishing, as the particulates accompanying sulfur dioxide consist of fly ash which contain traces of transition metals, such as vanadium, manganese, etc., which are commercially used catalysts in the manufacture of sulfuric acid. After inhalation, the fly ash particles are phagocytized and lodge in the cytoplasm of macrophages in the sinusoids, as long range studies on cynomolgus monkeys have shown (Abe, 1967).

2. Epidemiology

In fossile fuel burning electric power plants, and in some smelters, sulfur dioxide constitutes an unwanted by-product which is formed in mixture with enormous quantities of waste gases. Until very recently, these gases were commonly mixed and released into the air. Often special devices, such as tall stacks or stack gas heaters, were used to deliberately spread these obnoxious wastes over large areas. Thus, an acute local problem was transformed into a chronic regional nuisance. As a result, large segments of regional populations have been exposed to the effect of traces of poisonous or objectionable gases. During unfavorable weather conditions, the emission can be trapped and accumulates in inversion layers, reaching critical concentrations. During the last ten years, no acute epidemic episodes have occurred, but earlier such episodes used to be quite common. The incidents of the Meuse Valley of Belgium in 1930, in Donora, Pennsylvania in 1948, and in Tokyo, London, and New York during the 1950's and early 1960's are still well remembered. In several of these places, mortality above the statistically predicted rate could be demonstrated and autopsy data and cardio-respiratory distress in clinical patients (Severs, 1975) was well established (Higgins, 1971, 1973), and eventually lead to large scale legal action and to the study of the health effect of pollutants from combustion. Carnow (1973) showed a direct correlation between the frequency of acute chest illness and the incidence of sulfur dioxide concentrations exceeding 0.3 ppm. The threshhold for excessive deaths in London was established at 2 mg/m^3 smoke and 0.4 ppm (1.1 mg/m^3) sulfur dioxide (Burgess, 1959). Humidity was found to be a significant factor. British studies (Lawther, 1970) indicate that a reduction of smoke levels, without reduction of sulfur dioxide, can cause a marked decrease in the correlation of pollution with daily deaths. A careful dose-response correlation for the synergistic effect of sulfur dioxide and particulate was compiled by Higgins (1973), Table 10.1.

While a correlation between acute symptoms and acute episodes is now well established, the effects of chronic exposure to uncontrolled coal burning power plants and smelters are not yet clearly established. For example, pollution has not been conclusively proven to have increased the incidence of lung cancer. Such correlations are hard to establish because of the long range over which observations are necessary, and because of regional and national differences in the frequency—and diagnosis—of diseases. The U.S. Community Health and Environmental Surveillance System (CHESS) has indicated very interesting trends about the occurrence of various respiratory infections, differences in the lung function of children and the incidence of asthmatic attacks (Chapman, 1973; French, 1973) but the significance of these trends is not yet conclusively established.

An elaborate computer model has been developed to establish the social costs of health and material damage of pollution (North, 1975). It is based on statistical evidence which suggests that there is a direct correlation between sulfate and sulfur dioxide concentration and at least five forms of health defects: Chronic respiratory disease, aggravated heart and lung disease, asthma attacks, and children's lower respiratory disease. The epidemiological data indicates that with increasing concentration the health effects increase at a higher than linear rate. Table 10.4 shows the result of different sulfur dioxide and sulfate increases at an ambient total sulfur concentration of 16 micro gram/m^3. The first case, an increase of 0.145 micro gram/m^3 in sulfate and 0.35 micro gram/m^3 in sulfur dioxide, corresponds to the effect caused by a remote power plant; for example in the U.S. Northeast. The second case represents the effect of an urban power plant on a city, for example New York City, assuming an increase of 1.86 micro gram/m^3 in sulfate, and a 7.5 micro gram/m^3 increase in sulfur dioxide. Preliminary results of the effect of exposure length are shown in Fig. 10.6. The significance of these results is presently being hotly debated, and it is not clear whether the present understanding of

Table 10.4
ESTIMATED COST OF SULFUR EMISSION FROM REMOTE AND URBAN PLANTS ON A METROPOLITAN AREA

Effect	Cases/mil		Cost/case	Cost		Cost	
			$	% of total		$/year	
	remote	urban		remote	urban	remote	urban
Increased daily mortality	.3	3.5	30,000	2.5	2.5	0.40	1.30
Aggravation from heart & lung disease	5,100	65,000	20	25.0	28.0	5.10	15.10
Asthmatic attack	1,000	15,000	10	3.0	3.0	0.50	1.60
Lower respiratory disease in children	120	1,500	75	2.5	2.5	0.50	1.40
Chronic respiratory disease	510	6,500	250	32.0	35.0	6.4	18.9
Total health				65.0	72.0	12.90	38.30
Material & Aesthetics						6.00	13.40
Acid rain						1.40	1.40
Total Cost						20.00	53.00
Total Cost/lb Sulfur						.21	.55

remote: 10 tons S/hr causing an increase of 0.145 micro gram/m^3 SO_4^{-4} and 0.35 micro gram/m^3 increase of SO_2 in N.E. U.S., affecting 50,000,000 people.
urban: 10 tons S/hr cuasing an increase of 1.86 micro gram/m^3 SO_4^{-4} and 7.5 micro gram/m^3 increase of SO_2 in metropolitan area, affecting 11,500,000 people.

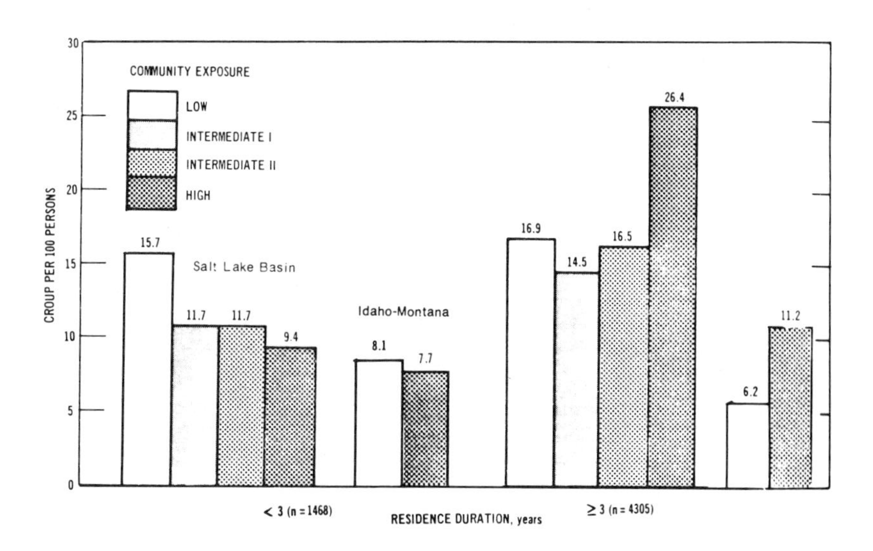

Fig. 10.6. Preliminary results of CHESS on the effect of exposure length on the incidence of croup in the Salt Lake basin and in Idaho-Montana (after Chapman, 1973)

the basic atmospheric and health effects is sufficient to justify the application of elaborate theoretical models and extended computer calculations (Neyman, 1977).

No meaningful quantitative conclusions can be drawn before we understand the synergistic effect of other pollutants which are emerging from power plants. Among these are carbon monoxide, nitric oxide, and particulates containing heavy metals, all of which are known to be far more toxic than sulfur dioxide. However, all work confirms that sulfur dioxide aggravates the influence of these pollutants, and it is now clearly established that sulfur dioxide in concentrations above 16 micro gram/m^3 measurably aggravates other, pre-existing conditions.

H. SULFURIC ACID

The dose-response curve for sulfuric acid is shown in Figure 10.1. The concentrated acid attacks the skin, burns clothing, and its fumes attack the lung. Ingenstion of 5-6 gr of 30% acid leads to the death of adults. Chronic exposure is now a rare event. The example from Modena, Chapter Nine, shows that in earlier times acid fumes used to be a common complaint. If concentrated acid is diluted from 100% to 50%, 500 Btu/lb heat is released; thus, acid can be heated to the boiling point by dilution. This increases the toxicity, and the rate of reaction. Furthermore, if water is poured into the acid, violent splashing can occur.

Chapter 11

Sulfur in Agriculture and Food

A. SULFUR IN AGRICULTURE

Sulfur is an essential component of animal and vegetable tissues. The human brain contains 1.1% sulfur and 1.5% cystine. Blood albumen contains 5% cystine. Egg white contains 1.6% sulfur and cystine. Table 11.1 shows other representative sulfur concentrations. In muscle tissue, i.e. meat, the average sulfur content is about 0.25%, of which three quarters is in organic form. The most important sulfur containing amino acids are cysteine, cystine, and methionine:

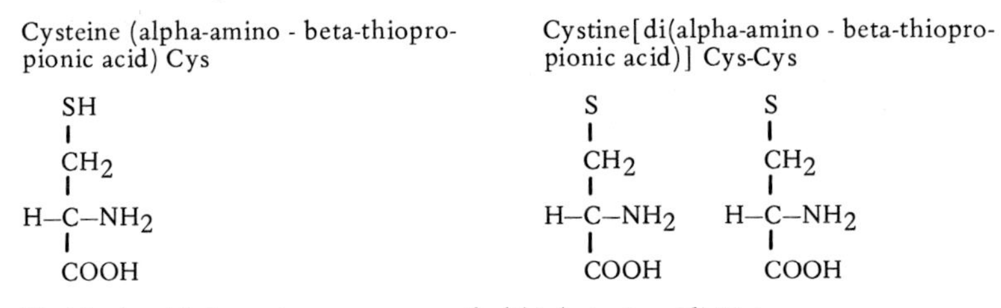

All non-ruminant animals need methionine in their food for the synthesis of cysteine and cystine. Ruminant animals can absorb both inorganic sulfates and sulfur containing amino acids. Chicks need 0.8% sulfur, and ruminants need either .17% sulfur in the form of amino acids, or .32% sulfur as sulfate (Albert, 1956). Sheep and cows need far more. A cow producing 20 kg milk per day excretes 10 g sulfur perday and needs about one g sulfur per kg dry weight of feed (Playne, 1975; Bouchard, 1972). Adult men need 1.1 g methionine per day (Rose, 1957). Thus, the world human dietary needs for the four billion people expected to live in 1980

Table 11.1
SULFUR CONTENT OF ORGANIC MATERIALS

Material	Sulfur Content	Material	Sulfur Content
Animal tissues	%	Plant tissue	%
Brain	1.10	Barley	0.25
Albumen	1.77	Oats	0.41
Milk	0.95	Clover	0.79
Egg white	1.60	Parsnip	0.61
Muscle	1.10	Peas	0.76
Kidney	1.00	Apple	0.45
Food Stuff	% Sulfur-Amino acid	Plant and Animal Residue	% Sulfur-Amino acid
Milk	1.16	Fish meal	0.70
Cheese	7.87	Milk powder	1.30
Egg	5.50	Fresh blood	0.008
Beef	3.74	Egg yellow	0.14
Chicken	3.07	Malt	0.15
Bread	1.05	Sugar pulp	0.65
Beans	4.88	Linseed	0.36
Peanut flour	7.78	Cabbage	0.31
Soybean flour	14.00	Rice	0.20
Wheat flour	4.90	Corn	0.25

After Lemoigne (1973), Allaway (1966), Beaton (1974).

will require amino acids containing 400 tons of sulfur daily or 150,000 tons per annum (Allaway, 1966).

The biochemistry and metabolism of sulfur containing compounds such as cysteine and cystine, glutathione, methionine, sulfurtransferase, sulfohydrolases, lipoic acid, thiamine, biotine, coenzyme A, insulin, and similar substances has been well reviewed (Greenberg, 1975; Allaway, 1966) but is far from being fully understood. The sulfur metabolism in plants has been reviewed by Thompson (1967). Siegel (1975) lists 473 references on the biochemistry of the sulfur cycle. Earlier work has been discussed by Peck (1962), Trudinger (1969), and Roy (1970). Oae (1976) discussed synthetic, bioorganic sulfur chemistry.

Plants form reduced sulfur compounds from sulfate, while animals do the reverse. Thus, the sulfur cycle differs from the nitrogen cycle in that animals excrete reduced nitrogen, but oxidize sulfur. Microorganisms are capable of both reduction and oxidation. During putrefaction, the microbial decomposition of plants and animal bodies, organic thiols are

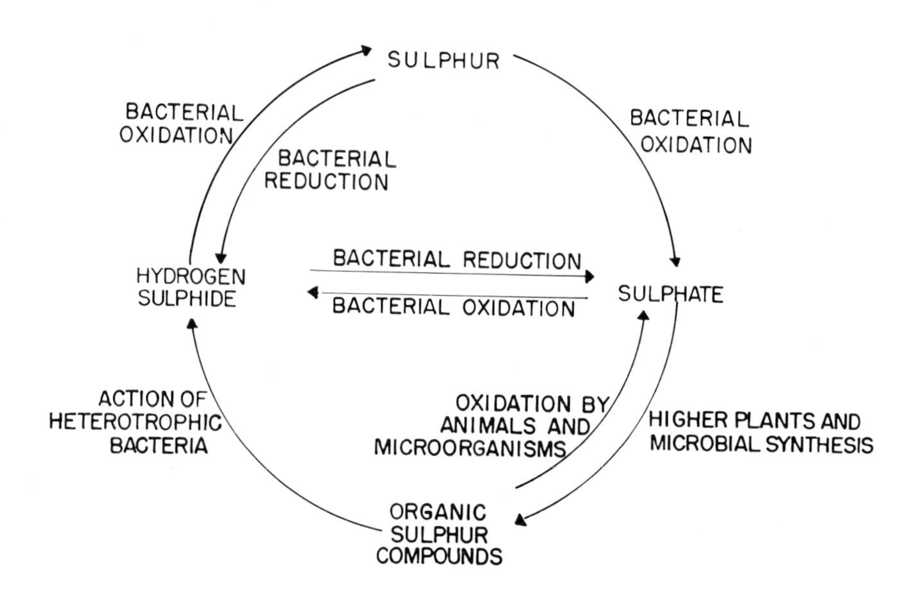

Fig. 11.1. Biological transformations of sulfur in soil (Burns, 1967).

metabolized and hydrogen sulfide is released. The genus of desulfovibrio of the small group of sulfate reducing bacteria uses sulfate as the terminal electron acceptor in the manner in which aerobic organisms use oxygen. Only little of the resulting hydrogen sulfide is utilized for synthesizing molecules, most of it is excreted into the surroundings as a gas. This respiratory process, the "dissimilatory sulfate" reduction and the sulfuretum, Chapter 9, constitute a primitive analogue of the oxygen cycle. This microbial sulfur cycle is over 3.5 billion years old, almost twice as old as the oxygen cycle. Ruminant animals host microorganisms which reduce sulfate for the synthesis of amino acids, and thus combine redox abilities of both animals and plants. The most important steps in the sulfur metabolism are shown in Fig. 11.1 (Allaway, 1966). Sulfur is involved both in structural and metabolic functions. Since sulfur exists primarily as part of proteins, nutritionists usually measure the adequacy of diets by the ratio N:S. However, measurement of this ratio in tissues is not a useful indicator because sulfur shortage prevents formation of the entire protein. In lactating cows fed a sulfur deficient diet, all 17 analyzed amino acids reduced

Table 11.2
SULFUR AND PRIMARY NUTRIENTS CONTAINED IN VARIOUS CROPS
(pounds)[a]

Crop	Yield/Acre	Sulfur	Nitrogen	Phosphorus	Potassium
Corn	250 bu.	45-55	500	65	275
Cotton	3.5 bales	28-35	250	50	150
Wheat	80-200 bu.	18-24	170	30	100
Alfalfa	10-16 tons	50-60	500	50	400
Clovers	5 tons	25-30	250	22	165
Grasses	6 tons	20-35	180	27	150
Cabbage	25 tons	30-60	140	20	125
Turnips	30 tons	40-55	130	25	190
Onions	25 tons	30-40	130	25	90

a) Expressed as elemental sulfur, nitrogen, phosphorus and potassium.

significantly (Jacobson, 1967), and milk production decreased within nine weeks by 35%. Table 11.1 shows how many grams of various plant or animal foods are needed to fulfill the daily human sulfur requirement.

1. Sulfur as Plant Nutrient

Animals derive their sulfur requirements from plants. Thus, sulfur is an essential nutrient for plants. Inorganic sulfate is used by plants for the synthesis of the amino acids cystine and cysteine which are components of proteins, and account for 90% of the organic plant sulfur. Smaller amounts are needed for the synthesis of vital biotin, thiamin (vitamin B), glutathione, and coenzyme A, for the formation of the disulfide linkages in protoplasm, formation of nitrogenase enzyme (the system involved in nitrogen fixation by microorganisms), the activity of ATP sulfurylase (an enzyme regulating the sulfur metabolism), the activation of some proteolytic enzymes, for example the papainases, and for the synthesis of the glucoside oils of oinions and cruciferous plants. Furthermore, sulfur is needed for the synthesis of chlorophyll, even though it does not appear in the final molecule. The sulfur content of some plants used as cattle feed is listed in Table 11.1. The analytical methods for determining sulfur content are discussed in Chapter 4. Sulfur nutrient requirements vary. Corn, sorghum, cabbage, turnips, and onions are well known for their high sulfur requirements. Alfalfa and other legumes, as well as peanuts, cotten, and tobacco require substantial amounts of sulfur, while small grains and grasses need less. Table 11.2 shows the sulfur requirement per acre, on the basis of typical yields, as indicated.

2. Sulfur in Soil

Field soils contain between 0.001 and 0.5% sulfur. Most soils average between 0.01 and 0.05% (Chapman, 1961). Methods for determining sulfur in soil are described by Beaton (1968) and in Chapter 4. To be fertile, soils must contain sulfur in the proper form (Freney, 1975) and proper concentration—not too high and not too low—for the chosen crop. Furthermore, they must possess an internal reservoir from which the seasonal need for sulfur during growth can be met. If this is not the case, sulfur must be supplemented at the correct time and in the correct amount.

In humid regions the sulfur occurs largely in the form of organic matter, which together with all inorganic sulfur is slowly oxidized to sulfate (Burns, 1967; Bloomfield, 1967; Harward, 1966; Li, 1966; Starkey, 1966) which makes the element available to plants. If the subsoils contain aluminum or iron hydroxides, the sulfate is absorbed and retained for plant use. In coarse-textured, sandy soils, the sulfate is quickly lost by leaching. However, in arid regions, sulfate and other salts accumulate. The same occurs in soils with poor water penetration; this can be overcome by pH adjustment. In anaerobic soil, sulfur is reduced to sulfide, which is toxic to normal plants. This process is discussed in Chapter 6.

The rate of oxidation of elemental sulfur in soil determines the rate at which sulfate is available to plants. A series of studies indicates that oxidation is preceeded by an incubation period of about two months, afterward the oxidation proceeds quickly and smoothly, if the grain size is below 100 mesh. Li (1966) determined the kinetics as a function of temperature, soil pH, organic content of soil, and particle size. The earlier literature was reviewed and the reaction steps analyzed by Starkey (1966). Bloomfield (1967) studied the effect of diammonium phosphate, triple superphosphate, and wax-coated fertilizer. Harward (1966) studied the influences of pH, ions, and oxidation on movement of inorganic sulfur in the soil, and Burns (1967) reviewed the entire literature on sulfur oxidation in soil. Other excellent reviews of our present knowledge of sulfur in soils can be found in Kelly (1972), Zajic (1969), and Roy (1970).

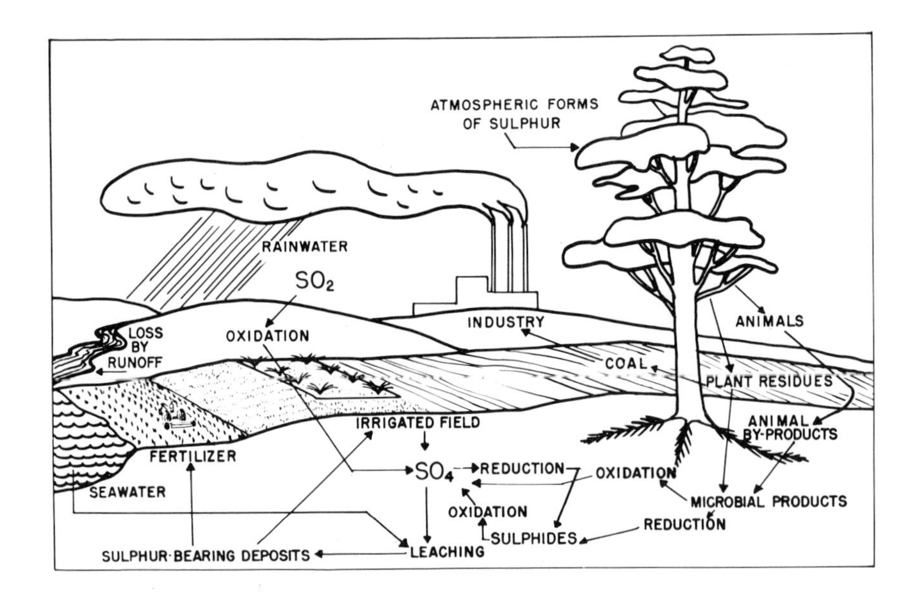

Fig. 11.2. Pedospheric sulfur cycle (Burns, 1967).

3. The Pedospheric Sulfur Cycle

In an undisturbed, natural environment, the consumption of sulfur by plants and animals forms part of a cycle in which sulfur from animal wastes and carcass decompotion returns to the soil, becomes available for plants, and in turn again reaches the animal. This cycle establishes equilibrium with the other sulfur cycles, described in Chapter 6, and reaches a steady state around which all rates adjust. Sulfur is added or received from the pedospheric cycle by sulfate gained from or lost to rivers and by precipitation. This effect is discussed in Chapter 6; typical sulfur balances are shown in Table 6.2 of that chapter. In addition, one to ten lbs per acre reach the earth each year in rain. It is often overlooked that both soil (Smith, 1973) and plants (Unsworth, 1972) are also very efficient at complementing the sulfur reservoir by absorbing sulfur dioxide from the atmosphere. The pedospheric sulfur cycle is shown in Fig. 11.2. Rural air normally contains about 10 ppb sulfur dioxide, i.e. 28.6 micro g/m^3. If pollution from electric power plants exceeds 500 ppb, visible damage is caused to sensitive plants. It is due to stomatal injury (Unsworth, 1972). The correlation between exposure time and damage for different sulfur dioxide concentrations is shown in Chapter 6.

The growth of crops is dependent upon the proper concentration of nitrogen, phosphorus, potassium, magnesium, sulfur, and trace elements in the soil. The growth rate is determined by the individual plant's efficiency in extracting these elements and by the continued availability of the trace elements. If seasonal plant nutrient needs can be met from the soil reservoir, the growth-decay cycles of plants can go on as long as all plant material remains and decays on site. This equilibrium continues to hold if resident foraging animals, or man, participate in the cycle.

It has been recognized for many hundred years that the addition of the proper fertilizer can enhance natural growth. In 1768 a pastor in Switzerland discovered that sulfur can be a growth-limiting factor, and that crop yields can be increased in some locations by the addition of gypsum. Benjamin Franklin dramatically demonstrated the same effect in Pennsylvania by using plaster to write "this land has been plastered" on a hillside. The fertilizer increased the growth of grass and made the writing visible.

Since the beginning of industrialization, the natural growth cycle has been depleted by increased removal of crops and animals from the soil to urban centers. In this unbalanced, open growth cycle, the concentration of nutrients in the soil decreases, and the concentration gradient between soil and plant decreases and becomes the rate determining step. Unless the nutrients are replaced, the plant growth slows down and eventually ceases. Already a hundred years ago, phosphorus, nitrogen, and potassium were recognized to be important elements, and fertilizers were formulated to replenish these three nutrients. In subsequent years potential sulfur deficiencies were masked because of the wide use of normal superphosphate, first produced by John Laws in 1840, and ammonium sulfate, which contain 12 and 24% sulfur respectively. However, since about 1900, sulfur deficiencies have been identified in the U.S. Pacific Northwest, and many other parts of the world, as indicated in Table 11.3. During the last two decades, the increasingly forced harvesting of high yield crops has put extreme stress on soils, and agribusiness, in the desire to minimize fertilizer application and distribution costs, resorted to yet "higher" phosphate and nitrogen analysis compounds which contain little or no sulfur. At the same time, modern, capital intensive agribusiness management has drastically increased the energy needed for producing food. Today, 15 calories of energy have to be invested to produce one calorie of meat on a cattle feed lot. In 1940 five calories of energy produced one calorie of food.

Table 11.3
AREAS WITH SULFUR DEFICIENCY[a]

UNITED STATES	Nebraska	EUROPE	BRAZIL
Alabama	North Carolina	Czechoslovakia	Chile
Alaska	North Dakota	France	Costa Rica
Arkansas	Ohio	Germany	Honduras
California	Oklahoma	Iceland	Venezuela
Colorado	Oregon	Ireland	AFRICA
Florida	South Carolina	Netherlands	Ghana
Georgia	South Dakota	Norway	Kenya
Hawaii	Tennessee	Poland	Malawi
Idaho	Texas	Spain	Nigeria
Indiana	Virginia	Sweden	Senegal
Iowa	Washington	Yugoslavakia	South Africa
Kansas	Wisconsin	AUSTRALASIA	Tanzania
Louisiana		Australia	Uganda
Maryland	CANADA	Fiji, Solomon Islands	Upper Volta
Michigan	Alberta	New Guinea	Zambia
Minnesota	British Columbia	New Zealand	ASIA
Mississippi	Manitoba	SOUTH AND	Ceylon
Missouri	Ontario	CENTRAL AMERICA	India
Montana	Saskatchewan	Argentina	Japan

a) After Beaton (1974).

In 1910 this ratio was 1. In developing nations, 0.05 calories suffice to produce one calorie worth of rice. The consequences of this have been recently discussed by Hirst (1974) and Steinhart (1974).

Modern agriculture operates on a negative sulfur balance. Careful estimates, using a variety of independent methods (Beaton, 1974) have shown that presently each year 3 to 4 million tons of sulfur are being removed or harvested from U.S. soils, whereas less than 1.1 million tons are being restored by the addition of fertilizer. The situation and impact vary locally and depend on the crop. Sulfur deficiencies have been identified in 13 U.S. states. In the East and North Central states, the U.S. "bread basket", field crops receive only 55% and 40%, respectively, of the required sulfur. Hay and pasture receive less than 15% of the needed replacement sulfur. On the average, the imbalance amounts to over 50% of the yearly sulfur turnover. Obviously, the natural sulfur cycles and industrial emissions are insufficient to adjust the balance. In recent years, soil sulfur deficiencies have become increasingly more visible and similar deficiencies have been recognized in many other countries (Odelien, 1966;

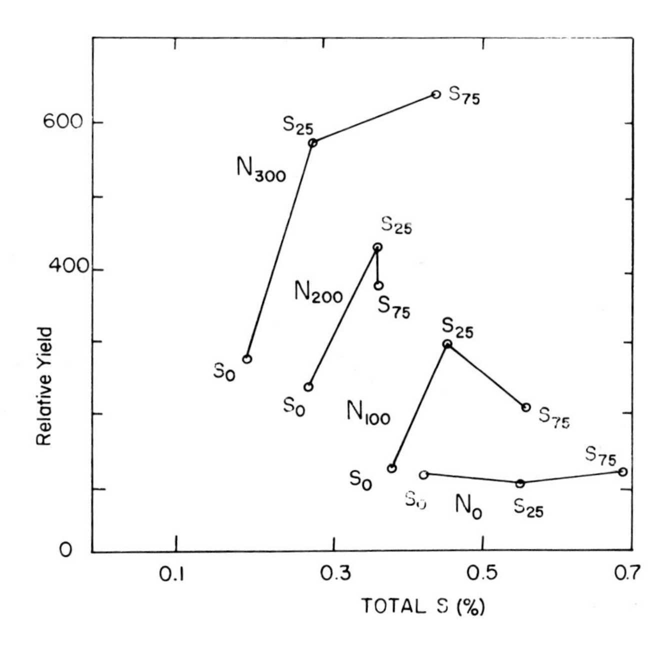

Fig. 11.3. Response of alfalfa to sulfur in combination with different nitrogen levels. (After Metson, 1973).

Coleman, 1966; Platou, 1972; Beaton, 1974). As a result, sulfur is now being added to fertilizers, and entirely new families of sulfur containing formulations are being developed, tested, and marketed.

4. Sulfur Demand

The sulfur demand of crops cannot be computed from the concentration in which it is present in the plant, because the efficiency of sulfur absorption differs from plant to plant and is often less than that for other elements. Generally, an N:S ratio of 5:1 has been found desirable, even though the plant proteins contain the element in the ratio N:S = 15:1. In the case of phosphorus, the natural ratio is P_2O_5:S = 2.3:1 and the best fertilizer ratio is 3:1. The discrepancy stems from the fact that many plants, such as soybeans, alfalfa, and legume crops in general, are efficient at extracting atmospheric nitrogen. Thus, they can by-pass the soil cycle. The response of alfalfa to sulfur as a function of nitrogen goes through a maximum, shown in Fig. 11.3, for four sets of nitrogen levels (Metson, 1973).

Sulfur deficiencies

The symptoms of sulfur deficiency are not always easily recognized, as the deficiencies result in generally reduced growth and yield. Often, plants are small and spindly with short, slender stalks. In cereals, the growth rate is retarded and maturity is delayed. Young leaves are light green to yellowish. In cotton, citrus, and tobacco, the symptoms resemble nitrogen deficiency. In corn and sorghum, they resemble the look of zinc and iron (Tisdale, 1967). In animals, sulfur shortage yields more specific symptoms, usually caused by lack of methionine. The best known effect is probably diathesis in chickens (Allaway, 1970). This, and similar deficiencies are often accompanied by selenium and other deficiencies (Whanger, 1970) of related elements. Sulfur and selenium are often jointly present, but their fate is different in the animal metabolism: While sulfur is oxidized, selenium is reduced and excreted via lungs and urine.

5. Sulfur Containing Fertilizers

The addition of sulfur to the soil can be achieved in a variety of ways. Conventional fertilizers ammonium sulfate and potassium phosphate inherently contain sulfur.

Solid Fertilizers

Enriched superphosphate with an analysis N:P:K:S = 0:30:0:7 can be made by partly substituting sulfuric acid for phosphate. In West Germany ammonium sulfate-nitrate with a composition 26:0:0:12 is marketed. TVA invented a fertilizer containing 79% ammonium nitrate, 21% ammonium sulfate, and 3% conditioner, which avoids the explosion hazard of other high nitrate materials. The widely used ammonium phosphate fertilizer 16:20:0:0 can be easily upgraded with ammonium sulfate to yield 16:20:0:14. Urea-ammonium sulfate mixtures with 35:0:0:10 are now available. Urea can also be mixed with gypsum. In this composition urea complexes the calcium ion, and replaces the water of crystallization. Several potassium fertilizers are available, among them a prilled material with the composition 20:0:20:7.

The use of elemental sulfur as a fertilizer additive has been reviewed by Slutskaya (1972). Elemental sulfur is well established as an insecticide and fungicide, and can be applied directly to the soil, but these grades are expensive. It was found (Beaton, 1970) that fertilizer sulfur can be quite

Table 11.4
SULFUR COATED UREA FORMULATIONS

J. Jung	BASF	Ger. P.	2,451,723	1976
A. R. Shirley	TVA	U.S. P.	3,903,333	1975
A. R. Shirley	TVA	U.S. Def. Publ.	268,520	1972
M. Berenbaum	Thiokol	U.S. P.	3,576,613	1967
G. M. Blouin	TVA	U.S. P.	3,342,577	1967
G. M. Blouin	TVA	U.S. P.	3,295,950	1967
B. O'Connor	Standard Oil	U.S. P.	3,206,297	1965
A. W. Green	Mississippi Chem.	U.S. P.	3,313,613	1964

An up-to-date summary of literature is provided by Jarrell (1977).

coarse, of the order of minus 100 mesh. Using coarse grades reduces costs by a factor of up to ten. The best and most convenient application of sulfur consists in blending it with the fertilizer before application (Dardinalovkaya, 1975). Urea and sulfur can be fused together before grinding.

In order to avoid dust problems, fertilizer is often prilled, granulated and flaked. Such materials must be augmented with an additive to hasten disintegration in the soil. For this, sodium bentonite surfactants, gypsum, and calcium lignosulfate (gorlac) are used.

In recent years, as a result of the influence of agribusiness, which emphasizes high productivity with high capital investment, controlled-release fertilizers have become popular, and substantial research efforts are underway to provide elaborately formulated materials. Ideally, such a fertilizer should in one application provide the continued release of plant nutrient concentrations matching the daily changing needs of growing crops. While the theoretical advantages are not yet fully translated into practical realization, it has been discovered that sulfur, which always has served as a chemical "slow release" source for sulfate, also makes excellent coatings (Scheib, 1976; McClellan, 1975), and thus can be used in a multiple function as plant nutrient, binding agent, and coating. Table 11.4 lists some patent compositions. The sulfur coated urea (SCU) is by far the best established mixture, and has been extensively described (Shirley, 1975; Davies, 1974; TVA, 1974; McClellan, 1973; Chekhosvskiks, 1974 and Morozow, 1976), and successfully tested on a variety of soils and crops (Allen, 1971; Liegel, 1976; Jaramillo, 1976; Sharma, 1976), and marketed (Davies, 1974). Jarrell (1977) very carefully studied the release of urea from sulfur coated granules, and developed a model for the release mechanism which quantitatively predicts the release at various

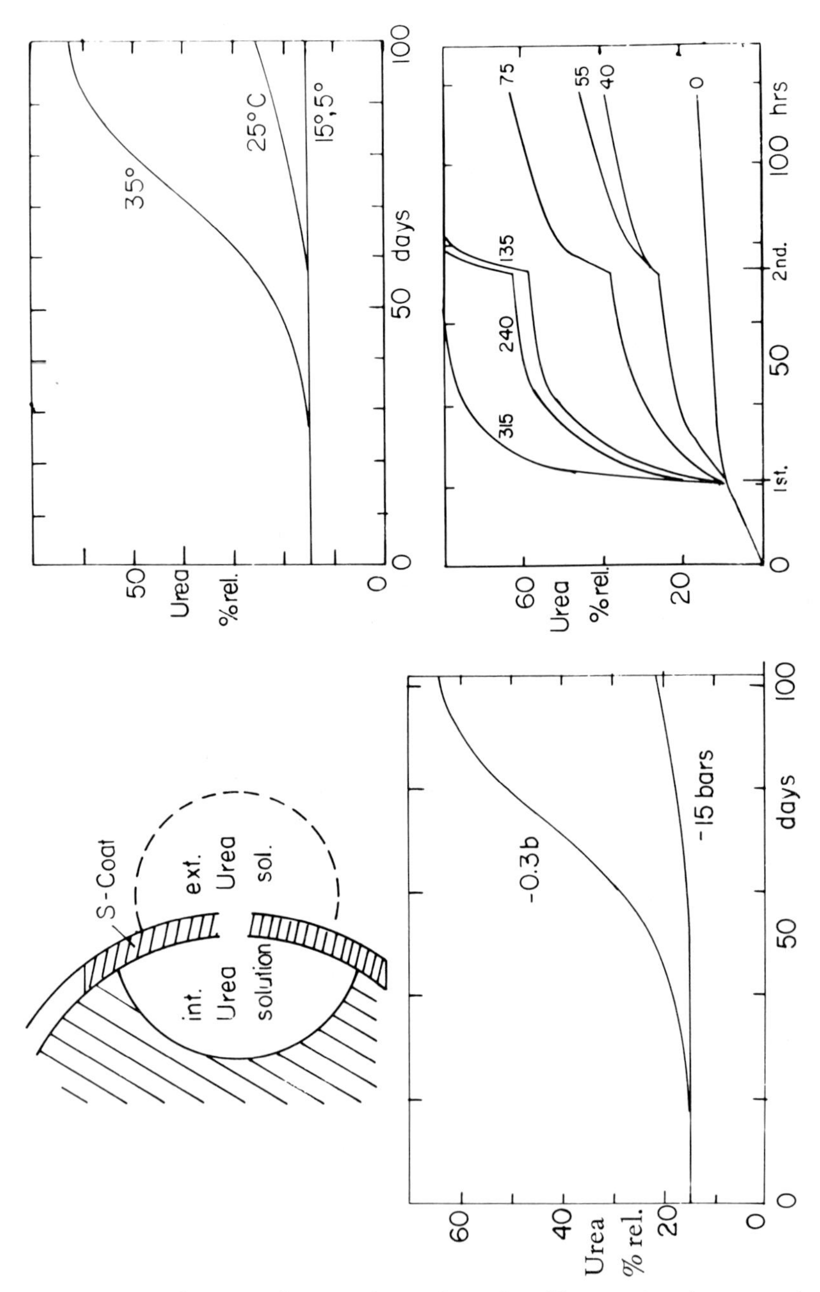

Fig. 11.4. a) Mechanism of urea release through sulfur coating; b) urea release at 5, 15, 25, and 35°C; c) urea release as a function of soil water potential; and d) urea release from SCU, extracted with benzene for 44, 55, 75, 135, 240, and 315 seconds. (After Jarrell, 1977)

Table 11.5

SOLUBILITY AND SULFUR CONTENT OF FERTILIZERS

Material	% S	Solubility in Water at 32°F	
		Parts salt per 100 parts H_2O	% S in sat. sol.
Ammonia bisulfite, NH_4HSO_3	32.3	267	23.5
Ammonium thiosulfite, $(NH_4)_2S_2O_3$	43.1	148	26.0
Ammonium bisulfate, NH_4HSO_4	27.8	100	13.9
Ammonium sulfate, $(NH_4)_2SO_4$	24.2	70.6	10.0
Ammonium sulfite, $(NH_4)_2SO_3H_2O$	23.8	32.4	5.8
Potassium thiosulfate, $K_2S_2O_3$	33.6	96.1	16.4
Potassium bisulfate, $KHSO_4$	23.5	36.3	6.2
Potassium magnesium sulfate, $K_2SO_4{\cdot}MgSO_4{\cdot}6H_2O$	15.9	19.26	2.6
Potassium sulfate, K_2SO_4	18.4	7.35	1.3
Sodium thiosulfate, $Na_2S_2O_3.5H_2O$	25.8	79.4	11.4
Sodium sulfide, Na_2S	41.0	15.4 (10°C)	5.4
Sodium sulfite, $Na_2SO_3{\cdot}7H_2O$	12.7	32.8	3.1
Sodium sulfite, Na_2SO_3	25.4	12.5	2.8
Sodium sulfate, $Na_2SO_4{\cdot}10H_2O$	10.0	11.0	1.0
Sodium sulfate, Na_2SO_4	22.5	4.76	1.0

temperatures, Fig. 11.4. He also predicted and measured the effect of lime on soil EMF and pH, and the release, and the effect of this fertilizer on the soil pH.

Liquid Fertilizers

In many parts of the world, especially in small farm areas, the spreading of liquid fertilizer and manure has remained popular. Recently, liquid fertilizers have been increasingly used for farming of high yield crops in arid regions. For this purpose, irrigation waters are spiked with fertilizer which is metered into the sprinklers, as needed. The solubility of various fertilizers is listed in Table 11.5. Fertilizers must be carefully formulated to prevent corrosion of the metal piping systems.

In the U.S. over 50% of all nitrogen fertilizer is presently applied in liquid form. Ammonium sulfide is well soluble in aqueous fertilizer. It has a pH of 10 and cannot be mixed with acids. Ammonium bisulfite, which is a disulfite with the formula $(NH_4)_2S_2O_5$, as discussed in Chapter Three, has been extensively used, as is ammonium thiosulfate and the sulfates. The latter are not very soluble. This makes distribution difficult,

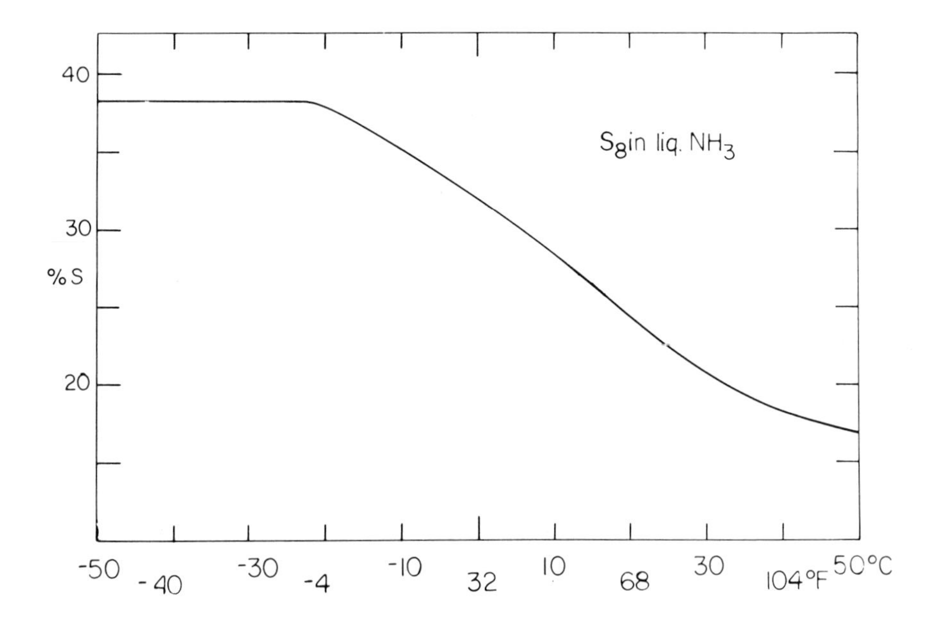

Fig. 11.5. Solubility of sulfur in ammonia

but enhances its residence time in soil. Elemental sulfur is not soluble in water, but it can be dissolved in anhydrous ammonia, Fig. 11.5, and directly applied to the soil. The solution constitutes a slowly reacting chemical system which has been studied for many decades but is not yet well understood (Lautenbach, 1969; Kerouanton, 1973).

Sulfur can also be introduced into aqueous fertilizer as a suspension. Eighty to 100 mesh (0.75 to 0.15 mm) sulfur is satisfactory for this purpose, and far cheaper than the pesticide grades which are minus 200 mesh (0.074 mm) and contain much of the material as 325 mesh (0.044 mm) and even 0.002 mm. Pesticide suspensions are usually called "colloidal sulfur" even though the size of even the smallest particle is at least an order of magnitude too large to constitute a true colloid (Chapter 3). In fertilizer solution, the sulfur tends to settle due to the ionic effect of the dissolved salts.

6. Sulfur in Animal Feed

That sulfur is an important component of the food proteins was discussed above. Only ruminant animals can synthesize these proteins from elemental sulfur (Virtanen, 1970). The progress of research in this field is continuously reported in the Journal of The Sulphur Institute. Elemental sulfur and inorganic sulfur compounds are not yet recognized animal nutrients. Man and most animals depend on cysteine and methionine as their sulfur sources.

B. SULFUR AS FUNGICIDE AND INSECTICIDE

Sulfur has long been known as a fungicide. Both elemental sulfur and sulfur dioxide have been used. Interest in the use of sulfur has gone through many cycles. The effect of sulfur is dependent on the surface area of the sulfur, i.e. particle size. The correlation between these two is shown in Fig. 10.1. The new and effective fine sulfur grades became available only after most users had already shifted to heavy metals and sophisticated organic insecticides. Many of these synthetic chemicals are not biodegradable, several of them are very poisonous, and some are even carcinogenic. This has helped renew interest in sulfur as a fungicide and insecticide.

There has been continued controversy about the reason for the fungicidal action of sulfur. Pollain (1875) was the first to recognize that hydrogen sulfide is toxic to mildew on grape leaves. During the following sixty years a lively debate continued over the relative effectiveness of hydrogen sulfide and pentathionic acid. Both are among the oxidation products resulting when sulfur is exposed to soil or biological material. McCallan (1931) published an excellent summary of early work. Young (1928) discussed the effect of pentathionic acid, while Marsh (1929) concentrated on hydrogen sulfide. Wilcoxon (1930) made experiments that indicated that pentathionate was not more effective than dilute sulfuric acid and that hydrogen sulfide was at least 200 times more poisonous to fungi than both of them. This explains why lime enhances the effectiveness of the element. Liming (1933) tested pentathionic acid on three types of bacteria and nine fungi, including apple scab and the bitter root of apple. In 1930 he had studied the relative toxicity of dithionic, trithionic, sulfurous, and sulfuric acid to spore of *sclerotimia cinerea* and found that pentathionic acid was ten to two hundred times more effective than sulfurous acid, depending on pH and other factors. He correlated his findings with his analysis of water filtrates from various grades of agricultural sulfur, and determined the effect of oxygen on acid-free sulfur, and the effect of weathering on the toxicity. In all cases, pentathionic acid accounted for all toxicity.

McCallan (1931) picked up earlier work and isolated glutathione from yeast and found that it could reduce sulfur to sulfur dioxide at room temperature. McCallan conducted extensive and careful experiments to determine the rate of reduction of finely dusted sulfur, and found that potted strawberry plants with a leaf surface of 460 cm could convert 0.252 mg sulfur per square meter during a 12 hour period at 35°C. The rate of production was greater at the very beginning; after about a day the plants showed symptoms of injury. He next studied fungus spores and found that the production of hydrogen sulfide by *Glomerella cingulata* and similar spores was of the order of 0.020 mg per 10^6 spores per hour. He found that after a short, initial induction period, the hydrogen sulfide release increased and reached a peak after about 5 hours. It decayed to about 10^{-3} mg/hr at 30°C after one day. From this data he determined the toxicity of hydrogen sulfide towards various specimens. He discovered that the sensitivity to hydrogen sulfide and elemental sulfur was not always following the same pattern. Among his most important findings was that direct contact between sulfur and the plant was not necessary, and that the equilibrium vapor pressure of sulfur was sufficient to account for the highest possible reaction rate.

Baldwin (1950) discussed the agricultural use of sulfur fungicides. At that time about 80,000 tons of dustable and wettable sulfur were used per annum, as were 2,000 tons of dithiocarbamates, $(R_2N\text{-}CS_2)_nMe$, fifty thousand tons of copper sulfate, and 70 tons of mercury compounds. The main applications of sulfur in the order of their relative importance were on apple scab (62%), peach scab and rot (17%), peanut leaf spots (7%), grape mildew and general orchard use (6%), with pecan leaf curl, pear scab, and cherry rot accounting for most of the remainder. Baldwin pointed out that the fungicidal agents in sulfur were not conclusively known, but that enzymatic action was involved. At that time interest in xanthates and dithiocarbonates had already overshadowed research on sulfur.

The recent compulsory pesticide registration program, which is jointly administered by the U.S. EPA and FDA, and which, according to the latest revised schedule is due to be completed in October, 1979, requires that toxicity tests have to be submitted for all of the 1500 active ingredients which are presently used in the over 35,000 products. This process will undoubtedly provoke renewed interest in the application of sulfur which is biodegradable, far less phytotoxic than most other fungicides, and finally, after use, converts to sulfate, which constitutes a vital plant nutrient. Sulfur is also a very important fungicide in developing countries, for example India (Soonderji, 1972; Mishra, 1973), where sulfur dust and wettable sulfur is now increasingly used as insecticide and fungicide on sugar beets (Paulus, 1975), tobacco, rubber, chili, and other cash crops, as well as on cotton, mangoes, peaches, etc. Sulfur dust is also used to treat jowar, wheat, and potato seeds. Reports of recent worldwide studies can be regularly found in the Sulphur Institute Journal. The Russian literature also reflects increasing interest in the application of fungicidal sulfur.

C. SULFUR IN THE FOOD INDUSTRY

Fumigation by *in situ* burning of sulfur has been used for thousands of years. When Eurycleia fumigated Ulysses house, her procedure was clearly an accepted practice, as was the sanitizing of wine casks. The first reference to food preservation stems from Evelyn (1664), who recommended that cider be stored in casks containing sulfur dioxide. An historic survey of its use as an antiseptic in food is given by Drummond (1939). Sulfur dioxide is popular for several reasons. Its strongest drawing point is clearly that it is far less toxic than any competing chemical. In the modern food industry, sulfur dioxide is mainly used as an antioxidant to prevent the browning of foods. It also is used in the baking industry, because it modifies bread and biscuit doughs by attacking the cystine S-S linkage, and thus improves baking characteristics.

Table 11.6 shows the permissible levels of sulfur dioxide in various foods in the United Kingdom; these are fairly representative of those accepted in other countries. Excess sulfur dioxide reduces nutritional values by degradation of thiamine.

Sulfur dioxide is applied as a gas, which dissolves in buffered aqueous food systems as bisulfite, HSO_3^- ion, or as disulfite salt, for example sodium disulfite or sodium pyrosulfite (Bourne, 1974). As discussed in Chapter 3, bisulfite exists only in solution, but not as a solid salt, because the pyrosulfite, $Na_2S_2O_5$, is far less soluble than the bisulfite, $NaHSO_3$, and the latter quantitatively converts to the pyrosulfite form. In all these compounds the active ingredient is sulfur in the oxidation state four, which can be readily oxidized to the more stable form S(VI), of the

Table 11.6

PERMITTED CONCENTRATIONS OF SULFUR DIOXIDE IN SOME FOODS

Food	Max. sulfur dioxide (ppm)
Beer, soft drinks and vinegar	70
Cabbage, dehydrated	2500
Candied or syruped peel	100
Cider and perry	200
Coffee extract, solid	150
Flavoring emulsions and syrups	350
Flour, biscuit	200
Fruit, dried	2000
Fruit or fruit pulp (other than tomato pulp) for manufacturing	3000
Fruit juices and concentrated soft drinks	350
Gelatin	1000
Horseradish, fresh grated	100
Jam	100
Pectin, liquid	250
Pickles	100
Potatoes, dehydrated	550
Sauces	100
Sausages or sausage meat	450
Starches, prepared	100
Starch hydrolysed (syrup)	450
Sugar or sugar syrups	70
Tomato pulp, paste or puree	350
Vegetables, dehydrated (other than cabbage or potato)	2000
Wine (including alcoholic cordials)	450
Others	50 or less

After Roberts (1972).

sulfate ion, SO_4^{2-}. Thus, sulfur dioxide is a latent reducing agent, i.e. a mild antioxidant. The four most common groups of food applications are inhibition of non-enzymic browning, inhibition of enzyme-catalyzed reactions, use as a reducing agent, and inhibition of microorganisms.

In many foods, sulfur dioxide acts in more than one way. Non-enzymic browning is caused by the polymerization of amino groups with ketones, or other active carbonyls (Hodge, 1953). This affects the appearance, and eventually the flavor of foods. Sulfur dioxide prevents this reaction. It is used for this purpose in canned, heat-processed foodstuff, for example in canned peas, carrots, pears, cabbage, and in pork rolls, suasages, luncheon meats, and chicken rolls. It also improves the color of canned herrings and other oily fish, especially if tomato sauce is used, or white sauces, for example in prepared convenience food. Twenty-nine percent pyrosulfite greatly increases the storage life of raw, iced fish.

Sulfur dioxide, sulfite, and pyrosulfite are very effective in preserving the appearance of dehydrated meats, fruits, and especially dry milk. Vegetables are best treated by immersion in a bath, while fruits are commonly treated with gaseous sulfur dioxide. Dried apples, peaches, bananas, raisins, currants, and apricots are all treated this way. Frozen food is dipped. White wines and fruit cordials are stabilized for storage in a sulfur dioxide atmosphere. In order to prevent the browning of beet and cane sugar, in the manufacturing process sulfur dioxide is bubbled into the clarified liquor after the lime has been precipitated with carbon dioxide.

Sulfur dioxide prevents the browning of unblanched fruits stored in sugar solutions, and the browning of peeled carrots, apples, onions, and potatoes. In all these systems, enzymic action is blocked, which causes the color change. In tomato puree, fruit cordials, dehydrated cabbage, and grape juice it also prevents the loss of ascorbic acid, and in tomato and carrot that of carotenoid, by antioxidant action. The same helps preserve the flavor of the citrus juices.

In comminuted meat products, for example raw sausages, sulfur dioxide prevents oxidation in air by reducing the brown Fe^{3+} in metmyoglobin to the red myoglobin (Fe^{2+}). The action of sulfur dioxide on pectins consists in depolymerization of pectin, while inhibiting the enzyme-catalyzed competing process. The effectiveness of sulfur dioxide in stopping mold has been discussed above. For treating *Botrytis* and *Cladosporium* attacks on grapes, raspberries, gooseberries, and cherries an aqueous solution is applied. In sausages, sulfite or pyrosulfite inhibits molds, yeast, and salmonellae. Sulfur dioxide is also an effective inhibitor of *E. Coli, Pseudomonas,* and other gram-negative growths, but it is not very effective against gram-positive rods (Dyett, 1966).

During the last twenty years, a number of chemicals have replaced sulfur dioxide in many of its traditional uses, because of their higher potency. Among these are p-hydrobenzoic acid, alpha, beta unsaturated acids, diethylpyrocarbonate, ascorbic acid, antibiotics, propionic acid. During the last ten years, such food additives were used quite indiscriminantly in prepared foods, often in such quantities that they affected the taste. The recently growing fear of potential long range effects of artificial food additives will likely revive interest in and stimulate increased use of sulfite.

Sulfur in Food Flavor

Sulfur compounds have strong, characteristic odors. Hydrogen sulfide exudes the "rotten egg" odor which is recognized by everyone. Volatile mercaptans have similarly objectionable odors. However, sulfur is also responsible for the odor of onions and other vegetables, and sulfur compounds are important components of many cosmetics and perfumes. Mabrouk (1976) studied the influence of sulfur compounds on the taste of red meats. He listed 36 compounds which have been identified in the pyrolysis products of sulfur containing amino acids. Among them are hydrogen sulfide, various thiophenes and thiazoles, aldehydes, amines, and ammonia. The reaction of sulfur containing amino acids in meat with glucose or pyruraldehyde leads to a large number of products of which 31 have been analyzed so far. Among them are thiophenes, thiazoles, picoline, pyrazine, furan, alcohols, and phenol.

Chapter 12

Industrial Uses of Sulfur and Its Compounds

Figure 1.2 shows the correlation between GNP and sulfur use. The present sulfur use is almost 100 lbs per capita per year in the U.S. The sulfur consumption world-wide is related to industrial production. Figures 1.1 and 12.1 show some of the end products for the production of which sulfur is an intermediate. Table 12.1 indicates how much sulfur and sulfuric acid are used to prepare 30 important materials. Several handbooks contain excellent reviews on the use of sulfur. The following constitutes only a short summary of this field, because several end uses are discussed in other chapters. Table 12.2 gives the weighted average for the expected U.S. end uses for the year 2000. As is seen, the estimated values do not differ significantly from the present, except for the field of new uses, discussed in Chapter 14.

A. ELEMENTAL SULFUR

The estimated total world production of sulfur during 1976 was about 80 million tons. About 60 million tons of this go into 120 m tons of sulfuric acid. Table 1.3 and Chapter 7 list the volume and price of earlier production. As mentioned above, because of its physical and chemical behavior, elemental sulfur is the preferred form for selling and shipping sulfur.

Since the middle 1950's, most Frasch sulfur from the Gulf of Mexico has been shipped in liquid form and stored in liquid form at the terminal for further distribution in heated tank cars. The safety problems caused by the slow conversion of trace impurities to hydrogen sulfide have been solved by purging and venting. Safety standards have been developed by the National Fire Protection Association and other organizations. With the advent of Claus sour gas sulfur, the shipment of solid sulfur has become more popular. Essentially all Canadian sulfur is shipped in solid form.

Table 12.1

ELEMENTAL SULFUR EQUIVALENT REQUIRED IN MANUFACTURE OF ONE TON OF PRODUCT (METRIC TON)

Product	Sulfur equivalent	Equivalent of H_2SO_4
Fertilizers		
Diammonium phosphate (DAP)	0.39	1.35
Granular triple superphosphate (GTSP)	0.28	0.95
P_2O_5 in 54% P_2O_5 wet phosphoric acid	0.85	2.88
Wet phosphoric acid (54% P_2O_5)	0.45	1.56
Granulated ammonium polyphosphate (GAPP)	0.49	1.65
Normal superphosphate (NSP)	0.11	0.37
Liquid fertilizer 11-37-0 grade	0.59	1.98
Sulfuric acid, 100%	0.31	1.00
Synthetic fibre intermediates		
Hydrogen cyanide (Modacrylic fibre)	0.07	0.25
Caprolactam (nylon 6 fibre)	0.92	3.12
Acetate rayon (fibres, photographic film, etc.)	0.03	0.10
Vulcanized synthetic rubber (SBR)	0.01	0.04
Carbon disulfide (fibres, cellophane, other chemicals	0.85	2.86
Paper pulp	0.10	0.33
Indigo dye	0.28	0.91
Phenol by sulfonation (plastics)	0.40	1.35
Explosives (nitrocellulose)	0.15	0.52
Lithopone paint pigment	0.10	0.32
Leather tanning (chrome tan)	0.07	0.23
Teflan (100%) (herbicide)	0.38	1.28
Alum, 17% Al_2O_3 (water treatment chemical)	0.13	0.46
Sodium dichromate (tanning, dyeing, paint pigments, etc.)	0.12	0.43
Uranium 235	16.4	55.34
Sodium sulfate (100%)	0.21	0.69
Ammonium sulfate (100%)	0.22	0.74

After R.N. Shreve, 'Chemical Process Industries,' McGraw-Hill, New York, 1967.

Much research is presently under way to determine the best size and shape of prills, slates, granules, and the like, and exploratory research is in progress for converting sulfur to a form suitable for transport in pipelines Berquin (1976). In the shipment of solid sulfur, moisture constitutes a problem. Not only does it add to shipping weight, but it enhances corrosion of shipping vessels, because nascent rust can react with elemental sulfur yielding pyrophoric iron sulfides, Section 3D.

Fig. 12.1 shows that acid is still the predominant intermediate. However, the uses of elemental sulfur as a nutrient and fungicide in agriculture is presently increasing. It will gain further importance in the

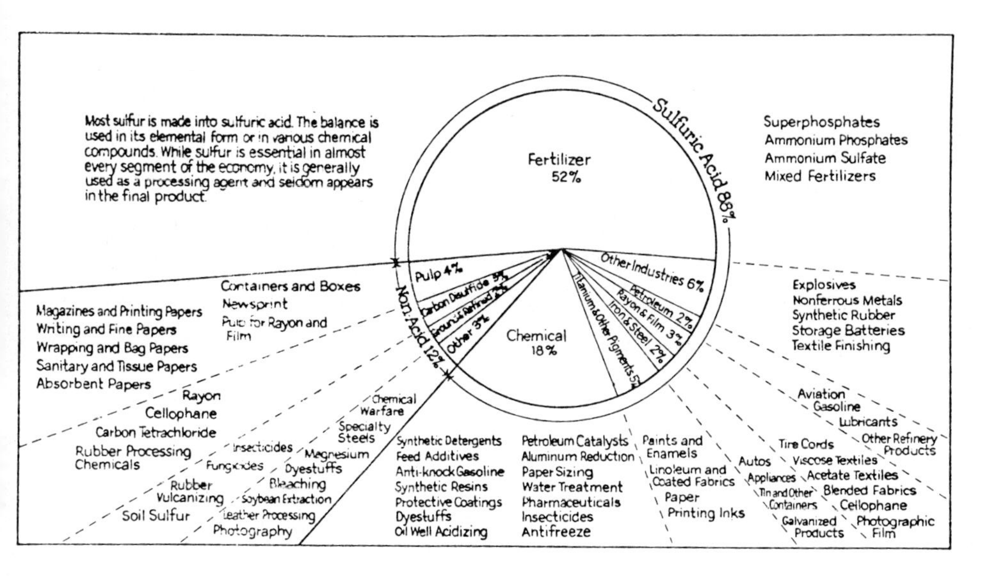

Fig. 12.1. End uses of sulfur (after Gittinger, 1975).

Table 12.2

U.S. PRESENT AND ESTIMATED FUTURE END USES OF SULFUR

End Use	1968 m tons	1968 %	Year 2000 Consumption Quantity (mt) Estimate	low	high	%
Fertilizers	4.550	50	16.800	12.000	22.300	56
Inorganic pigments	.500	5	-	-	-	5
Cellulose fibres	.570	6	2.000	1.500	2.000	5
Nonferrous metals	.300	3	1.050	1.050	2.100	5
Explosives	.250	2.6	.875	.700	1.050	3
Iron pickling	.200	2.2	-	-	.020	0.06
Petroleum refining	.180	2	.255	.325	.530	1.3
Alcohols	.134	1.4	.470	.400	.500	1.5
Pulp and paper	.540	6	1.900	.800	1.000	3
New & other uses	1.860	20	6.500	6.225	7.500	22
Total	9.085		30.000	23.000	37.000	

After Hyne (1975).

future both in high intensity farming in industrial nations and in developing nations such as India, Chapter 2. The use of sulfur in construction and other non-chemical applications is covered in Chapter 14. It is dependent on the fluctuation of the price correlation between sulfur and other bulk materials. These prices, in turn, depend on the world economy, but are also strongly susceptible to local conditions. During the next ten years Canada and the Middle East will have relatively cheap sulfur available. In France, the sulfur production is already decreasing and industry is switching to high value thioorganics. The use of sulfur in polymers, Chapter 13, is now well established. It is merely a matter of time before this use becomes better recognized.

B. SULFURIC ACID

Table 12.3 shows the current trend in end use in sulfuric acid. The price of sulfuric acid is strongly dependent on the purity and concentration. The chemistry of sulfuric acid is reviewed in Sections 3C1 and 3C4. In production catalysts are most important. Phillips of Bristol used platinum as catalyst (B.P. 6,096, 1831). In 1875 Winkler published his work on catalytic oxidation. In 1900 de Haen patented a vanadium catalyst (D.P. 123,616) which was 150 times less effective than platinum but far less sensitive to poisoning. Today, vanadium on a zeolite carrier is the most commonly used material. Despite the highly reliable catalysts available, catalyst poisoning constitutes an important economical factor. Such poisoning has plagued all sulfur dioxide abatement processes aimed at direct conversion of sulfur dioxide to acid.

The most thorough, reliable and useful summary of scientific and technical data on sulfuric acid is still Duecker and West's book (1959). Excellent information is also available in Ullmann's and Kirk-Othmer's handbooks, in publications by Fasullo (1965), and and reports by Gardy (1957), Sittig (1971), Slack (1971), and in pamphlets available from manufacturers. The old monograph by Lunge (1909) still contains much valuable information. Gmelin (1953) lists 17 other full sized books exclusively devoted to sulfuric acid.

Sulfur acid can be mixed with water in any ratio. As mentioned, the concentration is a very important factor in determining the value of acid. Acid with less than 80% is far more corrosive (Fasullo, 1965), Section 3C4, and, of course, more expensive to ship. Sulfuric acid cannot be economically shipped further than a few hundred miles. About half of

Table 12.3
TEN IMPORTANT END USES OF SULFURIC ACID (1970)

End Use	1000 t H_2SO_4 (100%)
Fertilizer	
Phosphoric acid products	13,750
Normal superphosphate	1,240
Petroleum alkylation	2,400
Alcohols	1,800
TiO_2	1,440
HF	880
Iron and steel pickling	800
Pulp and paper	600
Alum	600
Rayon	520

the world production of sulfuric acid is used by the producer. Unfortunately, much of the acid which can be produced with present generation sulfur dioxide abatement processes is only about 40% pure. Thus, it is useless for many applications either for chemical reasons or because of transportation cost.

An excellent graph correlating material suitable for handling sulfuric acid with temperature and acid strength has been prepared by Shell (Stauffer, 1974). Acids containing more than 100% sulfuric acid, i.e. the sulfuric acid to sulfur trioxide ratio is 1, are called fuming acid, or oleum. They are prepared by mixing sulfur trioxide and sulfuric acid. The properties of sulfur trioxide are described in Section 3C1. It is now sold and shipped as a pure liquid (Stauffer, 1975). Detailed instructions for safe shipping and use are now readily available.

Eighty percent of all acid goes into the production of phosphate fertilizer, which follows the over-all scheme:

$$2Ca_3(PO_4)_2 + 6H_2SO_4 + 12H_2O \rightarrow 6CaSO_4(H_2O)_2 + 4H_3PO_4$$

$$\underset{\text{Phosphate rock}}{Ca_3(PO_4)_2} + 4H_3PO_4 \rightarrow \underset{\text{Triple super phosphate}}{3Ca(H_2PO_4)_2}$$

C. SULFUR DIOXIDE

The properties of sulfur dioxide have been described in Section 3C2 and are summarized in Table 3.12. Normally, sulfur dioxide is only used as an intermediate for the production of sulfuric acid, because in other applications it can be substituted by elemental sulfur which is far easier to store and can be shipped at exactly half the weight of sulfur dioxide. Sulfur dioxide can be produced from hydrogen sulfide and elemental sulfur by burning. Sulfur dioxide is a large scale by-product of the smelting of sulfide ores and of combustion of all fossile fuels, but it is almost always converted to sulfuric acid or elemental sulfur, or released as a waste. A family of processes for preparing sulfur dioxide from smelter gas has been patented by ASARCO.

Recently, increased efforts have been made to find new uses for sulfur dioxide. A potential prime application is in leaching of ores (Habashi, 1976). Ramos (1976) estimates the yearly potential by 1985 of other uses of sulfur dioxide as follows: 30 mil tons for water treatment, 30 mil tons for the manufacture of sugar, 120 mil tons for making agricultural chemicals, and 25 mil tons for other uses, including wood pulping.

The toxic effects of sulfur dioxide are described in Chapter 10. Its use in the food industry is described in Section 11C.

D. CARBON DISULFIDE

W. A. Lappadius discovered carbon disulfide in 1796 while superheating charcoal with elemental sulfur. Clement and Desormes described the synthesis in 1803. In 1838 Schroeter set up the first continuous still. The modern synthesis of carbon disulfide is based on work by C.B. Thaker (U.S.P. 2,330,934, 1939), who found catalysts for the reaction:

$$CH_4 + S_4 \rightarrow CS_2 + 2H_2S$$

Belchetz patented a process for producing carbon disulfide from carbon particles suspended in a sulfur gas stream (U.S.P. 2,487,039, Stauffer). According to Fig. 7.4, the preheated vapor contains at 500°C comparable amounts of several sulfur species such as S_2, S_3, S_4, S_5, S_6, S_7, and S_8. Carbon disulfide freezes at -112.1°C and boils at 46.25°C. The heat of fusion is 13.8 kcal/kg, the heat of vaporization is 85.4 kcal/kg. Carbon disulfide has an index of refraction of 1,62546 at 10°C; this value is 20 times larger than that of water and gives the substance its characteristic appearance.

Carbon disulfide has a flash temperature of -30°C; between 0.8 and 50% carbon disulfide explodes; the optimal combustion concentration is 6.8 volume %, i.e. 219 g/cubic meter. Above 95°C carbon dioxide ignites spontaneously. The combustion of carbon disulfide has been well studied. The solubility of carbon disulfide in water decreases from 2.04 g/l at 0°C to 0.14 g at 50°. Below 03°C a white hydrate, $[2CS_2 \cdot H_2O]$, precipitates. The following reactions are well known:

$$CS_2 + H_2O \rightarrow COS + H_2S$$
$$CS_2 + 2H_2O \rightarrow CO_2 + 2H_2S$$

With alcohols a xanthate is formed:

$$CS_2 + ROH + NaOH \rightarrow H_2O + RO\text{-}CS\text{-}SNa$$

These xanthates are used in the flotation of ore, and in the manufacture of viscose. With chlorine, carbon tetrachloride is formed:

$$CS_2 + 3Cl_2 \rightarrow Cl_4 + S_3Cl_2$$

With alkali sulfides, trithiocarbonates are formed:

$$CS_2 + K_2S \rightarrow K_2CS_3$$

Carbon disulfide is used in the viscose industry to prepare cotton, rayon, and other materials. World production is about 1 mil tons per year.

Toxicology

Carbon disulfide attacks the central nervous system, and in acute cases acts as a narcotic which is swiftly followed by death. First symptoms of poisoning are visual perturbation, i.e. veiling of objects, and a crawling sensation in the skin. More than 10 ppm carbon disulfide induces chronic polyneural damage; 100 ppm can cause acute psychosis and neural damage; and more than 1000 ppm will cause delirious excitation. More than 3,000 ppm causes death.

E. CARBONYLSULFIDE, COS

Carbonylsulfide is not an important industrial chemical, but it constitues a steady by-product in the oil and gas industry. It melts at -138.2°C, boils at -50.2°C, and slowly reacts with the amines which are used as absorber for hydrogen sulfide. COS is viciously poisonous; more so than hydrogen sulfide, and it is more dangerous because it has no odor.

F. THIOPHOSGENE, $CSCl_2$

Thiophosgene melts at -110°C and boils at 73.5°C. The fuming, red liquid is formed by reaction of carbon tetrachloride and hydrogen sulfide at high temperature. It is used to prepare isocyanates which serve as insecticides.

G. HYDROGEN SULFIDE, H_2S

Hydrogen sulfide was discovered by Scheele (1777). Its composition was recognized by Berthollet (1798), who gave it its present name. Hydrogen sulfide is recovered from natural gas and is a by-product of ore refineries. It is also prepared by acidification of sodium sulfide. It melts at -85.6°C and boils at -60.38°C. The heat of melting is 568 cal/mole; the heat of vaporization is 4.46 kcal/mole. More physical properties are listed in Table 3.12. The solubility of hydrogen sulfide in water decreases from 6.87 g/l00 ml to 1.23 g/100 ml at 100°C. The solubility in octane is 1.33 g/l00 ml, in methanole 3.0 g/100 ml, and in benzene it is 25.1 g/100 ml. Hydrogen sulfide is too toxic and dangerous to be commercially used on a large scale (Section 10B). Thus, neither commercial production nor utilization are intentionally developed. Its properties are described in Chapter 3. Its occurence as an intermediate is extensively discussed in Chapter 5. The preparation of hydrogen sulfide from natural gas is described in Chapter 7. The transportation has been discussed by Geddes (1969).

H. SODIUM SULFIDE, Na_2S

Sodium sulfide is a yellow solid which is hard to prepare in pure form. The commercial form is $Na_2S \cdot 9H_2O$, a colorless solid which melts at 70°C. The substance is easily oxidized. The solubility increases from 16 g/100 ml at 20° to 40 g/100 ml at 90°C. Sodium sulfide is prepared by reduction of Na_2SO_4:

$$Na_2SO_4 \quad + \quad 2C \rightarrow ? \rightarrow Na_2S \quad + \quad 2CO_2 \quad - 48 \text{ kcal/mole}$$

It can also be produced by electrolysis. The latter product is far more pure.

Sodium Hydrosulfide

Sodium hydrosulfide is used more as a common commercial form than sodium sulfide. It is prepared by absorption of hydrogen sulfide in sodium hydroxide. It is shipped as a 45% solution with a specific gravity of 1.303 and a pH of 10.4. The viscosity is 7 centipoise. The solution tends to crystallize below 62°F (17°C).

Sodium sulfide is used in the leather industry, for the flotation of ores, in the paper industry, in oil refineries, to regenerate lead sulfite, and in the organic industry. Details about handling and properties are given in a brochure by Stauffer (1974).

I. SULFITE, HYDROSULFITE, PYROSULFITE, DISULFITE

The chemical properties of sulfite solutions have been reviewed in Section 3B. An excellent detailed review is given by Gmelin (1953), and in many other handbooks. Only commercial compounds are mentioned here:

1. $Al_2(SO_3)_3$

$Al_2(SO_3)_3$ is prepared by absorption of sulfur dioxide in solutions containing $Al(OH)_n^{n-3}$. It is also used to absorb sulfur dioxide in wash towers; see Section 8D.

2. $(NH_4)_2SO_3 \cdot H_2O$

$(NH_4)_2SO_3 \cdot H_2O$ obtains by absorption of sulfur dioxide in ammonia solution. At low temperature $(NH_4)_2SO_3$ is formed, at 100°C $(NH_4)_2S_2O_5$. Both oxidize in air to ammonium sulfate. The sulfite is used for pulping. Very large quantities of ammonium sulfite or pyrosulfite occur as intermediates in flue gas scrubbing, for example in the COMINCO process. From these solutions sulfur dioxide can be regenerated, or sulfate is produced for use as fertilizer.

3. $CaSO_3$, $CaHSO_3$, $Ca_2S_2O_5$

Calcium pyrosulfite is a major waste product in the present generation 'throw-away' processes which use limestone, lime, or half-calcined dolomite to remove sulfur dioxide. The product crystallizes poorly, partly oxidizes to sulfate, and must be stored in large, water tight ponds to reduce leaching.

The setting and drying of these waste sludges is slow, because a solid crust forms on the surface and prevents oxidation as well as crystallization. A major research effort is under way to solve the problems of handling such solutions, which constitute a major nuisance (Princiotta, 1976).

4. K_2SO_3, $KHSO_3$, $K_2S_2O_5$

Only potassium sulfite and pyrosulfite are known. The bisulfite is more soluble than the pyrosulfite, and is not known in solid form.

5. $K_2S_2O_5$

$K_2S_2O_5$ is also incorrectly labelled $KHSO_3$. It is prepared by saturating a strong solution of KOH with sulfur dioxide at 80°C. A solution with 60-62% $K_2S_2O_5$ can be obtained. Carefully crystallized, pure $K_2S_2O_5$ is used in the food industry as a preservative (Section 11C), in agriculture, and for sterilizing wine and brandy.

6. Na_2SO_3

Sodium sulfite, with the formula $Na_2SO_3{\cdot}7H_2O$ is prepared at 40°C by dissolving sulfur dioxide in NaOH. It also is recovered during the preparation of phenol. It is used in the food industry, photographic industry, as antichlor in textile bleaching, and for preparing sodium bisulfate.

7. $Na_2S_2O_5$

Sodium pyrosulfite, often mislabelled 'sodium bisulfite, $NaHSO_3$' is obtained from soda or sodium hydroxide solutions. It is used for tanning, to purify aldehydes, for making glues, in the paper industry, and to prevent the yellowing of foods and as a preservative.

8. $MgSO_3{\cdot}6H_2O$

This material crystallizes from the dolomite solutions used to treat flue gases. The disposal of $MgSO_3$ is a major problem, as discussed for $CaSO_3$. Regeneration of the calcium oxide by thermal decomposition can be achieved at 1000°C, but the energy needed is exorbitant.

9. $2ZnSO_3{\cdot}5H_2O$

This material is obtained if ZnO is used to strip flue gases. The salt decomposes at 260°C; thus, regeneration is feasible.

K. SODIUM THIOSULFATE, $Na_2S_2O_3$

Sodium thiosulfate was discovered by L. N. Vauquelin in 1802 in the residue of Leblanc-soda solutions. It was originally called 'hyposulfite.' The name has been preserved in the photography industry and among photographers who use it to fix the image in film developing. It acts by complexing the unreacted silver ion. For this purpose, it must be very pure and must be free of heavy ions and sulfite. Normal commercial thiosulfate is 98% pure. It is used to remove excess chlorine in paper and textile bleaching ('antichlor'), to extract silver from ore, to prepare matches, to preserve soap, and in organic industry. Annual world production is about 50,000 tons. It can be prepared by oxidation of sulfides, the reaction of sulfur dioxide on sulfides, and the reaction of sulfite with elemental sulfur between 60 and 100°C. The latter method can be used to prepare concentrated solutions. Originally, it was largely obtained from $Ca(HS)_2$, which is a by-product of the Leblanc-soda process.

$(NH_4)_2S_2O_3$

This material is prepared by reaction of sodium thiosulfate with ammonium chloride, or by the reaction of ammonium sulfite with elemental sulfur. It is used as a rapid fixer in photography.

L. POTASSIUM POLYTHIONATES, $K_2S_nO_6$

Potassium polythionates occur as intermediates in some sulfur dioxide abatement reactions. The structure and reaction are discussed in Section Three C. The only practical use is as a substitute for colloidal sulfur in dermatology, Chapter 10.

M. SODIUM DITHIONITE, $Na_2S_2O_4 \cdot 2H_2O$

Sodium dithionites are also called 'hydrosulfite.' They acquired the name before their structure was known. The chemistry of these compounds is discussed in Section 3B. The solubility increases from 20g/100 ml at 15°C to 44 g/100 ml at 75°C. However, the dithionite decomposes rapidly above 50°C. The decomposition is catalyzed by thiosulfate and polysulfide. If air is excluded and the pH is kept at 8-9, dithionite solutions are stable for several days at room temperature.

Dithionite is prepared with zinc, with sodium amalgam, or with formic acid, Chapter 3B. It is used as a reducing agent and as a bleaching agent for wool, cotton, sugar, cooking oil, and gelatin. Common stabilizers are the polyphosphates. Annual world production is about 100,000 tons.

N. SODIUM HYDROXYMETHANESULFINATE

$NaSO_2{\cdot}CH_2OH{\cdot}2H_2O$ is prepared by reaction of dithionite with formaldehyde in the presence of alkali. This compound, with the tradename 'Rongalite,' is used in printing and for forming polymers, Chapter 13.

O. DISULFUR DICHLORIDE, S_2Cl_2

'Sulfur monochloride,' i.e. disulfur dichloride ('sulfur monochloride, SCl_2, is monosulfur dichloride) is used as an additive to lubricants, in drilling muds, and in organic reactions. Its technical use is described in a brochure (Stauffer, 1976). It reacts with water, yielding sulfur dioxide and sulfur. With ammonia N_4S_4 is formed. In organic chemistry it acts as a chlorinating agent. Disulfur dichloride can react with sulfur and give a mixture of chlorosulfanes, S_xCl_2 with a value of x between 2 and 8, see Section 3A. The main use is for vapor phase vulcanization of rubber, Section 13F. The physical properties are described in Table 3.12.

Organosulfur compounds are becoming increasingly important. Among the many important compounds are methylmercaptan, methyl mercaptopropionaldehyde, which is used to synthesize methiomine, Chapter 11, heavy linear mercaptans with C_8 to C_{12} which are used as transfer chain agents in styrene and butadiene polymerization, thioglycolic acid which is used to make intermediates for pharmaceuticals and cosmetics, and polysulfides which are used as asphalt binders, sealants, and for other purposes, Chapter 13. All these compounds are still primarily specialty chemicals with prices up to $2 per pound, because the usefulness of bulk thioorganic chemicals is not yet fully recognized. It can be anticipated that in the not too distant future thioorganics will become more important and more widely used.

Chapter 13

Sulfur Polymers

Sulfur polymers and the entire plastics industry are comparatively young. Sulfur has been used for the vulcanization of rubber for about a hundred years, but sulfur is not a large fraction of the product and its chemical role is still quite obscure, Chapter 12. The first commerically viable polymer containing a large percentage of sulfur was introduced in 1929 by Patrick. The consumption of polymeric materials has increased at an incredible rate, as shown in Fig. 13.1 (Platzer, 1973). Sulfur is a well established, important component in several high performance polymers, but the potential for making cheap sulfur-based polymers has not even been superficially tapped. Among the most commonly known examples are probably the Thiokol-type polymers which are used for low and high temperature applications, for *in situ* casting of automobile windshields, and similarly demanding purposes; the polysulfones, for example the polyaryl ether Bakelite, which have excellent electrical and high temperature properties, and the polyphenylene sulfides, for example Ryton, which find increasing acceptance as coatings for household and cooking wares and as lubricants. In the following attention is focused on easy to prepare polymers with high sulfur content.

A. POLYMERIC ELEMENTAL SULFUR

Above 160°C liquid sulfur autopolymerizes, because S_8 rings undergo slow thermal scission, and the resulting radical chains attack the residual S_8 rings and form long chains. Fig. 13.2 shows the average chain length and the total concentration of polymer as a function of temperature. At 190°C about 20% of the sulfur is polymeric, and the average chain length is at a maximum of about 10^5. The polymer formation and the physical properties have been extensively studied and reviewed (Tobolsky, 1962, 1965, 1966). The viscous, honey-like properties of the polymeric liquid are demonstrated in most beginning chemistry classes. If the molecular

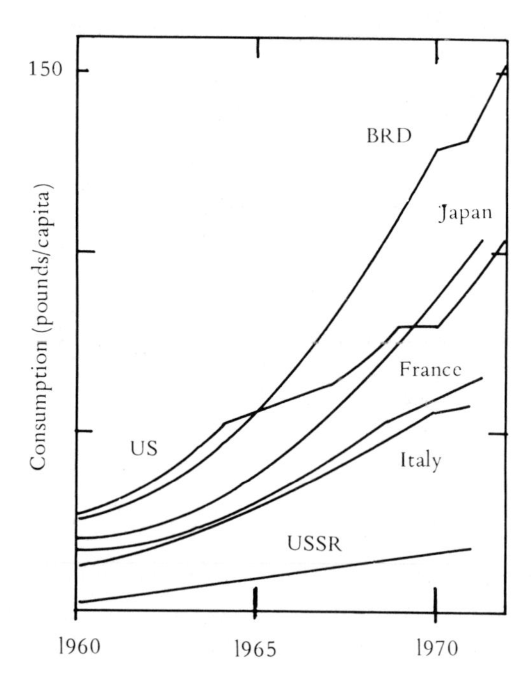

Fig. 13.1. Plastic consumption in six countries, 1960-1975 (after Platzer, 1973).

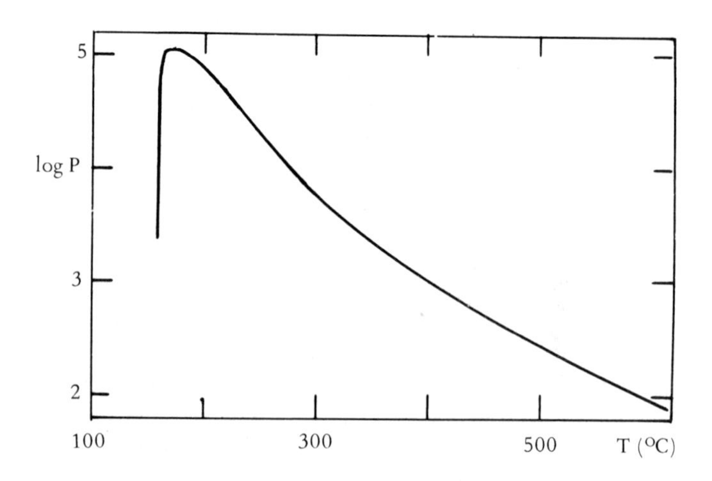

Fig. 13.2. Average chain length of sulfur between 130 and 600°C (after Tobolsky, 1965).

composition could be preserved at room temperature, the material would have very attractive properties. However, upon cooling, the liquid depolimerizes and crystallizes in the original orthorhombic form. Quickly quenched polymer yields a rubbery product which can be stretched up to twenty times its length and easily forms long fibres which contain long S_8 helices. Unfortunately, the fibrous form is metastable at room temperature, and slowly converts to the thermodynamically stable orthorhombic form. During the process, the polymeric chains spontaneously break and convert via an intramolecular ring closure to S_8 which forms brittle crystals. The conversion rate is about three months at 25°C, but it depends upon various factors. Light and mechanical stress strongly accelerate the conversion. In order to make polymeric sulfur useful, the material must be modified.

1. Modified Elemental Sulfur

It is well known that various additives change the composition of hot liquid polymeric sulfur. Fig. 3.8 shows that hydrogen sulfide reduces the average chain length, and thus reduces the viscosity. Arsenic and phosphorus influence the viscosity likewise. These trifunctional additives can form three dimensional molecules of the type $P(S_n)_3$. Schmidt and Wieber (1961) prepared a prototype of such a monomer at low temperature in ether by the reaction of tetrathiophosphoric acid, H_3PS_4, with chlorosulfane, S_xCl_2, which yields $SP[(S_x)P(SH)_2S]_3$

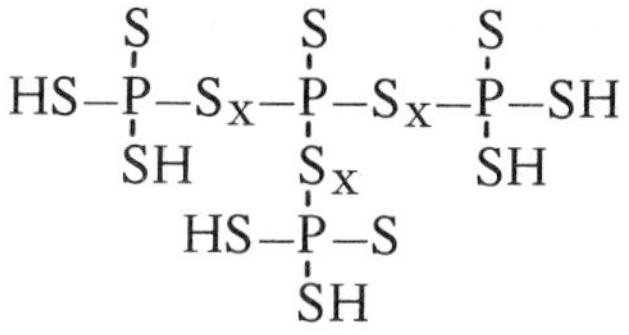

$$
\begin{array}{ccccccc}
 & S & & S & & S & \\
 & | & & | & & | & \\
HS- & P & -S_x- & P & -S_x- & P & -SH \\
 & | & & | & & | & \\
 & SH & & S_x & & SH & \\
 & & & | & & & \\
 & & HS- & P & -S & & \\
 & & & | & & & \\
 & & & SH & & &
\end{array}
$$

The structure of these molecules and their solution will become increasingly better studied, because the electric conductivity of the S-P system makes it attractive for use in lithium-sulfur bulk energy storage batteries. As polymers, these systems are not attractive, because the stabilization of polymeric sulfur is limited, and these compounds are all poisonous and exude a pungent odor. The same holds for other similar inorganic systems.

Several organic substances are known to stabilize elemental sulfur to some degree. Dicyclopentadiene is the most frequently used modifier, and a variety of substances have been used to stabilize polymeric sulfur for use in paints and coatings. Table 14.1 lists some of the tested and

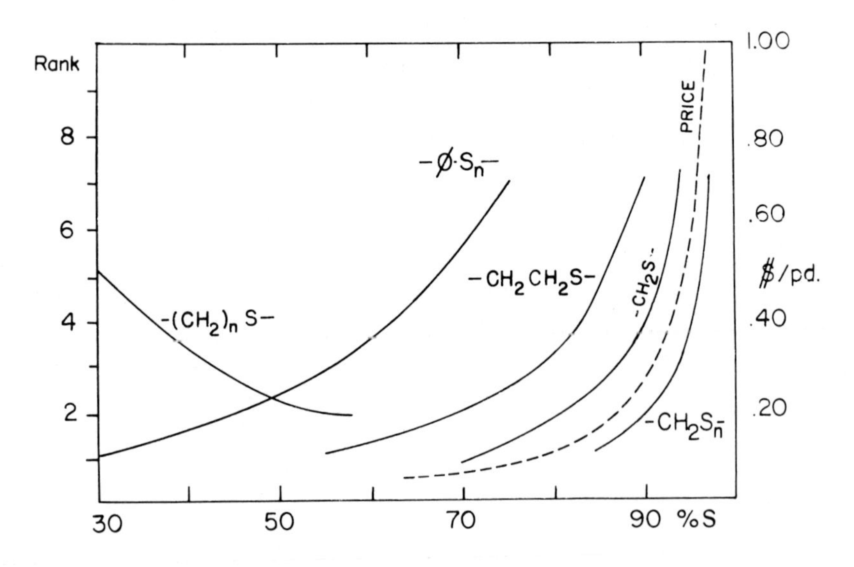

Fig. 13.3. Sulfur weight content of polymers as a function of rank and building block. The price of the polymer is correlated both to sulfur rank and sulfur content.

proposed chemicals. However, there is presently no truly satisfactory modifier available. Chapter 15 shows that a cheap and reliable modifier would immediately open new applications for sulfur, and would make it a viable material for construction and other vast markets. Therefore, research leading towards the development of modifiers is important for all who are forced to recover unwanted elemental sulfur for which marketing in present applications is difficult or impossible.

2. Sulfur Modifiers

In order to be practical, a modifier must be cheap and effective in small concentrations. Fig. 13.3 shows the sulfur content, in weight percent, of a series of possible modifiers. Some of these have been discussed by Jin (1974). The dashed line indicates the composition at which the sulfur and the modifier both constitute 50% of the value of the mixture as a function of the price of the modifier. At a value of 10 to 40 U.S. cents per pound, which is representative for cheap bulk chemicals, only 7 to 20% of modifier can be economically used. Assuming that the material consists mainly of carbon, this corresponds to a ratio of two to four sulfur atoms per carbon atom. Thus, in a pure polymer each monomer unit would have to contain two to four sulfur atoms for each carbon. This could be achieved by forming linear chains with tetrasulfide linkage, or by two or three dimensional branched mono- or disulfide linkage:

$$-[-CH_2-S-S-]_n- \qquad\qquad -[CH\begin{smallmatrix} \diagup S-]- \\ \diagdown S-]- \end{smallmatrix}$$

It will be shown later that such polymers are expected to have attractive properties. For many applications it is not necessary to make pure and well defined polymers, because sulfur ranks up to 6 are known to be tolerably stable. Furthermore, polysulfides are known to be good solvents for S_8; thus, up to 20% unmodified sulfur can be dissolved in such polymers. As a matter of fact, our experiments have shown that some polysulfides, for example sulfanes H_2S_n, are only stable in the presence of dissolved excess sulfur. Removal of the solute sulfur leads to degradation of the sulfane, accompanied by a corresponding reduction in rank, until the solute sulfur is replenished. Apparently, the dissolved sulfur acts as a plasticizer. This also holds for poly(polysulfides) (Tobolsky, 1964) and elemental polymeric sulfur, which, as shown in Fig. 13.2 and explained in Chapter 3, consists of a solution of polymeric chains containing substantial amounts of dissolved rings.

In order to be effective, a modifier must prevent conversion of the polymeric sulfur into orthorhombic sulfur. Unfortunately, the conversion mechanism is not yet known. Several conversion mechanisms can be imagined; all involve S-S bond scission followed by reaction of the newly formed reactive site. Polymers remain stable as long as supersaturation of the polymer solution can be prevented, spontaneous chain scission can be hampered, and if the intramolecular conversion of chains to S_8 can be evaded. Physical modifiers are additives which are soluble and compatible with both rings and chains and increase the solubility of both while preventing crystallization of S_8. Other useful agents are those which dissolve S_8 and form an eutectic, or co-crystallize in an unstable rather than an orthorhombic form. Chemical modifiers have the goal to capture free radical ends formed during spontaneous polymer chain scission, or react with the polymer during its synthesis, and become part of its skeleton and to interrupt the sulfur chains in such a manner that the sequence of eight atoms which is necessary for forming the stable eight-ring cannot form.

The presently known modifiers probably are quite efficient at the physical tasks. Dicyclopentadiene is well miscible with both polymeric and cyclic sulfur, and, thus, forms a ternary system which is bound to have a low melting eutectic and to increase the solubility of S_8. Furthermore, this reagent contains double bonds which can terminate "loose" chain ends, or at high temperature enter the polymer skeleton as a regular building block. The relative contribution of these effects depends on the concentration, temperature, and reaction time during the synthesis of the polymer. It has been experimentally determined that the addition of 6 to 10% of dicyclopentadiene at 150°C with stirring for about 5 minutes to 1 hour gives good results. Higher concentrations lead to high-carbon polymers which separate; higher temperature leads to excessive chemical reaction which forms an insoluble crust and leaves insufficient modifier to solubilize the polymer. For each modifier, the best working conditions are determined by its efficiency in the above described processes. Further improvement might be achieved by introducing a second modifier component which compensates for the shortcoming of the first. For example, one might want to add a chain stopper, such as a sulfide, to a modifier which acts primarily as a solvent, say an aromatic.

The above shows that "modifying" elemental sulfur can be a far more complex process than is generally assumed. This field has not yet been given the scientific attention which it deserves in view of the tremendous applications which seem feasible.

B. INORGANIC POLYMERS

This section deals with polymeric polysulfides, polymeric binary sulfur compounds, and monomeric polysulfides.

1. Polymeric Polysulfides

Only two inorganic polysulfide polymer groups are known; the $(PS_x)_n$ and $(CS_x)_n$ which can be viewed as derivatives of thiophosphoric acid and thiocarbonic acid, respectively. The first are obtained by reaction of S_xCl_2 with H_3PS_4 which was first prepared by Schmidt and Wieber in 1961. They form a light yellow plastic with a sulfur rank of 1 to 7. All are insoluble in water, but are readily decomposed by alkali. In an inert atmosphere, they decompose above 220°C and decompose at room temperature in ambient air giving off hydrogen sulfide.

The second group is formed from thiocarbonic acid, which is prepared by the reaction of barium hydrogensulfide with carbon disulfide followed by reaction with hydrogen chloride in an organic solvent. Polymers form if this acid is reacted with disulfur dichloride in ether at -78°C. The compounds $(CS_x)_n$ with a rank x of 4 to 9 are all orange powders which are insoluble in organic solvents or water; they decompose at 200°C to carbon disulfide and sulfur. The viscoelastic properties of copolymers of the latter were studied by Tobolsky (1962). So far no well defined silicon, polysulfide, or polysilicansulfide has been prepared.

2. Other Binary Inorganic Sulfur Polymers

Only two other compounds constitute truly binary sulfur polymers; $(N_2S_2)_n$ and $(SO_3)_n$. Poly(sulfur nitride) was first prepared by Goehring (1956), who discovered the diamagnetic properties of this semiconductor, which is best prepared by spontaneous polymerization of S_2N_2, or by sublimation. The latter method has been used by MacDiarmid (1975), Street (1975), and others to produce bronze colored single crystals which promise to offer novel material for the electronic industry. Their optical properties have been studied by Moeller (1976). The electronic band structure has been calculated with a theoretical, SCF-LCAO *ab initio* model by Merkel (1976).

Sulfur trioxide can polymerize in two forms, both of which derive from the gamma-form, a cyclic dimer. In the presence of traces of water, either it transforms into the alpha-form, which melts at 62°C, or the beta-form which melts at 32°C. Alpha-sulfur trioxide has an asbestos-like structure. Since 1945 sulfur trioxide has been sold in stabilized trimeric form, which contains traces of boron, antimony, phosphorus, silicium, or some sulfur compounds (Yodis, Ger. P. 2,545,603, 1976).

3. Monomeric Polysulfide

A variety of materials stabilize polysulfide monomer chains. The terminal groups H, F, Cl, Br, SO_3H, and the alkali and alkaline earth metals all form linear $X\text{-}S_x\text{-}X$.

Most preparation methods yield mixtures with an average sulfur chain length between 4.0 and 6.5. These are discussed in Section 3B. Hydrogen polysulfides, which according to their analogy to the alkanes are now officially called sulfanes, are obtained by reacting elemental sulfur with alkali sulfides, as described in Section 3C. Condensation of sulfanes with chlorosulfanes, obtained by the reaction of sulfur or hydrogen

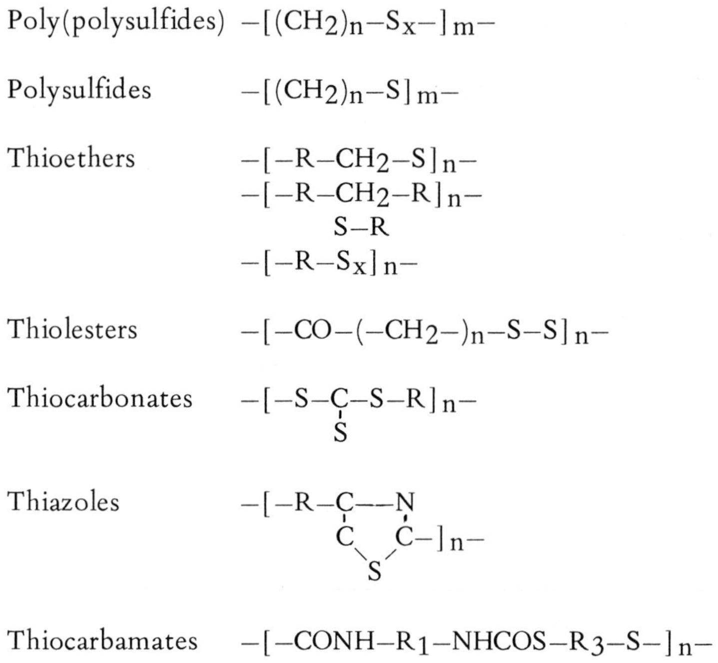

Poly(polysulfides)	$-[(CH_2)_n-S_x-]_m-$
Polysulfides	$-[(CH_2)_n-S]_m-$
Thioethers	$-[-R-CH_2-S]_n-$ $-[-R-CH_2-R]_n-$ (with S–R substituent) $-[-R-S_x]_n-$
Thiolesters	$-[-CO-(-CH_2-)_n-S-S]_n-$
Thiocarbonates	$-[-S-C(=S)-S-R]_n-$
Thiazoles	$-[-R-C$ (thiazole ring: C—N, C, C, S) $C-]_n-$
Thiocarbamates	$-[-CONH-R_1-NHCOS-R_3-S-]_n-$

Fig. 13.4. Organic sulfur building blocks.

sulfide with chlorine, Section 3B, produces heavy oils with the composition $H\text{-}S_x\text{-}H$, where x can be as high as 354 (Deines, 1928; Feher, 1951). However, these long polysulfides are not very stable. The preparation of monomeric organic polysulfides is described below. These substances can now be analyzed with the help of chromatography (Hiller, 1976). This undoubtedly will stimulate basic research of their chemistry.

Disulfides occur in the cysteine-cystine system and are very important in biochemical systems (Greenberg, 1975; Benesch, 1959; Kharasch, 1967). Cyclic polysulfides have been identified in the red algae *Chondria Californica* by Wratten (1976).

C. ORGANIC POLYMERS

Fig. 13.4 shows the different types of building blocks discussed in this section. The chemical reactions and the preparation of the starting materials have been reviewed by Kharasch (1967) and Pryor (1970). This chapter emphasizes those polymers which can be prepared by potentially simple and economic methods. A more general view of the various types of sulfur containing organic polymers can be found elsewhere (Goethals, 1968, 1970; Schmidt, 1962, 1970; Berenbaum, 1958, 1968; Bertozzi, 1968, Panek, 1962; Fettes, 1950; Warson, 1970).

1. Polythiols

Polymeric polythiols can be used as ion exchangers and redox resins; they act in biological systems as radiation antidotes, and have important functions in enzymes. Their general structure is:

$$-[R-\underset{\displaystyle SH}{\underset{|}{C}H}-R-]_n-$$

They can be obtained by polymerization of thiol-containing monomers (in which their group is masked to prevent inhibition), or by introduction of the thiol into the polymer. Overberger (1964) produced poly(vinyl thiol) from S-vinyl-o-t-butyl thiocarbonate which he hydrolyzed with HBr. Berger (1956) prepared poly-L-cysteine by polymerization of S-carbon-benzoxy-N-carboxy-L-cysteine:

$$H[NH-\underset{\underset{\displaystyle SH}{\underset{|}{CH_2}}}{\underset{|}{C}}-CO]OH$$

Overberger (1960) prepared thiol-containing polyamides, and other authors made monomolecular polythiols from polystyrene or styrene-divinyl-benzene copolymers. Liquid thiokols with the formula R-SH are discussed in the next section together with their parent compounds.

2. Poly(polysulfides)

The history of these compounds goes back to 1834, and has been documented by Berenbaum and Panek (1958) with over 800 references. Patrick's work (1924-1949) formed the basis of the commercial class of compounds which is now commonly called Thiokols. The standard synthetic techniques are:

$$Na_2S_x \quad + \quad Cl\text{-}R\text{-}Cl \quad \rightarrow \quad R\text{-}S_x\text{-}R \quad + \quad NaCl$$

The first commercial product was sold in 1929 under the name Thiokol A. It is described in the original patent by Patrick (BP 302,270, 1929); it suffered from an unpleasant and strong odor. The use of bis(2-chloroethyl)ether yielded the disulfide Thiokol D, and tetrasulfide, Thiokol T. Liquid thiokols were developed by Patrick (U.S.P. 2,466,963, 1949) and Jorczak (1951) by reduction of polysulfides with NaHS. Mild oxidizing agents can be used to convert these liquid compounds at room temperature to high quality rubbers. This makes possible their industrial use as *in situ* polymerizing agents. The most common synthesis consists of a condensation (Fettes, 1950).

$$Na_2S_x + Cl\text{-}R\text{-}Cl \rightarrow (R\text{-}S_x)_n + SNaCl$$

Schmidt and Blättner (1959) studied the reactions of sodium polysulfide with methylene chloride, CH_2Cl_2, benzal chloride, $C_6H_5CHCl_2$, and ethylidene chloride, CH_3CHCl_2, with Na_2S_x or H_2S_x with x values between 1 and 8. They found that CH_2Cl_2 reacts with hydrogen sulfide in an analogous way to formaldehyde, and yields pure thioformaldehydes, among them $(CH_2S)_n$, and the cyclic trithiane $(CH_2S)_3$ and tetrathiane $(CH_2S)_4$, a white crystalline substance which melts at 42°C and consists of an eight membered ring, analogous to S_8, but with an endo structure. In liquid hydrogen sulfide, in the presence of triethylamine, a polymer $(CH_2S_{1\cdot 1})_n$ is obtained, in which 10% disulfide bridges are present. Schmidt also obtained the polysulfide polymers $(CH_2S_x)_n$ with x values of 2 to 8, by starting with the corresponding polysulfide. The compound with x = 5 was very soft, those with x = 2 and 8 were brittle; all others were rubbery. All decompose above 150°C. Benzalchlorides with the formula $(C_6H_5CHS_x)_n$ with x values between 2 and 8 yielded red rubbery solids; unreacted benzal chloride acted as an effective softener. The molecular weight of these compounds, which are soluble in carbon disulfide, is about 10,000, which corresponds to an average of 50 monomer units per molecule. The reactivity of the chlorides followed in descending order: $C_6H_5CHCl_2$, CH_2Cl_2, and CH_3CHCl_2. Schmidt (1962) also prepared substituted compounds.

The commercial thiokol synthesis normally begins with ethylene dichloride, 1,2,3 trichloropropene, bis-2-chloroethyl formal, bis-4-chlorobutyl ether, or bis-4-chlorobutyl formal. All these monomers are liquids with boiling points between 84 and 150°C. Patrick described a method to purify the polymers in an aqueous dispersion (U.S. P. 1,950,744, 1934), developed a curing method (U.S. 2,195,380, 1940), and in 1949 developed a liquid polymer which can be crosslinked (U.S. P2,466,963). It has been conclusively established that sulfur chains in all polysulfide polymers are linear, i.e. contain divalent sulfur. Fettes (1954) showed that commercial polymers contain terminal R-OH groups. The molecular weight of many of these substances is very high, up to 500,000. The quality of these rubbers is best when tetrasulfides are formed, but the number and distance between other noncarbon atoms, such as Cl and O, also affects the properties. The most used rubber is made by copolymerization of dichloroethylene with bis(2-chloroethyl)formal and is called Thiokol FA. The glass transition temperature of these compounds lies between -80 and -30°C (Berenbaum, 1968). Their properties have been well reviewed (Berenbaum, 1959; Bertozzi, 1968).

Similar polymers can be obtained by using thiosulfate instead of sodium disulfide (Wiss, 1936). Instead of the organic halides, glycols can be used. For example, bis(2-hydroxyethyl)disulfide, $HO\text{-}CH_2\text{-}CH_2\text{-}S\text{-}S\text{-}CH_2\text{-}CH_2\text{-}OH$, i.e. dithiodiglycol can be made from $(CH_2)_2O$ with disodium disulfide (Bertozzi, 1968). Thiokol WD2 is obtained from bis(chloroethyl) formal and Na_2S_2. It has been found that hydroxyl groups in the beta-position to divalent sulfur are very reactive.

Cyclic di- and polysulfides have been polymerized by Dainton (1957), Tobolsky (1948), and others with various Lewis acids, alkaline salts, and iodine. The same method was used to polymerize fluorinated cyclic polysulfides (Krespan, 1962), but the latter have low molecular weights. The reactions of cyclic bis(arylene tetrasulfide) have been reviewed by Hiatt (1973).

Marvel (1957) and others have polymerized dimercaptans by oxidation in aqueous medium, using oxygen, air, or iodine. Selenious acid is a good catalyst for these simple reactions. Cu-amimine complexes were used by Hay (1966). This type of reaction deserves further detailed study.

The reaction of formaldehyde with sodium polysulfide was studied by Baer (U.S. P. 2,039,206, 1936), Walker, U.S. P. 2,429,859, 1947), Credali (1965, 1966), Ellis (U.S. P. 1,665,186, 1934), Delephine (1898), who reacted formalin solutions with ammonium polysulfide, Bannan (1890), who reacted formalin with hydrogen sulfide at room temperature, Hills (U.S. P. 2,195,248, 1939), who used sodium polysulfide and formaldehyde, Marks (1935) who reacted the same reagents at 50°C, Patrick who reacted formaldehyde with di- and tetra-sulfides (1940, 1941), and the author. It has been little used because of the difficulty in obtaining products with well controllable physical properties (Kawano, 1939; Kohno, 1941). The substances range from oils to brittle plastics.

3. Polythioethers

Polymers with the basic skeleton -[R-CH_2-S–] have been well studied. They can be obtained by several routes. One starts with episulfides, i.e. thiiranes. Delphin (1920) discovered that ethylene sulfide polymerizes in the presence of acids, bases and metal salts to form a white, insoluble substance. Ballard (1949) developed a patent based on the copolymerization of alkylene sulfides with alkylene oxides. In 1954 Marvel described the polymerization of propylene sulfide. For the formation of vinyl polymers, based on styrene, methylstyrene or methylmethacrylate sodium naphthalene or butyllithium have been found to make excellent catalysts. Cyclohexene sulfide can be polymerized with trimethyl oxonium fluoroborate (Stille, 1967). An entire series of patents is devoted to the polymerization of ethylene sulfide, propylene sulfide or similar materials with metal-organic compounds, base salts or acids (Osborn, 1966; Goethals, 1970). Thiethanes and thiolanes also form polythioethers. Likewise, thiols in the presence of peroxides combine with olefins to form thioethers. Without peroxide the reaction follows the Markovnikov rule and yields alpha, alpha-addition. Marvel (1950) prepared a family of copolymer dithiols-diolefins in various media, in bulk or in an aqueous emulsion at pH 3.5. Aromatic dithiols can be treated likewise.

Polythioethers can also be obtained by the condensation of thiols with halides, as described in the poly(poly sulfides) section. The same method also yields poly(phenylsulfides), as is described below (Macallum, 1948). Dithiols react similarly, i.e. in the manner described in Section 3C, where the formation of long sulfane chains from H_2S_x and S_yCl_2 is described. Thithiols containing more than five carbon atoms per unit react directly with aldehydes and ketones, as described by Marvel (1950). This corresponds to the reaction of formaldehyde with hydrogen sulfide, discussed above. Polythioacetone and similar polymers can be obtained in the same way. However, the reactions must be carefully guided at low temperature, because the monomeric thioaldehydes and thioketones are not very stable (Ettinghausen, 1966). Terminated polythioether can be obtained in the same way (Middleton, 1965) at -70 to -50°C. Polythiolesters have been reviewed by Bührer (1973).

Poly(thiocarbonates), the organic analogue to the carbonic acid polymers $(CS_x)_n$ described earlier, can be obtained by the reaction of dithiols with carbon disulfide in the presence of a tertiary amine (Braun 1965).

Polythiazoles, having good high temperature stability, can be synthesized by various methods (Sheehan, 1965).

4. Polysulfones

The sulfones form a large class of very useful and important polymers, in which sulfur plays an important chemical role; but they contain very little sulfur and are, therefore, not important in the context of this book. The reader is referred to reviews by Fettes (1957, 1962), Goethals (1968), Dawans (1961), and others.

Sulfones can be produced by reaction of olefins with sulfur dioxide (Dainton, 1958). Copolymerization of various olefins is very popular, as it yields alternating copolymers. Unsymmetrical olefins form either "head to tail" or "head to head, tail to tail" polymers, or a mixture of both, depending on the structure of the monomer. The reaction is often conducted under pressure. Polysulfones can also be prepared by oxidation of polythioethers, or by polycondensation of dihalides with the sodium salts of disulfinates, such as $C_6H_4(SO_2)Na_2$. Dichlorodiphenyl sulfone reacts with bisphenols in an alkaline medium forming rigid, strong, thermoplastic products containing 50 to 80 units. This material is made on an industrial scale (Union Carbide, Neth. P. 6,605,730, 1966) and has been well reviewed by Walton (1967), Springler (1966), and others. It contains linear structure of the type $[Av\text{-}SO_2\text{-}Av\text{-}O]_n$. Its physical properties have been reviewed by Johnson (1968).

5. Polysulfonic Acids

Polymers containing sulfonic acid groups are usually water soluble, and are used as tanning agents, adhesives, emulgators, and soil conditioners. Water insoluble polymers are used as ion exchange resins. They can be obtained by polymerization of the sulfonated monomer, for example by UV light, or by polycondensation of sulfonic acid containing compounds. Phenol sulfonic-formaldehyde resins have been extensively studied (Kressman, 1949). They form an entire family of compounds which are widely manufactured and used (Am. Cyanamid, B.P. 648,281, 1952; I.G. Farben, Ger. P. 733,679 & 734,279, 1943; Allied Chemical, U.S. P. 2,590,449, 1952; I.C.I., B.P. 829,704, 1960; Dow Chemical, U.S. P. 2,763,634, 1956; Signer, U.S. P. 2,604,456, 1952; Henkel Co., U. S. P. 2,764,576, 1953). They also can be produced by introduction of a sulfonic acid group into the polymer. Polysulfonamides are used as films or fibres which can be spun (Morgan, 1965). Further sulfur containing polymers can be obtained by the reaction of alkene oxides with isothiocyanates (Etlis, 1964).

6. Poly(phenylene sulfide)

It has been known since 1890 (Onufrovitch) that benzene and sulfur chlorides react in the presence of iodine or aluminum chloride and form phenylsulfides. The same polymer was obtained by oxidation of dithiohydroquinon by Leuckart (1890). Friedel and Crafts obtained them as by-products in 1888. Chloroaromatics react with sodium sulfides at a temperature above 250°C and at high pressure. Macallum (1948) synthesized a polyphenylene polysulfide mixture by using p-dichlorobenzene, elemental sulfur, and sodium carbonate:

$$p\text{-}C_6H_4Cl_2 \quad + \quad Na_2CO_3 \quad + \quad S \quad \rightarrow \quad [\text{-}C_6H_4\text{-}S]_n$$

The canary yellow powders can be plasticized with sulfur and form most attractive materials. The synthetic reaction has since been extensively studied (Lenz, 1959, 1961) and improved. It is now known that the air curing of the final product is crucial. It involves a combination of chain extension, oxidative cross linking, thermal cross linking, and oxidation followed by loss of sulfur dioxide (Hawkins, 1976). The crystal structure of polyphenylene monosulfide was studied by Lenz (1960). The orthorhombic unit cell contains four monomeric units in a volume of a = 8.67 A, b = 5.61 A, and c = 10.26·n A. The sulfur atoms of the molecular zig-zag chain lie in the (100) plane. The phenyl groups are off-set at an angle of 45° (Tabor, 1971), as indicated in Fig. 13.5.

This mono-sulfide polymer has found a wide market, because of its superior high temperature performance and the excellent mechanical properties. Some of the properties of the commercial polymer Ryton are summarized in several references. It can be used for casting or coating in such exposed applications as cooking ware or in ball bearings. The applications have been reviewed by Bailey (1974), Short (1972), Hill (1972, 1976), and Allcock (1974).

Phenol and sulfur react in the presence of NaOH at a temperature above 110°C. The kinetics of this reaction have been studied by Cherubim (1966). The sulfur rank of the polymer depends on the ratio of sulfur to

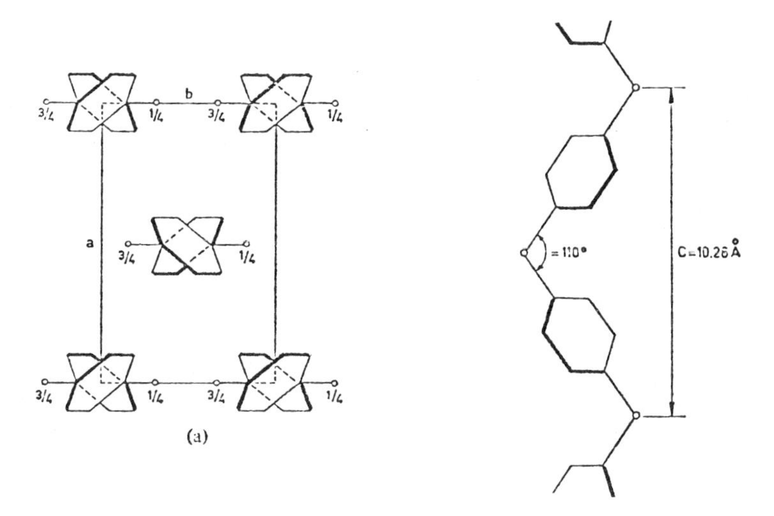

Fig. 13.5. Structure of polyphenylene sulfide unit cell.

phenol and on the reaction conditions. The product mixture is plasticized by excess sulfur dissolved in the polymer (Cherubim, 1966). The initial reaction produces diphenyl disulfide (Neale, 1969). Substituted phenols have been studied by Hay (1976), Popov (1968), Yukel'son (1971), and Geering (U.S. 3,494,966, 1970).

D. POLYMER MIXTURES AND BLENDS

It was stated above that sulfur can act as a plasticizer in many polymers. In such a system, excess sulfur modifies the properties of the polymers. The best known, but very complex, system is vulcanized rubber, discussed in the next section. Similar, crude reaction systems can be obtained by mixtures of sulfur and waxes. Depending on the reaction temperature, part of the two components reacts and releases hydrogen sulfide. This makes the system yet more complex. A carefully studied example of such a system was discussed by Bocca (1973). The asphalt-sulfur mixture will be discussed below.

Vulcanized Rubber

Rubber has been known for many hundreds of years, but the Aztec rubber balls, the Mayan rubber dolls, and the Haitian playballs observed by Columbus in 1493 did not contain sulfur. Despite this, the oldest known rubber samples, found in a 55 million year old Eocene lignite deposit in Germany, contained 2% sulfur. However, this sulfur was extracted by the rubber from adjacent material, and not added by man.

The beneficial effect of sulfur was discovered by Hayward (1838), who found that sunlight rendered a rubber-sulfur mixture non-tacky. The following year, Goodycar (1856) discovered that rubber heated with sulfur gives a mechanically superior product. In 1841, Hancock (1857) independently discovered the same effect in England, and proposed the term 'vulcanization'.

Vulcanization induces cross linkage in rubber and converts plastic natural rubber to an elastic material which is insoluble and resists oxidation. Rubber with 10% sulfur but no additives has to be cured at 140°C for six hours. The addition of 1% thiocarbanilide reduces the curing time to one hour. Thiazole treated material needs only 30 minutes; thiocarbamate reduces the time to one minute. Since 1846, disulfur dichloride has been used for gas phase curing. In 1918 a combination of sulfur dioxide and hydrogen sulfide was introduced, and since then a large number of materials, among them polysulfides, thiokols, and others, has been successfully used (Blokh, 1968). The vulcanization process was long a mystery and a matter of great controversy (Henriques, 1893; Weber, 1894; Bernstein, 1912). It is still very much an art, but we know today that elemental sulfur linkage initially has a rank of about 50. During curing, or by addition of agents, the rank gradually reduces to about 1.5, i.e. the products are a mixture of mono- and disulfides. They include episulfides and other cyclic compounds.

During the last thirty years, the rubber applications have become very sophisticated, and it has been found necessary or convenient to premix natural rubber with the vulcanizing agent and to set the mixture aside. Since the sulfur solubility drops from 10% at 150°C to 5% at 100° and 0.8% at 50°, such uncurred mixtures 'bleed' sulfur which crystallizes. This difficulty has been overcome by the use of polymeric insoluble sulfur with the tradename Crystex (Stauffer, U.S.P. 2,419,309, 1944; 2,419,324 and 2,460,365, 1945; 2,513,524 and 2,524,063, 1948, and 2,757,075,

1954). The advantages of Crystex and its application have been reviewed by Bartzsch (1974). The heat transfer and kinetics of vulcanization have been discussed by Hills (1971). Organic polysulfides have also been used for vulcanization (Patrick, U.S.P. 2,235,621; 2,278,128; 2,402,977; Hendry, 1,989,513; Forman, 3,038,883; Merrifield, 3,124,556; Ruppert, 3,455,851; Shim, 3,725,481; 3,862,923; Wilson, 3,957,738). Inorganic additives have been repeatedly proposed (Baer, U.S.P. 2,039,206; Bacon, U.S.P. 2,410,595; SNPA, Fr. P. 1,427,662). Obviously, vulcanization of rubber constitutes an entire field of research all by itself.

Chapter 14

Sulfur Containing Materials

During the last fifteen years significant progress has been made in formulating elemental sulfur compositions suitable for the construction industry. Many of the basic ideas for these uses were reviewed fifty years ago by Kobbe (1924) in a paper entitled: 'New Uses for Sulfur.' He mentioned six areas of non-chemical uses, among them: Portland cement saturated with 17% liquid sulfur yielding moisture and acid resisting products with ten fold increase in tensile strength; sewer pipes suitable to resist the alkaline Western U.S. soil; sandstone impregnated sulfur for use in grindstones; transite board consisting of asbestos, Portland cement, and sulfur for flooring, billiard table tops, dye vats, stair treads, etc.; high density wood-fibre board containing 50% sulfur; wood-pulp pipe impregnated with 4 times its own weight of sulfur, suitable for acid and electroplating tanks and battery boxes; 40% sulfur, 60% sand, coke, or aggregate compositions resisting nitric acid, phosphoric acid, hot ferric chloride, and even hydrofluoric acid, and suitable for making plaster casts or sulfur molds for casting bronze; and, finally, sulfur impregnated wood for use in railway ties or telephone poles.

This work was neglected during the Depression until after World War Two, when Duecker and others reactivated it. The recent revival and progress is based on work done during the early 1960's, especially by a French and U.S. companies. The present status has been summarized in proceedings of several conferences edited by West (1975), Carpentier (1976), and Fike (1977), and in reports and articles by many others.

Sulfur has long been used to set nails and poles in cement or concrete. The first documented used of sulfur mortar as a sealant is in a water pipe in Kentucky, which was placed before the U.S. Civil War (Sheppard, 1975).

This section is organized according to the presently estimated potential volume of the sulfur market, Table 14.1. The problems

Table 14.1
POTENTIAL NEW SULFUR USES

Section	Product	Potential Sulfur Use in 1980 (mt)
14 A	Asphalt	9
B	Concrete	4
C	Foam	2
D	Cardboard	.3
E	Forest products	.3
F	Batteries	.3
G	Ceramics	.25
	Other	.25
	Total	16.4 mt/yr

After Sundheim (1976).

connected with the end use in the construction industry are discussed in Section 14H.

A. SULFUR-ASPHALT

Asphalt can dissolve up to 14% sulfur. If more sulfur is present, it slowly crystallizes. This process, called blooming, is familiar to all producers and users of road or roofing asphalts. If sulfur-asphalt mixtures are heated, two effects can be observed: At a temperature between 115 and 140°C (240-300°F), sulfur slowly reacts and forms polysulfide resins which dissolve unreacted sulfur and yield highly desirable mechanical properties to the final product (Sadtler, U.S. P. 1,830,486, 1931; Bacon, U.S. P 2,182,837, 1939; Bencowitz, 1938). Above 200°C (400°F), sulfur dehydrogenates asphalt (Section 3C) and makes it hard and brittle, similar to air blown asphalt (van Ufford, 1965, 1962; Petrossi, 1972). The chemical composition and the mechanical properties of asphalt vary widely. The specifications for asphalts by highway departments vary accordingly, and Western and Eastern U.S. States use significantly different grades.

The use of sulfur in asphalt can considerably improve the quality of the product, it can reduce its price, and can even reduce the energy consumed in melting the mixture. The disadvantage of sulfur-asphalt mixtures is that normally special mixing equipment is necessary, and in most of the present processes, blending must be achieved in the field on the construction site. The fear that hydrogen sulfide release would seriously hamper the application has not proven to be justified, except in the roofing applications, where the temperature is frequently not well controlled, and occasionally is kept intentionally very high in order to yield an easy flowing, thin coat.

In road applications asphalt is mixed with aggregate in various ratios which depend on the climate, the road use, thickness of the pavement, the local price of the ingredients, and the intended quality. Since the 'energy crisis', the basics of road surfacing and asphalt application have been reviewed thoroughly by suppliers as well as by government agencies, among them the Federal Highway Administration. Recent asphalt-sulfur research was started by Speer (U.S. P. 3,239,361, 1961; Standard Oil) and Metcalf (B.P. 1,076,866, 1965; Shell Oil). U.S. research is now extensively coordinated by the Texas Transportation Institute at the University of Texas (Gallaway, 1974-1977). In 1975 a test road section of 1.1 km length at Lufkin, Texas, was paved by SNPA in cooperation with 9 companies and the U.S. F.H.A. using a method described in 1973 (SNPA, 1973). The properties of SNPA asphalt are described in detail (SNPA Munich meeting, 1973; Vincent, 1976; Nicolan, Ger. P. 2,016,568, 1970; Garrigues, U.S. P. 3,810,857, 1972, Ger. P. 2,422,469, 1974; SNPA, BP 1,303,318, 1973). In December 1975, a test site in Pau, France, followed (Medin, 1976). It has a traffic exposure of 6000 cars per day.

Gulf Canada developed another method which was patented by Kennepohl (Ger. P. 2,551,929, 1976). 20-50% Sulfur is used and the mixture is prepared on site. Differential thermal analysis shows that up to 19% sulfur stays in the asphalt. Shell Canada developed a 'Thermopave' process which uses aggregate, asphalt, and sulfur in the ratio 82:6:12. Detailed specifications are available (Shane, 1976; Deme, Ger. P. 2,434,417, Shell Int., Neth. Appl. 6,511,212, 1967, Fr. P. 1,444,629, 1965; Fenjiu, 1974). Polish researchers have also worked in this field (Gatarz, 1973). Roofing applications on paper and directly on plywood have been studied by Bryant (1976). Early work was described by Wellings (BP 289,737, 1926, and BP 1,303,318.) Other important asphalt formulations were described by Bacon (U.S.P. 2,182,837), Bencowitz (U.S.P. 2,343,860), Bradshaw (U.S. P. 2,602,029), Rappleyea (U.S.P. 2,780,607), Speer (U.S. P. 3,239,361), Leonards (U.S.P. 3,250,188), Bonitz (U.S.P. 3,634,293), Bennett (U.S. P. 3,721,578), Kopvillem (U.S.P. 3,738,853), Vondrak (U.S.P. 3,745,120). For more details, the reader is referred to the above mentioned references, including Burgess (1975), Saylak (1975), and

Garrigues (1975). Sulfur-asphalt concretes have been described by Pronk (1975).

B. SULFUR-CONCRETE

Two different types of products fall under this title: Sulfur impregnated concrete, and sulfur bonded aggregates.

1. Sulfur Impregnated Concrete

Bacon (U.S.P. 1,561,767, 1925) and Kobbe (U.S.P. 1,561,767, 1925) impregnated concrete with liquid sulfur and found that 10-20% sulfur increased the tensile strength tenfold from 7 to 70 bars. In recent years, vacuum impregnation has been used. In two hours, concrete absorbs 13% sulfur at 125°C. The resulting tensile strength is up to 700 bar, i.e. six to ten times that of the original specimen. Dutruel (1976) found that the strength depends on the age of the concrete and on many other factors. The properties of the product are so attractive that sulfur concretes are already widely accepted worldwide. The impregnation of concrete pipe (Bates, 1926; Saenger, 1933; Ludwig, 1971; Malhotra, 1975, 1976; Thaulow, 1975) imparts not only additional strength, but moisture resistance and chemical resistance. 8% Sulfur yields optimum conditions. Ludwig (1971) explained the cause for corrosion of sewer pipe. Dale and Ludwing have produced over a mile of different pipes which are now being tested in Texas and Delaware in practical application (Ludwig, 1971-1977). Wood (U.S.P. 3,677,994, 1972) demonstrated the use of poly(arylene sulfide) as a concrete filler; Inderwich (U.S.P. 3,954,480, 1976) developed a different cement penetration method.

2. Sulfur Bonded Aggregates

The patent literature on the use of sulfur as a binder goes back to 1859 (Wright, U.S.P. 25,074). This patent describes a clay-glass-quartz and sulfur mixture suitable for making ornamentation which can be colored. Bassett (U.S.P. 47,273, 1865) used clay, graphite, and sulfur in a ratio of 20:20:60 to line iron kettles for the storage of oils, which, at that time, were not yet desulfurized. Caduc (U.S.P, 80,856, 1868) used sand and gravel to make a street pavement. Gray (U.S.P. 84,877, 1868) pressed sand-sulfur slabs and used the tile for floors and sidewalks. Brault (Can.P. 9,642, 1879) used marble dust to make artificial stone. Herbert

(U.S.P. 273,527, 1883) made tile; Barnes (Can.P. 24,505, 1886) cast medallions, picture frames, and moldings; Broadbent (Can.P. 32,973, 1889) manufactured medallions, and Butty (U.S.P. 1,351,617, 1920) used 40% sulfur and 60% Portland cement to make imitation china. Kobbe obtained 7 patents related to sulfur cements: U.S. 1,508,144 (1924) describes sulfur containing structures; Can. 243,365 (1924) describes mixtures containing 40% sulfur and 60% cellular material or volcanic pumice. U.S. 1,551,573 (1925) describes smooth materials made from diatomacious earths; U.S. 1,594,417 (1926) deals with fibrous materials; U.S. 1,612,869 (1927) describes mixture containing coke for building electrolytic tanks; U.S. 1,647,528 (1927) deals with impregnated wood fibres, and U.S. 1,655,504 (1928) deals with sandstone, asbestos, limestone, Portland cement, and other compositions.

Duecker (Can.P. 356,181, U.S.P. 2,135,747, Can.P. 361,013, 1936) added 5% olefin polysulfide as plasticizer to aggregate mixtures and used melted sulfur-aggregate mixtures as replacement for concrete. He also used 50% sulfur, 40% silicate, and 5% carbon black. The high speed of setting and obtaining full mechanical strength were the reason why sulfur aggregate mixtures were employed during World War II for repairing air strips. The recent success of sulfur aggregates such as Sulfurcrete (Vroom, 1977), from which a variety of products and shapes can be cast, is due to the modifier (Dale, 1967; Crow, 1970; McBee, 1975; Vroom, 1976).

The most successful modifiers are dicyclopentadiene (DCPD), dipentene, methylcyclopentadiene, , styrene, polysulfides, and some olefins (Dale, 1965, 1967; Rennie, 1971; Currell, 1975, 1976; Sullivan, 1975; Diehl, 1976; Leutner, Ger.P. 2,461,483, 1976). DCPD is by far the most popular additive, not only because of the high compressive strength obtained (up to 700 kg/cm^2), but also because it mixes easily and readily with sulfur and increases wetting of surfaces by lowering the viscosity. Furthermore, the product is among the least odorous of all compositions. Diehl (1976) discussed the influence of reaction temperature and reaction time on the compressive strength of sulfur concretes with sulfur to sand ratios of 20:80 to 40:60, and of a sulfur foam with a density of 0.30 g/cm^3 obtained by adding talcum and an organic blowing agent, 'Porofor D33', to sulfur DCPD mixtures. He discusses in detail the weathering properties and the erosion by organic solvents, salts, acids, and alkalies. Other work has been summarized by Sullivan (1975), Schwartz (1975), Gamble (1975) and Dale (1975).

C. SULFUR FOAM

The first sulfur foam patent was filed by Mock (U.S. 1,297,583, 1919), who mixed sulfur with salt and leached the resulting structure free of salt. He also tried air injection. Dale and Ludwig (U.S.P. 3,337,355, 1967) prepared a series of well working foaming mixtures based on chemical results of earlier work sponsored by M.D. Barnes. 2% Sulfur was used as an additive to foams by van Raamsdonk (B.P. 1,107,237, 1968), because it proved to fire-retard polystyrene. Toland, Chevron Chemical Co., initiated commercial development of Dale's patent and formulated a range of densities and qualities of foams (Woo, Ger.P. 2,504,111, 1976; U.S.P. 3,887,504; U.S.P. 3,892,686; U.S.P. 3,954,685). Smith (U.S.P. 3,804,702, 1974) made a foam by mixing sulfur with expanded polystyrene; Yamamoto made heat insulating walls for liquid air storage tanks using sulfur foam (U.S.P. 3,870,588, 1975). Kajimur (Ger.P. 2,602,762, 1976) prepared a polysulfide foam; Mitchell (U.S.P. 2,814,600, 1957) made a thiomethane polysulfide foam, and Toone (U.S.P. 3,053,778, 1962) used sulfur, sulfite, thiosulfate, etc. to prevent discoloration of polymethane foams. Hunter (U.S.P. 3,114,723, 1963) prepared a cellular polysulfide rubber; Dahm added sulfur to polystyrene foam (U.S.P. 3,222,301, 1965), and Warne (U.S.P. 3,505,251, 1970) prepared a dimercaptan sponge rubber. Jacquelin (U.S.P. 3,787,276, 1975) used sulfur in corrugated cardboard.

The key to testing foam was the construction of field foaming equipment such as designed by Dale (1966). Sulfur foams have since been tested as collapsible liners in ICBM rocket silos, as thermal insulators for arctic housing, airstrips, and as road insulators in arctic regions and on permafrost (Science Dimensions, 1973). The latter work was initiated as a result of the Styropor experiments of Dow Co. on the Trans Canada Highway in Manitoba in 1962. The U.S. Army sponsored sulfur-polystyrene research in Alaska (Pazsint, 1972; Smith, 1973; Karalius, 1973). Chevron now markets foam under the trade names Sufoam and Furcoat based on their own development of Dale's patent (Toland, 1975). Progress in this field has been reviewed by Fike (1976). Today, foam with densities between 5 and 60 lbs. per cubic ft. with K-factors between 0.25 and 0.45 Btu·in./hr·ft^2·^{o}F and a compression strength of 50 to 300 psi are available. Four cubic meter of foam with a density of 0.16 g/cm^3 can be extruded at a rate of 0.3 m^3/min. The use of sulfur foam in insulating boards has also been tested.

D. CARDBOARD

Sulfur has been successfully used to impregnate cardboard and give it a printable surface (Jacquelin, U.S. P. 3,787,276). McKee (U.S. P. 2,568, 349) developed a machine for making corrugated cardboard using sulfur according to a method which he had patented in 1934. This invention has great potential if the waste disposal problem can be solved.

E. WOOD-SULFUR PRODUCTS

Sulfur is chemically and mechanically compatible with wood. Kobbe (U.S.P. 1,599,135-6, 1926) developed a method for impregnating wood by soaking in liquid sulfur. The wood releases moisture, but gains up to 100% of its weight in sulfur. The resulting product has greatly increased tensile and compressive strength, but is quite brittle, even though the wood resin partly mixes and dissolves in the sulfur. The compatibility of the two materials has been exploited by some 29 inventors to prepare new products. In most of these, wood chips are combined with liquid sulfur and shaped into various articles such as iron casting molds, radio cabinets, and similar high density products. It has been demonstrated that elemental sulfur can be used to make products with a density close to that of solid wood (Meyer, 1977). Elemental sulfur and polysulfides can also be used to modify traditional exterior and interior wood adhesives. Sulfur is chemically and physically compatible with traditional interior and exterior glues. Under proper conditions it can contribute to bond strength (U.S.P.Appl., 1976). If the demonstration work proves satisfactory weathering and aging behavior, sulfur could become a viable adhesive, a component of an adhesive, or a filler or extender for widely used wood products such as particle board and plywood (Meyer, 1976).

F. BATTERIES

Several U.S. government agencies, the Electric Power Research Institute, and several industries sponsor battery research which now approaches the commercial stage. Among the leading candidates are the sodium and lithium sulfur cells. The prime applications would be batteries for electric cars and for bulk electricity storage to provide energy during peak loads and the storage of excess energy during slack hours.

Electrochemical cells are classified according to their rechargeability. Primary cells convert Gibbs' free energy of the chemical reactions between anode and cathode into electricity. The discharge of primary batteries is intrinsically irreversible. Fuel cells are primary batteries in which hydrogen or other derivatives from fossile fuels are continuously fed at the anode and oxygen is added at the cathode. Secondary cells can be recharged electrically. Well known examples are the lead-acid battery and nickel-cadmium cells. Their usefulness depends on the number of charge-discharge cycles that can be expected. Car batteries can survive several thousand cycles. The lifetime depends on current density (current per unit of electrode area) and the depth of the discharge (the fraction of stored energy that is removed). Current batteries lack sufficient storage capacity per unit weight, expressed in Whr/kr, and power per unit weight, expressed in W/kg, for use in automobiles and for bulk power storage. Only nickel-cadmium and silver-zinc batteries can deliver a specific power of 200 W/kg, and no commercially available secondary cell produces the 200 Whr/kg which is necessary for these applications. The lithium-sulfur battery has a theoretical specific energy of 2700 Whr/kg and an equivalent weight of about 30 g and is the unchallenged leader among the 16 leading battery candidates. The sodium sulfur battery ranks seventh with a theoretical specific energy of 1000 Whr/kg and an equivalent weight of 60 g.

The functioning of the sulfur cathode is not yet well understood, because liquid sulfur is complex, as explained in Section 3A, and forms immiscible liquid phases in the presence of alkali sulfides, described in Section 3B. Sodium forms the stable sulfides Na_2S, Na_2S_2, Na_2S_4, and Na_2S_5, Fig. 3.12. In a sodium sulfur cell operated at 300-470°C, the fully discharged electrode consists of Na_2S_2 which melts at 475°C. During discharge excess sulfur reacts with sodium and forms first slightly soluble sodium pentasulfide which quickly forms a second liquid phase. After excess sulfur is consumed, the rank of the polysulfides is reduced by reaction with additional sodium. Details of this process are described by Cairns (1973), Sudworth (1972), and other references listed in this book. The electromotive force of the sodium cell changes during discharge from about 2.2 V to 2.07 V, and drops when the liquid sulfur phase is consumed to 1.72 V. The kinetics and mechanism of the sulfur electrode are not well known. The reduction of sulfur to polysulfide proceeds at a current density of about 0.5 A/cm^2. The pentasulfide reduction to disulfide can reach a current density of 5 A/cm^2. In cells containing molten salts as electrolyte, the chemistry is further complicated by the solubility of sulfur in the electrolyte. This system involves species such as the radical ion S_3^-, Section 3A. The phase diagram, the EMF of the lithium battery, and the voltage current density curves have been carefully measured (Cairns, 1973). Practical problems are encountered with sulfur cathodes because of the poor conducitivity of sulfur and the high vapor pressure at the operating temperature. The present cell design involves porous graphite filled with sulfur and molybdenum foam filled with electrolyte. The latter confines sulfur within the cathode because of interfacial tension. The porous graphite serves as a wick and keeps sulfur in contact with the current collector. The presently best working models use iron sulfide and lithium-silicon alloy electrodes. The current experimental lithium-iron sulfide batteries have an energy density of about 80 Whr/kg; this could be increased to 150 Whr/kg, i.e. a value five fold that of the lead-acid battery. Lithium-sulfur batteries already have lifetimes of more than 1000 hours and can endure several hundred recharge cycles. A peak current density of 3 W/cm^2 can be safely maintained for one minute (Birk, 1975). The sodium cells developed by Ford Motor Co. for use in automobiles and locomotives operate at 300°C, have a current density of 0.4 Whr/cm^2, and use a solid beta-alumina electrolyte in a stainless steel container.

Both the lithium and sodium sulfur cells still suffer problems, especially from corrosion and anode penetration by the electrolyte, but the presently achieved high power densities, high capacity densities, and good cycle lives indicate that it is only a matter of time before commercial vehicles can be driven with this advanced generation of batteries. Their viability as economic peak bulk power storage devices in the electricity generating industries will depend on the span of the present interest in this application.

G. SULFUR IMPREGNATED CERAMICS

Vacuum impregnation of ceramics and tiles makes them highly resistant to moisture, corrosion, and temperature shock. Thus, interior ceramic qualities can be upgraded for exterior use. This application has long been explored in Italy. In recent years it has been carefully studied and developed (Fike, 1973), and is now successfully applied in industry.

Table 14.2
PROPERTIES OF BUILDING MATERIALS AS RELATED TO COMBUSTION AND FIRE

Property	Wood	Thermoplastics (Asphalt, etc.)	Sulfur Based Material
Melting Point, °C	-	176	120
°F	-	350	246
Pilot Ignition, °C	288	300-500	175
°F	550	570-930	350
Heat of combustion, Btu/lb	7,000-8,000	12,000-16,000	4,000
Cal/g	3,900-4,400	6,600-8,900	2,200
Smoke Generation	Yes	Yes	No
Toxic Fumes	Yes	Yes	Yes

After Dale (1975).

H. APPLICATION OF SULFUR COMPOSITIONS

1. Sulfur as a Building Material

The attractive mechanical properties and the low cost have made sulfur a prime candidate as a construction material for low cost housing. Ortega (1976) and Rybczynski (1976) have worked in this field for many years. Ludwig and Dale (1969) have built several houses in Guatemala, two houses in Bogota, Colombia, 42 houses in Cartagena, Colombia, 4 houses in Tanzania, Africa, and 4 houses and a school in Botswana, Africa. The original sulfur house built in 1964 in San Antonio, Texas is still in excellent shape after a recent renewal of the original paint coating. Turner, Ludwig, Dale, and Corredor (1975) have prepared a manual, 'Techniques for Sulfur Surface Bonding for Low Cost Housing,' for the U.S. Agency for International Development. They discussed in detail architectural considerations of six types of simple housing styles, the construction sequence, the sulfur formulations, their applications, the fire hazards, and political-economical considerations. Typical formulations are 100 parts sulfur, 7 parts talcum, 3 parts filler, and 3 parts DCPD. The comparative fire hazards of sulfur are shown in Table 14.2.

Several different sulfur construction techniques have been proposed and tested. Dale's method consists of stacking normal bricks and brushing joint sections with liquid sulfur (Barnes, U.S.P. 3,306,000, 1967). An alternate method is to brush only the surface of stacked building sections. In the latter method glass or other fibre (Harris, U.S.P. 3,183,143, 1965), or burlap is used to facilitate the build-up of a surface layer and give it tensile strength. Tensile strength and quick setting are the properties which set the sulfur building method apart from all other traditional methods. It is possible to pre-assemble wall sections on site and lift them into place within minutes after completion. Other construction methods consist of pouring or casting sulfur, either into bricks (Rybczynski, 1973) or into slabs and walls (Sullivan, 1975).

2. Coatings

Sulfur coatings have not only high tensile strength, but also are acid proof. Kobbe (U.S.P. 1,508,144, 1923) built electroytic acid tanks and the surrounding structure. Sullivan (U.S.P. 1,808,081) patented a similar coating. Wellings (B.P. 289,737, 1926) used sulfur-bitumen coatings for tennis courts. Patrick (B.P. 456,351, 1936) used polysulfides for coating and impregnating surfaces and fabrics. Tieszen (Fr.P. 2,038,332, 1971) developed polyarylene coatings, Chapter 13, which are cured at 300°C onto metal surfaces. Tominaga (Jpn.P. 73 101,393) developed a non-flammable sulfur coating containing 3% styrene and 3% triphenyloxide. Wiekiera (Pol.P. 49,620, 1965) used 2% epoxy to plasticize sulfur for making putty. Milanoff (U.S.P. 1,305,146, 1919) used straw. Kobbe (U.S.P. 1,689,394 and 1,741,555) used felt or cotton panel to insulate pipe; Sullivan (U.S.P. 1,808,081) added 5% metallic coloring to sulfur-Portland cement before coating wooden tanks, pipes, and roofs. Duecker (U.S.P. 2,046,871) added up to 2.5% phosphorous halide and olefin sulfide to make sulfur coatings and lutes (seals); Payne (U.S.P. 2,134,837, 1938 and 2,227,228, 1940), Bacon and Bencowitz (U.S.P. 2,182,837), Patrick (U.S.P. 2,195,380), Gamson (U.S.P. 2,447,004-6, 1948), Stone (U.S.P. 3,053,816, 1962), Dresher (U.S.P. 3,256,106), SNPA (U.S.P. 3,384,609, 1966), Williams (U.S.P. 3,453,125, 1969), Berg (U.S.P. 3,549,571, 1970), Bennett (U.S.P. 3,619,258, 1971 and 3,721,678, 1972), Tieszen (U.S. 3,622,376, 1971), and Dale (U.S.P. 3,823,019, 1974) are only a few who have demonstrated the value of sulfur coatings. Chevron is presently commercializing a sulfur coating under the Trademark Sulcoat for use in acid pits, water sheds, and on other large surface areas. In the building field, the usefulness of coatings depends in some applications on the fire resistance.

3. Fire Resistance

Marx (U.S.P. 1,619,357, 1927) added chlorinated hydrocarbons to reduce the flammability of sulfur. Kobbe (U.S.P. 1,853,818, 1932) used chlorinated diphenyls; Darrin (U.S.P. 1,962,005, 1934) used selenium in combination with halogenated organic materials; SNPA (Neth.P. 6,606,699, 1966) used organic phosphate and bromates, and Allen (U.S.P. 3,440,064, 1969) added 3% styrene and alkene mercaptans, maleic acid, and similar compounds to fireproof sulfur. Van Raamsdonk (U.S.P. 3,542,701, 1970) discovered that sulfur fireproofed polymethanes. Tominaga (Jpn.P. 73 101,393) found that 3% styrene and 3% triphenyloxide fireproofs sulfur and plasticizes it at the same time.

4. Adhesives

Sulfur adhesives work well on almost all types of surfaces, including wood, brick, concrete, and metals. On the latter, a chemical reaction at the interface establishes surface contact. This effect can be used to bond rubber to metal. Hills (U.S.P. 2,195,248) used alkali sulfides; Patrick (U.S.P. 2,206,643) used alkane polysulfides; Wright (U.S.P. 2,331,951) and Ruggeri (U.S.P. 2,711,383, 1955) used a phenolic compound; Seymour (U.S.P. 2,799,593, 1957) used organic disulfides; Horning (U.S.P. 2,910,922, 1959) used liquid polysulfides; Simpson (U.S.P. 3,450,790, 1969) lead dioxide, peroxides, and thiols; Signouret (U.S.P. 3,465,064, 1969) used poly(thiomethylene alkanol); Reinhard (U.S.P. 3,617,361, 1970) used ethyl acrylate; Bertozzi (U.S.P. 3,635,873) used polythiol polymers; Lamboy (U.S.P. 3,635,880) used mercapto-terminated polysulfides; Kenton (U.S.P. 3,764,372) used polyisocyanate and propylene oxide; Paul (U.S. 3,770,678, 1973) used a polysulfide latex; Villa (U.S. 3,817,930, 1974) used thiomalic acid; Smith (U.S.P. 3,659,896, 1973) used benzyl phthalate, and Brossel (Ger.P. 2,545,910, 1976) used bitumen to impart adhesion to sulfur containing formulations.

5. Plasticized Sulfur

Sections 3A and 13A and B explained the need for additives to preserve the desirable properties of polymeric sulfur at room temperature. Research in this area has always been directly related to application. Basic research in this area is necessary and can be expected to yield prompt rewards, because organic polymer chemistry is a sophisticated art and could be quickly applied. Two groups of additives are presently known: The polysulfides, which form malodorous compounds, and the unsaturated organics such as styrene monomer, dicyclopentadiene (DCPD).

Duecker (Can.P. 361,013) and Hamor (U.S.P. 1,959,026 and 1,981,232) used arsenic, phosphorus, and olefins in combinations. These mixtures are quite poisonous. Patrick (U.S.P. 1,890,191) used sodium and calcium tetrasulfide and ethylenechloride; Ellis (U.S.P. 1,964,725) used methanol and polysulfide; Duecker (U.S.P. 2,039,070) used unsaturated hydrocarbon vapors at 120°C to saturate liquid sulfur; Hills (U.S. P. 2,174,000) and Patrick (U.S.P. 2,206,641) used formaldehyde; Patrick (U.S.P. 2,206,642) used tetramethyl thiuramsulfide, ethylene dichloride (U.S.P. 2,221,650), and methylene disulfide (U.S.P. 2,255,228), and SNPA (U.S.P. 3,374,206) used butadiene, dimercapto dithianononane and decane (Ger.P. 2,004,306), cyclic polysulfides (Ger.P. 2,004,305) to prepare moldable compositions.

Sulfur containing polymers were prepared by Tobolsky (U.S.P. 3,299,568), Fox (U.S.P. 3,311,587), Eckert (U.S.P. 3,317,488), Godfrey (U.S.P. 3,427,292), Louthan (U.S.P. 3,434,852, 3,674,525, 3,676,166), Signouret (U.S.P. 3,459,717, 3,560,451), Greco (U.S.P. 3,466,179, 3,734,753), Kane (U.S.P. 3,472,811, Fr.P. 1,517,897), Weesner (U.S.P 3,513,133), Hirsch (U.S.P. 3,576,590), Nitschmann (U.S.P. 3,591,550), Jing (U.S.P. 3,965,067), and Hwa (U.S.P. 2,968,077). A major impetus for this work was Hancock's work (1954) which showed that thiokols, wood resin, butyl polysulfides, chlorinated diphenyl, rubber, olive oil, dibutyl phthalate, p-dichlorobenzene, and alkyd resins are useful for plasticizing sulfur. He demonstrated that this material makes an excellent traffic marking composition, because it sticks to road surfaces and, due to the plasticizer, resists bacterial attacks. A substantial effort has been made to commercialize sulfur road paints, which can be dyed white. Road tests of such composition have shown that the sulfur polymer withstands road traffic better than conventional paints and that the paint deteriorates first in locations where it is not mechanically worked. This is probably connected with the free radical nature of the polymer.

As this chapter indicates, non-chemical uses of sulfur have now reached the threshold toward commercialization. The commercial pioneering work of high quality sulfur polymers which are now widely accepted should greatly aid acceptance of bulk sulfur containing construction products.

Chapter 15

Future Trends

This chapter consists of three parts. In the first part the trends in sulfur research, industrial production, use, and recovery are examined and related to trends in energy use and environmental attitudes. In the second part the future role of chemistry is discussed, as well as the influence of government and education. In the third part a brief attempt is made to look at the over-all developments which constitute the framework within which chemistry, science, sulfur, energy, and the environment must coexist.

A. SULFUR, ENERGY, and ENVIRONMENT

1. Trends in Sulfur Production, Use, and Recovery

In the 1880's two revolutionary inventions were patented: Frasch's production method which in one stroke made the Italian industry obsolete and made the U.S. the sole sulfur world power, and Claus' recovery method which, despite its amazing technical maturity, did not reach its full economic impact until seven decades later. Since 1970 (fig. 1.4) involuntary Claus sulfur production has exceeded voluntary sulfur production.

The sulfur consumption in leading industrialized nations approaches one hundred pounds per capita. If all countries were to approach this value, the world consumption of sulfur would increase fourfold. Table 5.1 shows that the world sulfur reserve, including gypsum, would be sufficient to cover such needs for several centuries. Thus, sulfur production is flexible. Fig. 15.1 shows, however, that the reserves quickly become expensive to produce. Presently, Frasch mines can produce a pure material (99.5% +) for about \$20/ton, a price similar to that of 1890 (Table 1.3). Today a Frasch plant requires about \$7/ton for investment and \$7/ton for production. Of the latter, about \$4/ton are necessary for energy, because about 3,000 gallons of water are needed to produce one ton of sulfur. The off-shore plants in the Gulf of Mexico need, of course, more energy. The cost

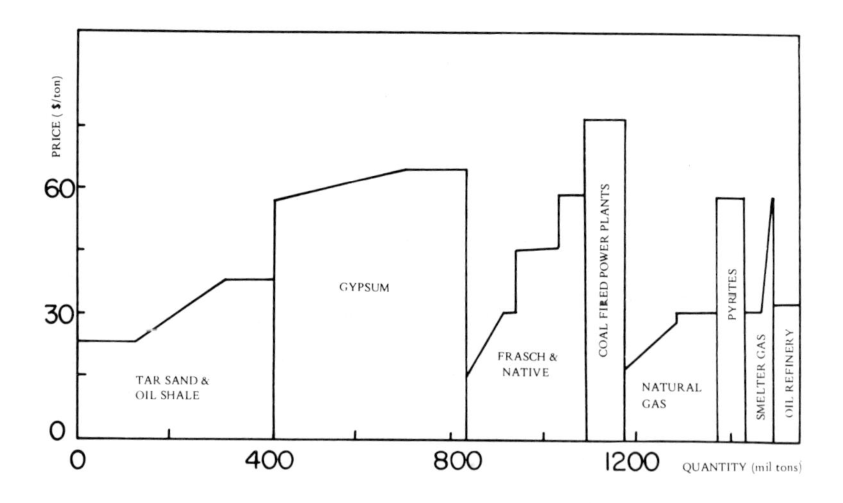

Fig. 15.1. Eight sulfur sources and the price of sulfur as a function of quantities produced.

of producing sulfur from sour gas is irrelevant, because its production increases in response and proportion to energy needs. Thus, sulfur production, energy production, industrial production, and GNP are related and the sulfur supply appears nearly balanced (fig. 1.2). However, regional factors disturb this over-all balance, because the transport of sulfur is far more costly than that of gas. Thus, Claus sulfur prices range from $15/ton in producing locations to $75/ton in distant world markets.

Since 1955 sulfur has been delivered in liquid form. In 1963, 90% of all shipments were liquid Frasch sulfur from Gulf Coast locations. Land-locked Claus sulfur caused a counter trend. A significant research effort is now expended at developing novel dust-free solid sulfur forms such as slates, prills, flakes, pop-corn, etc. A similar effort is aimed at producing sulfur in a form suitable for moving as a slurry in all-purpose pipelines.

All these efforts depend on an economic stimulus. As have all bulk chemical and mining industries, the sulfur industry has historically suffered from a well documented sequence of over-supply and over-demand cycles. Such cycles seem inevitable, because large-scale industry needs at least five years to develop new production sites, but has to respond to consumer demands, which in a free enterprise economy have a much shorter lead time.

Since 1920 it has been pointed out that coal combustion gases constitute an untapped source of sulfur. Industries and government have periodically expected that recovery of sulfur values from combustion gases might become a significant commercial factor. During the last ten years this expectation has significantly influenced attitudes and actions of the public, industry, and government. However, this abatement sulfur has not yet materialized. There are several reasons for this. For one, the extraction of small concentrations of sulfur dioxide from enormous gas volumes in the presence of other solid, liquid, and gaseous wastes is intrinsically, chemically, and thermodynamically inefficient and expensive. The chemistry is still poorly understood, and the necessary basic chemical research has not yet been sponsored. Furthermore, there is little hope that one comprehensive chemical and economical solution can be found, because the chemical composition of coal and the chemical composition and availability of potentially useful abatement reagents varies regionally. The abatement strategy also depends on the size and age of the plant and on whether it is located in an urban or a remote region. Thus, no universal process will serve all power plants equally. In addition, the market price of various abatement products and the cost of disposing of non-marketable wastes vary as a function of time and economic climate. Likewise, the need for energy varies daily, seasonally, and with economic cycles, so that the abatement reactions must be viable over a large range of working conditions.

In some locations the sulfur quantities involved are too small and accrue at too irregular a rate to ever earn a viable share in any commercial market. It is safe to say that less than half of all sulfur emitted from present generation plants is suitable for commercial recovery, and that sulfur from flue gas is not competitive with the present commercial brimstone. This situation could change in future generation plants, if sulfur were to be recovered before combustion. In order to recover sulfur in the near future, political incentives must be invoked. Table 8.7 demonstrates the U.S. situation in regard to seven potential abatement products. Fig. 15.2 shows the necessary 'social incentive.' For sulfuric acid, the social cost is larger than the sales value of the acid (the latter is called 'social gain'). Unfortunately, the political and social interests are dynamic and usually have a shorter lifetime than the amortization of a power plant. In order to overcome this handicap, the U.S. Environmental Protection Agency developed a series of strategies and technologies which are now in an advanced testing stage. The status of this work is being monitored in a series of excellent reports (PEDCo, 1977; Ponder, 1976). In Japan abatement technology has been tested in full-scale use for several years, and some products are commercially used, mainly as gypsum. Presently the problem of recovering nitric oxides and sulfur dioxide in each other's presence is being explored, as discussed in Chapter 8.

Fig. 15.3 shows the estimate for the sulfur supply and demand balance in eight world areas. Several detailed studies have been conducted. They all agree that the over-all system will remain closely balanced, with some temporary regional surplusses or shortages developing. However, a look at the sulfur production in Sicily from 1850 to 1950 (fig. 8.4) shows that the importance of large resources should not be overestimated: If new technologies develop and the international economic balance shifts, a supplier who produces at a constant cost cannot compete. Future sulfur production will be geared to energy production, and thus to energy consumption. Both reflect industrial and agricultural production, i.e. the health of the over-all economy. If coal gasification is introduced, sulfur from coal will eventually–maybe around the year 1990–become a viable by-product. It can be recovered from hydrogen sulfide via the Claus process or by recycling limestone through the coal bed using a recycling method yet to be developed. The impact of coal sulfur would be comparable to the present impact of sulfur from gas, but it would last far longer and would constitute a more dependable long range source.

2. Energy Consumption and Resources

There are presently two key questions regarding energy. One concerns the balance between consumption and reserves; the other concerns the choice of the candidate material which will most likely prove to be the cheapest replacement for oil and gas.

The worry over maintaining abundant power and energy is not new. In 1865 Jevons reviewed the English economy from 1650. Gladstone used his book, "The Coal Question," to prepare his budget presentation to Parliament in which he pleaded for assessment of an energy tax to pay off the national debt so as to compensate the future generation for the irreversible loss of natural resources. In regard to the English coal situation Jevons stated:

> The question concerning the duration of our present cheap supplies of coal cannot but excite deep interest and anxiety...The alternatives before us are simple: If

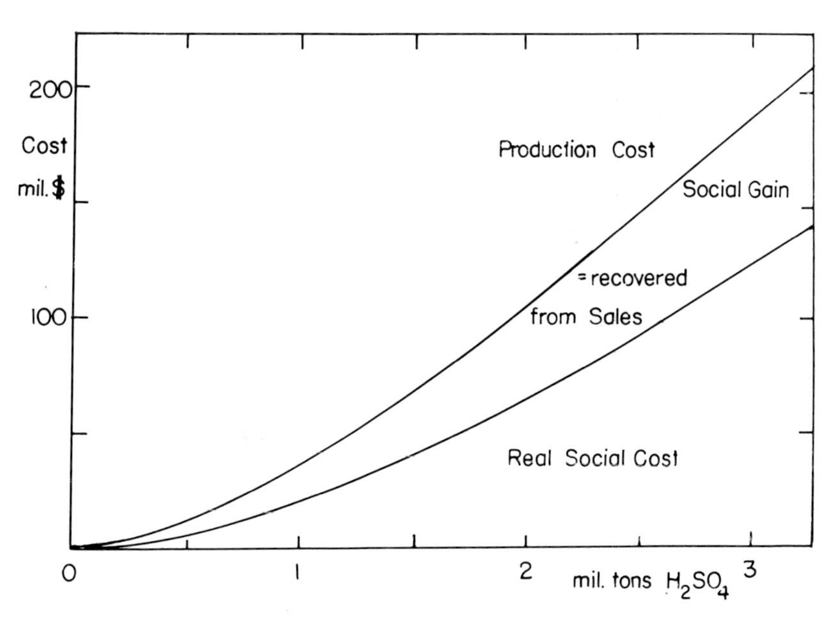

Fig. 15.2. Estimated 'Social Cost' of marketing recovered sulfuric acid as a function of quantity recoverd (after Bucy, 1976).

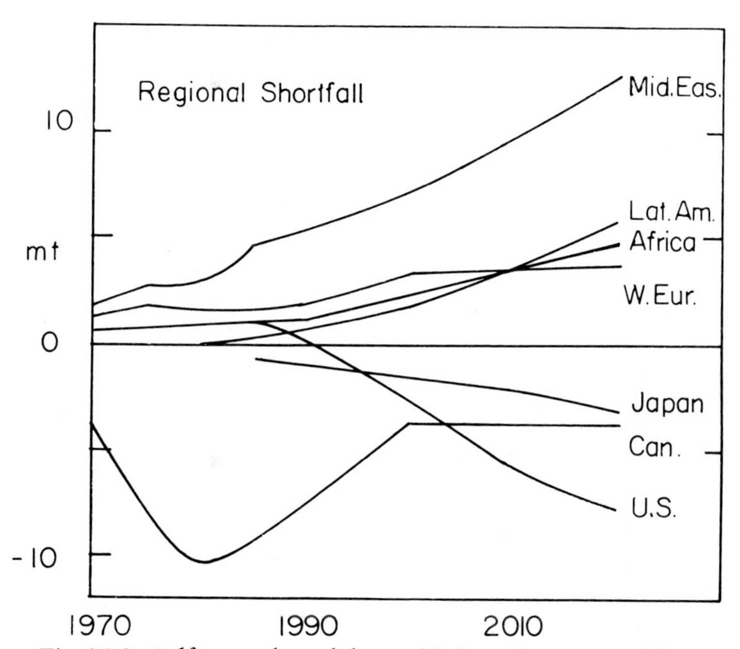

Fig. 15.3. Sulfur supply and demand balance in seven world areas (after Shell report for U.S. EPA, 1972).

we lavishly and boldly push forward in the creation and distribution of our riches, it is hard to overestimate the pitch of beneficial influence which we may attain in the present. But the maintenance of such a positon is physically impossible. We have to make the momentous choice between brief greatness and long, continued mediocrity.

In 1925, after 60 years of unforeseeable growth in population and per capital production and consumption of energy, Devine summarized the world situation differently:

All we have used thus far is less than one two-hundredth part of the original deposits...It seems reasonable to consider our coal reserves 'practically inexhaustible' in the sense that they may be expected to last through more centuries than it is profitable to undertake to forecast our needs.

In 1968 the bids were opened for oil lease in Alaska; the unexpectedly high offers surprised the public and the government, and the industry prepared to build a privately financed pipeline. In the following years environmental groups stopped the project, claiming it was not needed; in 1973 the 'energy crisis' caused hardship for industries and long waiting lines at gasoline stations. Congress ignored the Environmental Impact Statement and voted to build the Alaska pipeline as a crash project at a public expense of $6 billion. In December 1976 U.S. Government energy administrators predicted an impending 'oil glut' in the western U.S. and in February 1977 the new U.S. Administration called for a four-day work week to achieve a crash savings of energy. Despite the many expert reports and high-level government panels, there is wide confusion among all segments of the population about the cause, the extent, and the future of the situation. Everyone has looked for someone to blame, from the U.S. environmentalists to industrial oil concerns to the Arabian rulers (Inglis, 1974). In contrast, Fisher (1973) of the General Electric Co. stated:

The energy crunch of the 1970's was caused by factors that are so clearly identifiable, so clearly understood, and so readily compensatable through ordinary response mechanisms of government, industry, and technology that there can be little doubt of its transitory nature.

He suggested "a potential future in which all fuels must be refined, imports are restricted to protect marginal U.S. producers, but with sufficiently low fuel prices to allow the U.S. to continue its historic evolution; nominal fuel prices would increase because of coal gasification and other new technology, but the actual energy cost would drop below the present because of high inflation." Osborn (1974), a former director of the U.S. Bureau of Mines, also speaks of abundant reserves.

One problem in the evaluation of energy reserves is that the size of 'proven reserves,' both of energy and sulfur, has increased with consumption at an almost identical rate. The seemingly puzzling fact that the reserves increase with time is understood if one considers that exploration is normally undertaken with the purpose of production, and merely serves to establish whether the production time is sufficient to break even before a source is exhausted. Table 1.5 lists some approximate estimates of the presently proved reserves of various sources of energy. However, the main problem is not the size of the reserves, but the choice of a fuel which will be the cheapest during the coming 50 years. There is substantial ambiguity about commercial nuclear energy which fifteen years ago was hailed as the answer to all energy problems for the next five hundred years. The main problems with nuclear energy are still its cost, which is far higher than that of fossil fuels, and waste disposal, discussed in Section 9B. The latter is so severe that a panel of the U.S. National Academy of Sciences is presently in progress of examing the social impact of the commercial use of nuclear energy. Until it has been established which fuel will be cheapest and most acceptable, no one will want to invest in research and development of speculative methods and plants. This will cause more shortages and increased prices, until it becomes important and seems profitable to invest. Once the future balance between coal and nuclear energy becomes established, all utilities will rush to purchase new facilities. The future energy demand depends on attitudes toward, and the efficiency of, energy use.

It has been claimed that the U.S. energy use could be reduced from 100,000 to 70,000 Kwh/capita/yr (fig. 1.3) without changing lifestyles, if the public attitude toward efficiency could be changed to that of Sweden (Schipper, 1976). An English study indicates that the manufacturing industries could save an average of 20% of the energy they use if careful and modern management practices were followed. According to Short (1974), the leaders in energy waste are the iron and steel industries, metals and engineering industries, ceramics, chemicals, textiles, and food industries. The paper industry is an extreme example: In 1970 it used 39 Btu per ton of product. Modern plants could operate with 23.8 Btu/ton, but the thermodynamics of the process show that the industry could operate with no need for external energy subsidy whatsoever if it would find it economically feasible to employ the heat content of wastes. For aluminum production the theoretical thermodynamic minimum is 1/8 of what is

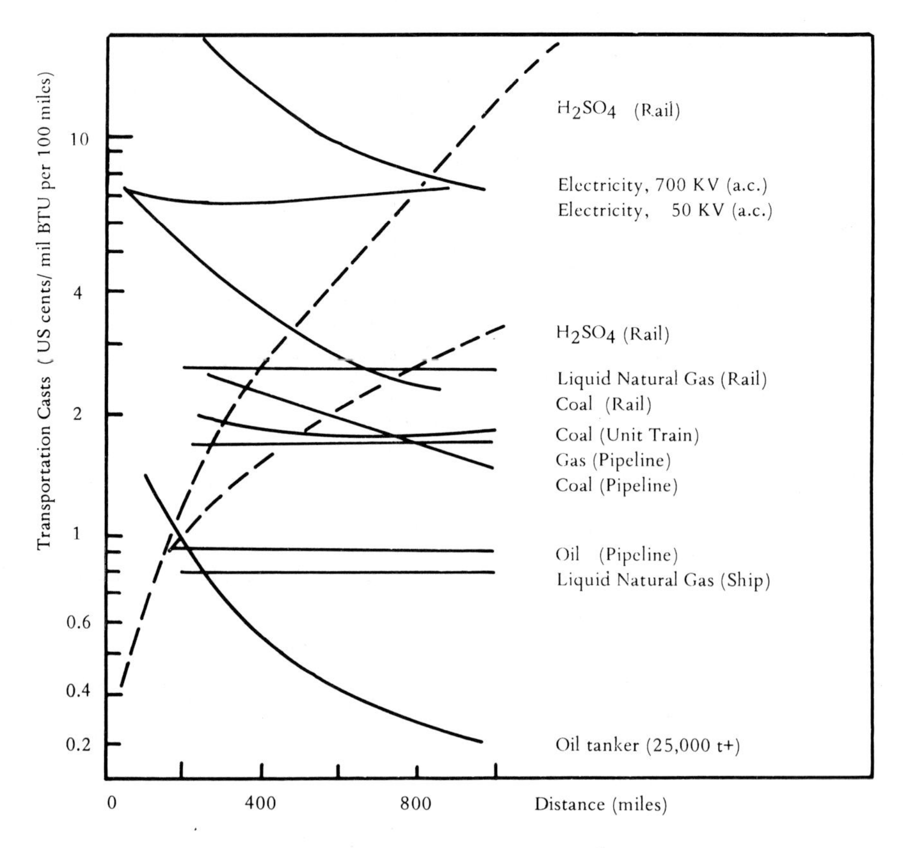

Fig. 15.4. Energy transportation costs (after Bucy, 1976).

now used; for steel the value would be 1/5, for petroleum refining 1/10, and for cement 1/10. Obviously, there is room for innovation, for finding a better chemical way, and for finding optimum engineering conditions to fulfill it. However, it should be pointed out that the present chemical energy production has an efficiency of 90%, while nuclear power has a value of 33%.

Modern industrialized agriculture consumes 100 times more energy for producing one calorie of food than simple cultures. Automobiles use ten times, airplanes 30 times, more energy than a pedestrian. Trucks use six times more energy than railroads or boats or pipelines. However, the aspiration of the present pluralistic governments is not to save energy, but to provide abundant, cheap energy to all its voting constituents.

Since energy sources are not evenly distributed, and since the cost of transporting energy to remote areas is substantial, Fig. 15.4, this goal

Table 15.1
PRODUCTS DERIVED FROM ONE TON OF BITUMINOUS COAL

High temperature carbonization:	Low temperature carbonization:
1440 lbs of coke	1500 lbs 'smoke less' coke
10^4 cft gas	5 gal fuel oil
22 lbs of $(NH_4)_2SO_4$	5 gal creosote
2½ gal. crude benzene	10 gal tar
9 gal. tar	1500 ft gas

cannot be achieved except by political means. Fortunately, most large nations have at least half of their dream fulfilled, i.e. they have abundant energy at a price. However, some regional shortages must be expected. The U.S. oil consumption rate is presently 16.5 million barrels per day; less than half of this is domestic production. Imports from the Middle East are 60% larger than during the oil embargo of 1973. Legislative action has not always been successful. In the long range, the balance between energy supply and the national economy will again be established. Furthermore, government sponsored research should succeed at finding a chemical sun-energy cycle which is shorter than that producing oil or coal. At the 1976 ACS Centennial meeting in San Francisco, Melvin Calvin (who won a Nobel prize for his work on photosynthesis in plants) proposed that certain trees and plants might be suitable for synthesizing gasoline-like products in a one year cycle. If the efficiency of sun conversion were only 0.01%, the present world energy demand could be indefinitely satisfied, and a growth or safety factor of one hundred would remain. However, again, distribution might well remain as a problem of gasoline plants as an energy source.

If coal is adopted as the predominant future source of electric power, sulfur will be removed by pre-combustion treatment as it was 150 years ago. Of the 50 U.S. companies which were technically skilled in the construction of such plants, none has maintained the know-how and the art has been virtually lost, except among some German, South African, and English companies. Modern large-scale technology could make coal plants not only highly efficient power sources, but also the source of cheap sulfur and of profitable coal-tar chemicals. Such chemicals are equivalent to and could become competitive with petrochemicals. Their viability was responsible for the wealth of industrial nations 150 years ago, Table 15.1. Lubin estimated that in 1927 the wasteful practice of direct combustion of coal was costing England 100,000,000*l*. sterling in loss of revenue from unrecovered tar.

3. Environment

Pollution is a serious problem. It is caused by chemical or other inefficiencies which increase when industries are strained. During the economic boom of the early 1960's in several urban and industrial locations sulfur dioxide repeatedly reached ambient air levels at which it acts as a fumigant and damages vegetation and health (fig. 10.3). Fig. 15.5 shows the change in sulfur dioxide concentration in the Chicago area during the last ten years. EPA has estimated that damage due to sulfur dioxide has dropped to 10% that of 1968 (Gilette, 1975). Similar progress has been made worldwide, because once sulfur dioxide is recognized as a serious, politically and socially unacceptable hazard, it can be abated quickly, and because sulfur dioxide settles within half a day after emission into the air (fig. 6.9), damage stops immediately. Today sulfur dioxide abatement constitutes an economic problem, not a chemical one; however, economic removal can only be achieved by improved basic chemistry.

The problems of the last fifty years were caused by the scale-up of processes which were never meant to be used on the present scale. Furthermore, when recognized during the early 1920's, the problem was not tackled chemically; instead it was disguised with tall stacks. The acid rain in Scandinavia and the U.S. has been caused by such stacks, which can interact with medium range metereology. The present approach to abating sulfur dioxide by recovery from combustion gases (Section 8B4) is likewise only palliative. However, the shifting of waste from a gaseous to a liquid form creates serious new dangers. The obnoxious sulfite slurries oxidize only slowly to sulfate, and stubbornly refuse to settle and instead form crusts. It is now expected that these sludges will linger for up to twenty years (Hammond, 1976). A further problem is caused by leaching of trace metals (Table 8.8) into ground water supplies (Princiotta, 1976). Solid wastes constitute an equally dubious solution: They constitute a permanent blight until they can be chemically converted or buried, at an additional expense. The reason for the present, seemingly irrational procedure is explained in Chapter 9. The basic problem is that the development of better basic chemical and engineering procedures takes fifteen to twenty years. Thus, the choice of abatement strategy becomes

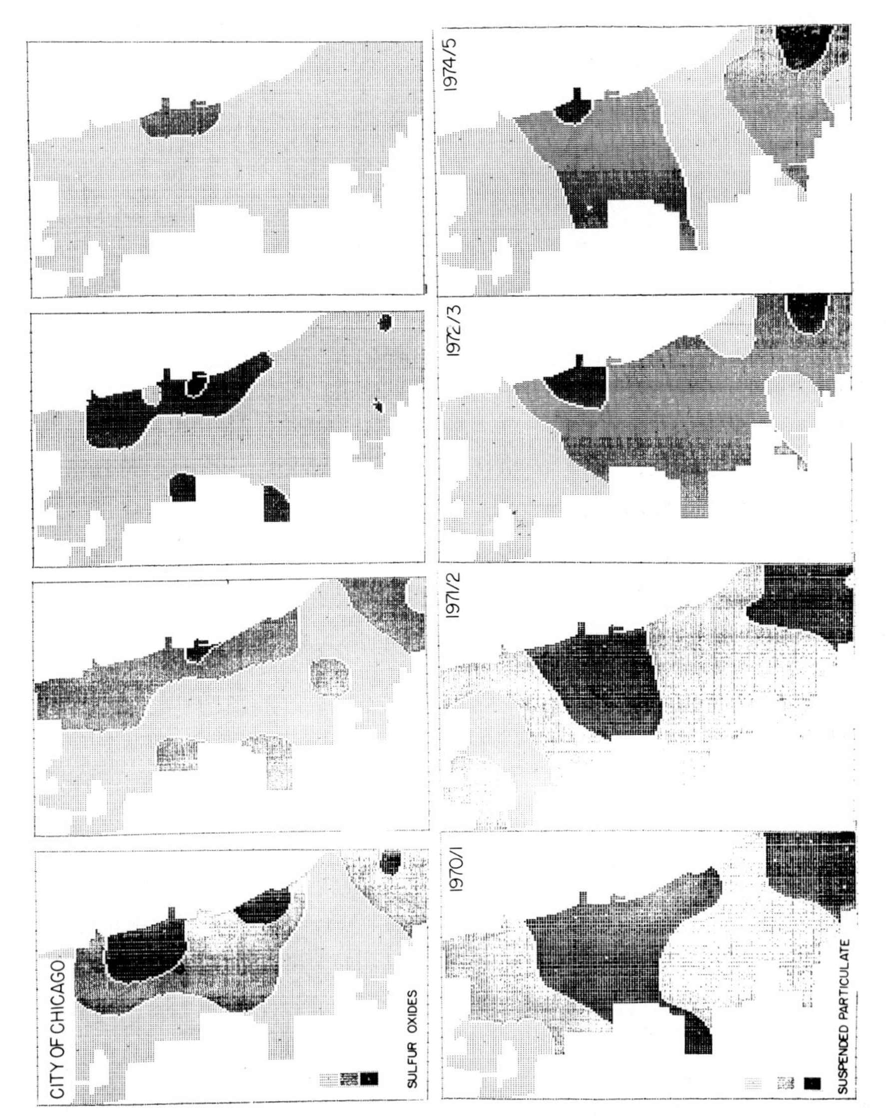

Fig. 15.5. Sulfur oxides and suspended particulates in the city of Chicago, 1970 to 1975.

not a chemical choice, but a matter of balancing long and short range 'social cost' and progress. The present situation is, however, transitory. Once the choice of the future fuel has been made, sulfur dioxide emission from power plants will cease within one or two decades. Coal gasification will make it far more economical to recover sulfur, because the recovery products are elemental sulfur and sulfate; both vital plant nutrients.

B. CHEMISTRY, GOVERNMENT, AND EDUCATION

1. Chemistry

Chemistry will influence future events in many forms: Present and future research will determine the limits of our ability to produce, use, and recover sulfur, the theoretical limits of coal's utilization, and our ability to make maximum use of the beneficial properties of sulfur, or to limit the damage caused by inadvertant by-products. Chemical knowledge will determine the limit of the engineer's ability to translate presently available chemistry into practical, commercially viable processes; and the judicious use of chemical facts by industry and government will determine the lowest possible energy price, and the physical quality of the environment.

Basic Research and Development

The function of sulfur chemistry has been discussed in 13 of the preceding chapters. Inorganic sulfur chemistry plays an important role in agriculture, coal combustion, coal gasification, atmospheric chemistry, environmental protection, sulfur production, corrosion, and many other fields.

In the following a few examples are listed to indicate the type of work that is now necessary. High on the list is the search for an efficient method of extracting sulfur from coal by solvent refining or hydrogenation. For this purpuse the reaction mechanisms and rates of coal gasification and coal combustion must be studied so that new efficient boilers which take advantge of the high temperature properties of modern alloys and refractory materials can be designed. Basic coal research should also include chemical study of sour water, coal tars, and their by-products. Among these is ammonia. The chemistry of ammonia and sulfur dioxide is poorly known (Section 3C).

Next on the list would be better chemical schemes for desulfurization of heavy residuals, oil shales, and tars. For this a more complete chemical analysis of these materials is necessary; similar to the outstanding data produced for crude oils by the Bureau of Mines in Bartlesville under sponsorship of the American Petroleum Institute. This is difficult because many of the organic sulfur compounds are reactive and exist in a dynamic equilibrium which shifts during attempts to separate the individual components. New catalysts are needed for refining because the present Co-Mo materials are still too slow and hydrogen requirements are too high to break the C-S and C-N bonds of heavy residuals and coals. For this, the surface chemistry of sulfur on catalysts must be explored. In this field valuable work has recently been performed in connection with the study of poisoning of automotive exhaust catalysts. The results of this research could make large, untapped petrochemical and fuel resources available. The Federal Highway Administration is well aware that this might disturb the supply of road and roof asphalts and is already looking for alternate surfacing materials; among them modified elemental sulfur.

High temperature chemistry and materials, a field pioneered by L. Brewer, can further contribute to our understanding of combustion chemistry, container materials, and corrosion. Much has been learned about the phase diagram of metal-metal sulfide systems (fig. 3.24) at high temperature. Materials resisting sea water have made possible the off-shore mining of Frasch sulfur in the Gulf of Mexico.

The analytical chemistry of sulfur rapidly reaches maturity. Instrumental methods, including Auger spectroscopy, low-energy high-resolution x-ray spectra, fluorescence, and UV fluorescence allow quantitative and rapid determination of as little as 10^{-12} g/sec of sulfur. Better and quicker methods are still needed to determine compounds with different oxidation states of sulfur in an equilibrium mixture. Furthermore, simpler routine methods for remote and field monitoring remain to be developed.

In the field of sulfur dioxide abatement, there is a basic need to improve the knowledge of inorganic aqueous and gaseous abatement reactions. Basic limestone chemistry (Boynton, 1966) should be updated and the fixation mechanism of sulfur dioxide must be better characterized. The original sulfur dioxide removal methods of the early 19th Century were clumsy but regenerable. The present trend to form throw-away products is counter-productive and not tenable. The key to recovery of reagents is a better knowledge of the reaction mechanism, the kinetics, and the thermodynamics of intermediate reactions and side reactions. Those who have successfully commercialized sulfur dioxide gas recovery from smelters state that, if given an opportunity to start anew, they would abandon the path chosen in the 1920's and would gamble on developing a new liquid-phase process rather than live with the intrinsic inefficiencies of the gas recovery systems. For this, equilibrium phase data on hydrogen sulfide and sulfur dioxide needs to be further improved.

Much of the chemistry of oxyacids of sulfur was described by Wackenroder (1846-1848). These compounds are involved in the oxidation of all aqueous sulfides, a field which has only sporadically found attention, mainly by pulp and paper researchers. Modern workers frequently overlook the oxyacids and their reactions, known to inorganic chemists of the 1850's, because their existence is not deducible from thermodynamics. The polythionate reactions should be studied in order to exploit their unique chemical potential for efficiently recovering Claus tail gases. Another example of knowledge lost in old, foreign language journals concerns the equilibrium between HSO_3^- and $S_2O_5^{2-}$ (Section 3C). The disulfites have been known since 1846; however, most chemical engineers use the Johnstone equation of 1939 which ignores this species and yields incorrect predictions. The oxyacids also hold the key to the oxidation of sludges. New instrumental methods should now make it possible to determine more accurately the mechanism and kinetics of sulfur dioxide absorption in water (Section 3C), its reactions, and the equilibrium composition of the solution. Solubility data and dissociation equilibria both in the gas and aqueous phases are inaccurate and incomplete (Tables 8.5 and 8.6, and fig. 8.6).

The chemistry of sulfur dioxide abatement products must be carefully explored to establish possible new end products. If sulfur is to be recovered, it should not be wasted; new uses for it should be found (Fike, 1974). Gypsum is the most prominent among the presently explored abatement products. The basic chemistry of gypsum has not been studied by inorganic research chemists for over fifty years. Surely modern research techniques could help find better data on the conversion of gypsum which is not a viable commercial outlet for abatement in the U.S. There are many other possible end products which should be explored. The most desirable would be elemental sulfur. A major effort should be made to develop the basic chemistryof new uses of sulfur. Sulfur-cement mixtures, for example, are well known to form extremely strong construction materials (Sections 14C & H). The reason for this is not known and the basic chemistry has not been studied.

Equal challenges await organic chemists. As stated above, the composition and analysis of tars and shales should be more fully explored. The chemistry of asphalt-sulfur mixtures is poorly understood. The reaction of sulfur with cheap monomers forming a stable, strong, non-odorous polymer containing 90% sulfur (Chapter 13) has hardly been explored. The organic and physical chemistry of modified polymeric sulfur has not been explored by basic chemists since Tobolsky, even though these materials are cheap and capable of revolutionizing the construction of roads, housing, and mining pits and tunnels. Sulfur foams constitute another viable material. The challenge consists in finding organic reagents which, in small concentrations, stabilize long sulfur chains and increase foaming characteristics. Only one company has begun to prepare large scale formulations. All present work on new uses of sulfur is applied and is based on research of the 1920's mostly by three people working for one company. Almost all recent efforts can be traced to the initiative of two men, one a mechanical engineer by training. During the last few years, substantial efforts have been made, especially in the U.S. and French industries, to prepare commercially viable compositions suitable for preparing bulk chemicals. Almost certainly, basic chemistry would quickly result in a major breakthrough.

In the biochemical field the importance of sulfur is being increasingly recognized. Sulfur is an important nutrient for plants and animals, and it constitutes a harmless and effective 'slow release' fungicide and insecticide which is increasingly in vogue in developing countries as well as in highly industrialized nations. It also serves as a valuable soil conditioner in arid and alkaline soils. Sulfur research related to agriculture is now being conducted worldwide in universities and industries. Much of this widespread interest is due to the work of The Sulphur Institute which demonstrates the high impact modest funds can have if they are used with persistence. In medicine the effectiveness of sulfur as a palliative against acne and as a seborrheic in shampoos is unexcelled. After a 100 year pause, a revival of research on the active ingredient is overdue. Hopefully, modern abatement strategy will make epidemiological research of sulfur dioxide obsolete. In the environmental sciences, sulfur dioxide has stimulated substantial interest as a monitor of the anthropogenic environmental cycle. It has a residence time of less than a day in the air and quickly integrates into other cycles. In some locations anthropogenic air sulfur is an important link in minimizing the sulfur shortage in underfertilized and overproduced soils.

Applied Chemistry

Over a hundred and fifty years of progress in all basic fields of chemistry waits to be exploited for new processes and products. As stated before, most of the bulk inorganic processes such as sulfuric acid manufacturing, iron and copper smelting, and coal gas manufacturing reached 'near perfection' in 1849 (Liebig, Chapter 2). It is now time for chemists to reconsider all these processes and test modern materials, knowledge, theory, and attitudes.

It has been claimed repeatedly for over a hundred years that sulfuric acid is an obsolete material. It is true that modern technology has reduced the need for sulfuric acid in some areas: Viscose rayon and nylon 66 have been replaced by more advanced materials, wood pulp is treated with less sulfite, titanium dioxide is made with chloride, and steel is pickled with HCl instead of sulfuric acid. However, sulfuric acid is readily available and the waste product, the sulfates, is more easily absorbed by the environment than those of any of its replacements. Thus, the over-all use of sulfuric acid has further increased. In the future uranium and copper enrichment will further add to the need for sulfuric acid. In the area of abatement chemistry, the search for the 'ideal process' described by Sherwood (1971) should continue. For this, a better understanding of aqueous sulfur chemistry at elevated temperatures and at intermediate and ionic strengths of the type explored by Oestreich (1976) and Borgwardt (1965-77) is necessary.

In the polymer field the thiokols, which can be cured *in situ* and are fairly non-odorous, have not only influenced the space industry, but also labor practices in automobile assembly. The polyphenylene sulfides of Macallum are now beginning to replace Teflon in cooking ware, as coating for drill bits, and in similar exacting applications.

Engineering

During the past 100 years engineers have significantly improved the efficiency of the Claus process, the sulfuric acid process, the smelting process, and the sulfur dioxide recovery process. Several of these now achieve over 99% of the theoretical efficiency of the chemistry. Whenever practical demands arise, engineers have become amazingly efficient at developing processes and improvements as needed. Rarely have engineering problems delayed new projects more than five years. If engineering seems inefficient, as it might be in the sulfur dioxide abatement field, it is almost without exception because engineers were pressed into areas of basic chemical research. As an example in point is the above mentioned study of the Wackenroder reaction which caused substantial delays during construction of Claus tail gas treatment plants. This happened because sulfur chemists were not available and sulfur chemistry has been dropped from modern basic chemistry textbooks. Thus, many scientists and engineers believe that basic inorganic sulfur chemistry is fully explored and well known.

A key area for progress in engineering is the design of coal combustion and gasification equipment using modern steels and other materials capable of handling a large fuel flow at higher temperature (Squires, 1971). Such work depends on greater knowledge and understanding of combustion chemistry and materials sciences. The present plan is for the U. S. Energy Research and Development Agency to build six liquid gas plants by 1985 at a cost of $1 billion per year. Similar costs have been estimated for the construction of a 250 mil scft/day synthetic natural gas plant. If these costs are compared to the estimate of the Texas Railroad Commission that it would take $18 billion to convert Texas industries from oil to coal, it is evident that the government effort for coal utilization is insufficient to successfully compete with other energy programs, such as the nuclear program.

2. The Role of Government

During World War II the U. S. government established strong leadership in academic research and in advanced training and education. This change has been profound and lasting. Today science research and education at universities is dependent upon federal support because many leading state and private universities draw up to 50% and more of their operating funds from overhead charges levied against their faculty's research grants. As a result, Congress demands increasing accountability of academic research. Some of the present problems in federal funding are similar to those encountered when stockholders in a private corporation realize that research expenditures have begun to exceed tax deductible income: The Federal government limits expenditures and at the same time expands its influence on the remaining work. The American Institute of Chemical Engineers believes that the next ten years will bring,

> ...problems of communication and credibility. However, because of a lack of understanding of the problems on the part of the public and a continued lack of political leadership, solutions to this problem will only come as a result of continued crises and will essentially be piecemeal rather than bold steps forward. For this and other reasons, it appears that we will be living under considerable uncertainty.

(Timmerhaus, 1976). However, due to nuclear experience, many governments are now more knowledgeable and have better expertise available than is yet commonly realized. A major intrinsic problem with implementing any forward-looking proposals of the executive and legislative branches of government in the field of energy and environmental matters is that they almost invariably conflict with vested interests and, thus, with political reality. Already 150 years ago De Tocqueville observed that in democracies the majority tends to become despotic and that the informed leader has to yield to the demands of the majority of his constituency. No matter how forward-looking most public leaders might like to be, regressive action frequently prevails (Affleck, 1976). Industry responds to such legislation with self-protecting action. Other countries will encounter similar problems and may have far smaller domestic resources. It is expected that the U.S. oil industry will further de-emphasize fuel production during the next ten years. As a result, the U. S. economy will increasingly depend on fuel imports or nonpetroleum fuels. This might well lead to increased government control and increased domestic and international frictions, similar to those encountered by other nations before nationalization of their power industry.

Conservation vs. *Consumption*

In large, industrialized countries abundant energy is a political necessity. Since transportation costs (fig. 5.5) are high, this desire cannot be fulfilled by scientific means. The present trend of shipping oil from the Middle East and natural gas from the North Sea to remote points all over the world is due to the synergistic effect of modern media and politics, and is transitory. A major difficulty in bringing consumption and conservation back into some type of balance is that modern pluralistic governments, in anticipation of unexpected power shifts, must attempt to serve the interests of divergent groups. Therefore, various government branches are forced to pursue conflicting goals. Therefore, many countries have consolidated energy related functions. In the U.S. this new government branch will have Departmental status.

Environmental Concern

Air pollution legislation is in flux. This subject is covered in Chapter Nine. Presently all stationary sulfur dioxide sources, except power plants, have taken steps to recover sulfur dioxide (fig. 15.5) and almost all work toward eventual compliance with air quality laws. The coal plants comply by temporarily burning low sulfur coals. It remains to be seen whether the air quality can be maintained in all locations in view of local pressures, but the transitory character of pollution makes it virtually certain that the nuisance will further decrease in the future. A vigorous struggle is presently underway to establish the need and desirability of increased legislation (Schimmel, 1975; Megonell, 1975; Larson, 1970). In the U.S. this struggle has caused a continuous organizational change in the agencies involved, and their functions have been constantly redefined. Currently in-house research has been greatly cut in the form of contract research which is largely conducted in industry, rather that at universities.

Government Innovation

During the nuclear age, leading inventors received public recognition, wielded power in government, and were rewarded with government research grants, leadership positions in government funded labs, and with Nobel prizes. These rewards replaced the personal wealth and private influence earned by earlier inventors; thus, the nuclear age changed the public attitude toward the private inventor and entrepreneur. This attitude shows in the emerging patent laws of Europe and the U.S. Today a large number of influential people believe that in the modern state political leadershp should establish directions for science and should select or designate inventors and carry responsibility for their work. As a result, there is heavy emphasis on peer evaluation and evaluation by frequent progress reports. Today the climate for innovation is not ideal (Bradbury, 1977). If Herman Frasch were born today, he would probably not learn to admire industry, he would likely become impatient with the scheduled learning activities in college and not earn an advanced degree, would likely not find an industrial group whose majority vote would approve financing of his enterprise, would not obtain a satisfactory environmental impact statement, a construction license, nor union agreement as to what skill should take over the new activities. And yet, the present public attitude toward private inventors cannot stop progress, as the work of the many excellent scientists in industry and government shows.

3. The Role of Education

Education is involved in training specialists, in training decision makers to understand and seek scientific knowledge, and in educating the public to learn how to understand issues, arrive at sensible judgments, and to live in harmony with its resources and environment.

Specialist Education

The image of chemistry, science, and engineering has gone from one extreme to the other. Alchemists were proud of their skill and their art. In 1792 Richter, the father of stoichiometry, proudly observed, "It was reserved for our century to make a science out of an art." However, skill and art remained an important part of chemistry. In Europe academic chemists retained a tradition of practice oriented laboratory work until after World War II. Today during the freshman training many leading U.S. universities offer 150 hours of large lecture sessions supported by movies and slides rather than demonstrations, and only a total of twenty hours of laboratory work. Most advanced laboratory work is purely organic and frequently the third year, advanced inorganic laboratory has been replaced by computer modeling and lecture classes on the theory of the chemical bond. As a result, the chemical profession attracts far more theoretically inclined young people than industry can employ and their education leaves them insufficiently prepared for supervising or evaluating laboratory work. The American Chemical Society's curriculum committee expects an increasing trend away from laboratory work because many schools find that in today's financial crunch a professor lecturing a large class and 'laboratories' with theoretical modeling are cheaper than experiments. In graduate schools similar trends are observed. Modern sciences involve expensive, complex instrumentation. Thus, universities have become dependent upon large scale, i.e. government, support; such schools produce highly specialized graduates with skills which make them dependent on working for very large organizations such as the government. Since the new curricula depend on highly specialized faculty, and faculty efficiency is measured by publication in specialist journals, the students often lack exposure to teachers who are familiar with industrial needs, and the rift between pure chemistry and industry will increase rather than narrow in the coming years. The main gap will be in the area of experimental laboratory work of the type necessary to develop basic new inorganic chemistry for finding better smelting methods, better sulfur removal methods for power plants, and other subjects of this book.

Public Education

Reflecting the current public confusion, the media present a wide spectrum of conflicting views on energy, environmental resources, and the need for conservation. Since most public schools operate under laws which require that students be prepared for adopting prevailing community standards, schools teach the social attitudes of the present adult generation which in turn draws its values from earlier times. This can perpetuate an obsolete attitude toward energy consumption (Section 9B) and can cause training for obsolete skills. However, during the last fifteen years the search for community values has led to sudden paradoxical discontinuities in educational goals. A generation of young teachers, trained to quickly introduce new math, new chemistry, and other new curricula attended the universities during a time of exploratory curricula dealing with acute social issues, such as the concern for protecting the endangered environment or for securing exotic forms of energy to insure continuity of energy supply. A substantial number of these former students has carried over such exploratory views into their positions as leaders in public schools and now prepare students for university studies in curricula which never reached implementation. This will haunt Western education in a more subtle but longer-lasting manner than the overreaction to anticipated needs for space and nuclear scientists of the late 50's.

C. CONCLUSIONS

The last 14 chapters have shown that sulfur, a traditionally leading industrial chemical, has adjusted to changing technology and the changing needs of society. Sulfur will likely remain a strategic chemical in most of its present applications, and it seems that sulfur will increasingly gain acceptance in non-chemical high volume construction uses. Since sulfur is so closely tied to GNP and industrial production, the future of sulfur production, use, and recovery is closely tied to development of industrial nations. This development depends on cheap energy. The present situation is characterized by uncertainty about the choice of the best fuel for the future, and uncertainty about the deadline for making that choice.

The energy industries are presently in the middle of a profound transition. Over the last several hundred years industry has made itself increasingly dependent on non-renewable sources, starting with coal mining in the 14th Century, the metallurgical art of Agricola in the 15th Century, and the commercial sulfuric acid production around 1750. The last century, characterized by the gold, oil, and natural gas rushes, yielded unprecedented wealth at an unprecedented speed with almost no effort. This created increasingly high expectations, topped in the early 1950's by a brief, but wide spread belief that nuclear energy would quickly yield perennial, essentially free, energy. Everybody quickly adjusted to the idea and many became overly dependent on energy as part of their life-style. When it was realized in the early 1960's that these expectations were premature and that substantial efforts had to be invested to make commercial nuclear energy work, discouragement followed and a mixture of foresight and greed have accelerated the realization that the exhaustion of oil and gas resources will eventually force adoption of a less convenient energy source. While a large segment of industry quickly prepared to provide new energy sources, a large fraction of the public in most nations was caught by surprise; some scientists, accustomed to using linear extrapolation, suddenly made totally erroneous predictions. In the scramble for adjustment, widely divergent, untested predictions emanated causing confusion and a loss of trust in the scientific method. For the consumers this dynamic period also caused severe personal, social, and economic problems which were specifically painful to the post-war generation in industrial societies which grew up with penicillin, Salk serum, and the birth control pill, and had learned to consider poverty, hunger, disease, and crime as painless entertainment on TV screens.

During the last ten years gifted and intelligent young students have gone through their education without a clear image of the structure of future society. Many received professional degrees before they established the limits of their ambitions or potential and before they formulated their professional goals. Thus, many of the people born since 1940 are now waiting for goals for fulfilling their dormant or delayed ambitions. For chemists, scientists, and engineers this situation creates new and rare opportunities of the type available only once every one or two hundred years.

They can now help develop new technology such as coal gasification, nuclear energy, and new sulfur production methods, which would be rejected during more stable times by those who have established interests and invested money and education in traditional systems and values.

The potential for future development is quite predictable, as discussed in earlier chapters. We presently have available 100 years of inorganic chemistry which can be applied quickly to current sulfur and energy technology. The number and the character of the scientists who will shape the future is also largely known, because the generation which will graduate from college in the year 2000 is being born this year. Those who will educate them are now being trained. The attitude of their teachers is now being established by current events and by the fact that human nature has remained essentially unchanged during the last 4000 years. And yet, the last fifteen years have demonstrated that during dynamic periods, simple linear extrapolation is not viable and leads to incorrect results. One reason for this is that science and engineering have created such a wealth of accumulated knowledge and skills that a gifted individual has an opportunity to make innovation in a form directly available to the public, making it difficult for religious or political leaders to control. Despite this, the reader will find that professional forecasts about the future energy and environmental conditions and the sulfur needs almost unanimously ignore future innovation. This is due to the intrinsic incapacity of science to deal with the unknown and unobserved, and to what Hildebrand (1976) called misused and misunderstood science.

Every parent and teacher can observe that many young people are ready to invent on every necessary scale. It can be safely expected that corrosion resistant steels, coal gasification furnaces, processes for the regeneration of limestone, calcium sulfite to sulfur, and similar goals can be reached. Thus, in many regions of the world commercially viable Claus sulfur from coal as an energy by-product is a virtual certainty. The timing will not be limited by technology alone, but will depend on the attitude and goals of the coming generation and on how strongly the present and next generation will knowingly and purposefully shape the developing technology before it has matured to the degree that it will dominate industrial acceptance for another hundred years. The purpose of this book is not to speculate further about the future science and technology, but to help make available presently known facts as an aid to those who now shape the technology which will prevail for the next hundred years.

Appendix

CONVERSION FACTORS FOR THE METRIC SYSTEM

1 U.S. gallon	3.785 liters
1 Imperial gallon	4.545 liters
1 cubic foot	28.316 liters
1 barrel (31.5 gal)	158.97 liters
1 short ton	0.907 tons
1 long ton	1.016 tons

Thermal conductivity:	1 Btu·inch/ft^2·sec·$^{\circ}$F = 518.87 joule/m·sec·$^{\circ}$K
Pressure:	1 psi = 53.8 Torr = 0.070 kg/cm^2
Heat capacity:	1 Btu/lb = 0.5556 cal/g
	1 Cal/gram·$^{\circ}$C = 0.996 Btu/lb·$^{\circ}$K

Energy

	kcal	Joule
1 Btu	0.252	1054
1 Hpr	641.	2.68×10^6
1 kWhr	860.0	3.59×10^6
1 cal	0.001	4.1846
1 kJ	0.23897	1000

SO_2 EMISSION NOMOGRAM

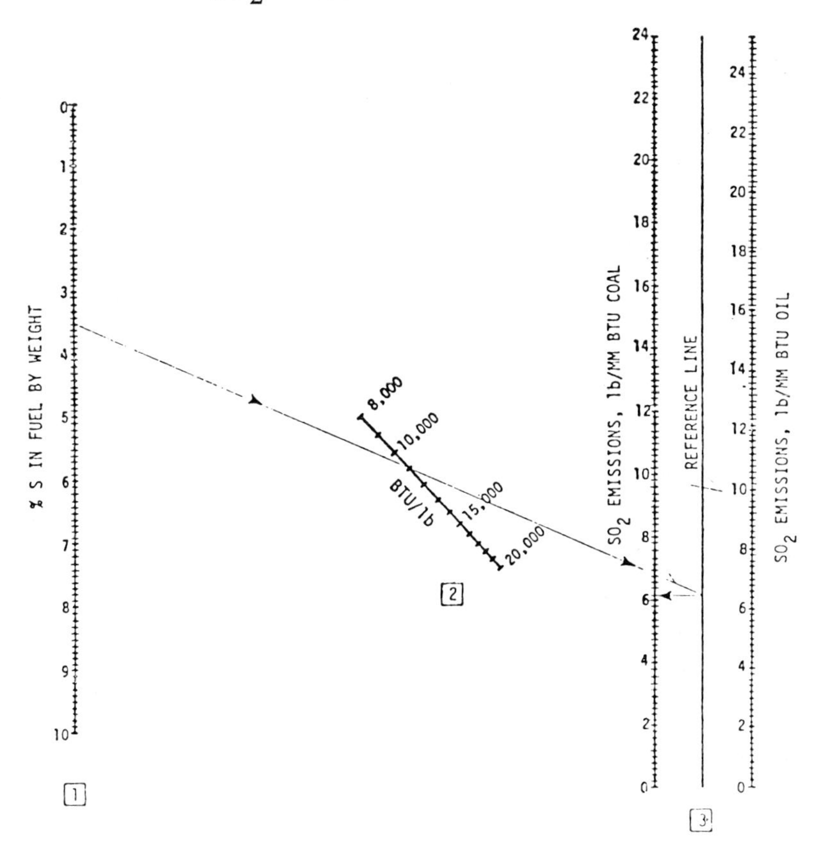

To Determine the SO_2 emissions (lb/MM Btu) in the flue gas, enter wt. % sulfur of fuel (oil or coal) on [1]; enter heating value of fuel (Btu/lb) on [2]. Connect [1] and [2] and extend to [3] and record and read: SO_2 emissions (lb/MM Btu) in flue gas–[3]. The assumptions are: (1) 95% of sulfur in coal is converted to SO_2 and (2) 100 % of sulfur in oil is converted to SO_2.

Bibliography

INTRODUCTION

This section contains references to all earlier chapters. Due to spatial considerations only 1600 references of over 5000 could be selected. Thus, many excellent articles had to be omitted. Recent documented specialist reviews and original research reports were given preference. Work before 1950 is included only if references are difficult to trace. Patent references are quoted in the text, but are not included in this bibliography. Government reports and industrial pamphlets are listed under the name of the company or government agency, as in the author index. U.S. government reports can be ordered from the National Technical Information Service, Department of Commerce, Springfield, Va., 22161, U.S.A. References are listed alphabetically according to the name of the first author. For each author the listing is chronological, with the oldest paper first. If a reference is missing, it is quoted in another paper by the same author, by co-authors cross referenced in the author index, or in the reviews listed in the same section of the text.

A

ABE, M., Bull. Tokyo Med. Dental Univ. 14 (1967) 415. 'Effects of Mixed NO_2-SO_2 Gas on Human Pulmonary Functions.'

ABERT, J. G., H. ALTER, and J. F. BERNHEISEL, Science 183 (1974) 1052s 'The Economics of Resource Recovery from Municipal Solid Waste.'

ABRAHAM, H., "Asphalts and Allied Substances," 6th Ed., Van Nostrand, New York, 1960.

ABRAHAMS, S. C., Acta Cryst. 8 (1955) 661. 'The Crystal and Molecular Structure of Orthorhombic Sulfur.'

AFFLECK, J. G., Chem. Eng. Progr., Aug. (1976) 17. 'The Monumental Challenge of Energy Supply.'

AGRICOLA, G., "De Re Metallica," translated from the 1st Latin Edition of 1556 by H. C. Hoover and L. M. Hoover, Dover Publications, New York, 1950.

AIZATULLIN, T. A., and A. V. LEONOV, Okeanologiya 15 (1975) 1020. 'Kinetics and Mechanism of the Oxidative Transformation of Inorganic Sulfur Compounds in Sea Water.'

ALARIE, Y., C. E. ULRICH, W. M. BUSEY, H. E. SWANN, and H. N. MACFARLAND, Arch. Environ. Health 21 (1970) 759. 'Long-Term Continuous Exposure of Guinea Pigs to Sulfur Dioxide.'

ALDERDICE, D. S., and R. N. DIXON, J. Chem. Soc., Faraday II, 72 (1976) 372. 'HeI photoelectron spectrum of SO_3.'

ALLAWAY, W. H., and J. F. THOMPSON, Soil Science 101 (1966) 240. 'Sulfur in the Nutrition of Plants and Animals.'

ALLAWAY, W. H., Sulphur Institute J. 6(3) (1970) 3. 'Sulphur-Selenium Relationships in Soils and Plants.'

ALLCOCK, H. R., Sci. Am. 230 (1974) 66. 'Inorganic Polymers.'

ALLEN, A. R., ACS 167th Nat. Conf., Los Angeles, April, 1974. 'Coping with the Oil Sands.'

ALLEN, E. R., R. D. MCQUIGG, and R. D. CADLE, Chemosphere 1 (1972) 25. 'The Photooxidation of Gaseous Sulfur Dioxide in Air.'

ALLEN, S. E., and D. A. MAYS, J. Agr. Food Chem. 19 (1971) 809. 'Sulfur-Coated Fertilizers for Controlled Release: Agronomic Evaluation.'

ALLEN, S. E., C. M. HUNT, and G. L. TERMAN, Agronomy J. 63 (1971) 529. 'Nitrogen Release from Sulfur-Coated Urea, as Affected by Coating Weight, Placement and Temperature.'

ALMAST, L., in "Sulfur in Organic and Inorganic Chemistry," Vol. 1, A. Senning, ed., Marcel Dekker, New York, 1971. 'The Sulfur-Phosphorus Bond.'

ALMGREN, T., and I. HAGSTRÖM, Water Research 8 (1974) 395. 'The Oxidation Rate of Sulphide in Sea Water.'

ALYEA, H. N., J. Chem. Educ. 46 (1969) A889. 'Colloids.'

AMBARTSUMYAN, R. V., N. V. CHEKALIN, YU. A. GOROKHOV, V. S. LETOKHOV, G. N. MAKAROV, and E. A. RYABOV, Lect. Notes Phys. 43 (1975) 121. 'Selective Photochemistry in an Intense Infrared Field.'

AMDUR, M. O., J. Air Poll. Contr. Assn. 19 (1969) 68. 'Toxicological Appraisal of Particulate Matter, Oxides of Sulfur and Sulfuric Acid.'

AMDUR, M. O., Arch. Environ. Health 23b (1971) 459. 'Aerosols Formed by Oxidation of Sulfur Dioxide.'

AMDUR, M. O., Proc. Contents on Health Effects of Air Pollutants, National Academy of Sciences National Research Council, Oct. 3-5, 1973, U.S. Gov. Print. Off., Washington, D.C., November 1973. 'Animal Studies.'

AMERICAN PHARMACEUTICAL ASSOCIATION, "The Pharmaceutical Recipe Book," 1st Ed., The Am. Pharmaceutical Assn., 1929; 2nd Ed., 1936.

ANDERSON, A., and Y. T. LOH, Can. J. Chem. 47 (1969) 879. 'Low Temperature Raman Spectrum of Rhombic Sulfur.'

ANDERSON, A., and P. G. BOCZAR, Chem. Phys. Letters 43 (1976) 3. 'Far Infrared Spectrum of Orthorhombic Sulfur.'

ANDERSSON, T., Tellus 21 (1969) 685. 'Small-Scale Variations of the Contamination of Rain Caused by Washout from the Low Layers of the Atmosphere.'

ANDO, J., Environ. Prot. Technol. Ser. EPA-650-2-73-038, (1973) 69. 'Status of Japanese Flue Gas Desulfurization Technology.'

ANDO, J., Proc. Symp. Flue Gas Desulfurization, New Orleans, March 1976, R. D. Stern and W. H. Ponder, eds., EPA Report 600/2-76-136, (1976). 'Status of Flue Gas Desulfurization and Simultaneous Removal of SO_2 and NO_x in Japan.'

ANDO, J. and G. A. ISAACS, Environ. Prot. Technol. Ser. EPA 600/-2-76-013a (1976). 'SO_2 Abatement for Stationary Sources in Japan.'

ANTIKAINEN, P. J., Suomen Khem. B31 (1958) 223. 'Infra-red Absorption Spectra of Inorganic Compounds in Aqueous Solution. 1. Some Oxyanions of Sulphur.'

APPEL, B. R., J. J. WESOLOWSKI, E. HOFFER, S. TWISS, S. WALL, S. W. CHANG, and T. NOVAKOV, Intern. J. Environ. Anal. Chem. 4 (1976) 169. 'An Intermethod Comparison of X-ray Photoelectron Spectroscopic (ESCA) Analysis of Atmospheric Particulate Matter.'

ARMITAGE, J. W. and C. F. CULLIS, Combust. Flame 16 (1971) 125. 'Studies of the Reaction Between Nitrogen Dioxide and Sulfur Dioxide.'

ASHWORTH, M. R. F., "The Determination of Sulphur-Containing Groups," Academic Press, New York, 1976.

ASINGER, F., Angew. Chem. 68 (1956) 413. 'Ueber die Gemeinsame Einwirkung von Schwefel und Ammoniak auf Ketone.'

ASPLUND, L., P. KELFVE, H. SIEGBAHN, O. GOSCINSKI, H. FELLNER-FELDEGG, K. HAMRIN, B. BLOMSTER, and K. SIEGBAHN, Chem. Phys. Lett. 40 (1976) 353. 'Chemical Shifts of Auger Electron Lines and Electron Binding Energies in Free Molecules. Sulfur Compounds.'

ASSOCIATION OF OFFICIAL AGRICULTURAL CHEMISTS, "Official Methods of Analysis of the Association of Official Agricultural Chemists," 8th Ed., Washington, D. C., 1955.

ATEN, A. H. W., Z. Phys. Chem. 88 (1914) 321. 'Herstellung und Eigenschaften von Hexaschwefel.'

ATKINS, R. S., Adv. Chem. 139 (1975) 120. 'The Commercialization of Lime/Limestone Flue Gas Scrubbing Technology.'

AUDRIETH, L. F., "Liquid Sulfur Dioxide," Stauffer Chemical Co., Westport, Conn., 1969.

AUSTIN, J. M., D. JENSEN, and B. MEYER, J. Chem. Eng. Data 16 (1971) 364. 'Solubility of Sulfur in Liquid Sulfur Dioxide, Carbon Disulfide, and Carbon Tetrachloride.'

AVERBUKH, T. D., N. P. BAKINA, A. A. RADIVILOV, L. V. ALPATOVA, and E. A. KRAVCHENKO, Int. Chem. Eng. 9 (1969) 317. 'Production of Elemental Sulfur by Reduction of Sulfur Dioxide with Natural Gas Methane Process.'

B

BAAS BECKING, L. G. M., Ann. Bot., Lond. 39 (1925) 613. 'Studies on the Sulfur Bacteria.'

BACON, R. F., and R. FANELLI, Ind. Eng. Chem. 34 (1942) 1043. 'Purification of Sulfur.'

BACON, R. F., and R. FANELLI, J. Am. Chem. Soc. 65 (1943) 639. 'The Viscosity of Sulfur.'

BADAR-UD-DIN and M. ASLAM, Pakistan J. Sci. Res. 5 (1953) 6. 'A Study of Ammonia-Sulphur Dioxide Reaction.'

BAGGIO, S., Acta Cryst. B27 (1971) 517. 'Crystal Structure of Ammonium Pyrosulfite.'

BAILEY, E. E., Adv. Chem. Ser. 139 (1975) 101. 'Continuing Progress for Wellman-Lord SO_2 Process.'

BAILEY, G. C., and H. W. HILL, JR., Am. Chem. Soc. Symp. Series 4 (1974) 83. 'Polyphenylene Sulfide: A New Industrial Resin.'

BAIN, R. L., and T. H. NORRIS, Proc. Int. Atomic Energy Agency, Vienna, 1965. 'Isotopic Exchange Reactions in Thionyl Chloride and Sulfuryl Chloride.'

BAKER, E. H., Mineral Proc. Extractive Metall. 80C (1971) 93. 'Vapour Pressure-Temperature Relation for Sulphur up to the Critical Point.'

BAKER, R. B., and E. E. REID, J. Am. Chem. Soc. 51 (1929) 1566. 'The Action of Sulfur on Normal-Heptane and Normal-Butane.'

BALARD, M., Ann. Chim. Phys. 32 (1826) 375. 'Du Bromure de Soufre.'

BALDWIN, M. M., Ind. Eng. Chem. 42 (1950) 2227. 'Sulfur in Fungicides.'

BALI, A., and K. C. MALHOTRA, Aust. J. Chem. 28 (1975) 983. 'Mixed Cations of Sulfur in Disulphuric Acid.'

BALINT, T., S. IGLEVSKI, and E. KERENYI, Hung. Sci. Instrum. 35 (1975) 13. 'Portable Instruments for Measuring the Concentration of Sulfur-Containing and Organic Air Pollutants in Workshops.'

BALLI, M. F., and G. DE NEGRI, Proc. Symp. New Uses for Sulphur and Pyrites, Madrid, May 1976, The Sulphur Institute, London, 1976. 'Laboratory Methods and First Results of Sulphur Impregnation Tests Carried out on Frost Non-Resistant Bodies Made in the Sassuolo Area.'

BANCHERO, J. T., and F. H. VERHOFF, J. Inst. Fuel, June (1975) 76. 'Evaluation and Interpretation of the Vapour Pressure Data for Sulphuric Acid Aqueous Solutions with Application to Flue Gas Dewpoints.'

BANERJEE, S., and R. K. DUTTA, Res. Ind. 20(4) (1975) 204. 'Rapid Method for Determination of Purity of Amorphous Varieties of Elemental Sulfur.'

BANISTER, A. J., L. F. MOORE, and J. S. PADLEY, in "Inorganic Sulphur Chemistry," G. Nickless, Ed., Elsevier, Amsterdam, 1968. 'Structural Studies on Sulphur Species.'

BANISTER, A. J., Nature Phys. Sci. 239 (1972) 69. 'Structures of Electron-Rich Cage Species Tetrasulphur Tetranitride and the Octasulphur Dication.'

BANISTER, A. J., "MTP Int. Rev. Sci., Inorg. Chem.," Ser. 2, Vol. 3, V. Gutmann, Ed., Butterworths, London, 1975. 'Cyclic Sulphur-Nitrogen Compounds.'

BARBIERI, R., La Ricerca Scientifica 2 (1962) 62. 'Indagini Sulla Reazione di Wackenroder, Nota I: Acidi Tiosolforici Quali Probabili Intermedi.'

BARBIERI, R., and G. FARAGLIA, Gazz. Chim. Hal. 92 (1962) 660. 'Indagini Sulla Reazione di Wackenroder, Nota II: Sul Meccanismo di Formazione Dello Zolfo Interno Degli Acidi Politionici.'

BARRIE, A., H. GARCIA-FERNANDEZ, H. G. HEAL, and R. J. RAMSAY, J. Inorg. Nucl. Chem. 37 (1975) 311. 'Investigations of Bonding in the Cyclic Sulphur Imides by X-ray Photoelectron Spectroscopy.'

BARRIE, L. A., and H. W. GEORGII, Atmosph. Environ. 10 (1976) 743. 'An Experimental Investigation of the Absorption of Sulphur Dioxide by Water Drops Containing Heavy Metal Ions.'

BARROW, R. F., and R. P. DUPARCQ, in "Elemental Sulfur," B. Meyer, Ed., Interscience, New York, 1965. 'Electronic Spectrum and Electronic States of S_2.'

BARROW, R. F., in "International Tables of Selected Constants," Vol. 17, B. Rosen, Ed., Pergamon Press, Oxford, 1970. 'Spectroscopic Data Relative to Diatomic Molecules.'

BARRY, C. B., Hydrocarbon Process. 51 (1972) 102. 'Reduce Claus Sulfur Emission.'

BARTHEL, Y., A. DESCHAMPS, S. FRANKOWIAK, P. RENAULT, P. BONNIFAY, and J. W. ANDREWS, Adv. Chem. Ser. 139 (1975) 100. 'Sulfur Recovery in Oil Refineries Using IFP Processes.'

BARTLETT, R. W., Report to EPA on Grant No. AP 00876 by Process Metallurgy Group, Dept. of Mineral Engineering, Stanford, 1972. 'Sulfation Kinetics in Sulfur Dioxide Absorption from Stack Gases.'

BARTZSCH, H., Gummi Asbest Kunstoffe 27 (1974) 72. 'Unlöslicher Schwefel und seine Anwendung in der Gummiindustrie.'

BASS, A. M., J. Chem. Phys. 21 (1953) 80. 'The Optical Absorption of Sulfur.'

BASSET, H., and A. J. HENRY, J. Chem. Soc. 1935 (1935) 914. 'The Formation of Dithionate by the Oxidation of Sulphurous Acid and Sulphites.'

BATAGLIA, O. C., Cienc. Cult. (Sao Paulo) 28 (1976) 672. 'Indirect Determination of Sulfur in Plants by Atomic Absorption Spectrophotometry.'

BATEMAN, L. C., E. D. HUGHES, and C. K. INGOLD, J. Chem. Soc. 1944 (1944) 243. 'Molecular Compounds between Amines and Sulphur Dioxide. A Comment on Jander's Theory of Ionic Reactions in Sulphur Dioxide.'

BATEMAN, L, and C. G. MOORE, "Organic Sulphur Compounds," Vol. 1, Pergamon Press, London, 1961.

BATES, D. V., Am. Rev. Res. Dis. 105 (1972) 1. 'Air Pollutants and the Human Lung.'

BATES, F. H., Zement 15 (1926) 373. 'Ueber die Widerstandsfaehigkeit von mit Schwefel vorbehandelten Zementdrainsteinen gegen die Einwirkung von Alkalien.'

BATTELLE MEMORIAL INSTITUTE, Report for U. S. Dept. of Health, Educ. & Welfare, 1966. 'Fundamental Study of Sulfur Fixation by Lime and Magnesia.'

BATTIGELLI, M. C., and J. F. GAMBLE, J. Occup. Med. 18 (1976) 334. 'From Sulfur to Sulfate: Ancient and Recent Considerations.'

BAULCH, D. L., D. D. DRYSDALE, and D. G. HORNE, "Evaluated Kinetics Data for High Temperature Reactions," Butterworths, London, 1973.

BEATON, J. D., G. R. BURNS, and J. PLATOU, Sulphur Institute Tech. Bull. 14 (1968). 'Determination of Sulphur in Soils and Plant Material.'

BEATON, J. D., R. H. LEITCH, R. E. MCALLISTER, and M. NYBORG, Preprint Can. Sulphur Symp., Calgary, May 1974. 'The Projected Need for Sulphur Fertilizers in Western Canada and Their Integration in Soil Fertility Programs.'

BEATON, J. D., D. W. BIXBY, S. L. TISDALE, and J. S. PLATOU, Sulphur Institue Tech. Bull 21 (1974). 'Fertilizer Sulphur, Status and Potential in the U.S.'

BEAVON, D. K., and N. FLECK, Adv. Chem. Ser. 139 (1975) 93. 'Beavon Sulfur Removal Process for Claus Plant Tail Gas.'

BECHER, W., and J. MASSONNE, Chem. Zeit. 98 (1974) 117. 'Neue Synthese von SF_4.'

BECKE-GOEHRING (see also GOEHRING)

BECKE-GOEHRING, M., Quart. Rev. 10 (1956) 437. 'Sulphur Nitride and its Derivative.'

BECKE-GOEHRING, M., Ber. 92 (1959) 855. 'Ueber die Schwefelnitride.'

BECKER, K. H., W. G. FILBY, and K. GUENTHER, "Chemical Reactions of Atmospheric Pollutants," Inst. f. Radiochemie Kernforschungszentrum Karlsruhe, 1974.

BEDDOME, J. M., Alberta Sulphur Res. Quart. Bull. 6 (1969) 1. 'Current Natural Gas Sweetening Practice; Solvent Systems for Removal of Hydrogen Sulphide and Carbon Dioxide from Natural Gas in Alberta.'

BEILKE, S., and H. W. GEORGII, Tellus 20 (1968) 435. 'Investigation on the Incorpuration of Sulfur-Dioxide into Fog- and Rain-Droplets.'

BEILKE, S., and D. LAMB, Tellus 26 (1974) 268. 'On the Absorption of SO_2 in Ocean Water.'

BEILKE, S., and D. LAMB, Am. Inst. Chem. Eng. J. 21 (1975) 402. 'Remarks on the Rate of Formation of Bisulfite Ions in Aqueous Solution.'

BEILKE, S., D. LAMB, and J. MÜLLER, Atmosph. Environ. 9 (1975) 1083. 'On the Uncatalyzed Oxidation of Atmospheric SO_2 by Oxygen in Aqueous Systems.'

BENESCH, R, et. al, Eds. "Sulfur in Proteins," Academic Press, New York and London, 1959.

BENSON, S. W., R. SHAW, and R. W. WOOLFOLK, Environ. Prot. Technol. Ser. EPA-600/2-76-152a (1976) 267. 'Estimation of Rate Constants as a Function of Temperature for the Reactions X + YZ, XY + Z, where X, Y, and Z are Atoms of the Elements Carbon, Hydrogen, Nitrogen, Oxygen, and Sulfur.'

BERENBAUM, M. B., and J. R. PANEK, "The Chemistry and Application of Polysulfide Polymers," Thiokol Corp., Trenton, N. J., 1958.

BERENBAUM, M. B., and R. N. Johnson, "Kirk-Othmer Encycl. Chem. Technol., 2nd Ed., 16 (1968) 253. 'Polymers Containing Sulfur.'

BERG, C. A., Science 181 (1973) 128. 'Energy Conservation through Effective Utilization.'

BERGER, A., J. NOGUCHI, and E. KATCHALSKI, J. Am. Chem. Soc. 78 (1956) 4483. 'Poly-L-cysteine.'

BERKOWITZ, J., and J. R. MARQUART, J. Chem. Phys. 39 (1963) 275. 'Equilibrium Composition of Sulfur Vapor.'

BERKOWITZ, J., and C. LIFSHITZ, J. Chem. Phys. 48 (1968) 4346. 'Photoionization of High-Temperature Vapors. II. Sulfur Molecular Species.'

BERKOWITZ, J., and W. A. CHUPKA, J. Chem. Phys. 50 (1969) 4245. 'Photoionization of High-Temperature Vapors. VI. S_2, Se_2, and Te_2.'

BERKOWITZ, J., J. Chem. Phys. 62 (1975) 4074. 'PES of High Temperature Vapors. VII. S_2 and Te_2.'

BERKOWITZ, J., Preprint, Symp. Chem. Elemental Sulfur and Its Binary Compounds, Am. Chem. Soc., New York, April 1976. 'Molecular Composition, Structure, and Spectra of Sulfur Molecules.'

BERLIE, E. M., Alberta Sulphur Res. Quart. Bull. 10 (1973) 1. 'H_2S to Sulphur–The Claus Sulphur Recovery Process.'

BERNARD, J. P., A. DE HAAN, and H. VAN DER POORTEN, Compt. Rend. 276 (1973) 587. 'Chimie Minerale–Anions non Oxygenes du Soufre dans l'Eutectique LiCl-KCl Fondu.'

BERNER, R. A., Am. J. Sci. 268 (1970) 1. 'Sedimentary Pyrite Formation.'

BERNER, R. A., J. Geophys. Res. 76 (1971) 6597. 'Worldwide Sulfur Pollution of Rivers.'

BERNER, R. A., in "The Changing Chemistry of the Oceans," Proc. 12th Nobel Symp., Aspenaesgarden, Lerum, and Goeteborg, Sweden (1971), D. Dyrssen and D. Jagner, Eds., Wiley Interscience, New York, 1972. 'Sulfate Reduction, Pyrite Formation, and the Oceanic Sulfur Budget.'

BERSTEIN, G., Z. Chem. Ind. Kolloide 11 (1912) 185. 'Studien ueber die Vulkanisation des Kautschuks. I. Beitrag zum Studium der Kalt-Vulkanisation des Kautschuks.'

BERQUIN, Y. F., Proc. Symp. New Uses for Sulphur and Pyrites, Madrid, 1976, The Sulphur Institute, London, 1976. 'New Developments in the Conditioning of Solid Sulphur.'

BERTHOLLET, C. L., Ann. Chim. Phys. (1)25 (1798) 233. 'Sur l'Hydrogene Sulfure.'

BERTOZZI, E. R., Rubbev Chem. Tech. 41 (1968) 114. 'Chemistry and Technology of Elastomeric Polysulfide Polymers.'

BERZELIUS, J., Ann. Phys. 38 (1811) 161. 'Erste Fortsetzung des Versuchs, die Bestimmten und Einfachen Verhältnisse Aufzufinden, nach Welchen die Bestandtheile der Unorganischen Natur mit Einander Verbunden Sind.'

BERZELIUS, J., Ann. Phys. 38 (1811) 249. 'Versuch, die Bestimmten und Einfachen, nach Welchen die Bestandtheile der Unorganischen Natur mit Einander Verbunden Sind.'

BESTOUGEFF, M., and A. COMBAZ, Adv. Org. Geochem. Proc. 6th Int. Meet. 1973 (Pub. 1974). 'Action of Hydrogen Sulfide and Sulfur on Some Actual and Fossil Organic Substances.'

BETTS, R. H., and R. H. VOSS, Can. J. Chem. 48 (1970) 2035. 'The Kinetics of Oxygen Exchange Between the Sulfite Ion and Water.'

BETTS, R. H., and S. LIBICH, Can. J. Chem. 49 (1971) 180. '18-O Transfer in the System Thiosulfate-Sulfite-Water: An Example of a Set of Consecutive Reversible First Order Rate Processes.'

BIENIEK, D., and F. KORTE, Naturwis. 59 (1972) 529. 'Organische Hochdruckchemie Neue Moeglichkeiten der Praeparativen Organischen Chemie.'

BIENSTOCK, D., L. W. BRUNN, E. M. MURPHY, and H. E. BENSON, U. S. Bureau of Mines Information Circular 7836 (1958). 'Sulfur Dioxide–Its Chemistry and Removal from Industrial Waste Gases.'

BIRK, J. R., and R. K. STEUNENBERG, Adv. Chem. 140 (1975) 186. 'Chemical Investigations of Lithium-Sulfur Cells.'

BISCHOFF, W. F., and Y. HABIB, Chem. Eng. Prog. 71 (May 1975) 59. 'The FW-BF Dry Adsorption System.'

BITUMINOUS COAL RESEARCH, INC., Reports to Nat'l Air Poll. Control Admin. on Contract PH-86-67-139, Sept. 1969 and Feb. 1970. 'An Evaluation of Coal Cleaning Processes and Techniques for Removing Pyritic Sulfur from Fine Coal.'

BITUMINOUS COAL RESEARCH, INC., Final Report to EPA on Contract CPA 70-26, April 1971. 'An Evaluation of Coal Cleaning Processes and Techniques for Removing Pyritic Sulfur from Fine Coal.'

BITUMINOUS COAL RESEARCH, INC., Report to EPA on Contract 68-02-0024, April 1972. 'An Evaluation of Coal Cleaning Processes and Techniques for Removing Pyritic Sulfur from Fine Coal.'

BIXBY, D. W., and J. D. BEATON, Sulphur Institute Tech. Bull. 17 (1970). 'Sulphur Containing Fertilizers, Properties and Applications.'

BLASIUS, E., and R. KRÄMER, J. Chromatog. 20 (1965) 367. 'Gegenueberstellung der Hochspannungspapierionophorese und der Papierchromatographie bie der Untersuchung einiger Reaktionen von Schwefelverbindungen.'

BLASIUS, E., G. HORN, A. KNÖCHEL, J. MÜNCH, and H. WAGNER, in "Inorganic Sulphur Chemistry," G. Nickless, Ed., Elsevier, Amsterdam, 1968. 'Analytical Chemistry of Sulphur Compounds.'

BLASIUS, E., W. NEUMANN, and H. WAGNER, in "Sulfur in Organic and Inorganic Chemistry," Vol. 3, A. Senning, Ed., M. Dekker, New York, 1972. 'Labeled Sulfur Compounds.'

BLASIUS, E., and H. WAGNER, Talanta 20 (1973) 43. 'Bestimmung 35-S-Markierter Polythionate nach Hochspannungsionophoretischertrennung.'

BLASIUS, E., and J. MUNCH, J. Chromatog. 84 (1973) 137. 'Hochspannungsionophoretische Trennung von Polysulfandisulfonaten auf Duennschiehtfolien in Temperatur-Kontrollierter Kammer.'

BLASIUS, E., C. SCHREIER, and K. ZIEGLER, Arch. Eisenhuettenwes. 45 (1974) 442. 'Polarographic Determination of Different Sulphur Compounds Dissolved in the Same Watery Solution.'

BLASIUS, E., and H. WAGNER, Chem. Zeit. 98 (1974) 154. 'Radiohochspannungsionophorese als Hilfsmittel zum Nachweis Neuer Langkettiger Schwefeverbindungen.'

BLOCK, E., J. Chem. Ed. 48 (1971) 814. 'Organic Sulfur Compounds in Organic Synthesis.'

BLOCK, S., and G. J. PIERMARINI, High Temp. High Press. 5 (1973) 567. 'The Melting Curve of Sulfur to 300°C and 12 Kbar.'

BLOKH, G. H., "Organic Accelerators in the Vulcanization of Rubber," Israel Program for Scientific Translation, Jerusalem, 1968.

BLOOMFIELD, C., Soil Sciences 103 (1967) 219. 'Effect of Some Phosphate Fertilizers on the Oxidation of Elemental Sulfur in Soil.'

BOCCA, P. L., U. PETROSSI, and V. PICONI, Chim. l'Ind. (Milan) 55 (1973) 425. 'Heavy Hydocarbons and Sulfur: Reactions and Reaction Products.'

BOCK, H., and G. WAGNER, Angew. Chem. 84 (1972) 119. ' 'Einsame' Elektronenpaare in Organischen Sulfiden und Disulfiden.'

BODEWIG, F. G., and J. A. PLAMBECK, J. Electrochem. Soc. 116 (1969) 607. 'Electrochemical Behavior of Sulfide in Fused LiCl-KCl Eutectic.'

BOGUN, E. F., L. D. ERMAK, and E. S. NENNO, Izv. Vyssh. Ucheb. Zaved., Khim. Khim. Technol. 17 (1974) 309. 'Kinetics of Sulfur Dioxide Absorption by Aqueous Suspensions of Manganese Dioxide and Ferric Oxide.'

BOHANNAN, D. L., Alberta Sulphur Res. Quart. Bull. 4(3) (1967) 18. 'Problems of Sulfur Deposition in the Sour Gas Wells, Gathering Systems, and Sulfur Plant at Harmattan Leduc Unit No. 1.'

BOLIN, B., G. ASPLING, and C. PERSSON, Tellus 26(1-2) 185 (1974). 'Residence Time of Atmospheric Pollutants as Dependent on Source Characteristics, Atmospheric Diffusion Processes, and Sink Mechanisms.'

BOLIN, B., and R. J. CHARLSON, Ambio 5(2) (1976) 47. 'On the Role of the Tropospheric Sulfur Cycle in the Shortwave Radiative Climate of the Earth.'

BOLOTOV, E. A., G. N. KLEPTZOVA, and P. S. MELNEKOV, Krystallografey (Russian) 16 (1971) 400. 'Two Types of Spherulites Formed During Crystallization of Molten Sulfur.'

BOONE, J. E., and J. H. TURNER, Environ. Prot. Technol Ser. EPA-650/2-73-012 (1973). 'Properties of Ammonium Sulfate, Ammonium Bisulfate, and Sulfur Dioxide Solutions in Ammonia Scrubbing Processes.'

BOR, P., Chem. Listy (Czech.) 69 (1975) 276. 'Preparation of Very Pure Sulfur.'

BORELLI, S., and H. WEITGASSER, Med. Klinik 50 (1955) 464. 'Zur Dermatologish-Kosmetischen Behandlung der Akne Vulgaris.'

BORGWARDT, R. H., Environ. Sci. Technol. 4 (1970) 59. 'Kinetics of the Reaction of SO_2 with Calcined Limeston.'
BORGWARDT, R. H., Report EPA on Project 37, Program Element 21ACY, Draft May 1974. 'Sulfate Scale Control in Lime/Limestone Scrubbers by Unsaturated Operation.'
BORGWARDT, R. H., Report EPA Project AAZ 004, Program Element EHB 528, Draft March 1975. 'SO_2 Scrubber Studies Related to Improving Limestone Utilization.'
BORGWARDT, R. H., Preprint, 68th Annual Meet., Am. Inst. Chem. Eng., Los Angeles, Nov. 1975. 'Increasing Limestone Utilization in FGD Scrubbers.'
BORGWARDT, R. H., Symp. EPA Flue Gas Desulfurization, New Orleands, 1976, Environ. Prot. Technol. Ser. 600/2-76-136, (1976). 'IERL-RTP Scrubber Studies Related to Forced Oxidation.'
BORGWARDT, R. H., Unpublished, 1976. 'EPA/RTP Pilot Studies Related to Unsaturated Operation of Lime and Limestone Scrubbers.'
BOTTENHEIM, J. W., and J. G. CALVERT, J. Phys. Chem. 80 (1976) 782. 'The Long-Lived Transient in the 2500-3500 A Flash Photolysis of Gaseous Sulfur Dioxide.'
BOTTS, W. V., and D. C. GEHRI, Adv. Chem. Ser. 139 (1975) 164. 'Regenerative Aqueous Carbonate Process for Utility and Industrial Sulfur Dioxide Removal.'
BOUCHARD, I., and H. R. CONRAD, J. Dairy Sci. 56 (1972) 1276. 'Sulphur Requirement of Lactating Dairy Cows. 1. Sulfur Balance and Dietary Supplementation.'
BOUCHARDAT, A., Ann. Therapeut. Matiere Med. de Pharm. et de Toxicologie (1852) 186. 'New Method for Administering Sulfur.'
BOULEGUE, J., and G. MICHARD, Compt. Rend. Acad. Sci. (France) 277D (1973) 2613. 'Geochemie–Formation de Polysulfures dans les Conditions Physico-Chimiques de l'Eau de Mer.'
BOURNE, D. W. A., T. HIGUCHI, and I. H. PITMAN, J. Pract. Chem. 63 (1974) 865. 'Chemical Equlibria in Solutions of Bisulfite Salts.'
BOWEN, J. S., and R. E. HALL, Environ. Prot. Technol Ser. EPA-600/2-76-152a (1976). 'Proceedings of the Stationary Source Combustion Symposium, Vol. I. Fundamental Research.'
BOWERS, J. W., M. J. A. FULLER, and J. E. PACKER, Chem. and Ind., Jan. 8 (1966) 65. 'Autoxidation of Aqueous Sulphide Solutions.'
BOYNTON, R. S., "Chemistry and Technology of Lime and Limestone," Interscience, New York, 1966.
BRADBURY, F. R., Chem. Tech. (Am. Chem. Soc.) 7 (1977) 22. 'Constraints to Innovation.'
BRADLEY, E. B., M. J. MATHUR, and C. A. FRENZEL, J. Chem. Phys. 47 (1967) 4325. 'New Measurements of the Infrared and the Raman Spectrum of S_2Cl_2.'
BRADLEY, E. B., C. A. FRENZEL, and M. J. MATHUR, J. Chem. Phys. 49 (1968) 2344. 'Measurements of the Raman and the Far-Infrared Spectrum of Sulfur Bromide.'
BRAITHWAITE, E. R., Ed., "Lubrication and Lubricants." Elsevier, New York, 1967.
BRAEKKE, F. H., Ed., Research Report FR 6/76, SNSF Project, Oslo Norway, 1976. 'Impact of Acid Precipitation on Forest and Freshwater Ecosystems in Norway.'
BRAKEL, J., Ann. Gembloux 81(1) (1975) 35. 'Microbiological Problems in Relation to the Sanitary Engineer.'

BRATSCHIKOV, G. G., ZH. A. KOVAL, and M. D. BABUSHKINA, Sb. Tr., Tsentr. Nauchno-Issled. Inst. Bum. 9 (1974) 227. 'Absorbers for the Recovery of Sulfur Dioxide from Waste Gases of Sulfite Pulp Production.'

BRAUN, D., and M. KIESEL, Monatschr. 96 (1965) 631. 'Ueber Darstellung und Eigenschaften Einiger Cyclischer und Polymer Trithiokohlensaeureester.'

BRETSZNAJDER, S., and J. PISKORSKI, Bull. Acad. Polon. Sci., Ser. Sci. Chim. 15 (1967) 93. 'Temperature Effect on the Solubility of Sulphur In Ammonium Sulphide.'

BRIMBELCOMBE, P., and D. J. SPEDDING, Nature 236 (1972) 225. 'Rate of Solution of Gaseous SO_2 at Atmospheric Concentrations.'

BRIMBLECOMBE, P., and D. J. SPEDDING, Tellus 26 (1974) 273. 'The Absorption of Low Concentrations of Sulphur Dioxide into Aqueous Solutions.'

BRINKE, R., Oel Gas Feuerungstech. 21(3) (1976) 22, 24. 'Coordination of Oil Burners to Combustion Chambers and Chimneys,' and 'Studies on the Acquisition of the Optimal Tuning of Combustion Chamber, Oil Burner, and Waste Gas Plant in Heating Plants.'

BRIQUET, J. C., and J. SOUSSEN-JACOB, Mol. Phys. 28 (1974) 921. 'Etude des Mouvements Moleculaires en Milieu Liquide a Partir du Profil des Bandes de Vibration dans l'Infrarouge et des Fonctions de Correlation, II. Dioxyde de Soufre.'

BRÖLLOS, K., and G. M. SCHNEIDER, Ber. Buns. Ges. 78 (1974) 296. 'Optical Absorption of Liquid Sulfur under High Pressure. Pressure Dependence of the Polymerisation of Liquid Sulfur.'

BROOKS, E. S., Proc. 26th Power Sources Symp., April 1974. 'Performance Characteristics of Lithium Primary Batteries.'

BROWDER, T. J., Chem. Eng. Progress 67(5) (1971) 45. 'Modern Sulfuric Acid Technology.'

BROWN, C., in "Sulfur in Organic and Inorganic Chemistry,' Vol. 3, A. Senning, Ed., Marcel Dekker, New York, 1973. 'NMR Spectra of Sulfur Compounds.'

BROWN, J. D., and B. P. STRAUGHAN, J. Chem. Soc. Dalton 1972 (1972) 1750. 'Reassignment of the Fundamental Frequencies of Sodium Sulfite.'

BRUNFAUT, J., "L'Exploitation des Soufres en Italie," Paris, 1874.

BRYCE, W. A., and C. HINSHELWOOD, J. Chem. Soc. 4 (1949) 3379. 'The Reaction Between Paraffin Hydrocarbons and Sulphur Vapour.'

BUCHLER, J., Angew. Chem. 78 (1966) 1021. 'Massenspectrometrische Untersuchung des Cyclododecaschwefels, S_{12}.'

BUCY, J. I., J. L. NEVINS, P. A. CORRIGAN, and A. G. MELICKS, Proc. Symp. Flue Gas Desulfurization, New Orleans, March 1976, R. D. Stern and W. H. Ponder, Eds., Environ. Prot. Technol. Ser. EPA 600/2-76-136 (1976). 'Potential Utilization of Controlled SO_x Emissions from Power Plants in Eastern United States.'

BÜHRER, H. G., and H. G. ELIAS, Adv. Chem. Ser. 129 (1973) 105. 'Poly(thiolesters).'

BUFALINI, M., Environ. Sci. Technol. 5 (1971) 685. 'Oxidation of Sulfur Dioxide in Polluted Atmospheres - A Review.'

BUILOV, V. V., Dokl. Akad. Nauk SSSR 229 (1976) 185. 'Problem of the Cycle of Sulfur Isotopes in Soils.'

BULKLEY, L. D., Trans. Am. Med. Soc. 31 (1880) 185. 'On the Uses of Sulfur and Its Compounds in Diseases of the Skin.'

BUONICORE, A. J., and L. THEODORE, "Industrial Control Equipment for Gaseous Pollutants," Vol. II., CRC Press, Inc., Cleveland, Ohio, 1975.

BURDICK, L. R., and J. F. BARKLEY, U.S. Bureau of Mines Inf. Circular 7064, April 1939. 'Effect of Sulfur Compounds in the Air on Various Materials.'

BURGESS, R. A., and I. DEME, Adv. Chem. Ser. 140 (1975) 85. 'Sulfur in Asphalt Paving Mixes.'

BURGESS, S. G., and C. W. SHADDICK, Royal Soc. Health 1 (1959) 10. 'Bronchitis and Air Pollution.'

BURLAMACCHI, L., G. GUARINI, and E. TIEZZI, Trans. Faraday Soc. 65 (1969) 496. 'Mechanism of Decomposition of Sodium Dithionite in Aqueous Solution.'

BURLAMACCHI, L., G. GUARINI, and E. TIEZZI, Can. J. Chem. 49 (1971) 1139. 'Comment: On the Decomposition of Aqueous Sodium Dithionite.'

BURNS, G. R., Sulphur Institute Tech. Bull. 13 (1967). 'The Oxidation of Sulphur in Soils.'

BUROW, D. F., in "Chemistry of Non-Aqueous Solvents," J. J. Lagowski, Ed., Academic Press, New York, 1970. 'Liquid Sulfur Dioxide.'

BURTON, K. W. C., and P. MACHMER, in "Inorganic Sulphur Chemistry," G. Nickless, Ed., Elsevier, Amsterdam, 1968. 'Sulphanes.'

BURTON, K. W. C., and G. NICKLESS, in "Inorganic Sulphur Chemistry," G. Nickless, Ed., Elsevier, Amsterdam, 1968. 'Amido- and Imido-Sulphonic Acids.'

BURWELL, J. T., Z. Kristallogr. 97 (1937) 123. 'The Unit Cell and Space Group of Monoclinic Sulphur.'

BUTCHER, S. S., and R. J. CHARLSON, "An Introduction to Air Chemistry," Academic Press, New York-London, 1972.

BUTLIN, K. R., and J. R. POSTGATE, in "Biology of Deserts," Proc. Symp. Biology of Hot and Cold Deserts, J. L. Cloudsley-Thompson, Ed., Institute of Biology, London, 1954. 'The Microbiological Formation of Sulphur in Cyrenaican Lakes.'

BUTTERWORTH, C. E., and J. W. SCHWAB, Ind. Eng. Chem. 30 (1938) 746. 'Sulfur Mining as a Processing Industry.'

C

CAIRNS, E. J., and H. SHIMOTAKE, Science 164 (1969) 1347. 'High-Temperature Batteries.'

CAIRNS, E. J., and R. K. STEUNENBERG, in " Progress in High Temperature Physics and Chemistry," Vol. 5, C. A. Rouse, Ed., Pergamon Press, New York, 1973. 'High Temperature Batteries.'

CALIFORNIA AIR RESOURCES BOARD, Calif. Air Qual. Data 7 (1975). 'Air Analysis Comments.'

CANESSON, P., and P. GRANGE, C. R. Hebd. Seances Acad. Sci., Ser. C. 281 (1975) 757. 'Photoelectron Spectroscopy of the Molybdenum (IV) Sulfide-Cobalt Sulfide (Co_9S_8) System. Hydrodesulfurization Catalysts with Low Cobalt Contents.'

CANTY, C., F. G. PERRY, JR., and L. R. WOODLAND, Tappi 56 (1973) 52. 'Economic Impact of Pollution Abatement on the Sulfite Segment of the U. S. Pulp and Paper Industry.'

CARDONE, M. J., in "The Analytical Chemistry of Sulfur and Its Compounds," Vol. 2, J. H. Karchmer, Ed., Wiley-Interscience, New York, 1972.

CARLSON, G. L., and L. G. PEDERSEN, J. Chem. Phys. 62 (1975) 4567. 'An *ab Initio* Investigation of S_8.'

CARNOW, B. W., and P. MEIER, Arch. Environ. Health 27 (1973) 207. 'Air Pollution and Pulmonary Cancer.'

CARON, A., and J. DONOHUE, J. Phys. Chem. 64 (1960) 1767. 'The X-Ray Powder Pattern of Rhombohedral Sulfur.'

CARON, A., and J. DONOHUE, Acta Crystallogr. 18 (1965) 562. 'Bond Lengths and Thermal Vibrations in Orthorhombic Sulfur.'

CARPENTER, H. D., and P. L. COTTINGHAM, U. S. Bureau of Mines Inf. Circular 8156 (1963). 'A Survey of Methods for Desulfurizing Residual Fuel Oil.'

CARPENTIER, A., Proc. Symp. New Uses for Sulphur and Pyrites, Madrid, 1976, The Sulphur Institute, London, 1976. 'Introductory Remarks on Foamed Sulfur.' and 'Study of the Impregnation of Ceramic Tiles under Vacuum.'

CASATI, P., and P. VIGANO, Atti. Soc. Ital. Sci. Nat. Mus. Civ. Stor. Nat. Milano 115 (1975) 251. 'Sulfurated Water Near the Milan Arena.'

CASTILLO JUSTO, R., Bol. Soc. Geol. Peru 46 (1975) 53. 'The Sulfur Deposits of Tacna Department, Peru.'

CASTLEMAN, A. W. JR., H. R. MUNKELWITZ, and B. MANOWITZ, Nature 244 (1973) 345. 'Contribution of Volcanic Sulphur Compounds to the Stratospheric Aerosol Layer.'

CASTLEMAN, A. W. JR., H. R. MUNKJ LWITZ, and B. MANOWITZ, Tellus 26(1-2) (1974) 222. 'Isotopic Studies of the Sulfur Component of the Stratospheric Aerosol Layer.'

CASTLEMAN, A. W. Jr., R. E. DAVIS, H. R. MUNKELWITZ, I. N. TANG, and W. P. WOOD, in "International Journal of Chemical Kinetics, Symp. No. 1, Chemical Kinetics Data for the Upper and Lower Atmosphere," John Wiley & Sons, 1975. 'Kinetics of Association Reactions Pertaining to H_2SO_4 Aerosol Formation.'

CAVALLARO, J. A., M. T. JOHNSTON, and A. W. DEURBROUCK, Environ. Prot. Technol. Ser. EPA-600/2-76-091 (1976). 'Sulfur Reduction Potential of U. S. Coals: A Revised Report of Investigations.'

CHAMBERLAIN, D. L. JR., Final Report Stanford Research Institute to The Sulphur Institute, SRI Project No. P-3812 (1962). 'Oxidation of Sulfur.'

CHANDLER, R. H., and R. A. C. ISBELL, "The Claus Process. A Review 1970-75," London, 1976.

CHANG, S. G., and T. NOVAKOV, Lawrence Berkeley Rep. 3068, (1974). 'Formation of Pollution Particulate Nitrogen Compounds by NO-Soot and NH_3-Soot Gas Particle Surface Reactions.'

CHANG, S. G., and T. NOVAKOV, Lawrence Berkeley Rep. 4446 (1975). 'Infrared and Photoelectron Spectroscopic Study of SO_2 Oxidation on Soot Particles.'

CHANG, S. G., and T. NOVAKOV, Atmosph. Environ. 9 (1975) 495. 'Formation of Pollution Particulate Nitrogen Compounds by NO-Soot and NH_3-Soot Gas-Particle Surface Reactions.'

CHAPMAN, D., R. J. WARN, A. G. FITZGERALD, and A. D. YOFFE, Trans. Faraday Soc. 60 (1964) 294. 'Spectra and the Semi-Conductivity of the $[SN]_x$ Polymer.'

CHAPMAN, R. S., C. M. SHY, J. F. FINKLEA, D. E. HOUSE, H. E. GOLDBERG, and C. G. HAYES, Arch. Environ. Health 27 (1973) 138. 'Sporadic Respiratory Disease.'

CHARLSON, R. J., A. H. VANDERPOL, D. S. COVERT, A. P. WAGGONER, and N. C. AHLQUIST, Science 184 (1974) 156. 'Sulfuric Acid-Ammonium Sulfate Aerosol: Optical Detection in the St. Louis Region.'

CHASE, H. B., Physiol. Rev. 34 (1954) 113. 'Growth of Hair.'

CHAUDHARI, M. A., F. AKHTAR, and S. P. SHAHID, J. Nat. Sci. Math. 13(1) (1973) 199. 'The Determination of Sulfur in Organic Compounds Using N-Bromo-succinimide.'

CHEKHOVSKIKH, A. I., T. K. MIKHALEVA, and S. I. VOL'FKOVICH, Soviet Chem. Ind. 6 (1974) 3. 'Use of Elemental Sulfur in Fodder and Fertilzer Production.'

CHEKHOVSKIKH, A. I. , T. K. MIKHALEVA, and S. I. VOL'FKOVICH, J. Applied Chem. 36 (1974) 196. 'Applications of Elemental Sulfur in Fodder and Fertilizer Production.'

CHEKHOVSKIKH, A. I., T. K. MIKHALEVA, and S. I. VOL'FKOVICH, Khim. Prom. (Moscow) 3 (1974) 196. 'Use of Elemental Sulfur in the Production of Feed Supplement and Fertilizers.'

CHEN, K. Y., and J. C. MORRIS, Environ. Sci. Technol. 6 (1972) 529. 'Kinetics of Oxidation of Aqueous Sulfide by O_2.'

CHEN, K. Y., and S. K. GUPTA, Environ. Lett. 4 (1973) 187. 'Formation of Polysulfides in Aqueous Solution.'

CHEN, W. F., H. C. MEHTA, and R. G. SLUTTER, J. Test. Eval. 4 (1976) 283. 'Sulfur- and Polymer-Impregnated Brick and Block Prisms.'

CHERUBIM, M., Kaut. Gummi 19 (1966) 676. 'Zur Kinetik der Reaktion zwischen Schwefel und Phenol.'

CHERUBIM, M., Kunstoff-Rundschau 13 (1966) 235 & 291. 'Herstellung, Eigenschaften und Anwendung Schwefelmodifizierter Phenolharze.'

CHEVRON CHEMICAL CO., Brochure, San Francisco, 1976. 'Chevron SUCOAT Coating Compounds.'

CHI, C. T., E. C. EIMUTUS, W. H. HEDLEY, M. V. JONES, R. JONES, and L. B. MOTE, Environ. Prot. Technol. Ser. EPA-600/2-75-045 (1975). 'A Method for Evaluating SO_2 Abatement Strategies.'

CHILTON, T. H., Chairman, Nat'l Research Council Panel, 1970. 'Abatement of Sulfur Oxide Emissions from Stationary Combustion Sources.'

CHILTON, T. H., Chem. Eng. Progress 67 (1971) 69. 'Reducing SO_2 Emission from Stationary Sources.'

CHISTONI, A., Boll. Soc. Ital. Biol. Speriment. 7 (1932) 1266. 'Sul Comportamento Dello Zolfo Colloidale Nell'organismo in Rapporto Alla Reazione Attuale del Sangue.'

CHIVERS, T., and I. DRUMMOND, Chem. Comm. 1971 (1971) 1623. 'Blue Solutions of Sulphur in Hexamethylphosphoramide.'

CHIVERS, T., and I. DRUMMOND, Inorg. Chem. 11 (1972) 2525. 'Characterization of the Trisulfur Radical Anion S_3^- in Blue Solutions of Alkali Polysulfides in Hexamethylphosphoramide.'

CHIVERS, T., and I. DRUMMOND, J. Chem. Soc. Dalton 1974 (1974) 632. 'Formation of the Blue Trisulphur Radical Anion, S_3^-, in Solutions of Alkali Polysulphides in Dimethylformamide, and from Elemental Sulphur and Piperidyl-lithium in Hexamethylphosphoramide.'

CHIVERS, T., Nature 252 (1974) 32. 'Ubiquitous Trisulphur Radical Ion S_3^-.'

CHRISTIANSEN, B., and T. H. CLACK, JR., Science 194 (1976) 578. 'A Western Perspective on Energy: A Plea for Rational Energy Planning.'

CHUNG, K., J. G. CALVERT, and J. W. BOTTENHEIM, Int. J. Chem. Kinetics 7 (1975) 161. 'The Photochemistry of Sulfur Dioxide Excited within Its First Allowed Band (3130 A) and the 'Forbidden' Band (3700-4000 A).'

CHURBANOV, M. F., Tr. Khim. Khim. Tekhnol. (1) (1967) 181. 'Nature of Organic Impurities in Commercial Sulfur.'

CIESIELSKI, C. A., Ph.D. Thesis, Pennsylvania State Univ., University Park, Pa., 1974. 'Sulfur Contents in Asphalts and Asphaltenes.'

CLARK, D., J. Chem. Soc. 1952 (1952) 1615. 'The Structure of Sulfur Nitride.'

CLARK, G. L., Science 120 (1954) 40. '35-S in Wool.'

CLARK, L. B., Ph.D. Thesis, Univ. of Washington, Seattle, Wa., 1963. 'Calculations of Electronic Spectra.'

CLEAVER, B., and A. J. DAVIES, Electrochim. Acta 18 (1973) 741. 'Properties of Fused Polysulphides–IV. Cryoscopic Studies on Solutions of Alkali Metal Polysulphides in Potassium Thiocyanate.'

CLEGHORN, H. P., and M. B. DAVIES, J. Chem. Soc. 1970A (1970) 137. 'Thermal Decomposition of Oxysulphur Salts and The Infrared Spectra of their Products.'

COCKE, D. L., G. ABEND, and J. H. BLOCK, J. Phys. Chem. 80 (1976) 574. 'Mass Spectrometric Observation of Large Sulfur Molecules from Condensed Sulfur.'

COGBILL, C. V., Masters Thesis, Cornell University, Ithaca, N.Y., 1975. 'Acid Precipitation in the Northeastern U.S.'

COLE, E. R., and R. F. BAYFIELD, in "Sulfur in Organic and Inorganic Chemistry," Vol. 2, A. Senning, Ed., Marcel Dekker, New York, 1972. 'Chromatographic Techniques in Sulfur Chemistry.'

COLEMAN, R., Soil Sci. 101 (1966) 230. 'The Importance of Sulfur as a Plant Nutrient in World Crop Production.'

COMBUSTION INSTITUTE, THE, Int. Symposia on Combustion, Pittsburgh, Pa., 1969-1977.

COMPANION, A. L., Theor. Chim. Acta 25 (1972) 268. 'Molecular Orbital Study of Bond Energies in Compounds of Sulfur and Fluorine.'

CORTELYOU, C. G., Chem. Eng. Progress 65 (1969) 69. 'Commercial Processes for SO_2 Removal.'

COUTANT, R. W., R. E. BARRETT, E. H. LOUGHER, Am. Soc. Mech. Engineers Manuscript No. 69-WA/APC-1 (1969). 'SO_2 Pickup by Limestone and Dolomite.'

COUTANT, R. W., R. H. CHERRY, H. ROSENBERG, J. GENCO, and A. LEVY, Final Report NAPCA Contract PH 86-68-84 (1969). 'Investigation of the Limestone SO_2 Wet Scrubbing Process.'

COX, D., Phil. Trans. 9 (1674) 66. 'A Continuation of the Discourse Concerning Vitriol.'

COX, R. A., and S. A. PENKETT, Nature 229 (1971) 486. 'Photo-Oxidation of Atmospheric SO_2.'

COX, R. A., J. Photochem. 2 (1973/74) 1. 'The Sulphur Dioxide Photosensitized Cis-Trans Isomerization of Butene-2.'

COX, R. A., Tellus 26 (1974) 235. 'Particle Formation from Homogeneous Reactions of Sulphur Dioxide and Nitrogen Dioxide.'

CRAIG, D. P., and T. THIRUNAMACHANDRAN, J. Chem. Phys. 43 (1965) 4183. 'd-Orbital Sizes in Sulfur.'

CRAIG, N. L., A. B. HARKER, and T. NOVAKOV, Atmosph. Environ. 8 (1974) 15. 'Determination of the Chemical States of Sulfur in Ambient Pollution Aerosols by X-Ray Photoelectron Spectroscopy.'

CREDALI, L., L. MORTILLARO, G. GALIAZZO, N. DEL FANTI, and G. CARAZZOLO, J. Appl. Polymer Sci. 9 (1965) 2895. 'The Water-Formaldehyde-Hydrogen Sulfide System. I. Solubility of Hydrogen Sulfide in Aqueous Solutions of Formaldehyde and First Solid Product of the System.'

CREDALI, L., L. MORTILLARO, M. RUSSO, and C. DE CHECCHI, J. Appl. Polymer Sci. 10 (1966) 859. 'Water-Formaldehyde-Hydrogen Sulfide System. II. Formation and Growth of Thiomethylenic Chains.'

CREED, E. R., D. R. LEES, and J. G. DUCKETT, Nature 244 (1973) 276. 'Biological Method of Estimating Smoke and Sulphur Dioxide Pollution.'

CRISS, C. M., and J. W. COBBLE, J. Am. Chem. Soc. 86 (1964) 5390. 'The Thermodynamic Properties of High Temperature Aqueous Solutions; The Calculation of Ionic Heat Capacities up to 200°C; Entropies and Heat Capacities above 200°C.'

CROCKER, T. T., Ed., Conf. Health Eff. Atmos. Salts, Gases of Sulfur, Nitrogen in Assoc. with Photochemical Oxidant, Vol. 2, (NTIS, Springfield, Va.), 1974.

CROW, L. J., and R. C. BATES, U. S. Bureau of Mines R.I. 7349 (1970). 'Strengths of Sulfur-Basalt Concretes.'

CRUICKSHANK, D. W. J., B. C. WEBSTER, and D. F. MAYERS, J. Chem. Phys. 40 (1964) 3733. 'd-Orbitals of Sulfur.'

CRUICKSHANK, D. W. J., and B. C. WEBSTER, in "Inorganic Sulphur Chemistry," G. Nickless, Ed., Elsevier, Amsterdam, 1968. 'Orbitals in Sulphur and Its Compounds.'

CULLIS, C. F., and M. F. R. MULCAHY, Combust. Flame 18 (1972) 225. 'The Kinetics of Combustion of Gaseous Sulphur Compounds.'

CUNLIFFE, R. S., Alberta Sulphur Res. Quart. Bull. 6 (1970) 1. 'Optimizing Conversion of Hydrogen Sulphide to Sulphur.'

CUNNINGHAM, P. T., S. A. JOHNSON, and E. J. CAIRNS, J. Electrochem. Soc. 119 (1972) 1448. 'Phase Equilibria in Lithium-Chalcogen Systems, II. Lithium-Sulfur.'

CURRELL, B. R., and A. J. WILLIAMS, Thermochim. Acta 9 (1974) 255. 'Thermal Analysis of Elemental Sulphur.'

CURRELL, B. R., A. J. WILLIAMS, A. J. MOONEY, and B. J. NASH, Adv. Chem. 140 (1975) 1. 'Plasticization of Sulfur.'

CURRELL, B. R., and A. J. WILLIAMS, Proc. Anal. Div. Chem. Soc. 12(3) (1975) 87. 'Differential Scanning Calorimetry of Elemental Sulfur.'

CURRELL, B. R., Proc. Symp. New Uses for Sulphur and Pyrites, Madrid, 1976, The Sulphur Institute, London, 1976. 'The Importance of Using Additives in the Development of New Applications for Sulphur.'

CUSACHS, L. C., and D. J. MILLER, Adv. Chem. Ser. 110 (1972) 1. 'Semiempirical Molecular Orbital Calculations on Sulfur Containing Molecules and their Extension to Heavy Elements.'

CYVIN, S. J., Acta Chem. Scand. 24 (1970) 3259. 'Normal Coordinate Analysis and Mean Amplitudes of Vibration for Octasulphur.'

CZELNY, Z., and J. BANDROWSKI, Inz. Apar. Chem. 14(6) (1975) 16. 'Mathematical Model of the Tower System of a Sulfuric Acid Plant.'

D

DACRE, B., Trans. Faraday Soc. 63 (1967) 604. 'Oleum Solutions. III. Cryoscopic Behavior of Acids, Bases, and Salts in Disulfuric Acid.'

DAINTON, F. S., J. A. DAVIES, P. P. MANNING, and S. A. ZAHIR, Trans. Faraday Soc. 53 (1957) 813. 'The Polymerization of 1-OXA-4,5 Dithio-cycloheptane.'

DAINTON, F. S., and K. J. IVIN, Quart. Rev. (London) 12 (1958) 61. 'Some Thermodynamic and Kinetic Aspects of Addition Polymerization.'

DALE, J. M., and A. C. LUDWIG, in "Elemental Sulfur," B. Meyer, Ed., Interscience, New York, 1965. 'Mechanical Properties of Sulfur.'

DALE, J. M., Sulphur Institute J. 1(1) (1965) 11. 'Sulphur-Fibre Coatings.'

DALE, J. M., and A. C. LUDWIG, Unclassified Document AD 628 064, N.B.S., Institute for Applied Technology (1966). 'Development of Equipment and Procedures for Producing Large Quantities of Foamed Sulphur in the Field.'

DALE, J. M., and A. C. LUDWIG, Sulphur Institute J. 2(3) (1966) 6. 'Rigid Sulphur Foams.'

DALE, J. M., and A. C. LUDWIG, J. Materials 2(1) (1967) 131. 'Fire Retarding Elemental Sulfur.'

DALE, J. M., Sulphur Institute J. 3(4) (1967-68) 19. 'Mechanical Properties of Sulphur Suggests Many Potential Uses.'

DALE, J. M., and A. C. LUDWIG, Civ. Eng. 37(12) (1967) 66. 'Sulphur-Aggregate Concrete.'

DALE, J. M., Mining Eng. 25(10) (1973) 49. 'Utilizing Sulfur-Based Spray Coatings.'

DALE, J. M., and A. C. LUDWIG, Adv. Chem. Ser. 140 (1975) 167. 'Cold Region Testing of Sulfur Foam and Coatings.'

DALTON, J., "Ein Neues System des Chemischen Teils der Naturwissenschaft," Vol. 2, Berlin (1812).

DALY, F. P., and C. W. BROWN, J. Phys. Chem. 80 (1976) 480. 'Raman Spectra of Rhombic Sulfur Dissolved in Secondary Amines.'

DAVENPORT, S. J., and G. G. MORGIS, U. S. Bureau of Mines Bulletin 537 (1954). 'Air Pollution. A Bibliography.'

DAVIES, C. G., R. J. GILLESPIE, J. J. PARK, and J. PASSMORE, Inorg. Chem. 10 (1971) 2781. 'Polyatomic Cations of Sulfur. II. The Crystal Structure of Octasulfur Bis(hexasfluoroarsenate), $S_8(AsF_6)_2$.'

DAVIES, L. H., Sulphur Institute J. 10(2) (1974) 6. 'Gold-N, a Sulphur Coated Urea from ICI.'

DAVIS, A. J. III, and T. F. YEN, Adv. Chem. Ser. 151 (1976) 137. 'Feasibility Studies of a Biochemical Desulfurization Method.'

DAVIS, C. S., and J. B. HYNE, Thermochim. Acta 15 (1976) 375. 'Thermomechanical Analysis of Elemental Sulphur: The Effects of Thermal History and Ageing.'

DAVIS, D. D., G. SMITH, and G. KLAUBER, Science 186 (1974) 733. 'Trace Gas Analysis of Power Plant Plumes Via Aircraft Measurment: O_3, NO_x, and SO_2 Chemistry.'

DAVIS, P. R., E. BECHTOLD, and J. H. BLOCK, Surface Sci. 45 (1974) 585. 'Sulfur Surface Layers on Tungsten Investigated by Field Ion Mass Spectroscopy.'

DAVIS, R. D., in "Inorganic Sulphur Chemistry," G. Nickless, Ed., Elsevier, Amsterdam, 1968. 'Mechanisms of Sulfur Reactions.'

DAVY, H., Phil. Trans. 1 (1809) 59. 'Analytical Experiments on Sulphur.'

DAVY POWERGAS INC., Brochure, 1976. 'Clean Stacks...Clean Air: The Wellman-Lord SO_2 Recovery Process.'

DAWANS, F., and G. LEFEBVRE, Rev. Inst. Franc. Petrole Ann. Combust. Liquides 16 (1961) 941.

DEBAERDEMAEKER, T., and A. KUTOGLU, Naturwissensch. 60 (1973) 49. 'The Crystal and Molecular Structure of a New Elemental Sulfur, S_{18}.'

DEBAERDEMAEKER, T., and A. KUTOGLU, Cryst. Struct. Comm. 3 (1974) 611. 'Cyclooctadecasulfur, S_{18} (beta).'

DEBUS, H., J. Chem. Soc. 53 (1888) 278. 'Chemical Investigation of Wackenroder's Solution, and Explanation of the Formation of its Constituents.'

DE HAAN, Y. M., Physica 24 (1958) 855. 'The Crystal Structure of Gamma-Sulfur.'

DEINES, v. O., Liebig's Ann. Chem. 440 (1924) 213. 'Polythionates.'

DELAMAR, M., J. Electroanal. Chem. Interfacial Electrochem. 63 (1975) 339. 'Reduction Mechanism of Elemental Sulfur in Dimethyl Sulfoxide.'

DELEPINE, M., Bull. Soc. Chim. France 27 (1920) 740. 'Sur le Sulfure d'Ethylene C_2H_2S.'

DE MAINE, P. A. D., J. Chem. Phys. 26 (1957) 1036. 'Interaction between Sulfur Dioxide and Polar Molecules. I. Systems Containing Aliphatic Alcohols, Ethers, or Benzene in Carbon Tetrachloride.'

DEME, I., Proc. Tech. Sessions, Assoc. Asphalt Paving Technologists, Williamsburg, Va., 1974. 'Processing of Sand-Asphalt-Sulphur Mixes.'

DENIS, W., and L. REED, J. Biol. Chem. 72 (1927) 385. 'The Action of Blood on Sulfides.'

DEPOY, P. E., and D. M. MASON, Faraday Symp. Chem. Soc. 9 (1974) 47. 'Periodicity in Chemically Reacting Systems. A Model for the Kinetics of the Decomposition of $Na_2S_2O_4$.'

DERRETT, C. J., and C. BROWN, Atmosph. Environ. 10 (1976) 303. 'A Fast Response Sulphur Dioxide Meter for Laboratory Studies.'

DEUSTER, v. E., and R. E. CONNICK, in Preparation, 1977. 'UV and Raman Spectra of Sulfur Dioxide, Disulfite, and Sulfite in Aqueous Solution.'

DEUSTER, v. E., Unpublished, 1976. 'The Raman Spectra of 34-S_8.'

DEVILLE, C. S. C., Ann. Chim. Phys. 47(3) (1856) 94. 'Modifications du Soufre sons l'Influence de la Chaleur et des Dissolvants.'

DEVILLE, C. S. C., and L. TROOST, Ann. Chim. Phys. 58 (1860) 257. 'Les Densities de Vapeur a des Temperatures Tres Elevees.'

DEVINE, E. T., "Coal, Economic Problems of the Mining, Marketing, and Consumption of Anthracite and Soft Coal in the U.S." Ann. Reveiw Service Press, Bloomington, Ind., 1925.

DHARMARAJAN, V., R. L. THOMAS, R. F. MADDALONE, and P. W. WEST, Sci. Total Environ. 4 (1975) 279. 'Sulfuric Acid Aerosol.'

DIAMOND, R. A., Ed., "Energy Crisis in America," Congressional Quarterly, Washington, D. C., 1973.

DIEHL, L., Proc. Symp. New Uses for Sulphur and Pyrites, Madrid, 1976, The Sulphur Institute, London, 1976. 'Dicyclopentadiene [DCPD]-Modified Sulphur and Its Use as a Binder, Quoting Sulphur Concrete as an Example.'

DIETZEL, R., and S. GALANOS, Z. Elektrochem. 31 (1925) 466. 'Optische Untersuchungen ueber die Schweflige Saeure und Ihre Alkalisalze, Insbesondere das Kalium- und Ammoniumpyrosulfit.'

DIOT, M., J. M. LETOFFE, M. PROST, and J. BOUSQUET, Bull. Soc. Chim. 1972 (1972) 4490. 'K_2S_5; I—Heat Capacity; II—Crystal Structure.'

DIVERS, E., and M. OGAWA, J. Chem. Soc. 77 (1900) 335. 'Products of Heating Ammonium Sulfites, Thiosulphites, and Trithionate.'

DOCHINGER, L. S., and R. A. SELIGA, Eds., Proc. 1st Int. Symp. Acid Precipitation and the Forest Ecosystem, USDA Forest Service Gen. Tech. Report NE-23, Northeastern Forest Experiment Station, Upper Darby, Pa., 1976.

DODD, J. L., R. G. WOODMANSEE, W. K. LAUENROTH, R. K. HEITSCHMIDT, and J. W. LEETHAM, U.S. Environ. Prot. Agency, Off. Res. Dev. EPA-600/3-76-013 (1976). 'Effects of Sulfur Dioxide and Other Coal-Fired Power Plant Emissions on Producer, Invertebrate Consumer, and Decomposer Structure and Function in a Southeastern Montana Grassland.'

DÖBEREINER, J., J. Chem. Phys. 47 (1826) 119. 'Aus Einem Brief des Ritter und Geheimrates Döbereiners. Das Neue Pneumatische Versuch.'

DOI, T., Rev. Phys. Chem. Japan 37 (1967) 62; 35 (1965) 1, 11 & 18. 'Physico-Chemical Properties of Sulfur, I. Pressure Effects on Viscosity of Liquid Sulfur, Addendum; II. Effects of Different Types of Reagent on Viscosities of Liquid Sulfur; III. Dissolved State of Sulfur Polymers in Liquid Sulfur; IV. Critical Polymerization Equilibrium Constants of Sulfur.'

DONBAVAND, M. H., and H. J. MÖCKEL, Z. Naturf. 29b (1974) 742. 'Formation of Radical Intermediates in Sulphur Dioxide Solutions of Tertiary Aliphatic Amines.'

DONOHUE, J., A. CARON, and E. GOLDISH, J. Am. Chem. Soc. 83 (1961) 3748. 'The Crystal and Molecular Structure of S_6.'

DONOHUE, J., "The Structures of the Elements," John Wiley & Sons, New York, 1974.

DORABIALSKA, A., and A. PLONKA, Zes. Nauk Polit. Lodz. Chemia 24 (1973) 139. 'Radiometryczne Badania Kinetyni Rozpuszczalnosci Siarki Polimerycznej.'

DRAEMEL, D. C., Environ. Prot. Technol. Ser. EPA-R2-73-186 (1973). 'Regeneration Chemistry of Sodium-Based Double-Alkali Scrubbing Process.'

DRAEMEL, D. C., Preprint, May 1973. 'An EPA Overview of Sodium-Based Double Alkali Processes. Part I. A View of the Process Chemistry of Identifiable and Attractive Schemes.'

DRAHOWZAL, F. A., Alberta Sulphur Res. Quart. Bull. 1(3) (1965) 1. 'Hydrogen Sulphide in Organic Chemistry.'

DUECKER, W. W., and E. W. EDDY, Chem. Ind. 50 (1942) 174. 'Sulfur's Role in Industry.'

DUECKER, W. W., and J. R. WEST, "Manufacture of Sulfuric Acid," Am. Chem. Soc. Monograph Series 144, 1959.

DUMAS, A., Ann. Chim. Phys. 50 (1811) 170. 'Dissertation sur la Densite de la Vapeur de Quelques Corps Simples.'

DUMAS, J., Poggendorff's Ann. Phys. Chem. 4 (1825) 474. 'Zerlegung des Chlorschwefels.'

DUMAS, J., Ann. Chim. Phys. 36 (1827) 83. 'Observation sur Quelque Proprietes du Soufre (Liquide).'

DUMAS, J., Ann. Chim. Phys. 49 (1832) 204. 'Sur les Chlorures de Soufre.'

DUMAS, J., Ann. Chim. Phys. 50 (1832) 170. 'Le Densite de la Vapeur de Quelques Corps Simples.'

DUNHAM, K., Adv. Sci. 18 (Sept.) (1961) 284. 'Black Shale, Oil, and Sulphide Ore.'

DUNITZ, J. D., Acta Cryst. 9 (1956) 579. 'Structure of Sodium Dithionite, and the Nature of the Dithionite Ion.'

DUTRUEL, F., Proc. Symp. New Uses for Sulphur and Pyrites, Madrid, 1976, The Sulphur Institute, London, 1976. 'Impregnation of Concrete with Sulphur.'

DVORACEK, L. M., Corrosion 32(2) (1976) 64. 'Pitting Corrosion of Steel in H_2S Solutions.'

DZIEWIATKOWSKI, D. D., in "Mineral Metabolism, An Advanced Treatise," Vol. 2, C. L. Comar and F. Bronner, Eds., Part B, The Elements, Academic Press, New York, 1962. 'Sulfur.'

E

EARL, C. B., and W. J. OSBORNE, Power Eng. 78(11) (1974) 46. 'A Method of Meeting EPA's SO_2 Regulations.'

ECHTMAN, J., J. Am. Inst. Homeopathy 23 (1930) 1222. 'The Effect of Homeopathic Sulphur in Diabetes: A Preliminary Report.'

EIGEN, M., K. KUSTIN, and G. MAASS, Z. Phys. Chem., Neue Folge 30 (1961) 130. 'Die Geschwindigkeit der Hydratation von SO_2 in Waessriger Loesung.'

EINBRODT, H. J., Staub Reinhalt. Luft 36(3) (1976) 122. 'Combinations of Effects of the Fine Dusts in the Lung.'

EISENBERG, A., Macromolecules 2 (1969) 44. 'The Viscosity of Liquid Sulfur. A Mechanistic Reinterpretation.'

EKLUND, R. B., Ph.D. Thesis, Royal Institute Tech., Uppsala, 1956. 'The Rate of Oxidation of Sulfur Dioxide with a Commercial Vanadium Catalyst.'

ELAM, C. J., Proc. AFMA Nutr. Counc. 34 (1974) 12. 'Sulfur Requirement of Ruminants.'

ELECTRIC POWER RESEARCH INSTITUTE, Res. and Develop. Projects (Listing) February 5, 1976.

ELLIOT, A. J., and F. C. ADAM, Can. J. Chem. 52 (1974) 102. "Identification of Intermediates in the Photolysis of Mercaptans in Dilute Glass Matrices. II.'

EMELEUS, H. J., Ed., Adv. Inorg. Chem. and Radiochem., Vol. 1-19, 1958-1977.

EMERALD, R. L., R. E. DREWS, and R. ZALLEN, Phys. Rev. 14B (1976) 808. 'Polarization-Dependent Optical Properties of Orthorhombic Sulfur in the Ultraviolet.'

EMERSON, H. T., Pap. 3rd Int. Pollut. Eng. Congr., Pollut. Eng. Tech., A. Smith, Ed., Clapp & Poliak, Inc. New York, 1974. 'Sulfur Dioxide Recovery. Emission Control in Heavy Chemicals Plants.'

ENGEL, M. R., Compt. Rend. Acad. Sci., Paris 112 (1891) 866. 'Sur Deux Nouveaux etats du Soufre.'

ENVIRONMENTAL PROTECTION AGENCY, U.S., Technology Series, EPA-600/2-76-136a & b, (1976), R. D. Stern and W. H. Ponder, Eds. 'Proceedings: Symposium on Flue Gas Desulfurization, New Orleans, March 1976, Volumes I and II.'

ENVIRONMENTAL PROTECTION AGENCY, U.S., Progress Report 26, July 1976. 'Limestone Scrubbing of SO_2 at EPA/RTP Pilot Plant.'

EPHRAIM, F., and C. AELLIG, Helv. Chim. Acta 6 (1923) 37. 'Ueber Komplexe mit Schwefeldioxyd.'

EPSTEIN, M., H. N. HEAD, S. C. WANG, and D. A. BURBANK, Proc. Symp. Flue Gas Desulfurization, New Orleans, 1976, R. D. Stern and W. H. Ponder, Eds., Environ. Prot. Technol Ser. EPA 600/2-76-136 (1976). 'Results of Mist Elimination and Alkali Utilization Testing at the EPA Alkali Scrubbing Test Facility.'

ERÄMETSÄ, O., and L. NIINISTÖ, Suomen Kemistilehti B42 (1969) 471. 'Untersuchungen ueber die Allotropie des Schwefels XI. Die Entstehung des Monoklinen beta-Schwefels.'

ERDEY, L., J. SIMON, S. GAL, and G. LIPTAY, Talanta 13 (1966) 67. 'Thermoanalytical Properties of Analytical-Grade Reagents-IVA.'

ERDMAN, D. A., Proc. Am. Chem. Soc. Meet., Atlantic City, 1974. 'Removal of SO_2 in Stack Gases.'

ERDÖS, E., Coll. Czech. Chem. Comm. 27 (1962) 1428 & 2273. 'Thermodynamic Properties of Sulphites. I. Standard Heats of Formation; II. Absolute Entropies, Heat Capacities, and Dissociation Pressures.'

ERDÖS, E., Coll. Czech. Chem. Comm. 27 (1962) 2152. 'Equilibria in the Systems SO_2-CO_2-M_mO.'

ERICKS, L. J., and R. L. POWELL, "Thermal Conductivity of Solids at Low Temperatures," U. S. Nat'l Bureau of Standards Monograph, 1971.

ERIKSEN, T. E., Chem. Eng. Sci. 24 (1969) 273. 'Diffusion Studies in Aqueous Solutions of Sulfur Dioxide.'

ERIKSEN, T., and J. LIND, Acta Chem. Scand. 26 (1972) 3325. 'Spectrophotometric Determination of Sulphur Dioxide and Thiosulphate in Aqueous Solutions of Hydrogen Sulphite.'

ERIKSEN, T., J. Chem. Soc. Faraday I., 70 (1974) 208. 'pH Effects on the Pulse Radiolysis of Deoxygenated Aqueous Solutions of Sulphur Dioxide.'

ERIKSSON, E., Tellus 2 (1959) 375. 'The Yearly Circulation of Chloride and Sulfur in Nature; Meteorological, Geochemical, and Pedological Implications. Part I.'

ERIKSSON, E., Tellus 12 (1960) 63. 'The Yearly Circulation of Chloride and Sulfur in Nature; Meteorological, Geochemical, and Pedological Implications, Part II.'

ERIKSSON, E., J. Geophys. Res. 68 (1963) 4001. 'The Yearly Circulations of Sulfur in Nature.'

ETLIS, V. S., A. P. SINEOKOV, and G. A. RAZUVAEV, J. Gen. Chem. USSR (Engl. Transl.) 34 (1963) 4076. 'Reaction of Alkene Oxides with Isothiocyanates. I.'

F

FABIAN, J., in "Sulfur in Organic and Inorganic Chemistry," Vol. 3, A. Senning, Ed., Marcel Dekker, New York, 1973. 'The Quantum Chemistry of Sulfur Compounds.'

FALLER, N., K. HERWIG, and H. KUEHN, Plant & Soil 33 (1970) 283. 'Sulfur Dioxide-35-S Uptake from the Air. II. Uptake, Metabolism, and Distribution in Plants.'

FANG, H. Y., in Proc. Symp. New Horizons in Construction Materials, Vol. I., Lehigh Univ. Press, Bethlehem, Pa. 1976. 'Study of Sulphur Sand Treated Bamboo Pole.'

FARADAY, M., Ann. Chim. Phys. 7 (1817) 71. 'Sur le Sulfure de Phosphore.'

FARMER, M. H., and R. R. BERTRAND, Environ. Prot. Agency APTD 1069, Nov. 1971. 'Long Range Sulfur Supply and Demand Model.'

FARWELL, S. O., and R. A. RASMUSSEN, J. Chromatogr. Sci. 14(5) (1976) 224. 'Limitations of the FPD and ECD in Atmospheric Analysis. A Review.'

FASULLO, O. T., "Sulfuric Acid Use and Handling," McGraw-Hill, New York, 1965.
FEELY, H. W., and J. L. KULP, Bull. Amer. Assoc. Petrol. Geol. 41 (1957) 1802. 'Origin of Gulf Coast Salt-Dome Sulphur Deposits.'
FEHER, F., and G. WINKHAUS, Z. Anorg. Allg. Chem. 288 (1956) 123. 'Ueber die Darstellung der Sulfane H_2S_5, H_2S_6, H_2S_7, und H_2S_8.'
FEHER, F., K. NAUSED, and H. WEBER, Z. Anorg. Allg. Chem. 290 (1957) 303. 'Ueber die Darstellung und Eigenschaften der Chlorsulfane S_3Cl_2, S_4Cl_2, S_5Cl_2, und S_6Cl_2.'
FEHER, F., and G. WINKHAUS, Z. Anorg. Allg. Chem. 292 (1957) 210. 'Zur Thermochemie der Sulfane: Bildungsenthalpien und Bindungsenergien.'
FEHER, F., G. KRAUSE, and K. VOLGEBRUCH, Chem. Ber. 90 (1957) 1570. 'Zur Kenntnis der Dailkyl- und Diarylsulfane.'
FEHER, F., and W. KRUSE, Z. Anorg. Allg. Chem. 293 (1958) 302. 'Ueber die Reaktion des Schwefelwasserstoffes mit Chlorsulfanen bezw. Chlor oder Brom.'
FEHER, F., and S. RISTIC, Z. Anorg. Allg. Chem. 293 (1958) 307. 'Ueber die Syntheses der Chlorsulfane S_5Cl_2, S_6Cl_2, S_7Cl_2, und S_8Cl_2.'
FEHER, F., and S. RISTIC, Z. Anorg. Allg. Chem. 293 (1958) 312. 'Ueber die Darstellung Definierter Bromsulfane S_nBr_2 durch Umwandlung Entsprechender Chlorsulfane S_nCl_2 mit Hilfe von Bromwasserstoff.'
FEHER, F., and G. HITZEMANN, Z. Anorg. Allg. Chem 294 (1958) 50. 'Verdampfungsenthalpien, Dampfdrucke, Siedepunkte, Kritische Temperaturen und Drucke sowie Troutonsche Konstanten von Sulfanen.'
FEHER, F., and E. HELLWIG, Z. Anorg. Allg. Chem. 294 (1958) 63. 'Zur Kenntnis des Fluessigen Schwefels.'
FEHER, F., and H. MÜNZNER, Chem. Ber. 96 (1963) 1150. 'Zur Kenntnis der Jodsulfane.'
FEHER, F., in "Handbook of Preparative Inorganic Chemistry," Vol. 1, 2nd Ed., Georg Brauer, Ed., Academic Press, New York, 1963. 'Sulfur, Selenium, Tellurium.'
FEHER, F., and H. D. LUTZ, Z. Anorg. Allg. Chem. 333 (1964) 216. 'Die Elektrische Leitfaehigkeit des Fluessigen Schwefels.'
FEHER, F., R. SCHLAFKE, and A. MÜLLER, Z. Naturforsch. 22b (1967) 221. 'Die UV-Spektroskopische Untersuchung des Systems S_2F_2.'
FEHER, F., and H. KULUS, Z. Anorg. Allg. Chem. 364 (1969) 241. 'Sulfur Chemistry. XCIV. Preparation of Chlorosulfanes in Carbon Tetrachloride.'
FEHER, F., Chem. Non-Aqueous Solv. 3 (1970) 219. 'Liquid Hydrogen Sulfide.'
FEHER, F., G. P. GOERLER, and H. D. LUTZ, Z. Anorg. Allg. Chem. 382 (1971) 135. 'Schmelzwaerme und Spezifische Waerme des Fluessigen Schwefels.'
FENIJN, J., Preprint Can. Sulphur Symp., Calgary, 1974. 'Elemental Sulphur in Asphaltic Paving Mixes.'
FENIMORE, C. P., Symp. Int. Combust. Proc. 1972, 14 (1973) 955. 'Two Modes of Interaction of Sodium Hydroxide and Sulfur Dioxide in Gases from Fuel-Lean Hydrogen Air Flames.'
FERRIS, B. G. JR., I. T. T. HIGGINS, M. W. HIGGINS, and J. M. PETERS, Environ. Health 27 (1973) 179. 'Sulfur Oxides and Suspended Particulates.'
FETTES, E. M., and J. S. JORCZAK, Ind. Eng. Chem. 42 (1950) 2217. 'Polysulfide Polymers.'
FETTES, E. M., J. S. JORCZAK, and J. R. PANEK, Ind. Eng. Chem. 46 (1954) 1539. 'Mechanism of Vulcanization of Polysulfide Rubbers.'

FETTES, E. M., and F. O. DAVIS, in "High Polymers," Vol. 13, N. G. Gaylord, Ed., Wiley-Interscience, New York, 1962. 'Chemical Reactions of Polymers.'

FIKE, H. L., and C. RIXON, "Current Academic Research in Sulphur Chemistry," The Sulphur Institute, Washington, D. C., 1974.

FIKE, H. L., and M. CONITZ, in Proc. Symp. New Horizons in Construction Materials, Vol. I., Lehigh Univ. Press, Bethlehem, Pa., 1976. 'Surface Bond Construction.'

FIKE, H. L., Proc. Symp. New Uses for Sulphur and Pyrites, Madrid, 1976, The Sulphur Institute, London, 1976. 'Sulphur Coatings, a Review and Status Report.'

FINK, U., H. P. LARSON, and T. N. GAUTIER III, Icarus 27 (1976) 439. 'New Upper Limits for Atmospheric Constituents on Io.'

FINLAYSON, B. J., and J. N. PITTS, JR., Science 192 (1976) 111. 'Photochemistry of the Polluted Troposphere.'

FISHER, G. L., D. P. Y. CHANG, and M. BRUMMER, Science 192 (1976) 553. 'Fly Ash Collected from Electrostatic Precipitators: Microcrystalline Structures and the Mystery of the Spheres.'

FISHER, J. C., Physics Today, Dec. (1973) 40. 'Energy Crises in Perspective.'

FISHER, J. C., "Energy Crisis in Perspective," John Wiley & Sons., New York, 1974.

FITTIPALDI, F., and L. PAUCIULO, J. Appl. Phys. 37 (1966) 4292. 'Electrical Conductivity Measurements for Good Insulators.'

FLIS, I. E., G. P. ARKHIPOVA, and K. P. MISHCHENKO, Zh. Prikl. Khim. 38 (1965) 1494. 'Equilibrium in Aqueous Solutions of Sulfites at 10-35°.'

FLIS, I. E., G. P. ARKHIPOVA, and K. P. MISHCHENKO, J. Appl. Chem. USSR 541 (1965) 121, 546 (1965) 224. 'Investigation of the Equilibrium in Aqueous Solutions of Sulfite under the Temperatures 10-35°.'

FMC CORP., Bulletin 25300, 1976. 'Boiler Emissions Control.'

FOCK, A., and K. K. KLÜSS, Chem. Ber. 23 (1890) 3149. 'Pyroschwefligsaures Ammonium.'

FOERSTER, F., F. LANGE, O. DROSSENBACH, and W. SEIDEL, Z. Anorg. Allg. Chem. 128 (1923) 245. 'Ueber die Zersetzung der Schwefligen Saure und Ihrer Salze in Waessriger Loesung.'

FOERSTER, F., and G. HAMPRECHT, Z. Anorg. Allg. Chem. 158 (1926) 277. 'Beitraege zur Kenntnis der Schwefligen Saeure und Ihrer Salze. V. Das Verhalten der Pyrosulfite in der Hitze.'

FOGLEMAN, W. W., D. J. MILLER, H. B. JONASSEN, and L. C. CUSACHS, Inorg. Chem. 8 (1969) 1209. 'The 3d Orbitals of Phosphorus and Sulfur.'

FORCHHAMMER, M., Compt. Rend. 5 (1837) 395. 'Action de l'Acide Sulfureux sur l'Ammoniaque.'

FORD, G. P., and V. K. LA MER, J. Am. Chem. Soc. 72 (1950) 1959. 'Vapor Pressure of Supercooled Liquid Sulfur.'

FORREST, J., and L. NEWMAN, J. Air Poll. Control Assoc. 23 (1973) 761. 'Ambient Air Monitoring for Sulfur Compounds. A Critical Review.'

FORYS, M., A. JOWKO, and I. SZAMREJ, J. Phys. Chem. 80 (1976) 1035. 'Rare Gas Sensitized Radiolysis of Hydrogen Sulfide in the Low Concentration Range.'

FOSS, O., and O. TJOMSLAND, Acta Chem. Scand. 10 (1956) 424. 'Solvates of Barium Pentathionates with Acetone and Tetrahydrofuran.'

FOSS, O., and O. TJOMSLAND, Acta Chem. Scand. 12 (1958) 44. 'Structure of a Solvate of Barium Pentathionate with Acetone.'

FOWLER, J. M., and K. E. MERVINE, Environmental Resource Packet Project, Dept. of Physics and Astronomy, Univ. of Maryland, College Park, Md., 1974. 'Energy and the Environment.'

FRANCK, H. G., Brennstoff Chem. 45 (1964) 5. 'Entwicklungsmoeglichkeiten des Steinkohlenteers als Chemische Rohstoffquelle.'

FRANK, R., Conf. Health Eff. Atmos. Salts Gases Sulfur Nitrogen, Assoc. Photochem. Oxid. 2 No. II (1974). 'Sulfur Oxides and Particles: Effects on Pulmonary Physiology in Man and Animals.'

FRANKISS, S. G., J. Mol. Struct. 2 (1968) 271. 'Vibrational Spectra and Structures of S_2Cl_2, S_2Br_2, Se_2Cl_2, and Se_2Br_2.'

FRASCH, H., J. Ind. Eng. Chem. 4 (1912) 134. 'Perkin Medal Award Address of Acceptance.'

FREIBERG, J., Atmosph. Environ. 9 (1975) 661. 'Mechanism of Iron-Catalyzed Oxidation of Sulfur Dioxide in Oxygenated Solutions.'

FRENCH, J. C., D. LAWRIMORE, W. C. NELSON, J. F. FINKLEA, P. ENGLISH, and M. HERTZ, Arch. Environ. Health 27 (1973) 129. 'The Effects of Sulfur Dioxides and Suspended Sulfates on Acute Respiratory Disease.'

FRENEY, J. R., G. E. MELVILLE, and C. H. WILLIAMS, Soil Biol. Biochem. 7 (1975) 217. 'Soil Organic Matter Fractions as Sources of Plant-Available Sulphur.'

FRIEND, J. P., in "Chemistry of the Lower Atmosphere," S. I. Rasool, Ed., Plenum Press, New York-London, 1973. 'The Global Sulfur Cycle.'

FRIEND, J. P., R. LEIFER, and M. TRICHON, J. Atmosph. Sci. 30 (1973) 465. 'On the Formation of Stratospheric Aerosols.'

FROHLIGER, J. O., and R. KANE, Science 194 (1976) 647. 'Acid in Rain Water.'

FUJII, T., Bunseki Kiki 14 (1976) 83. 'A Study on the Analysis of Sulfur Compounds by FPD.'

FURUYA, H., J. Dermatol. 3(3) (1976) 73. 'Effect of Hot Spring Waters on the Growth of Trichophyton Mentagrophytes and Trichophyton Rubrum.'

G

GALL, R. L., and E. J. PIASECKI, Chem. Eng. Progr. 7(May) (1975) 72. 'The Double Alkali Wet Scrubbing System.'

GALLOWAY, J. E., G. E. LIKENS, and E. S. EDGERTON, Science 194 (1976) 722. 'Acid Precipitation in the Northeastern United States: pH and Acidity.'

GAMBLE, B. R., J. E. GILLOTT, I. J. JORDAAN, R. E. LOOV, and M. A. WARD, Adv. Chem. Ser. 140 (1975) 154. 'Civil Engineering Applications of Sulfur-Based Materials.'

GAMSON, B. W., and R. H. ELKINS, Chem. Eng. Progr. 49 (1953) 203. 'Sulfur from Hydrogen Sulfide.'

GARDNER, D. M., and G. K. FRAENKEL, J. Am. Chem. Soc. 78 (1956) 3279. 'Paramagnetic Resonance of Liquid Sulfur: Determination of Molecular Properties.'

GARDNER, L. R., Geochim. Cosmochim. Acta 37 (1973) 53. 'Chemical Models for Sulfate Reduction in Closed Anaerobic Marine Environments.'

GARDNER, M., and A. ROGSTAD, J. Chem. Soc. Dalton 1973 (1973) 599. 'Infrared and Raman Spectra of Cycloheptasulphur.'

GARDY, H., "La Production Directe de l'Anhydride Sulfurique," Annales du Genie Chimique, Toulouse, 1957.

GAREEV, S. Z., S. N. GANZ, V. A. KRASNOVSKAYA, and Z. N. GAREEVA, Izv. Vyssh. Uchebn. Zaved., Khim. Khim. Tekhnol. 19 (1976) 618. 'Study of the Kinetics of Sulfur Dioxide Absorption by Aqueous Solutions of Alkalies.'

GARLAND, J. A., W. S. CLOUGH, and D. FOWLER, Nature 242 (1973) 256. 'Deposition of Sulphur Dioxide on Grass.'

GARLAND, J. A., and J. R. BRANSON, Atmosph. Environ. 10 (1976) 353. 'The Mixing Height and Mass Balance of Sulfur Dioxide in the Atmosphere Above Great Britain.'

GARRELS, R. M., and C. R. NAESER, Geochim. Cosmochim. Acta 15 (1958) 113. 'Equilibrium Distribution of Dissolved Sulphur Species in Water at 25°C and 1 atm Total Pressure.'

GARRELS, R. M., and M. E. THOMPSON, Amer. J. Sci. 260 (1962) 57. 'A Chemical Model for Seawater at 25°C and one Atmosphere Total Pressure.'

GARRELS, R. M., and F. T. MACKENZIE, Marine Chem. 1 (1972) 27. 'A Quantitative Model for the Sedimentary Rock Cycle.'

GARRELS, R. M., and E. A. PERRY, JR., in "The Sea," Vol. 5, E. D. Goldberg, Ed., Wiley-Interscience, New York, 1974. 'Cycling of Carbon, Sulfur, and Oxygen Through Geologic Time.'

GARRIGUES, C., and P. VINCENT, Adv. Chem. Ser. 140 (1975) 130. 'Sulfur/Asphalt Binders for Road Construction.'

GATARZ, Z., Prozem. Chemi. 52 (1973) 415. 'Siarka w Drogownictwie.'

GAUTIER, G., and M. DEBEAU, Spectrochim. Acta 30A (1974) 1193. 'Spectres de Vibration d'un Monocristal de Soufre Orthorhombique.'

GAUTIER, G., and M. DEBEAU, Spectrochim. Acta 32A (1976) 1007. 'Spectres de Diffusion Raman du Soufre beta-Monoclinique.'

GAY, E., R. K. STEUNENBERG, J. E. BATTLES, and E. J. CAIRNS, Proc. Intersoc. Energy Convers. Eng. Conf., AIAA, New York, 1973. 'The Development of Lithium/Sulfur Cells for Application to Electric Automobiles.'

GAY, E. C., W. W. SCHERTZ, F. J. MARTINO, and K. E. ANDERSON, Proc. 9th Intersoc. Energy Convers. Eng. Conf., San Francisco, 1974. 'The Development of Lithium/Sulfur Cells for Applications to Electric Automobiles.'

GAY, E. C., F. J. MARTINO, and Z. TOMCZUK, Proc. 10th Intersoc. Energy Convers. Eng. Conf., Newark, 1975. 'Development of High-Performance Iron Sulfide Electrodes with Porous Current Collector Structures.'

GAYDON, A. G., "The Composition of Flames," 2nd Ed., Chapman and Hall, London, 1974.

GEDDES, J. H., Alberta Sulphur Res. Quart. Bull. 5(4) (1969) 5. 'Manufacture and Transportation of Liquid Hydrogen Sulphide.'

GEHRI, D. C., and R. D. OLDENKAMP, Proc. Symp. Flue Gas Desulfurization, New Orleans, 1976, R. D. Stern and W. H. Ponder, Eds., EPA 600/2-76-136 (1936). 'Potential Utilization of Controlled SO_x Emissions from Power Plants in Eastern United States,' and 'Status and Economics of the Atomics International Aqueous Carbonate Flue Gas Desulfurization Process.'

GELLER, S., and M. D. LIND, Acta Crystallogr. 25B (1969) 2166. 'Indexing of the psi-Sulfur Fiber Pattern.'

GEORGE, Z. M., Adv. Chem. Ser. 139 (1975) 75. 'Effect of Basicity of the Catalyst on Claus Reaction.'

GEORGIEVA, Z., Stoit. Mater. Silikat. Prom. 13(5) (1972) 10. 'Sulfur Cement.'

GERARD, M., G. DUMAS, and F. VOVELLE, Proc. 12th Eur. Congr. Mol Spectrosc., 1975, M. Grosmann, S. G. Elkomoss, and J. Ringeissen, Ed., Elsevier, Amsterdam, 1976. 'Raman Scattering of Molecular Solid–Solid Phase Transition.'

GERDING, H., and K. ERIKS, Recueil 69 (1950) 724. 'The Raman Spectra of Di-, Tri-, and Tetrathionate Ions in Aqueous Solutions.'

GERMERDONK, R., Chem. Ing. Tech. 47 (1975) 897. 'Desulfurization of Waste Gases Using a Wet Claus Process.'

GERNEZ, D., Compt. Rend. Acad. Sci. France 82 (1876) 1151. 'Sur la Determination de la Temperature de Solidification des Liquides et en Particulier du Soufre.'

GERSTLE, R., T. DEVITT, and F. K. ZADA, Proc. Annu. AIChE Southwest Ohio Conf., Energy Environ., 1973, A. J. Buonicore, E. J. Rolinski, and D. E. Earley, Eds., Am. Inst. Chem. Eng., Dayton, Ohio, 1974. 'Sulfur Dioxide Emission Control. Lime/Limestone Scrubbing Problem Areas and Solutions.'

GIBBONS, D. J., Mol. Cryst. Liq. Cryst. 10 (1970) 137. 'Electronic Structure of the S_8 Molecule and its Ions and the Transport Properties in the Crystalline Solid and the Liquid.'

GIGGENBACH, W., Inorg. Chem. 10 (1971) 1308. 'The Blue Solutions of Sulfur in Salt Melts.'

GIGGENBACH, W., Inorg. Chem. 10 (1971) 1333. 'Optical Spectra of Highly Alkaline Sulfide Solutions and the Second Dissociation Constant of Hydrogen Sulfide.'

GIGGENBACH, W., Inorg. Chem. 11 (1972) 1201. 'Optical Spectra and Equilibrium Distribution of Polysulfide Ions in Aqueous Solution at 20°C.'

GIGGENBACH, W., Inorg. Chem. 13 (1974) 1724. 'Equilibria Involving Polysulfide Ions in Aqueous Sulfide Solutions up to 240°C.'

GIGGENBACH, W., Inorg. Chem. 13 (1974) 1730. 'Kinetics of Polysulfide-Thiosulfate Disproportionation up to 240°.'

GILLESPIE, R. J., and E. A. ROBINSON, in "Non-Aqueous Solvent Systems," Academic Press, London, 1965. 'Sulphuric Acid.'

GILLESPIE, R. J., in "Inorganic Sulphur Chemistry," G. Nickless, Ed., Elsevier, Amsterdam, 1968. 'Sulphur Acid as a Solvent System.'

GILLESPIE, R. J., and J. PASSMORE, Accounts Res. 4 (1971) 413. 'Polycations of Group VI.'

GILLESPIE, R. J., and J. PASSMORE, Int. Rev. Sci. Inorg. Chem. Ser. II 3 (1975) 121. 'Polyatomic Cations of Sulphur, Selenium, and Tellurium.'

GITTINGER, L. B., JR., in "Industrial Mineral Rocks," 4th Ed., S. J. Lefond, Ed., Am. Inst. Min. Eng., New York, 1975.

GLEMSER, O., Z. Naturforsch. 31b (1976) 610. 'Recent Investigations on Cyclic Sulfur-Nitrogen-Halogen Compounds.'

GLOOR, M., I. PAPE, and H. C. FRIEDERICH, Fette Seif. Anstrichmit. 75 (1973) 550. 'Ueber den Effekt von Schwefel- und Teerzusaetzen zu Kopfwaschmitteln auf die Seborrhoea capitis.'

GLUUD, W., "International Handbook for the By-Product Coking Industry," Amer. Ed., D. L. Jacobsen, Transl., Chem. Cat. Co., New York, 1931.

GLUUD, W., Ber. Ges. Kohletechnik 3 (1931) 466. 'Schwefelwasserstoff-Ammoniakwaesche.'

GMELIN, L., "Handbook of Chemistry," 1st Ed., translated by H. Watts for the Cavendish Society, London, 1849.

GMELIN, L., "Handbook der Anorganischen Chemie," 8th Ed. publ. between 1950 and 1970; 9th Ed. in publ. since 1970, Verlag Chemie, Weinheim. 'Schwefel,' 9. Systemnummer. (Six parts published in different years. For sulfite, disulfites, sulfates, etc., see also the volumes on 'Natrium,' Kalium,' 'Calcium,' 'Magnesium,' etc.)

GOEHRING (see also BECKE-GOEHRING)

GOEHRING, M., and H. W. KALOUMENOS, Z. Anorg. Allg. Chem. 263, (1951) 138. 'Ueber die Einwirkung von Schwefeldioxyd auf Ammoniak.'

GOEHRING, M., H. W. KALOUMENOS, and J. MESSNER, Z. Anorg. Allg. Chem. 264 (1951) 48. 'Ueber das Amid der Imidodisulfinsaeure.'

GOEHRING, M., and D. VOIGT, Z. Anorg. Allg. Chem. 285 (1956) 181. 'Ueber Dischwefel-Dinitrid, S_2N_2, und Polyschwefelstickstoff, $(SN)_x$.'

GOETHALS, E. J., in "Encycl. Pol. and Pol. Science," Vol. 13, H. F. Mark, Ed., Interscience, 1970. 'Sulfur-Containing Polymers.'

GOLD, V., and F. L. TYE, J. Chem. Soc. 1950 (1950) 2932. 'The State of Sulphur Dioxide dissolved in Sulphuric Acid.'

GOLDHABER, M. B., and I. R. KAPLAN, in "The Sea," Vol. 5, E. D. Goldberg, Ed., Wiley-Interscience, New York, 1974. 'The Sulfur Cycle.'

GOLDHABER, M. B., and I. R. KAPLAN, Marine Chem. 3 (1975) 83. 'Apparent Dissociation Constants of Hydrogen Sulfide in Chloride Solutions.'

GOLDING, R. M., J. Chem. Soc. 1960 (1960) 3711. 'Ultraviolet Absorption Studies of the Bisulphite-Pyrosulphite Equilibrium.'

GOLDSCHMIDT, K., Fortschritt Ber. VDI-Zeitschr., Ser. 6(21), VDI-Verlag BmbH, Duesseldorf, 1968. 'Experiments in the Use of White Lime Hydrate and Dolomite Lime Hydrate to Desulfurize Flue Gases from Oil- and Pulverized Coal-Fired Furnaces.'

GOLDSTEIN, B., see National Academy of Science Report, 1975, pg. 37.

GOLLMER, K. D., F. H. MÜLLER, and H. RINGSDORF, Makramol. Chem. 92 (1966) 122. 'Untersuchung von O-, S-, und N-Vinylverbindungen. IV. Darstellung und Polymerisation von Formaldehyde-S-Vinylmercaptalen.'

GOODYEAR, C., "Gum Elastic and its Varieties, with Detailed Account of its Application and Uses and the Discovery of Vulcanization," New Haven, Conn., 1856.

GORDON, C. C., U.S. Environ. Prot. Agency, Off. Res. Dev. Rep. EPA-600/3-76-103 (1976). 'Investigations of the Impact of Coal-Fired Power Plant Emissions Upon Plant Disease and Upon Plant-Fungus and Plant-Insect Systems.'

GORIN, E., and H. E. LEBOWITZ, Alberta Sulphur Res. Quart. Bull. 10 (1973) 2. 'The Removal of Sulfur and Mineral Matter from Coal.'

GORMAN, P. G., Environ. Prot. Agency Rep. EPA-600/2-76-120 (1976). 'Control Technology for Asphalt Roofing Industry.'

GOSHI, Y., O. HIRAO, and I. SUZUKI, Adv. X-Ray Anal. 18 (1975) 406. 'Chemical State Analyses of Sulfur, Chromium and Tin by High Resolution X-Ray Spectrometry.'

GOULDEN, J. D. S., and D. J. MANNING, Spectrochim. Acta 23A (1967) 2249. 'IR Spectroscopy of Inorganic Materials in Aqueous Solution.'

GOVOROV, V. V., N. S. AVRAMENKO, and A. V. GLADKII, Zh. Vses. Khim. O-va. 20 (1975) 468. 'Equilibrium Pressure of Sulfur Dioxide Vapors over Calcium Sulfite and Bisulfite Solutions.'

GOYPIRON, A., J. DE VILLEPIN, and A. NOVAK, Spectrochim. Acta 31A (1975) 805. 'Spectres de Vibration des Acides H_2SO_4 et D_2SO_4 a l'etat Cristallise.'

GRANAT, L., H. RODHE, and R. HALLBERG, in "Nitrogen, Phosphorus, and Sulphur Global Cycle," Ecol. Bull., B. H. Svensson, Ed., (Stockholm) 22 (1976) 89. 'The Global Sulphur Cycle.'

GREENOUGH, K. F., and A. B. F. DUNCAN, J. Am. Chem. Soc. 83 (1961) 555. 'The Fluorescence of Sulfur Dioxide.'

GREENBERG, D. M., "Metabolism of Sulfur Compounds," (Vol. VII of Metabolic Pathways), Academic Press, New York, 1975.

GREENE, K. T., Transp. Res. Rec. 564 (1976) 21. 'A Setting Problem Involving White Cement and Admixture.'

GREENE, R. L., G. B. STREET, and L. J. SUTER, Phys. Rev. Lett. 34 (1975) 577. 'Superconductivity in Polysulfur Nitride $(SN)_x$.'

GREENGARD, H., and J. R. WOOLLEY, J. Biol. Chem. 132 (1940) 83. 'Studies on Colloidal Sulfur–Polysulfide Mixture; Absorption and Oxidation after Oral Administration.'

GREY, D. C., and M. L. JENSEN, Science 177 (1972) 1099. 'Bacteriogenic Sulfur in Air Pollution.'

GRINENKO, V. A., and L. N. GRINENKO, V sb., XI Mendeleevsk. S'ezd po Obshch. i Prikl. Khimii. Ref. Dokl. i Soobshch. (1) (1975) 306. 'Some High-Temperature Reactions for Separation of Sulfur Isotopes.'

GROTEWOLD, G., Gwf-gas/Erdgas 113 (1972) 73. 'Schwefel in deutschen Erdgasen und seine Entfernung.'

GRUEN, D. M., R. L. MCBETH, and A. J. ZIELEN, J. Am. Chem. Soc. 93 (1971) 6691. 'Nature of Sulfur Species in Fused Salt Solutions.'

GRÜNERT, A., and G. TÖLG, Talanta 18 (1971) 881. 'Zur Elementaranalyse des Schwefels im Nanogramm-Bereich.'

GUNNING, H. E., in "Elemental Sulfur," B. Meyer, Ed., Interscience, New York, 1965. 'The Reactions of Atomic Sulfur.'

GUNNING, H. E., and O. P. STRAUSZ, Adv. Photochem. 4 (1966) 143. 'The Reactions of Sulfur Atoms.'

GUTENMANN, W. H., C. A. BACHE, W. D. YOUNGS, and D. J. LISK, Science 191 (1976) 966. 'Selenium in Fly Ash.'

GUYOL, N. B., "The World Electric Power Industry," University of California Press, Berkeley and Los Angeles, 1969.

H

HAAS, L. A., U.S. Bureau of Mines Info. Circular 8608, 1973. 'Sulfur Dioxide: Its Chemistry as Related to Methods for Removing it from Waste Gases.'

HABASHI, F., Mining Congr. J. 55 (1969) 38. 'Processes for Sulfur Recovery from Ores.'

HABASHI, F., Sulphur Institute J. 12(3-4) (1976) 15. 'Sulfur Dioxide in Metallurgy.'

HADEISHI, T., Proc. 2nd Int. Conf. Nucl. Methods Environ. Res., J. R. Vogt and W. Meyer, Eds., NTIS, Springfiled, Va., 1974. 'Isotope Shift Zeeman Technique for Detection of Atoms and Molecules.'

HALL, N. F., and O. R. ALEXANDER, J. Am. Chem. Soc. 62 (1940) 3455. 'Oxygen Exchange between Anions and Water.'

HALLIWELL, B. Proc. Symp. New Uses for Sulphur and Pyrites, Madrid, 1976, The Sulphur Institute, Lond, 1976. 'The Sodium Sulphur Battery.'

HAMADA, S., Y. NADAZAWA, and T. SHIRAI, Bull. Chem. Soc. Japan 43 (1970) 3096. 'Nucleation in Liquid Sulfur Droplets.'
HAMADA, Y., and A. J. MERER, Can. J. Phys. 53 (1975) 2555. 'Rotational Structure in the Absorption Spectrum of Sulfur Dioxide between 3000 A and 3300 A.'
HAMMICK, D. L., and M. ZVEGINTZOV, J. Chem. Soc. 1928 (1928) 1785. 'Pseudo-ternary Systems Containing Sulphur. Part III. The System of Sulphur-Sulphur Monochloride.'
HAMMOND, A. L., W. D. METZ, and T. H. MAUGH II, "Energy and the Future," American Assoc. for the Advancement of Science, Washington, D.C., 1973.
HAMMOND, A. L., Science 189 (1975) 128. 'Cleaning up Coal: A New Entry in the Energy Sweepstakes.'
HAMMOND, A. L., Science 194 (1976) 172. 'Coal Research (IV): Direct Combustion Lags its Potential.'
HAMPTON, E. M., and J. N. SHERWOOD, Phil. Mag. 29 (1973) 763. 'Self-Diffusion in alpha Sulphur Crystals.'
HANCOCK, C. K., Ind. Eng. Chem. 46 (1954) 2431. 'Plasticized Sulfur Compositions for Traffic Marking.'
HANCOCK, G., J. D. CAMPBELL, and K. H. WELGE, Optics Comm. 16 (1976) 177. 'Sulfur Isotope Enrichment in SF_6 by High Intensity CO_2 Laser Radiation.'
HANCOCK, T., "Personal Narrative of the Origin and Progress of the Caoutchouc and India-Rubber Manufacture in England," London, 1857.
HANNICK, D. L., and M. ZVEGINTZOV, J. Chem. Soc. 1928 (1928) 1785. 'Solubility of Sulfur in $CHCl_3$, S_2Cl_2, and Benzene.'
HANSON, W. C., Alberta Sulphur Res. Quart. Bull. 12(3) (1975) 14. 'Proposed Method for Determination of Hydrogen Sulphide in Sulphur.'
HAPPEL, J., and M. A. HNATOW, U.S. Environ. Prot. Agency EPA-600/2-73-020, (1973). 'Catalytic Oxidation of Sulfur Dioxide Using Isotopic Tracers.'
HARIMA, M., Chem. Econ. Eng. Rev. 6 (1974) 13. 'Production of Clean Energy and Sulfur Recovery.'
HARKER, A. B., P. J. PAGNI, and T. NOVAKOV, Chemosphere 6 (1975) 339. 'Manganese Emissions from Combustors.'
HARRER, T. S., in Kirk-Othmer "Encyclopedia of Chemical Technology," Vol. 19 John Wiley, New York, 1969. 'Sulfuric Acid.'
HARRINGTON, R. E., Int. J. Sulfur Chem. 7B (1972) 57. 'Current Status of Sulfur Dioxide Control Technology.'
HARRIS, C. L., F. MARASHI, and E. B. TITCHENER, Nucleic Acids Res. 3 (1976) 2129. 'Increased Isoleucine Acceptance by Sulfur-Deficient Transfer RNA from Escherichia Coli.'
HARRIS, R. E., J. Phys. Chem. 74 (1970) 3102. 'The Molecular Composition of Liquid Sulfur.'
HARRIS, R. E. Private Communication, 1971.
HARRISON, A. G., and H. G. THODE, Trans. Faraday Soc. 53 (1957) 1648. 'The Kinetic Isotope Effect in the Chemical Reduction of Sulphate.'
HARTLER, N., J. LIBERT, and A. TEDER, I&EC Proc. Design Develop. 6(4) (1967) 398. 'Rate of Sulfur Dissolution in Aqueous Sodium Sulfide.'

HARTLER, N., J. LIBERT, and G. AKERLUND, Sv. Papperstidn. 75 (1972) 673. 'Dissolution of Sulfur in Alkaline Liquors. Degree of Conversion from Sulfur to Polysulfide.'

HARTLEY, E. M., JR., and M. J. MATTESON, Ind. Eng. Chem. Fundam. 14 (1975) 57. 'Sulfur Dioxide Reactions with Ammonia in Humid Air.'

HARWARD, M. E., and H. M. REISENAUER, Soil Sci. 101 (1966) 326. 'Reactions and Movements of Inorganic Soil Sulfur.'

HASINSKI, S., Chem. Anal. (Warsaw) 20 (1975) 1135. 'Gas Chromatographic Determination of Traces of Phosphorus Compounds. New Construction of a Nonquenching Flame Photometric Detector (FPD).'

HATA, T., and S. KINUMAKI, Nature 203 (1964) 1378. 'Reactions of Ammonia and Aliphatic Amines with Sulphur Dioxide.'

HATA, T., Sendai, Tohoku Univ., J. Sci. Res., Ser. A, 14 (1964) 5. 'Infrared Spectra of Reaction Products of Gaseous Ammonia and Sulfur Dioxide.'

HATFIELD, J. D., and A. V. SLACK, Adv. Chem. Ser. 139 (1975) 130. 'Lime-Limestone Scrubbing: Factors Affecting the Concentration of Sulfur Dioxide-Absorbing Species in Solution.'

HAWKINS, R. T., Macramolecules 9 (1976) 189. 'Chemistry of the Cure of Poly(p-Phenylene Sulfide).'

HAY, A. S., J. Org. Chem. 41 (1976) 1710. 'Reaction of Sulfur with 2,6-Disubstituted Phenols.'

HAYNES, B. S., and N. Y. KIROV, Combust. Flame 23 (1974) 277. 'Nitric Oxide Formation During the Combustion of Coal.'

HAYNES, W., "Brimstone, The Stone That Burns," McGraw Hill, New York, 1959.

HAYON, E., A. TREININ, and J. WILF, J. Am. Chem. Soc. 94 (1972) 47. 'Electronic Spectra, Photochemistry, and Autoxidation Mechanism of the Sulfite-Bisulfite-Pyrosulfite Systems. The SO_2^-, SO_3^-, SO_4^-, and SO_5^- Radicals.'

HAZLETON, J. E., "The Economics of the Sulphur Industry," Resources for the Future, Inc., John Hopkins Press, Baltimore and London, 1970.

HAZUCHA, M., and D. V. BATES, Nature 257 (1975) 50. 'Combined Effect of Ozone and Sulphur Dioxide on Human Pulmonary Function.'

HEAL, H. G., in "Inorganic Sulphur Chemistry," G. Nickless, Ed., Elsevier, Amsterdam, 1968. 'The Nitrides, Nitride-Halides, Imides, and Amides of Sulphur.'

HEINTZE, K., F. BRAUN, and A. FRICKER, Ind. Obst-Gemueseverwert. 59 (1974) 452. 'Stability of Sulfur Dioxide in Aqueous Solutions.'

HENDERSON, J. M., and J. B. PFEIFFER, Min. Eng. 26(11) (1974) 36. 'How ASARCO Liquifies SO_2 Off-Gas at Tacoma Smelter.'

HENDERSON, J. M., and J. B. PFEIFFER, Adv. Chem. Ser. 139 (1975) 35. 'Remuction of Sulfur Dioxide to Sulfur: The Elemental Sulfur Pilot Plant of ASARCO and Phelps Dodge Corp.'

HENRIQUES, H., Chem. Zeit. 17 (1893) 707. 'Beitraege zur analytischen Untersuchung von Kautschukwaaren. II.'

HENRIQUES, R., Chem. Zeit. 17 (1893) 634 and 707. 'Beitraege zur Kenntniss der Kautschuksurrogate.'

HENSEL, G., and H. G. TRUEPER, Arch. Microbiol. 109 (1976) 101. 'Cysteine and S-Sulfocystein Biosynthesis in Phototrophic Bacteria.'

HERBER, R. H., and T. H. NORRIS, J. Am. Chem. Soc. 76 (1954) 3849. 'Isotopic Exchange Reactions in Triethylamine-Liquid Sulfur Dioxide Solutions.'

HEREDY, L. A., S. C. LAI, L. R. MCCOY, and R. C. SAUNDERS, Adv. Chem. Ser. 140 (1975) 203. 'Metal Sulfide Electrodes for Secondary Lithium Batteries.'
HERLINGER, A. W., and T. V. LONG II., Inorg. Chem. 8 (1969) 2661. 'An Investigation of the Structure of the Disulfite Ion in Aqueous Solution Using Raman and Infrared Spectroscopies.'
HESTER, R. E., and R. A. PLANE, Inorg. Chem. 3 (1964) 769. 'Raman Spectrophotometric Comparison of terionic Association in Aqueous Solutions of Metal Nitrates, Sulfates, and Perchlorates.'
HEYDEN CO., Pamphlet, New York, 1937. 'Sulfidal, Colloidal Sulfur.'
HIATT, N. A., Adv. Chem. Ser. 129 (1973) 92. 'Polymerization of Cyclic Bis(Arylene Tetrasulfide).'
HIGGINS, I. T. T., Arch. Env. Health 22 (1971) 581. 'Effects of Sulfur Oxides and Particulates on Health.'
HIGGINS, I. T. T., and B. G. FERRIS, JR., Proc. Conf. Health Effect of Air Pollutants, Nat'l Academy of Sci., Nat'l Res. Council (1973) 227. 'Epidemiology of Sulphur Oxides and Particles.'
HILDEBRAND, J. H., Chem. Eng. News, Sept. 13 (1976) 26. 'From Then to Now.'
HILL, A. C., and E. M. CHAMBERLAIN, JR., Proc. 1974 Symp. Atmos.-Surf. Exch. Part. Gaseous Pollut., Environ. Res. Develop. Agency Symp. Ser. (1976) 153. 'The Removal of Water Soluble Gases from the Atmosphere by Vegetation.'
HILL, H. W., JR., and J. T. EDMONDS, JR., Adv. Chem. Ser. 129 (1973) 80. 'Properties of Polyphenylene Sulfide Coatings.'
HILLER, K. O., B. MASLOCH, and H. J. MOECKEL, Z. Anal. Chem. 280 (1976) 293. 'Liquid-chromatographische Trennung von aliphatischen Polysulfiden.'
HILLS, D. A., "Vulcanization of Rubber," Elsevier, Amsterdam, 1971.
HIRST, E., Science 184 (1974) 134. 'Food-Related Energy Requirements.'
HISATSUNE, I. C., and JULIAN HEICKLEN, Can. J. Chem. 53 (1975) 2646. 'Infrared Spectroscopic Study of Ammonia-Sulfur Dioxide-Water Solid State System.'
HISSONG, D. W., K. S. MURTHY, and A. W. LEMMON, JR., Environ. Prot. Agency Rep. EPA-600/2-76-130 (1976). 'Reductant Gases for Flue Gas Desulfurization Systems.'
HITCHCOCK, D. R., J. Air. Pollut. Control Assoc. 26 (1976) 210. 'Atmospheric Sulfates from Biological Sources.'
HITCHCOCK, L., and A. K. SCRIBNER, Ind. Eng. Chem. 23 (1931) 743. 'Anhydrous Liquid Sulfur Dioxide.'
HO, T. Y., M. A. ROGERS, H. V. DRUSHEL, and C. B. KOONS, Am. Assoc. Petroleum Geol. Bull. 58 (1974) 2338. 'Evolution of Sulfur Compounds in Crude Oils.'
HOERING, T. C., and J. W. KENNEDY, J. Am. Chem. Soc. 79 (1957) 56. 'The Exchange of Oxygen between Sulfuric Acid and Water.'
HOEY, W. A., B. W. NORTON, and K. W. ENTWISTLE, Proc. Aust. Soc. Anim. Prod. 11 (1976) 377. 'Preliminary Investigations into Molasses and Sulfur Supplementation of Sheep Fed Mulga (Acacia Aneura).'
HOFFMAN, H., Hydrocarbon Proc., Int. Ed., 2(1) (1976) 77. 'Sulfur Split Among Products.'
HOFFMANN, M. R., and J. O. EDWARDS, J. Phys. Chem. 79 (1975) 2096. 'Kinetics of the Oxidation of Sulfite by Hydrogen Peroxide in Acidic Solution.'
HOFMANN, H. J., and K. ANDRESS, Z. Anorg. Allg. Chem. 284 (1956) 234. 'Ramanspektroskopische Untersuchungen an Schwefel-Stickstoff-Sauerstoff-Verbindungen.'

HOFMANN, K. A., and F. HÖCHTLEN, Chem. Ber. 36, (1903) 3090. 'Kyrstallisirte Polysulfide von Schwermetallen.'

HOHENBERG, P. M., "Chemicals in Western Europe: 1850-1914, An Economic Study of Technical Change," Rand McNally & Co., Chicago, Ill., 1967.

HOLLANDER, J., Ed.,"Annual Review of Energy," Vol I., Annual Reviews, Palo Alto, Ca., 1976.

HOLSER, W. T., and I. R. KAPLAN, Chem. Geol. 1 (1966) 93. 'Isotope Geochemistry of Sedimentary Sulfates.'

HOLT, B. D., P. T. CUNNINGHAM, and A. G. ENGELKEMEIR, Int. Conf. Stable Isot., New Zealand, 1976. 'Application of 16-O Analysis of the Study of Atmospheric Sulfate Formation.'

HOLT, E. L., K. C. BACHMAN, W. R. LEPPARD, E. E. WIGG, and J. H. SOMERS, SAE Tech. Pap. 750683 (1975). 'Control of Automotive Sulfate Emissions.'

HOLZER, W., W. F. MURPHY, and H. J. BERNSTEIN, J. Mol. Spectrosc. 32 (1969) 13. 'Raman Spectra of Negative Molecular Ions Doped in Alkali Halide Crystals.'

HOMBERG, M., Histoire de L'Academie Royale des Sciences (1703) 47. 'Sur L'Analise du Souffre Commun.'

HOMBERG, M., Histoire de L.AcademieRoyale des Sciences, Avec les Memoires de Mathematique & de Physique, pur la mem Annee 1710, (1710) 225 'Sur les Matieres Sulphureuses & sur la Facilite de les Changer d'une Espece de Souffre en une Autre.'

HOODLESS, R. A., M. SARGENT, and R. D. TREBLE, Analyst (London) 101 (1976) 757. 'Sulfur Resonse of the Alkali Flame-Ionization Detector.'

HOPKINS, A. G., F. P. DALY, and C. W. BROWN, J. Phys. Chem. 79 (1975) 1849. 'Infrared Spectra of Matrix Isolated Disulfur Monoxide Isotopes.'

HORSEMAN, F., "World Sulphur Supply and Demand (1960-1980)," for U.N. Ind. Dev. Org., published by UNESCO, New York, 1970.

HOWARD, H. C., in "Chemistry of Coal Utilization, Supplementary Volume," H. H. Lowry, Ed., Wiley, New York, 1963. 'Pyrolytic Reactions of Coal.'

HUBBERT, M. K., Sci. Amer. 225(Sept.) (1971) 60. 'The Energy Resources of the Earth.'

HUEBERT, B. J., Science 194 (1976) 646. 'Acid in Rain Water.'

HUNTER, W. D., JR., J. C. FEDORUK, A. W. MICHENER, and J. E. HARRIS, Adv. Chem. Ser. 139 (1975) 23. 'The Allied Chemical Sulfur Dioxide Reduction Process for Metallurgical Emissions.'

HYNE, J. B., E. MÜLLER, and T. K. WIEWIOROWSKI, J. Phys. Chem. 70 (1966) 3733. 'Nuclear Magnetic Resonance of Hydrogen Polysulfides in Molten Sulfur.'

HYNE, J. B., Alberta Sulphur Res. Quart. Bull. 3 (1967) 15. 'Sulphur 'Solubility' in Sour Gas.'

HYNE, J. B., Alberta Sulphur Res. Quart. Bull. 7 (1970) 2. 'Removal of SO_2 from Stack Gases–A Revieh of Methods.'

HYNE, J. B., Alberta Sulphur Res. Quart. Bull. 7(4) (1971) 23. 'Instrumentation in Sour Gas Processing.'

HYNE, J. B., Alberta Sulphur Res. Quart. Bull. 8 (1972) 1. 'Desulfurisation of Effluent Gas Streams–Review and Comparison of Techniques.'

I

IAMMARTINO, N. R., Chem. Eng. 82 (1975) 48. 'New Batteries are Coming.'
IGUMNOV, S. A., and N. B. PONER, Ezheg., Inst. Geol. Geokhim., Akad. Nauk SSR, Ural. Nauchn. Tsentr 1974 (Pub. 1975) 114. 'Experimental Study of Isotope Exchange Between Dissolved Sulfate and Sulfide Sulfur under Hydrothermal Conditions.'
IGUMNOV, S. A., Geokhimiya (4) (1976) 497. 'Experimental Study of Isotope Exchange between Sulfide and Sulfate Sulfur in Hydrothermal Solution.'
IKEDA, S., S. SATAKE, T. HISANO, and T. TERAZAWA, Talanta 19 (1972) 1650. 'Potentiometric Argentimetric Method of the Successive Titration of Sulphide and Dissolved Sulphur in Polysulphide Solutions.'
INGLIS, K. A. D., Ed., "Energy: From Surplus to Scarcity?" Halsted Press, John Wiley & Sons, New York-Toronto, 1974.
ITO, Y., and Y. YASUMOTO, Nippon Kagaku Kaishi (7) (1975) 1160. 'Solubility Equilibrium of the Sodium Imidobisulfate-Sodium Hydroxide-Water and Potassium Imidobissulfate-Potassium Hydroxide-Water Systems and Preparation of Sodium and Potassium Imidobissulfates.'
IVANOV, M. V., "Microbiological Processes in the Formation of Sulfur Deposits," S. I. Kuznetsov, Ed., S. Nemchonok, Transl., Israel Program for Scientific Translations, Jerusalem, 1968.
IVINS, R. O., A. A. CHILENSKAS, V. M. KOLBA, W. L. TOWLE, and P. A. NELSON, Proc. IEEE Reg. 3 Conf., Vol. I, 3D-2-1, Charlotte, 1975. 'Design of a Lithium/Sulfur Battery for Load Leveling on Utility Networks.'
IWAKURA, Y., and M. SAKOMOTO, J. Polymer Sci. 47 (1960) 277. 'Polyaddition Reaction of Polymethylene and Polymethylene Dimercaptans.'
IWAKURA, Y., and M. SAKOMOTO, J. Polymer Sci. 52 (1964) 881. 'Polymethane Sulfides.'

J

JACKSON, J. A., and H. TAUBE, J. Phys. Chem. 69 (1965) 1844. 'Chemical Shifts in the Nuclear Magnetic Resonance Absorption for Oxygen-17 in Oxy Ions.'
JACOBSON, D. R., J. W. BARNETT, S. B. CARR, and R. H. HATTON, J. Dairy Sci. 50 (1967) 1248. 'Voluntary Feed Intake, Milk Production, Rumen Content, and Plasma-Free Amino Acid Levels of Lactating Cows on Low Sulfur and Sulfur-Supplemented Diets.'
JANDER, G., "Die Chemi in Wasseraehnlichen Loesungsmitteln," Springer-Verlag, Berlin, 1949.
JANDER, J., "Chemistry in Anhydrous Liquid Ammonia," Vol. I, Part 1, Interscience Publishers, New York-London, 1966.
JANEF, "Thermodynamic Tables," 1974 Suppl., Dow Chemical Co., Midland Mich., J. Phys. Chem. Ref. Data 3 (1974) 471.
JANSSEN, M. J., Ed., "Organosulfur Chemistry," Interscience, New York, 1967.
JANZ, G. J., E. RODUNER, J. W. COUTTS, and J. R. DOWNEY, Inorg. Chem. 15 (1976) 1751, 1755. 'Raman Studies of Sulfur-Containing Anions in Inorganic Polysulfides. Barium Trisulfide. Potassium Polysulfides.'

JANZ, G. J., J. R. DOWNEY,JR., E. RODUNER, G. J. WASILCZYK, J. W. COUTTS, and A. ELUARD, Inorg. Chem. 15 (1976) 1759. 'Raman Studies of Sulfur-Containing Anions in Inorganic Polysulfides. Sodium Polysulfides.'

JANZEN, E. G., J. Phys. Chem. 76 (1972) 157. 'Electron Spin Resonance Study of the SO_2^- Formation in the Thermal Decomposition of Sodium Dithionite, Sodium Potassium Metabisulfite, and Sodium Hydrogen Sulfite.'

JARAMILLO CELIS, R., and R. BAZAN, Turrialba 26(1) (1976) 90. 'Effect of Urea and Urea-Sulfur on 'Giant Cavendish' Banana Production in Guapiles, Costa Rica.'

JARRELL, W. M., Ph.D. Thesis, Oregon State Univ., Corvallis, Or., 1977. 'Nitrogen Release from Granules of Sulfur-Coated Urea.'

JEFFERS, R. E., and S. H. BAUER, Combust. Flame 17 (1971) 432. 'Equilibrium Compositions at High Temperatures in Selected Mixtures of SO_2, CO, Oxides of Nitrogen, and Hydrocarbons.'

JELLINEK, F., in "Inorganic Sulphur Chemistry, G. Nickless, Ed., Elsevier, Amsterdam, 1968. 'Sulphides.'

JENSEN, D., in "Selected Values of Thermodynamic Properties of the Elements,' D. T. Hawkins, M. Gleiser, and K. K. Kelly, Eds., American Society of Metals, New York, 1973. 'Selected Values of Thermodynamic Properties of Sulfur.'

JENTSCH, C., in "Ullmann's Encyckl. der technischen Chemie," 4th Ed., Verlag Chemie, 1974, pg. 696. 'Erdoelverarbeitung.'

JEVONS, W. S., "The Coal Question," MacMillan and Co., London, 1865.

JEWELL, D. M., R. G. RUBERTO, E. W. ALBAUGH, and R. C. QUERY, Ind. Eng. Chem. Fundam. 15 (1976) 206. 'Distribution and Structural Aspects of Sulfur Compounds in Petroleum Residuals (Kuwait).'

JIN, J. I., Am. Chem. Soc. Meet. Abstracts, Chem. Eng. Section, Los Angeles (1974) 234. 'Chemistry of Plasticized Sulfur.'

JOHNSON, P. Y., Chem. Comm. 1971 (1971) 1083. 'A Convenient Synthesis of 2-substituted 1,3,5-Trithians.'

JOHNSON, W. H., Science 192 (1976) 629. 'Social Impact of Pollution Control Legislation.'

JOHNSTONE, H. F., Ind. Eng. Chem. 27 (1935) 587. 'Recovery of Sulfur Dioxide from Waste Gases; Equilibrium Partial Vapor Pressures over Solutions of the Ammonia-Sulfur Dioxide-Water System.'

JOHNSTONE, H. F., and D. B. KEYES, Ind. Eng. Chem. 27 (1935) 659. 'Recovery of Sulfur Dioxide from Waste Gases; Distillation of a Three-Component System Ammonia-Sulfur Dioxide-Water.'

JOHNSTONE, H. F., and A. D. SINGH, Ind. Eng. Chem. 29 (1937) 286. 'Recovery of Sulfur Dioxide.'

JOHNSTONE, H. F., Ind. Eng. Chem. 29 (1937) 1396. 'Recovery of Sulfur Dioxide from Waste Gases; Effect of Solvent Concentration on Capacity and Steam Requirement of Ammonium Sulfite-Bisulfite Solutions.'

JOHNSTONE, H. F., Pulp Paper Mag. Can. (Mar.) (1952) 105. 'Recovery of Sulphur Dioxide from Dilute Gases.'

JOHNSTONE, S. J., and M. G. JOHNSTONE, "Minerals for the Chemical and Allied Industries," Chapman and Hall, London, 1961. 'Sulphur and Pyrites.'

JONES, G. G., and R. L. STARKEY, Appl. Microbiol. 5 (1957) 111. 'Fractionation of Stable Isotopes of Sulfur by Microorganisms and Their Role in Deposition of Native Sulfur.'

JONES, R. V., and H. W. HILL, JR., Adv. Chem. Ser. 140 (1975) 174. 'Polyphenylene Sulfide–A New Item of Commerce.'

JORCZAK, J. S., and E. M. FETTES, Ind. Eng. Chem. 43 (1951) 324. 'Polysulfide Liquid Polymers.'

JORDAAN, I. J., J. E. GILLOTT, R. E. LOOV, and J. B. HYNE, Mater. Sci. Eng. 26 (1976) 105. 'Effect of Hydrogen Sulfide on the Mechanical Strength of Sulfur and of Sulfur Mortars and Concretes.'

JOST, D., Tellus 26 (1974) 206. 'Aerological Studies on the Atmospheric Sulfur Budget.'

JUNGE, G., and M. OTTNAD, Chem. Zeit. 98 (1974) 147. 'Spektroskopische Methoden zur Untersuchung organischer Thiole, Sulfide, und Disulfide.'

JUNGE, C. E., and T. G. RYAN, Quart. J. Royal Met. Soc. 84 (1958) 46. 'Study of the SO_2 Oxidation in Solution and its Role in Atmospheric Chemistry.'

JUNGE, C. E., "Air Chemistry and Radioactivity," Vol. 4, Geophysical Series, Academic Press, New York, 1963.

JUNGE, C. E., and G. SCHEICH, Atmosph. Environ. 3 (1969) 423. 'Studien zur Bestimmung des Saeuregehaltes von Aerosol-Teilchen.'

JUNGE, C. E., Quart. J. Royal Met. Soc. 98 (1972) 711. 'The Cycle of Atmospheric Gases–Natural and Man Made.'

JURASZYK, H., Chem. Zeit. 98 (1974) 126. 'Reaktionen des elementaren Schwefels mit organischen Verbindungen.'

K

KAGARISE, R. E., U.S. Naval Res. Lab. Report 6394, May 1966. 'Intensity Measurements of Some Fundamental Absorption Bands of H_2O, D_2O, and CS_2.

KAHLERT, H., and B. KUNDU, Mater. Res. Bull 11 (1976) 967. 'The Preparation and Growth of Large Single Crystals of Polysulfur Nitride.'

KALRA, H., D. B. ROBINSON, and T. R. KRISHNAN, J. Chem. Eng. Data 22 (1977) 1. 'The Equilibrium Phase Properties of the Ethane-Hydrogen Sulfide System at Subambient Temperatures.'

KAO, J., and N. L. ALLINGER, Inorg. Chem. 16 (1977) 35. 'Conformational Analysis of Elemental Sulfur.'

KAPLAN, I. R., and S. C. RITTENBERG, in "Biogeochemistry of Sulfur Isotopes," M. L. Jensen, Ed., Yale Univ. Press, New Haven, Conn., 1962. 'The Microbiological Fractionation of Sulfur in Isotopes.'

KAPLAN, I. R., K. O. EMERY, and S. C. RITTENBERG, Geochim. Cosmochim. Acta 27 (1963) 297. 'The Distribution and Isotopic Abundance of Sulphur in Recent Marine Sediments off Southern California.'

KAPLAN, N., Environ. Prot. Technol. Ser. EPA-650-2-73-038 (1973) 1019. 'EPA Overview of Sodium-Based Double Alkali Processes. II. Status of Technology and Description of Attractive Schemes.'

KAPLAN, N., Proc. Symp. Flue Gas Desulfurization, New Orleans, 1976, R. D. Stern and W. H. Ponder, Eds., EPA Report 600/2-76-136 (1976). 'Introduction to Double Alkali Flue Gas Desulfurization Technology.

KAPLINA, E. G., and N. A. PROKOPENKO, Koks Khim. (2) (1968) 35. 'Preparation of Colloidal Sulfur.'

KARALIUS, J., and N. SMITH, Cold Regions Res. Eng. Lab. Tech. Note, Feb. 1973. 'Construction of an Expedient Road Test Section Using Sulfur/Foamed Polystyrene Bead Insulation Composite.'

KARCHMER, J. H., Ed., "Analytical Chemistry of Sulfur and Its Compounds," Wiley-Interscience, New York, 1970.

KARDINALOVSKAYA, R., and A. V. LAZURSKII, Khim. Sel'sk. Khoz. 13 (1975) 272. 'Sulfur Containing Fertilizers.'

KASAHARA, M., and K. TAKAHASHI, Atmosph. Environ. 10 (1976) 475. 'Experimental Studies on Aerosol Particle Formation by Sulfur Dioxide.'

KATZ, M., and R. J. COLE, Ind. Eng. Chem. 42 (1950) 2258. 'Recovery of Sulfur Compounds from Atmospheric Contaminants.'

KAWADA, I., and E. HELLNER, Angew. Chem. 82 (1970) 390. 'Zur Struktur von Cycloheptaschwefel.'

KAWAKAMI, T., N. KUBOTA, and H. TERNI, Technol. Rep. Iwate Univ. 1971 (1971) 77. 'Solubility of Sulfur in H_2O-$(CH_3)_2SO$, Ethanol, Acetone, and NH_3-$(CH_3)_2SO$.'

KAWANO, T., J. Soc. Rubber Ind. Japan 12 (1939) 9. 'Rubber-Like Substances Derived from Formalin. I.'

KEETON, M., and D. P. SANTRY, Chem. Phys. Lett. 7 (1970) 105. '*Ab Initio* Calculations for the 3d-Exponent of Phosphorus and Sulfur.'

KELLOG, W. W., R. D. CADLE, E. R. ALLEN, A. Z. LAZRUS, and E. A. MARTELL, Science 175 (1972) 587. 'The Sulfur Cycle.'

KELLY, D. P., Int. Symp. Sulphur in Agric., Inst. National Rech. Agronomique (1972) 217. 'Transformations of Sulphur and its Compounds in Soils.'

KENNEDY, J. H., and F. ADAMO, J. Electrochem. Soc. 116 (1969) 1518. 'Electrochemistry of Sulfur in LiCl-KCl Eutectic.'

KENNEPOHL, G. J. A., A. LOGAN, and D. C. BEAN, Proc. Assoc. Asphalt Pav. Tech. 44 (1975) 485. ' 'Conventioanl' Paving Mixes with Sulphur-Asphalt Binders.'

KENNEPOHL, G. J. A., Proc. Symp. New Uses for Sulphur and Pyrites, Madrid, 1976, The Sulphur Institute, London, 1976. 'The Gulf Canada Sulfur-Asphalt Process for Pavements.'

KENNEPOHL, G. J. A., Energy Process. Can. 68(6) (1976) 24. 'A Novel Method of Incorporating Sulfur in Conventional Paving.'

KENT, R. L., and B. EISENBERG, Hydrocarbon Process. 2(1) (1976) 87. 'Better Data for Amine Treating.'

KEROUANTON, A., M. HERLEM, and A. THIEBAULT, Anal. Lett. 6 (1973) 171. 'The Behavior of Sulfur in Liquid Ammonia.'

KERR, C. P., Am. Inst. Chem. Eng. J. 20 (1974) 1213. 'Aqueous Sodium Sulfite, Bisulfite, and Sulfate Equilibrium.'

KERTAMUS, N. J., M. A. PAISLEY, and W. L. SAGE, Hydrocarbon Process. 53 (1974) 95. 'Process for SO_2/Char Reaction.'

KETTNER, H., Bull. World Health Organiz. 32 (1965) 421. 'The Removal of Sulfur Dioxide from Flue Gases.'

KHALAFALLA, S. E., and L. A. HAAS, Adv. Chem. Ser. 139 (1975) 60. 'Dual-Catalyst Beds to Reduce Sulfur Dioxide to Elemental Sulfur in the Presence of Water Vapor.'

KHARASCH, N., "Organic Sulfur Compounds," Pergamon Press, London, 1961.

KHARASCH, N., C. M. BUESS, and R. B. LANGFORD, Quart. Rep. Sulfur Chem. 2 (1967) 241. 'Sulfur-Containing Reagents. I.'

KHARE, B. N., and C. SAGAN, Science 189 (1975) 722. 'Cyclic Octatomic Sulfur: A Possible Infrared and Visible Chromophore in the Clouds of Jupiter.'

KICE, J. L, in "Sulfur in Organic and Inorganic Chemistry," Vol. 1, A. Senning, Ed., Marcel Dekker, New York, 1971. 'The Sulfur-Sulfur- Bond.'

KIRK-OTHMER, "Encyclopedia of Chemical Technology," Vol 19, J. Wiley, New York, 1969.

KITABATAKE, M., Taiki Osen Kenkyu 10 (1976) 700, 712, & 718. 'The Effect of Air Pollutants on Experimental Provocation of Asthma Attacks in Guinea Pigs, I. The Effects of Sulfur Dioxide and Sulfuric Acid Mist. II. The Effects of Sulfur Dioxide, Nitrogen Dioxide and Ozone. III. The Effects of Nitrogen Monoxide and Nitrogen Dioxide.'

KITTRELL, J. R., and N. GODLEY, Environ. Prot. Agency Report 600/2-76-161a (1976). 'Impact of SO_x Emissions Control on Petroleum Refining Industry, Volume 1: Study Results and Planning Assumptions.'

KLABUNDE, K. J., and P. S. SKELL, J. Am. Chem. Soc. 93 (1971) 3807. 'The Chemistry of Atomic Carbon. Desulfurization.'

KLEMENT, W., JR., J. Chem. Phys. 45 (1966) 1421. 'Thermodynamics of the Lambda Transition in Liquid Sulfur.'

KLEMENT, W., JR., J. Polymer Sci. 12 (1974) 815. 'The Gamma Transition in Liquid Sulfur.'

KLING, C. S., D. STAUFFER, and V. A. MOHNEN, Nature, Phys. Sci. 244 (1973) 53. 'Possibilities for Atmospheric Aerosol Formation involving NH_3.'

KLOEPPEL, F. W., Muenchener Med. Wochenschr. 78 (1931) 151. 'Therapeutische Mitteilungen. Ein wirksamer kolloidaler Schwefelpuder (Sulfoderm-Heyden).

KLUGE, W., and G. MANIG, Energietechnik 26(1) (1976) 32. 'Position and Problems of Flue Gas Desulfurization in the GDR.'

KOBAYASHI, M., Nenryo Oyobi Nensho 42 (1975) 229. 'Development of Simultaneous Treatment for Sulfur Oxides (SO_x) and Nitrogen Oxides (NO_x) by the Wet Method.'

KOBBE, W. H., Ind. Eng. Chem. 16 (1924) 1026. 'New Uses for Sulfur in Industry.'

KOHL, A. L., and F. C. RIESENFELD, "Gas Purification," McGraw Hill, New York, 1960.

KOHNO, T., J. Soc. Rubber Ind. Japan 14 (1941) 436. 'Rubberlike Substances Derived from Formalin. III.'

KOMIYAMA, H., J. M. SMITH, Am. Inst. Chem. Eng. J. 21 (1975) 664. 'Sulfur Dioxide Oxidation in Slurries of Activated Carbon.'

KONINGSBERGER, D. C., Ph.D. Thesis, Technical University of Eindhoven, 1971. 'On the Polymerization of Sulfur and Selenium in the Liquid State. An ESR Study.'

KONINGSBERGER, D. C., and T. DE NEEF, Chem. Phys. Lett. 14 (1972) 453. 'New Method in Calculating the Polymerization Parameters of Liquid Sulphur and Selenium.'

KÖPF, H., B. BLOCK, and M. SCHMIDT, Chem. Ber. 101 (1968) 272. 'Di-pi-cyclopentadienyl-titan(IV)-pentaselenid und -pentasulfid, zwei Hetero-cyclohexachalkogene in fixierter Konformation.'

KÖPF, H., and B. BLOCK, Z. Naturforsch. 23b (1968) 1534. 'Umsetzungen von Mono-pi-cyclopentadienyltitan(IV)-chlorid mit Thiophenol und mit Kaliumrhodanid.'

KÖPF, H., Angew. Chem. 81 (1969) 332. 'Ein Tetrasulfid-Chelat des Molybdaens.'
KOPP, H., "Geschichte der Chemie," F. Vieweg, Braunschweig, 1845.
KOPPANYI, T., and A. LINEGAR, Fed. Proc. Pharmacology 1 (1942) 155. 'Sulfides.'
KORN, J., H. W. PRINZLER, and D. PAPE, Erdoel Kohle 19 (1966) 651. 'The Reaction between Paraffinic Hydrocarbons and Sulfur.'
KOROSY, L., H. L. GEWANTER, F. S. CHALMERS, and S. VASAN, Adv. Chem. 139 (1975) 192. 'The Citrate Process.'
KOSEV., A., G. PEEV, D. ELENKOV, and G. GOCHEV., Metalurgiya (Sofia) 29(10) (1974) 20. 'Absorption of Sulfur Dioxide by Zinc Oxide Suspension in a Column with a Fluidized Bed.'
KOWAKA, M., Tetsu To Hagane 62 (1976) 1052. 'Corrosion of Iron and Steels in a Flue Gas.'
KREBS, H., and E. F. WEBER, Z. Anorg. Allg. Chem. 272 (1953) 288. 'Ueber die Struktur und Eigenschaften der Halbmetalle. VI. Die Allotropie des Schwefels.'
KREBS, H., and H. BEINE, Z. Anorg. Allg. Chem. 355 (1967) 113. 'Versuche zur Aufspaltung des pi-Schwefels durch Gegenstromverteilung.'
KRESPAN, C. G., and W. R. BRASEN, J. Org. Chem. 27 (1962) 3995. 'Fluorinated Cyclic Polysulfides and Their Polymers.'
KRESSMAN, T. R. E., and J. A. KITCHENER, J. Chem. Soc. 1949 (1949) 1190. 'Cation Exchange with a Synthetic Phenolsulphonate Resin. Part I. Equilibria with Univalent Cations.'
KRONICK, P. L., H. KAYE, E. F. CHAPMAN, S. B. MAINTHIA, and M. M. LABES, J. Chem. Phys. 36 (1962) 2235. 'Electronic Properties of Polysulfur Nitride.'
KUCZKOWSKI, R. L., J. Am. Chem. Soc. 86 (1964) 3617. 'The Microwave Spectrum of S_2F_2.'
KUGEL, R. W., Diss. Abstr. Int. B. 35 (1975) 5835, Ph.D. Thesis, Stanford Univ., Palo Alto, Ca., 1975. 'Matrix Isolation Studies of Reactions of Atomic Oxygen with Sulfur Dioxide and Sulfur Trioxide.'
KURAMOTO, Y., Y. WATANABE, and Y. INUISHI, Technol. Rep. Osaka Univ. 19 (1969) 571. 'High Field Effects in alpha-Sulfur.'
KURODA, Y., and K. MACHIDA, Bull. Chem. Soc. Japan 49 (1976) 1475. 'Transition Dipole Interaction and Intermolecular Potential of Carbon Disulfide and Sulfur Dioxide Crystals.'
KURTENACKER, A., Z. Anorg. Allg. Chem. 161 (1927) 201. 'Iodometrische Analyse von Sulfid, Sulfit, und Thiosulfat.'
KUTOGLU, A., and E. HELLNER, Angew. Chem. 78 (1966) 1021. 'Kristallstruktur von Cyclododecaschwefel, S_{12}.
KVET, R., Geokhimyia 4 (1973) 625. 'Origin of Hydrogen Sulfide of Natural Waters.'
KYLE, M. L., H. SHIMOTAKE, and R. K. STEUNENBERG, Proc. Intersoc. Energy Conversion Eng. Conf., SAE, 1971, New York. 'High Energy Storage Batteries.'

L

LAFON, G. M., and F. T. MACKENZIE, Soc. Econ. Paleontol. Mineral, Spec. Publ. 20 (1974) 205. 'Early Evolution of the Oceans—A Weathering Model.'
LALO, C., and C. VERMEIL, J. Photochem. 3 (1974) 141. 'Photochemistry of SO_2 in the Vacuum-UV. II. Luminescence Studies.'

LAMANTIA, C. R., R. R. LUNT, R. E. RUSH, T. M. FRANK, and N. KAPLAN, Proc. Symp. Flue Gas Desulfurization, New Orleans, 1976, R. D. Stern and W. H. Ponder, Eds., EPA Report 600/2-76-136 (1976). 'Operating Experience–CEA/ADL Duel Alkali Prototype System at Gulf Power/Southern Services, Inc.'

LAND, G. W., Preprint, Am. Chem. Soc. Meeting, Los Angeles, 1974. 'Synthetic Fluid Fuels from Coal.'

LANDOLT-BÖRNSTEIN, H. H., "Kalorische Zustandgroessen," Vol. 2, Springer-Verlag, Berlin, 1961.

LANDRETH, R., R. G. DE PENA, and J. HEICKLEN, J. Phys. Chem. 78 (1974) 14. 'Thermodynamics of the Reactions $(NH_3)_n{\cdot}SO_2(s) - nNH_3(g) + SO_2(g)$.'

LANDRETH, R., R. G. DE PENA, and J. HEICKLEN, J. Phys. Chem. 79 (1975) 1785. 'Thermodynamics of the Reaction of Ammonia and Sulfur Dioxide in the Presence of Water Vapor.'

LANG, C. J., R. S. SALTZMAN, and G. G. DE HAAS, Tappi 58 (1975) 88. 'Monitoring Volatile Sulfur Compounds in Kraft and Sulfite Mills.'

LARKIN, J. A., J. KATZ, and R. L. SCOTT, J. Phys. Chem. 71 (1967) 352. 'Phase Equilibria in Solutions of Liquid Sulfur. II. Experimental Studies in Ten Solvents: Carbon Disulfide, Carbon Tetrachloride, Benzene, Toluene, o-Xylene, Naphthalene, Biphenyl, Triphenylmethane, cis-Decalin, and trans-Decalin.'

LARSEN, R. I., J. Air Pollut. Control Assoc. 20 (1970) 214. 'Relating Air Pollutant Effects to Concentration and Control.'

LARSON, T., R. J. CHARLSON, E. J. KNUDSON, G. CHRISTIAN, and H. HARRISON, Water Air Soil Pollut. 4 (1975) 319. 'The Influence of a Sulfur Dioxide Point Source on the Rain Chemistry of a Single Storm in the Puget Sound Region.'

LASTER, L. L., Environ. Prot. Agency Rep. EPA-650/2-73-046 (1973). 'Atmospheric Emissions from the Asphalt Industry.'

LAUR, P. H., in 'Sulfur in Organic and Inorganic Chemistry,' Vol. 3, A. Senning, Ed., Marcel Dekker, New York, 1973. 'Steric Aspects of Sulfur Chemistry.'

LAUTENBACH, D., Ph.D. Thesis, Technical University, Berlin, 1969. 'Untersuchungen zur Solvolyse der S-S-S-Bindung in fluessigem Ammoniak.'

LAVOISIER, A., "Elements of Chemistry," R. Kerr, Transl., Creech & Son, Edinburgh, 1794.

LAVRINENKO, R. F., Tr. Gl. Geofiz. Observ. 207 (1968) 87. 'Sulfur Content in Rainfall.'

LAWTHER, P. J., R. E. WALKER, and M. HENDERSON, Thorax 25 (1970) 525. 'Air Pollution and Exacerbations of Bronchitis.'

LAWTHER, P. J., A. J. MACFARLANE, R. E. WALLER, and A. G. F. BROOKS, Enc. Res. 10 (1975) 355. 'Pulmonary Function and Sulphur Dioxide, Some Preliminary Findings.'

LEBEVRE, A., A. ETIENNE, J. COQUELIN, and C. JACQUOT, Bull. Soc. Chim. France 1973 (1973) 210. 'Acides Sulfocarboxyliques et Derives.'

LECLERCQ, R., in 'Sulfur in Organic and Inorganic Chemistry,' Vol. 2, A. Senning, Ed., Marcel Dekker, New York, 1972. 'Commercially Important Sulfur Compounds.'

LEITHE, W., "The Analysis of Air Pollutants," Ann Arbor Science Publishers, Ann Arbor, Mich., 1971.

LEE, D. Y., and F. W. KLAIBER, Proc. Symp. New Horizons in Constr. Materials, Vol. 1 Lehigh Univ. Press, Bethlehem, Pa., 1976. 'Fatigue Behavior of Sulfur Concrete.'

LEM, W. J., and M. WAYMAN, Can. J. Chem. 48 (1970) 776. 'Decomposition of Aqueous Dithionite. Part I. Kinetics of Decomposition of Aqueous Sodium Dithionite.'
LEM, W. J., and M. WAYMAN, Can. J. Chem. 48 (1970) 2778. 'The Decomposition of Dithionite III. Stabilization of Dithionite by Cations.'
LEMMON, W. A., Chem. Econ. Eng. Rev. 6 (1974) 30. 'Air Pollution Control in Canada.'
LEMOIGNE, M., "Le Soufre en Agriculture," L'Office de Publicite Generale Paris, Societe du Superphosphate, Paris, 1963.
LENZ, R. W., and W. K. CARRINGTON, J. Polymer Sci. 41 (1959) 333. 'Phenylene Sulfide Polymers. I. Mechanism of the Macallum Polymerization.'
LENZ, R. W., and C. E. HANDLOVITS, J. Polymer Sci. 43 (1960) 167. 'Phenylene Sulfide Polymers. II. Structure of Polymers Obtained by the Macallum Polymerization.'
LENZ, R. W., C. E. HANDLOVITS, and H. A. SMITH, J. Polymer Sci. 58 (1961) 351. 'Phenylene Sulfide Polymers. III. The Synthesis of Linear Polyphenylene Sulfide.'
LEONE, S. R., and K. G. KOSNIK, J. Chem. Phys. (in press)(1977). 'A Tuneable Visible and UV Laser on S_2 (Triplet B to Triplet X).
LESKOVSEK, H., J. MARSEL, ad L. KOSTA, Isotopenpraxis 10 (1974) 375. 'The Application of SF_6 in the Geochemical Research.'
LETOFFE, J. M., R. D. JOLY, J. THOUREY, G. PERACHON, J. BOUSQUET, J. Chim. Phys. 71 (1974) 427. 'Determination des Enthalpies de Formation des Polysulfures de Potassium.'
LEUCKART, R., J. prakt. Chem. 41 (1890) 179. 'Eine neue Methode zur Darstellung aromatischer Mercaptane.'
LEUNG, Y. C., J. WASER, S. v. HOUTEN, A. VOS, G. A. WIEGERS, and E. H. WIEBENGA, Acta Cryst. 10 (1957) 574. 'The Crystal Structure of P_4S_3.'
LEVENE, H. D., and J. W. HAND, Coal Min. Process 12(2) (1975) 46. 'Sulfur Stays in the Ash When Lignite Burns.'
LEVINE, C., Proc. 25th Power Sources Symp., 1972. 'Sodium-Sulfur Batteries.
LEVITT, B. P., and D. B. SHEEN, Trans. Faraday Soc. 63 (1967) 540. 'Light Emission from Shock-Heated Sulfur Dioxide. II. Ultraviolet Emission.'
LEVY, A., and E. L. MERRYMAN, Combust. Flame 9 (1965) 229. 'The Microstructure of Hydrogen Sulphide Flames.'
LEVY, A., E. L. MERRYMAN, and W. T. REID, Environ. Sci. Technol. 4 (1970) 653. 'Mechanisms of Formation of Sulfur Oxides in Combustion.'
LEWIS, J. S., Ann. Rev. 24 (1973) 339. 'Chemistry of the Planets.'
LEY, H., and E. KÖNIG, Z. Phys. Chem. 41 (1938) 365. 'Die Loesungsspektren von wichtigeren Saeuren der Elemente der Schwefelgruppe.'
LI, P., and A. C. CALDWELL, Soil Sci. Soc. Amer. Proc. 30 (1966) 370. 'The Oxidation of Elemental Sulfur in Soil.'
LIEBIG, J., "Letters on Chemistry," J. Gardner, Transl., Campbell, Philadelphia, Pa, 1843.
LIEGEL, E. A., and L. M. WALSH, Agron. J. 68 (1976) 457. 'Evaluation of Sulfur-Coated Urea (SCU) Applied to Irrigated Potatoes and Corn.'
LIKENS, G. E., N. M. JOHNSON, J. N. GALLOWAY, and F. H. BORMANN, Science 194 (1976) 643. 'Acid Precipitation: Strong and Weak Acids.'
LIKENS, G. E., Chem. Eng. News. (Nov. 22) (1976) 29. 'Acid Precipitation.'

LIMING, O. N., J. Agricult. Res. 40 (1930) 951. 'Toxicity of Sulphur to Spores of Sclerotinia Cinerea as Affected by the Presence of Pentathionic and Other Sulphur Acids.'

LIMING, O. N., Phytopathology 23 (1933) 155. 'The Preparation and Properties of Pentathionic Acid and its Salts; its Toxicity to Fungi, Bacteria, and Insects.'

LIND, M. D., and S. GELLER, J. Chem. Phys. 51 (1969) 348. 'Structure of Pressure-Induced Fibrous Sulfur.'

LINDQVIST, I., and M. MÖRTSELL, Acta Cryst. 10 (1957) 406. 'The Structure of Potassium Pyrosulfite and the Nature of the Pyrosulfite Ion.'

LINDQVIST, I., J. Inorg. Nucl. Chem. 6 (1958) 159. 'Relation between Bond Length and Hybridization.'

LINKE, S., Ed., Report of the Cornell Worshops on the Major Issues of a National Energy Research and Development Program, College of Engineering, Cornell Univ., Ithaca, N. Y., 1973.

LIPPINCOTT, E. R., and M. C. TOBIN, J. Chem. Phys. 21 (1953) 1559. 'Vibrational Spectra and Structure of S_4N_4.'

LIU, C. H., A. J. ZIELEN, and D. M. GRUEN, J. Electrochem. Soc. 120 (1973) 67. Electrochemical Generation and Measurement of Sulfide Ion in Molten LiCl-KCl Eutectic.'

LIU, S. H., and F. J. LIU, T'ai-wan Huan Ching Wei Sheng 6(2) (1974) 21. 'Air Pollution Control at Kaohsiung, Taiwan.'

LLOYD, D. R., P. J. ROBERTS, I. H. HILLIER, and I. C. SHENTON, Mol. Phys. 31 (1976) 1549. 'On the Photoelectron Spectrum of Sulphur Trioxide.'

LOENING, K. L., in "Sulfur in Organic and Inorganic Chemistry,' Vol. 3, A. Senning, Ed., Marcel Dekker, New York, 1973. 'The Nomenclature of Sulfur Compounds and Their Selenium and Tellurium Analogs.'

LOOMAN, C. M., and F. TUINSTRA, Physica 42 (1969) 291. 'Thermal Expansion of Fibrous Sulfur.'

LOOV, R. E., A. H. VROOM, and M. A. WARD, J. Prestressed Concr. Inst. 19(1) (1974) 86. 'Sulfur Concrete–A New Construction Material.'

LORENZEN, J. A., Adv. X-Ray Anal. 18 (1975) 568. 'Environmental Monitoring Device for X-Ray Determination of Atmospheric Chlorine, Reactive Sulfur and Sulfur Dioxide.'

LOVELOCK, J. E., R. J. MAGGS, and R. A. RASMUSSEN, Nature 237 (1972) 452. 'Atmospheric Dimethyl Sulphide and the Natural Sulphur Cycle.'

LOVELOCK, J. E., Nature 248 (1974) 625. 'CS_2 and the Natural Sulphur Cycle.'

LOVINS, A. B., Foreign Affairs J., Oct. 1976. 'Energy Strategy, The Road Not Taken.'

LOW, H. S., and R. A. BEAUDET, J. Am. Chem. Soc. 98 (1976) 3849. 'The Identification of S_5^+ as a Paramagnetic Species in Sulfur-Oleum Solutions by ESR.'

LOWELL, P. S., Final Rep. to EPA, Contract CPA 70-45, Radian Corp., Nov. 1, 1971. 'A Study of the Limestone Injection Wet Scrubbing Process.'

LU, C. S., and J. DONOHUE, J. Am. Chem. Soc. 66 (1944) 818. 'Electron Diffraction Investigation of Sulfur Nitride and Sulfur.'

LUBIN, I., and H. EVERETT, "The British Coal Dilemma," The MacMillan Co., New York, 1927.

LUCAZEAU, G., A. LAUTIE, and A. NOVAK, J. Raman Spec. 3 (1975) 161. 'Low Temperature Raman Spectra of NH_3SO_3 and ND_3SO_3 Single Crystals.'

LÜDEMANN, H. D., and E. U. FRANCK, Ber. Bunsenges. 72 (1968) 523. 'Absorptionsspecktren bei hohen Drucken. III. Waessrige SO_2-Loesungen bei 25°C bis zu 6 kbar.'

LUDWIG, A. C., United Nations Rep. No. TAO/GUA/4, July 14, 1969. 'Utilization of Sulphur and Sulphur Ores as Construction Materials in Guatemala.'

LUDWIG, A. C., and J. M. DALE, Environ. Prot. Agency Water Pollut. Control Res. Ser. 11024 EQE 06/71. 'Impregnation of Concrete Pipe.'

LUDWIG, S., Chem. Eng. 75(Jan. 29) (1968) 70. 'Antipollution Process Uses Absorbent to Remove SO_2 from Flue Gases.'

LUNGE, G., "Coal-Tar and Ammonia," 4th Ed., Part 1, Gurney & Jackson, London, 1909.

LYNN, S., R. E. RINKER, and W. H. CORCORAN, J. Phys. Chem. 68 (1964) 2363. 'ESR Spectrum of SO_2^- in Dithionite.'

LYONS, D., and G. NICKLESS, in "Inorganic Sulphur Chemistry," G. Nickless, Ed., Elsevier, Amsterdam, 1968. 'The Lower Oxy-Acids of Sulphur.'

M

MABROUK, A. F., Am. Chem. Soc. Symp. Ser. 26 (1976) 146. 'Nonvolatile Nitrogen and Sulfur Compounds in Red Meats and Their Relation to Flavor and Taste.'

MACALLUM, A. D., J. Org. Chem. 13 (1948) 154. 'A Dry Synthesis of Aromatic Sulfides: Phenylene Sulfide Resins.'

MACALUSO, P., in 'Kirk-Othmer Encyclopedia of Chemical Technology," Vol. 19 J. Wiley, New York, 1969. 'Sulfur Compounds.'

MACDONALD, D. D., D. DUNAY, G. HANLON, and J. B. HYNE, Albevta Sulphur Res. Quart. Bull. 7(1) (1970) 6. 'Properties of the N-Methyl-2-Pyrrolidinone/Water System.'

MACKNIGHT, W. J., and A. V. TOBOLSKY, in "Elemental Sulfur," B. Meyer, Ed., Interscience, New York, 1965. 'Properties of Polymeric Sulfur.'

MAGEE, P. S., in "Sulfur in Organic and Inorganic Chemistry," Vol. 1, A. Senning, Ed., Marcel Dekker, New York, 1971. 'The Sulfur-Bromine Bond.'

MAILLARD, D., M. ALLAVENA, and J. P. PERCHARD, Spectrochim. Acta 31A (1975) 1523. 'Spectres Vibrationnels du Dioxyde de Soufre dans une Matrice d'Argon, d'Azote et de Xenon.'

MAJEWSKA, J., Chem. Anal. (Warsaw) 13 (1968) 29. 'Sulfur Determination in Polymers and Copolymers for Fiber Production.'

MALANCHUK, M., Ann. Chim. Acta 56 (1971) 377. 'Thermal Analysis of Sodium Metabisulfite.'

MALHOTRA, V. M., J. Am. Concr. Inst. 72 (1975) 466. 'Development of Sulfur-Infiltrated High-Strength Concrete.'

MALHOTRA, V. M., J. A. SOLES, and G. CARETTE, Proc. Symp. New Uses for Sulphur and Pyrites, Madrid, 1976, The Sulphur Institute, London (1976). 'Research and Development of Sulphur-Infiltrated Concrete at Canmet, Canada.'

MALLAT, R. C., Alberta Sulphur Res. Quart. Bull. 7 (1971) 1. 'Air Pollution Standards in Perspective.'

MAMURO, T., Kagaku Kojo 20 (1976) 91. 'Continuous and Simultaneous Analysis of Gaseuus and Particle Sulfur Components in Air.'

MANOHARAN, S., "The Oil Crisis, End of an Era," S. Chand & Co., Ram Nagar, New Delhi, 1974.

MANTEL, A., Sci. Dimensions 5 (1973) 16. 'Insulated Highways for Canada's North.'
MARCOUX, L. S., and E. T. SEO, Ad. Chem. Ser. 140 (1975) 216. 'Sodium-Sulfur Batteries.'
MARSH, R. W., J. Pomology Horticult. Sci. 7 (1929) 237. 'Investigations on the Fungicidal Action of Sulphur. II. Studies on the Toxicity of Sulphuretted Hydrogen and on the Interaction of Sulphur with Fungi.'
MARTIN, R. P., W. H. DOUB, JR., J. L. ROBERTS, JR., and D. T. SAWYER, Inorg. Chem. 12 (1973) 1921. 'Further Studies of the Electrochemical Reduction of Sulfur in Aprotic Solvents.'
MARVEL, C. S., E. H. H. SHEN, and R. R. CHAMBERS, J. Am. Chem. Soc. 72 (1950) 2106. 'Polymercaptals and Polymercaptols.'
MARVEL, C. S., and G. NOWLIN, J. Am. Chem. Soc. 72 (1950) 5026. 'Polyalkylene Sulfides. IV. The Effect of pH on Polymer Size.'
MARVEL, C. S., and E. D. WEIL, J. Am. Chem. Soc. 76 (1954) 61. 'The Structure of Propylene Polysulfone.'
MARVEL, C. S., and L. OLSEN, J. Am. Chem. Soc. 79 (1957) 3089. 'Polyalkylene Disulfides.'
MASON, D. B., Ind. Eng. Chem. 30 (1938) 740. 'The Sulphur Industry, History and Development.'
MASON, T. J., F. W. MCKAY, R. HOOVER, W. J. BOLT, and T. F. FRAUMENI, "Atlas of Cancer Mortality for U.S. Counties: 1950-1969," NIH 75-780, U.S. Dept. of Health, Education, and Welfare, Washington, D.C., 1976.
MASSEY, M. J., Ph.D. Thesis, Carnegie-Mellon Univ., Pittsburgh, 1974. 'Environmental Control of Sulfur in Iron and Steelmaking.'
MATSON, R. F., T. K. WIEWIOROWSKI, D. E. SCHOF, JR., and R. A. GRIFFIN, Preprint, 1963. 'On-Stream Infrared Analysis of Molten Sulfur.'
MATSUSHIMA, T., and K. ONO, Tohoku Univ., Sendai, J. Sci. Res. 10 (1958) 58. 'The Interfacial Tension between Liquid Sulfur and Water.'
MATSUSHIMA, T., and K. ONO, Tohoku Univ., Sendai, J. Sci. Res. 10 (1958) 375. 'Some Interfacial Phenomena among Liquid Sulfur, Water, Solid and Gas.'
MAW, G. A., in "Sulfur in Organic and Inorganic Chemistry," Vol. 2, A. Senning, Ed., Marcel Dekker, New York, 1972. 'Metabolic Pathways of Organic Sulfur Compounds.'
MAYER, R., Z. Chem. 13 (1973) 321. 'Elementarer Schwefel als Syntheserohstoff aus der Sicht des Organikers.'
MAYLOR, R., J. B. GILL, and D. C. GOODALL, J. Chem. Soc. Dalton 1972 (1972) 2001. 'Tetra-n-alkylammonium Bisulphites: A New Example of the Existence of the Bisulphite Ion in Solid Compounds.'
MAYO de, P., Acc. Chem. Res. 9 (1976) 52. 'Thione Photochemistry, and the Chemistry of the S_2 State.'
MCAMISH, L. H., Diss. Abstr. B 34 (1974) 4254. 'Decomposition Kinetics in Solid Thiosulfate.'
MCBEE, W. C., and T. A. SULLIVAN, Sulphur Institute J. 11(3-4) (1975) 12. 'Sulphur Composite Material.'
MCBEE, W. C., Proc. Symp. New Horizons in Constr. Materials, Vol. 1, Lehigh Univ. Press, Bethlehem, Pa., 1976. 'Sulfur as a Partial Replacement for Asphalt in Bituminous Pavements.'

MCCAIN, J. E., K. M. CUSHING, and A. N. BIRD, JR., Environ. Prot. Agency Rep. EPA-650/2-73-035. (1973). 'Field Measurements of Particle Size Distribution with Inertial Sizing Devices.'

MCCALLAN, S. E. A., and F. WILCOXON, Boyce Thompson Institute 3 (1931) 13. 'The Fungicidal Action of Sulphur. II. The Production of Hydrogen Sulphide by Sulphured Leaves and Spores and its Toxicity to Spores.'

MCCLELLAN, G. H., and R. M. SCHEIB, Sulphur Institute J. 9(3-4) (1973) 8. 'Characteristization of Sulphur Coatings on Urea.'

MCCLELLAN, G. H., and R. M. SCHEIB, in "New Uses of Sulfur," J. R. West, Ed., Adv. Chem. Ser. 140 (1975) 18. 'Texture of Sulfur Coatings on Urea.'

MCCOWAN, P. K., and M. L. M. NORTHCOTE, Lancet 223 (1932) 237. 'Sulphur Therapy in the Psychoses.'

MCDONALD, W. S., and D. W. S. CRUICKSHANK, Acta Crystallogr. 22 (1967) 48. 'A Refinement of the Structure of S_3O_9.'

MCGLAMERY, G. G., H. L. FAUCETT, R. L. TORSTRICK, and L. J. HENSON, Proc. Symp. Flue Gas Desulfurization, New Orleans, 1976, R. D. Stern and W. H. Ponder, Eds., EPA 600/2-76-136 (1976). 'Flue Gas Desulfurization Economics.'

MCIVER, D. T., J. B. CHATELAIN, and B. A. AXELRAD, Chem. Ind. 30 (1938) 752. 'Re-Use of Bleed Water in Sulfur Mining.'

MCJILTON, C. E., Ph.D. Thesis, Univ. of Washington, Seattle, Wa., 1973. 'The Role of Relative Humidity on the Synergistic Effect of SO_2-Aerosol Mixtures in the Lung.'

MCLAREN, E., A. J. YENCHA, J. M. KOSHNIR, and V. A. MOHNEN, Tellus 26 (1974) 291. 'Some New Thermal Data and Interpretations for the System SO_2-NH_3-H_2O-O_2.'

MEDIO, J., Proc. Symp. New Uses for Sulphur and Pyrites, Madrid, 1976, The Sulphur Institute, London, 1976. 'Road Construction Near Pau, France, December 1975.'

MEGONNEL, W. H., J. Air Pollut. Control Assoc. 25 (1975) 9. 'Atmospheric Sulfur Dioxide in the United States: Can the Standards be Justified or Afforded?'

MEISEL, G. M., J. Metals 24 (1972) 31. 'Sulfur Recovery.'

MELBER, E., Proc. Symp. New Uses for Sulphur and Pyrites, Madrid, 1976, The Sulphur Institute, London, 1976. 'The Present Situation of Pyrite Roasting in Central and Eastern Europe.'

MERKEL, C., and J. LADIK, Phys. Lett. 56A (1976) 395. 'Ab Initio SCF LCAO Band Structure of Poly(SN).'

MERRITT, M. V., and D. T. SAWYER, Inorg. Chem. 9 (1970) 211. 'Electrochemical Reduction of Elemental Sulfur in Aprotic Solvents. Formation of a Stable S_8 Species.'

MERRYMAN, E. L., and A. LEVY, J. Air Pollut. Control Assoc. 17 (1967) 800. Kinetics of Sulfur-Oxide Formation in Flames: II. Low Pressure H_2S Flames.'

MERWIN, R. W., and W. F. KEYES, Minerals Yearbook (1973) 1173. 'Sulfur and Pyrites.'

METSON, A. J., Sulphur Institute Tech. Bull. 20 (1973). 'Sulphur in Forage Crops.'

MEYER, B., Chem. Rev. 64 (1964) 429. 'Solid Allotropes of Sulfur.'

MEYER, B., Ed., "Elemental Sulfur, Chemistry and Physics," Interscience, New York, 1965.

MEYER, B., in "Inorganic Sulphur Chemistry," G. Nickless, Ed., Elsevier, Amsterdam, 1968. 'Elemental Sulphur.'

MEYER, B., and J. J. SMITH, Alberta Sulphur Res. Quart. Bull. 6(1) (1969) 2. Liquid Hydrogen Sulfide as a Solvent.'

MEYER, B., T. V. OOMMEN, and D. JENSEN, J. Phys. Chem. 75 (1971) 912. 'The Color of Liquid Sulfur.'

MEYER, B., T. V. OOMMEN, B. GOTTHARDT, and T. R. HOOPER, Inorg. Chem. 10 (1971) 1632. 'Reactions of Lower Fluorides of Sulfur with Hydrogen Sulfide.'

MEYER B., and J. M. HOLLANDER, Eds., Proc. Symp. SO_2 Abatement Chemistry, Berkeley, 1971, Int. J. Sulfur Chem. B 7 (1972). 'Quarterly Reports on Sulfur Chemistry.'

MEYER, B., D. JENSEN, and T. OOMMEN, in "Sulfur in Organic and Inorganic Chemistry," Vol. 2, A. Senning, Ed., Marcel Dekker, New York, 1972. 'Diatomic Species Containing Sulfur.'

MEYER, B., and K. SPITZER, J. Phys. Chem. 76 (1972) 2274. 'Extended Hueckel Calculations on the Color of Sulfur Chains and Rings.'

MEYER, B., M. GOUTERMAN, D. JENSEN, T. OOMMEN, and T. STROYER-HANSEN, Adv. Chem. Ser. 110 (1972) 53. 'The Spectrum of Sulfur and Its Allotropes.'

MEYER, B., in "Encyclopaedia Britannica," 15th Ed., 1974. 'Sulfur Products and Production.'

MEYER, B., Adv. Inorg. Chem. 18 (1976) 287. 'The Structures of Elemental Sulfur.'

MEYER, B., Chem. Rev. 76 (1976) 367. 'Elemental Sulfur.'

MEYER, B., and B. MULLIKEN, Lawrence Berk. Lab. Report 4159, 1977. 'Wood-Sulfur Bonding.'

MEYER, B., L. PETER, and K. SPITZER, Inorg. Chem. 16 (1977) 27. 'Trends in the Charge Distribution of Sulfanes, Sulfanesulfonic Acids, Sulfanedisulfonic Acids, and Sulfurous Acid.'

MEYER, B., C. SHASKEY-ROSENLUND, and L. PETER, Lawrence Berk. Lab. Rep. 4160, 1977. 'Isotopic Spectra of Disulfites.'

MEYER, B., K. SPITZER, and L. PETER, Proc. Int. Conf. Homonuclear Polyatomics, Plattsburgh, N. Y., 1976, Elsevier Publishers, 1977. 'Charge Distribution in Elemental Sulfur Molecules and Ions.'

MEYER, E., Chem. Ing. Tech. 41 (1969) 1056. 'Schwefeldioxid-Emission und Smog-Bildung.'

MEYER, J. W., W. J. JONES, and M. M. KESSLER, Environ. Prot. Agency Rep. EPA-600/2-76-044a (1976). 'Energy Supply, Demand/Need, and the Gaps Between; Volume I—An Overview.'

MEYER, K. H., and Y. GO, Helv. Chim. Acta 17 (1934) 1081. 'Sur le Soufre Filiforme et la Structure.'

MIDDLETON, W. J., H. W. JACOBSON, R. E. PUTNAM, H. C. WALTER, D. G. PYE, and W. H. SHARKEY, J. Polymer Sci. A3 (1965) 4115. 'Fluorothiocarbonyl Compounds. V. Polymerization of Thiocarbonyl Compounds.'

MIKULSKI, C. M., P. J. RUSSO, M. S. SARAN, A. G. MACDIARMID, A. F. GARITO, and A. J. HEEGER, J. Am. Chem. Soc. 97 (1975) 6358. 'Synthesis and Structure of Metallic Polymeric Sulfur Nitride, $(SN)_x$, and Its Precursor, Disulfur Dinitride, S_2N_2.'

MILLER, D. J., and L. C. CUSACHS, Chem. Phys. Lett. 3 (1969) 501. 'Semi-Empirical Molecular Orbital Calculations on the Bonding in Sulfur Compounds. I. Elemental Sulfur, S_6 and S_8.'

MILLER, D. J., Ph.D. Thesis, Tulane Univ., New Orleans, La., 1970. 'A Theoretical Investigation of the Nature of the Bonding in Sulfur-Containing Molecules.'

MILLER, D. J., and L. C. CUSACHS, in "The Jerusalem Symposia on Quantum Chemistry and Biochemistry. II. Quantum Aspects of Heterocyclic Compounds in Chemistry and Biochemistry," Israel Acad. Sci. & Humanit., Jerusalem, 1970. 'Semi-Empirical Molecular Orbital Calculations on the Bonding in Sulfur Compounds. 1. Elemental Sulfur, S_6 and S_8.'

MILLER, D. J., and T. K. WIEWIOROWSKI, Eds, Adv. Chem. Ser. 110 (1972). 'Sulfur Research Trends.'

MILLER, G. W., J. Appl. Polym. Sci. 15 (1971) 1985 'Thermal Analysis of Polymers. VIII. Diatometric and Thermal Optical Behavior of Sulfur.'

MILLER, H. E., Arch. Dermatol. Syphilol. 31 (1935) 516. 'Colloidal Sulfur in Dermatology.'

MILLS, G. A., J. Am. Chem. Soc. 62 (1940) 2833. 'Oxygen Exchange between Water and Inorganic Oxy-anions.'

MINKWITZ, R., and R. MANZEL, Habilitationsschrift, Free Univ., Berlin, 1976. 'Disulfur Diiodide.'

MISHIN, V. F., Y. P. LUZIN, A. L. ZEN'KOVICH, M. F. PODOROZHNYI, V. N. NIKIFOROV, and V. M. KROVTSOV, Metall. Gornorudn. Prom-st. (6) (1975) 78. 'Physicochemical Characteristics of Dust-Gas Streams during the Operation of a Furnace on Chlorine-Containing Sinter Cake.'

MISHRA, M. M., and S. R. VYAS, J. Sci. Ind. Res. 32 (1973) 481. 'Microbial Production of Sulphur.'

MITA, A., T. AKABANE, and T. YAMAMOTO, Chem. Econ. Eng. Rev. 7 (1975) 36. 'New Pollution-Free Techniques for Pulp Manufacture.'

MITCHELL, B. R., "European Historic Statistics, 1750-1970," Columbia Univ. Press, New York, 1971.

MITOFF, S. P., and J. B. BUSH, JR., Proc. 9th Intersoc. Energy. Conver. Eng. Conf., ASME, New York, 1974. 'Characteristics of a Sodium-Sulfur Cell for Bulk Energy Storage.'

MITTSCHERLICH, H., Pogg. Ann. Phys. Chem. 29 (1833) 193. 'Ueber das Verkaeltnis des spezifischen Gerichts der Gasarten zu den chemische Proportionen.'

MIZOGUCHI, T., S. TSUGIO, and T. OKABE, Hakko Kogaku Zasshi 54 (1976) 181. 'New Sulfur-Oxidizing Bacteria Capable of Growing Heterotrophically, Thiobacillus Rubellus nov. sp. and Thiobacillus Delicatus nov. sp.'

MOECKEL, H. J., Fresenius' Z. Anal. Chem. 279 (1976) 199. 'FID Response Factors for Aliphatic Sulfur Compounds at Higher Concentration Levels.'

MOELLER, W., H. P. GESERICH, and L. PINTSCHOVIUS, Solid State Comm. 18 (1976) 791. 'Optical Properties of Orientated Polymeric Sulfur Nitride ($(SN)_x$) Films in the Visible and Near Infrared Region.'

MOGILEVSKI, B. M., and A. F. CHUDNOVSKII, Proc. Int. Conf. Phys. Semi-Cond., 2 (1968) 1241. 'Thermal Conductivity of Solids.'

MOISEYEV, N., and I. PLATZNER, J. Chromat. Sci. 14 (1976) 143. 'Isotope Effect in Gas-Solid Chromatography of SF_6.'

MOL, A., B. DE MOET, and J. DRAAISMA, Chem. Econ. Eng. Rev. 13 (1976) 7. 'How to Cope with Sulfur in Steam Cracker Feedstocks.'

MONTGOMERY, R. L., Science 184 (1974) 562. 'Monoclinic Sulfur: Heat Capacity Anomaly at 198°K Caused by Disordering of the Crystal Structure.'

MONTGOMERY, R. L., Ph.D. Thesis, Univ. of Oklahoma, Norman, Okla.,1975. 'The Heat Capacity of Monoclinic Sulfur.'

MOORE, C. E., "Atomic Energy Levels as Derived from Analysis of Optical Spectra," Vol. I., U.S. Government Printing Office, Washington, D.C., 1949-1958.

MORAY, R., Phil. Trans. 1 (1666) 45. 'Of the Mineral of Liege, yielding both Brimstone and Vitriol, and the way of extracting them out of it, used at Liege.'

MORGAN, P. W., in "Polymer Reviews," Vol. 10, H. F. Mark and E. H. Immergut, Eds., Wiley-Interscience, New York, 1965. 'Condensation Polymers: By Interfacial and Solution Methods.'

MORGENSTERN, L., U. S. NTIS, PB Rep. PB-246081 (1975). 'Summary Report on Modeling Analysis of Selected Power Plants in 128 AQCRs for Evaluation of Impact on Ambient Sulfur Dioxide Concentrations.'

MORGENSTERN, L., U. S. NTIS, PB Rep. PB-246146 (1975). Summary Report on Modeling Analysis of Power Plants for Fuel Conversion.'

MORICE, J. A., L. V. C. REES, and D. T. RICKARD, J. Inorg. Nucl. Chem. 31 (1969) 3797. 'Moessbauer Studies of Iron Sulphides.'

MOROZOV, V. I., N. IVANOV, E. G. KARKHALEVA, and O. F. SHERANOVA, Agrokhimiya 4 (1976) 56.

MORTH, A. H., and E. E. SMITH, Am. Chem. Soc. Div. Fuel Chem. Preprints 10 (1966) 83. 'Kinetics of the Sulfide-to Sulfate Reaction.'

MORTITA, T., Soda To Enso 26(3) (1975) 77. 'Desulfurization Processes Using Sodium Sulfite.'

MORTON, J. R., J. Phys. Chem. 71 (1967) 89. 'Identification of Some Sulfur-Containing Radicals Trapped in Single Crystals.'

MULLER, E., and J. B. HYNE, Alberta Sulphur Res. Quart. Bull. 4(3) (1967) 22. 'Hydrogen Bonding in Sulphanes.'

MULLER, E., A. PARTHASARATHY, H. J. LANGER, Alberta Sulphur Res. Quart. Bull. 5(4) (1969) 17. 'The Reaction of H_2S with Sulphur.'

MULLER, E., and J. B. HYNE, J. Am. Chem. Soc. 91 (1969) 1907. 'Thermal Decomposition of Tri- and Tetrasulfanes.'

MULLER, H., and H. HEEGN, Acta Chim. Acad. Sci. Hung. 53 (1967) 67. 'Die Electronengasmethode in Anwendung auf zyklische gleichkernige Molekuele.'

MURPHY, T. J., Science 194 (1976) 645. 'Acid in Rain Water.'

MURRAY, R. W., and S. L. JINDAL, J. Org. Chem. 37 (1972) 3516. 'Photosensitized Oxidation of Dialkyl Disulfides.'

MURTHY, K. S., U. S. NTIS, PB Rep. PB-248602 (1974). 'Characterization of Sulfur Recovery in Oil and Natural Gas Production.'

MURZAEV, P. M., Tr. Geol. Inst. Kazan (26) (1968) 67. 'Bacteria in the Sulfur Cycle During Formation of Weathering of Sulfides, Sulfur, and Carbonates.'

MUTHMANN, W., Z. Kristallogr. 17 (1890) 336. 'Zur Kristallstruktur des Schwefels.'

N

NAEGLE, J. A., Ed., Adv. Chem. Ser. 122 (1973). 'Air Pollution Damage to Vegetation.'

NAGARAJAN, G., E. LIPPINCOTT, and J. STUTMAN, J. Phys. Chem. 69 (1965) 2017. 'Mean Amplitudes of Vibration, Bastiansen-Morino Shrinkage Effect, Thermodynamic Functions, and Molecular Polarizability of Sulfur Trioxide.'

NAKAGAWA, H., F. MATSUDA, and T. SENDA, Trans. Jpn. Weld. Soc. 5 (1974) 39. 'Effect of Sulfur on Solidification Cracking in Weld Metal of Steel (Report 1). Fundamental Investigation on Sulfides in Iron-Sulfur Binary Alloy Steel.'

NAKAGAWA, H., F. MATSUDA, T. SENDA, T. MATSUZAKA, and K. WATANABE, Trans. Jpn. Weld. Soc. 5 (1974) 134. 'Effect of Sulfur on Solidification Cracking in Weld Metal of Steel (Report 3). Applicability of New Parameter Manganese^3/Sulfur to Weld Metal of Iron-Sulfur-Manganese Alloy Steel.'

NAKAI, N., and M. L. JENSEN, Geochim. Cosmochim. Acta 28 (1964) 1893. 'The Kinetic Isotope Effect in the Bacterial Reduction and Oxidation of Sulfur.'

NALDRETT, A. J., and J. R. CRAIG, Carnegie Inst. Wash., Yearbook 66 (1968) 436. 'The Iron-Nickel-Sulfur System. Partial Pressure of Sulfur in the Vapor Co-existing with the $Fe_{1-x}S$-$Ni_{1-x}S$ Solid Solution at 600 and 400°.'

NATESAN, K., Corrosion 32 (1976) 364. 'Corrosion-Erosion Behavior of Materials in a Coal-Gasification Environment.'

NATIONAL ACADEMY OF ENGINEERING, Committee on Technology Transfer and Utilization, J. H. Newman, Chairman, Report to Nat'l Sci. Found. PB-232 123, Feb. 1974. 'Technology Transfer and Utilization: Recommendations for Redirecting the Emphasis and Correcting the Imbalance.'

NATIONAL ACADEMY OF SCIENCE, Nat'l Res. Council, Committees on Pollution Abatement and Control, "Abatement of Sulfur Oxide Emissions from Stationary Combustion Sources," Washington, D. C., 1970.

NATIONAL ACADEMY of SCIENCE, Nat'l Res. Council, Nat'l Academy of Eng., Report by the Commission on Natural Resources prepared for the Committee on Public Works, U. S. Senate, Serial No. 94-4, March 1975. 'Air Quality and Stationary Source Emission Control.'

NATIONAL COAL ASSOCIATION, Proc. Coal Utilization Symp. Focus on SO_2 Emission Control; Coal and Environnment Tech. Conf., Louisville, Kentucky, 1974., National Coal Association, Wash., D.C., 1974.

NATUSCH, D. F. S., Anal. Chem. 44 (1972) 2067. 'Sensitive Method for Measurement of Atmospheric Hydrogen Sulfide.'

NEALE, A. J., P. J. S. BAIN, and T. J. RAWLINGS, Tetrahedron 25 (1969) 459. 'Some Observations of the Reactions between Phenol and Sulphur.'

NELEN, I. M., V. I. GORYACHKIN, V. I. SADIKOV, and M. V. MIRONOV, Kim. Prom. 7 (1975) 1669. 'Removal of Arsenic from Sulfur.'

NELSON, P. A., A. A. CHILENSKAS, and R. K. STEUNENBERG, ANL-8075, Argonne Nat'l Lab., Nov. 1974. 'The Need for Development of High-Energy Batteries for Electric Automobiles.'

NERSASIAN, A., and D. E. ANDERSEN, J. Appl. Polymer Sci. 4 (1960) 74. 'The Structure of Chlorosulfonated Polyethylene.'

NESS, H. M., E. A. SONDREAL, and P. H. TUFTE, Proc. Symp. Flue Gas Desulfurization, New Orleans, 1976, R. D. Stern and W. H. Ponder, Eds., EPA Rep. 600/2-76-136 (1976). 'Status of Flue Gas Desulfurization Using Alkaline Fly Ash from Western Coals.'

NETTLETON, M. A., and R. STERLIN, Proc. 12th Int. Symp. on Combustion, U. of Poitiers, 1968, The Combustion Institute, Pennsylvania, 1969. 'Formation and Decomposition of Sulfur Trioxide in Flames and Burned Gases.'

NEUMANN, P., and F. VÖGTLE, Chem. Ziet. 98 (1974) 138. 'Aktuelle Forschungsrichtungen der organischen Schwefelchemie.'

NEWBY, R. A., D. L. KEAIRNS, and E. J. VIDT, Chem. Eng. Progr. 71(5) (1975) 77. 'Residual Oil Gasification/Desulfurization.'

NEYMAN, J., Science 195 (1977) 754. 'Public Health Hazards from Electricity-Producing Plants.'

NGUYEN BA CUONG, B. BONSANG, J. L. PASQUIER, and G. LAMBERT, Proc. Int. Symp. Chem. Sea/Air Particulates Exchanges Processes, Nice, 1973, J. de Recherches Atmospheriques, July-Dec. 1974. 'Composantes Marine et Africaine des Aerosols de Sulfates dans L'Hemisphere Sud.'

NGUYEN BA CUONG, B. BONSANG, and G. LAMBERT, Tellus 26(1-2) (1974) 241. 'The Atmospheric Concentration of Sulfur Dioxide and Sulfate Aerosols over Antarctic, Subantarctic Areas and Oceans.'

NGUYEN BA CUONG, B. BONSANG, G. LAMBERT, and J. L. PASQUIER, Pageoph. 113 (1975) 489. 'Residence Time of Sulfur Dioxide in the Marine Atmosphere.

NICKLESS, G., Ed., "Inorganic Sulphur Chemistry," Elsevier Publishing Co., Amsterdam, 1968.

NIELSEN, H., Tellus 26(1-2) (1974) 213. 'Isotopic Composition of the Major Contributors to Atmospheric Sulfur.

NIMON, L. A., V. D. NEFF, R. E. CANTLEY, and R. O. BUTTLAR, J. Mol. Spectrosc., 22 (1967) 105. 'The Infrared and Raman Spectra of S_6.'

NISHIJIMA, C., K. KANAMARU, and K. KIMURA, Bull. Chem. Soc. Japan 49 (1976) 1151. 'Primary Photochemical Process of Sulfur in Solution.'

NISSEN, W. I., D. A. ELKINS, and W. A. MCKINNEY, Proc. Symp. Flue Gas Desulfurization, New Orleans, 1976, R. D. Stern and W. H. Ponder, Eds., Environ. Prot. Agency Rep. EPA-600/2-76-136 (1976). 'Citrate Process for Flue Gas Desulfurization, A status Report.'

NOBLE, K. E., and W. T. DAVIS, "Power Generation: A. P. Monitoring and Control," 1971.

NONHEBEL, G., Ed., "Gas Purification Processes for Air Pollution Control," Newnes-Butterworths, London, 1972.

NORDIN-FOSSUM, G., and A. TEDER, Svensk Papperstidning 77 (1974) 211. 'Scrubbing of Hydrogen Sulfide with Hydrogen Sulfite Solution.'

NORDSIEK, K. H., and K. VOHWINKEL, Kaut. Gummi Kunststoffe 18 (1965) 566. 'Das Verhalten von Schwefel in cis-1,4-Polybutadien.'

NORTH, D. W., Nat'l Academy of Sci. Report prepared for the Committee on Public Works, U. S. Senate pursuant to U. S. Senate Resolution 135, Aug. 2, 1973, pg. 580 (1975). 'Air Quality and Stationary Source Emission Control.'

NORTHMORE, J. W., Ann. Phys. 30 (1808) 295. 'Versuch 17: Verdichtung von Schweflig-Saurem Gas.'

NOSENKOVA, A. N., and O. N. PONOMAREVA, Tr. Perm. Farm. Inst. No. 3 (1969) 175. 'Rheological Studies of 33% Sulfur Ointment.'

NOVAKOV, T., Proc. Conf. Sensing Environ. Pollutants, Washington, D. C., 1973. 'Chemical Characterization of Atmospheric Pollution Particulates by Photoelectron Spectroscopy.'

NOVAKOV, T., and S. G. CHANG, Lawrence Berk. Lab. Rep. 2693, 1974. 'Catalytic Oxidation of SO_2 on Carbon Particles.'

NOVAKOV, T., Lawr. Berk. Lab. Rep. 3035, 1974. 'Sulfates in Pollution Particulates.'

NOVAKOV, T., S. G. CHANG, and A. B. HARKER, Science 186 (1974) 259. 'Sulfates as Pollution Particulates: Catalytic Formation on Carbon (Soot) Particles.'
NOVAKOV, T., S. G. CHANG, R. L. DOD, and H. ROSEN, Lawrence Berk. Lab. Rep. 5215, 1976. 'Chemical Characterization of Aerosol Species Produced in Heterogeneous Gas-Particle Reactions.'
NOVAKOV, T., R. L. DOD, and S. G. CHANG, Lawrence Berk. Lab Rep. 5217, 1976. 'Study of Air Pollution Particles by X-ray Photoelectron Spectroscopy.'
NOVITSKAJA, N. N., R. V. KUNAKOVA, L. K., YULDASHEVA, E. E. ZAEV, E. B. DMITRIEVA, News USSR Acad. Sci., Chem. Ser. 2 (1976) 384. 'Examination of the Interaction of Halogenized Sulfur with Unlimited Combinations.'
NRIAGU, J. O., Ed., "Environmental Biogeochemistry, Vol. 1: Carbon, Nitrogen, Phosphorus, Sulfur and Selenium Cycle," Ann Arbor Sci. Publishers, Ann Arbor, Mich., 1976.

O

OAE, S., Y. TSUCHIDA, and N. FURUKAWA, Bull. Chem. Soc. Japan 46 (1973) 648. 'The Reaction of Elemental Sulfur with Organic Compounds. V. Reactions of Optically Active Sulfoximine with Elemental Sulfur and with Diphenyl Disulfide.'
OAE, S., S. MAKINO, and Y. TSUCHIDA, Bull. Chem. Soc. Japan 46 (1973) 650. 'The Reaction of Elemental Sulfur with Organic Compounds. VI. S-Tracer Study of Aromatic Displacement Reaction of Thianthrene, Phenoxathiin, Dibenzothiophene and Their Oxidation Compounds by Sulfur.'
OAE, S., D. FUKUSHIMA, and K. WATANABE, Kagaku No Ryoiki, Zokan 113 (1976) 127. 'Bioorganic Reactions and Synthetic Processes around Sulfur and Nitrogen Atoms.'
ODELIEN, M., Sulphur Institute J. 2(2) (1966) 13. 'Sulphur and Crop Quality.'
ODEN, S., Z. Kolloide 8 (1911) 186. 'Ueber die Darstellung kolloider Schwefelloesungen von verschiedenem Dispersitaetsgrad durch fraktionierte Koagulation.'
OEI, D. G., Inorg. Chem. 12 (1973) 438. 'The Sodium-Sulfur System. II. Polysulfides of Sodium.'
OESTREICH, D. K., Environ. Prot. Agency Rep. EPA-600/2-76-279 (1976). 'Equilibrium Partial Pressure of Sulfur Dioxide in Alkaline Scrubbing Processes.'
OIL, PAINT & DRUG REPORTER, 195(Feb. 17) (1969) 5. 'European See Sulfur Price Cut as a Reflection of Rising Supply from Poland, Other New Sources.'
OKABE, H., Anal. Chem. 48 (1976) 1487. 'Fluorescence Quenching of Sulfur Dioxide by Source Emission Gases.'
O'NEILL, E. P., D. L. KEAIRNS, and W. F. KITTLE, Thermochim. Acta 14 (1976) 209. 'A Thermogravimetric Study of the Sulfation of Limestone and Dolomite—The Effect of Calcination Conditions.'
ONO, K., and T. MATSUSHIMA, Tohoku Univ., Sendai, J. Sci. Res. 9 (1957) 309. 'The Surface Tension of Liquid Sulfur.'
ONUFROWICZ, S., Ber. 23 (1890) 3355. 'Ueber Sulfide des beta-Naphtols.'
OOMMEN, T. V., Ph.D. Thesis, Univ. of Washington, Seattle, Wa., 1970, Diss. Abstr. Int. B., 31 (1970) 3904. 'Spectra and Reactions of Elemental Sulfur.'
OPPENHEIMER, M., and A. DALGARNO, Astrophysical J. 187 (1974) 231. 'The Chemistry of Sulfur in Interstellar Clouds.'

ORR, W. L., and A. G. GAINES, JR., Adv. Org. Geochem., Sept. (1973) 791. 'Observations on Rate of Sulfate Reduction and Organic Matter Oxidation in the Bottom Waters of an Estuarine Basin: The Upper Basin of the Pettaquamscutt River (Rhode Island).

ORR, W. L., in "Handbook of Geochemistry," Vol. 2, Part 4, Springer-Verlag, Berlin, 1974. 'Sulfur.'

ORR, W. L., Am. Assoc. Petrol. Geol. Bull. 58 (1974) 2295. 'Changes in Sulfur Content and Isotopic Ratios of Sulfur during Petroleum Maturation—Study of Big Horn Basin Paleozoic Oils.'

ORR, W. L., Preprint 7th Int. Meet. Organic Geochem., Madrid, 1975. 'Geologic and Geochemical Controls on the Distribution of Hydrogen Sulfide in Natural Gas.'

ORTEGA, A., Proc. Symp. New Uses for Sulphur and Pyrites, Madrid, 1976, The Sulphur Institute, London, 1976. 'Experiences in the United Arab Emirates on the Use of Sulphur.'

ORR, W. L., Proc. Am. Chem. Soc. Div. Petrol. Chem., San Francisco, 1976. 'Sulfur in Petroleum and Related Fossil Organic Materials.'

OSBORN, E. F., Science 183 (1974) 477. 'Coal and the Present Energy Situation.'

OSBORNE, W. J., Adv. Chem. 139 (1975) 158. 'Recent Experience of the Wellman-Lord SO_2 Recovery Process.'

OSMOLSKI, T., Kwartalnik Geologiczny 17 (1973) 310. 'Problemy genezy i wieku koncentracji siarki.'

OSWALD, W. J., Paper Symp. Clean Fuels Biomass, Sewage, Urban Refuse, Agric. Wastes, Inst. Gas Technol., Chicago, Ill., 1976. 'Gas Production from Micro Algae.'

OVERBERGER, C. G., and H. ASCHKENASY, J. Org. Chem. 25 (1960) 1648. 'Preparation of Polymeric Condensation Products Containing Functional Thiol Side Chains. Polyamides.'

OVERBERGER, C. G., and W. H. DALY, J. Am. Chem. Soc. 86 (1964) 3402. 'S-Vinyl-o-t-butyl Thiolcarbonate. A New Route to Polymercaptans.'

OZAKI, S., M. IGUCHI, H. ATAKE, M. MATSUMOTO, R. KOIKE, and Y. NAGASHIMA, Chem. Econ. Eng. Rev. 8 (1976) 22. 'Development of New Coke Oven Gas Desulfurization Process, 'Ammonia Takahax-Wet Oxidation Process.' '

OZIN, G. A., J. Chem. Soc. A, 1969 (1969) 116. 'The Single-crystal Raman Spectrum of Rhombic Sulphur.'

OZIN, G. A., Chem. Comm. 1969 (1969) 1325. 'Gas-phase Raman Spectroscopy of Phosphorus, Arsenic, and Saturated Sulphur Vapours.'

P

PACOR, P., Anal. Chim. Acta 37 (1967) 200. 'Applicability of the Du Pont 900 DTA Apparatus in Quantitative Differential Thermal Analysis.'

PALMA, A., and N. V. COHAN, Rev. Inst. Mex. Pet. 2 (1970) 100.

PALMA, A., and N. V. COHAN, Rev. Mex. Fis. 19 (1970) 15.

PALMER, H. B., and D. J. SEERY, Ann. Rev. Phys. Chem. 24 (1973) 235. 'Chemistry of Pollutant Formation in Flames.'

PANCHESHNIKOVA, R. B., and K. T. DANILOV, Khim. Vysokomol. Soedin Neftekhim. 71 (1975). 'Kinetic Principles of the Reaction of Sodium Sulfide and Formaldehyde.'

PANEK, J. R., in "High Polymers," Vol. 13, N. G. Gaylord, Ed., Wiley-Interscience, New York, 1962. 'Polysulfide Polymers: II. Applications.'

PARFENOVA, M. A., E. S. BRODSKII, K. I. ZIMINA, N. K. LYAPINA, V. S. NIKITINA, G. G. KAKABEKOV, A. A. SIMEONOV, and E. M. KALAMASHVILI, Neftekhimiya 15 (1975) 902. 'Study of Structural Group Composition of Petroleum Organosulfur Compounds by Mass Spectrometric and Molecular Spectroscopic Methods.'

PARKER, A. J., and N. KHARASCH, Chem. Rev. 59 (1959) 584. 'The Scission of the Sulfur-Sulfur Bond.'

PARUS, J., and J. KIERZEK, J. Radioanal. Chem. 31 (1976) 309. 'Determination of of Sulfur Content in Drill Core Samples by Backscattered Beta-Particles and Gamma-Ray Absorption.'

PASHKOV, I. A., and E. A. RUBINOVICH, Nauka i Tekhn. u Mis'k. Gospod. Resp. Mizhvid. Nauk. tekhn. Zb. (29) (1975) 16. 'Special Purpose Binders and Concretes Based on Sulfur Compositions.'

PATEL, W., and L. B. BORST, J. Chem. Phys. 54 (1971) 822. 'First-Order Lambda Transition in Liquid Sulfur.'

PAUKOV, I. E. P., E. Y. TONKOV, and D. S. MIRINSKI, Dokl. Akad. Nauk. SSSR 164 (1965) 588.

PAULUS, A. O., O. A. HARVEY, J. NELSON, and V. MEEK, Plant Disease Rep. 59 (1975) 516. 'Fungicides and Timing for Control of Sugarbeet Powdery Mildew.'

PAVLOV, V. I., and L. N. KIRILLOV, Tr. Tol'yattinsk. Politekh. Inst. 1 (1969) 78.

PAZSINT, D. A., and N. SMITH, U.S. Army Cold Reg. Res. Eng. Lab. Tech. Note, May 1972. 'Laboratory Development of a Sulfur/Foamed Polystyrene Insulation Composite.'

PEABODY ENGINEERING SYSTEMS, "The Holmes-Stretford Process for Removal of Hydrogen Sulfide from Gas Streams," 1974.

PEARCE, T. J. P., in "Inorganic Sulphur Chemistry," G. Nickless, Ed., Elsevier, Amsterdam, 1968. 'Sulphuric Acid: Physico-Chemical Aspects of Manufacture.'

PECK, D. H., Bacteriol. Ref. 26 (1962) 67. 'Biochemistry of the Sulfur Cycle.'

PEDCo-ENVIRONMENTAL SPECIALISTS, Quart. Summary Reports to EPA on Contract No. 68-02-1321, 1965-Present. 'Summary Report—Flue Gas Desulfurization Systems.'

PEDROSO, R. I., Proc. EPA Symp. Flue Gas Desulfurization, New Orleans, 1976. 'An Update of the Wellman-Lord Flue Gas Desulfurization Process.'

PELLETIER, B., Ann. Chim. 4 (1790) 1. 'Memoire sur le Phosphore, dans Lequel il est Traite de sa Combinaison avec le Soufre.'

PENATI, A., and M. PEGORARO, J. Polymer Sci. Symp. 53 (1975) 151. 'Cross-linking with S_2Cl_2 of some EPDM Rubbers and their Stress-Strain Behaviors.'

PENKETT, S. A., and J. A. GARLAND, Tellus 26(1-2) (1974) 284. 'Oxidation of Sulphur Dioxide in Artificial Fogs by Ozone.'

PERALTA, L., Y. BEVTHIER, and J. OUDAR, Surface Sci. 55 (1976) 199. 'Adsorption of Sulphur on the (110) Face of Molybdenum.'

PEREIRA, J., "The Elements of Materia Medica and Therapeutics," Vol. 1, J. Carson, Ed., Blanchard and Lea, Philadelphia, Pa., 1852. 'Sulphur and its Compounds with Oxygen, Hydrogen, and Carbon.'

PERRINI, E. M., "Oil From Shale and Tar Sands 1975," Noyes Data Corp. , Park Ridge, New Jersey, 1975.

PERRY, H., and J. A. DECARLO, Mech. Eng. 89 (1967) 22. 'The Search for Low-Sulfur Coal.'

PERRY, H., Scientific Amer. 230(3) (1974) 19. 'The Gasification of Coal.'
PERRY, R. A., R. ATKINSON, and J. N. PITTS, JR., J. Chem. Phys. 64 (1976) 3237. 'Rate Constants for the Reactions $OH+H_2S-H_2O+SH$ and $OH+NH_3-H_2O+NH_2$ over the Temperature Range 297-427°K.'
PETER, S., and H. WOY, Chem. Ing. Technik 41 (1969) 1. 'Gewinnung von Schwefel aus Schwefelwasserstoff nach dem Claus-Verfahren.'
PETROSSI, U., P. L. BOCCA, and P. PACOR, Ind. Eng. Chem. Prod. Res. Develop. 11 (1972) 214. 'Reactions and Technological Properties of Sulfur-Treated Asphalt.'
PETRUSHOVA, N. I., E. I. VORONIN, Protection of Flora 3(June) (1971) 355. 'Measures for Containing Powdery Mildew of Peach with Colloidal Sulfur.'
PHILLIPS PETROLEUM CO., Tech. Bull. TSM-266, "Ryton PPS," 1976. 'Physical, Chemical, Thermal and Electric Properties.'
PICKERING, T. L., and A. V. TOBOLSKY, in "Sulfur in Organic and Inorganic Chemistry," Vol. 3, A. Senning, Ed., Marcel Dekker, New York, 1973. 'Inorganic and Organic Polysulfides.'
PISKORSKI, J., W. KAWECKI, and A. OSUCHA, Chemia Stosowana 16 (1972) 133. 'Cieplo Tworzenia Wielosiarczku Amonowego w Roztworze.'
PITZER, K. S., J. Phys. Chem. 26 (1976) 2863. 'Thermodynamic Properties of Dilute Sulfuric Acid.'
PLATOU, J., Sulphur Institute J., 7(2) (1971) 2. 'Brimstone and Fire.'
PLATOU, J., Hydrocarb. Process. 51 (1972) 86. 'Use Sulfur in Fertilizers.'
PLATTNER, H., "Praeperative Methoden der Organischen Chemie," Verlag Chemie, Berlin, 1943. 'Dehydrienungen mit Schwefel, Selen und Plattinmetallen.'
PLATZER, N. A., Adv. Chem. Ser. 129 (1973) 1. 'Preface.'
PLAYNE, M. J., Sulphur Institute J. 11(2) (1975) 4. 'Sulphur Sources for Cattle.'
PLYLER, E. L., and M. A. MAXWELL, Eds., Environ. Prot. Agency Technol. Ser. EPA-650/2-73-038 (1973). 'Proceedings: Flue Gas Desulfurization Symposium, 1973.'
POLLACCI, E., Gazzetta Chim. Italiana 5 (1875) 451. 'Ella Ragione per cui il Sulfo Uccide l'Oidio della Vite, e Sulla Emissione d'Idrogeno Libero Dalle Piante.'
POLLARD, F. H., J. F. W. MCOMIE, and D. J. JONES, J. Chem. Soc. 1955 (1955) 4337. 'The Analysis of Inorganic Compounds by Paper Chromatography. Part VIII. The Separation of the Thionic Acids by a New Paper-chromatographic Technique.'
PONDER, T. C., JR., L. V. YERINO, V. KATARI, Y. SHAH, and T. W. DEVITT, Environ. Prot. Agency Rep. EPA-600/2-76-150 (1976). 'Simplified Procedures for Estimating Flue Gas Desulfurization System Costs.'
PONDER, W. H., Proc. Thermal Power Conf., Pullman, Wa., 1974. 'Status of Flue Gas Desulfurization Technology for Power Plant Pollution Control.'
PONDER, W. H., R. C. STERN, and G. G. MCGLAMERY, Preprint 3rd Ann. Int. Conf. Coal Gasification and Liquefaction, Pittsburgh, 1976. 'SO_2 Control Technologies—Commercial Availabilities and Economics.'
POPOV, I. F., and V. F. MATRENKIN, Sb. Nauch. Tr. Sredneaziat. Nauch.-Issled, Proekt. Inst. Tsvet. Met. 1 (1968) 89. 'Synthesis of Selective Sulfur-Containing Anion Exchanger. 1. Reaction of Phenol with Elemental Sulfur (Production of Sulfides.'
POSTGATE, J. R., in "Inorganic Sulphur Chemistry," G. Nickless, Ed., Elsevier, Amsterdam, 1968. 'The Sulphur Cycle.'
POTTER, A. E., Am. Ceramic Soc. Bull 48 (1969) 855. 'Sulfur Oxide Capacity of Limestones.'

POTTER, B. H., and C. B. EARL, Nat'l Eng., Aug. (1973) 10. 'The Wellman-Lord SO_2 Recovery Process.'

POWELL, R. W., C. Y. HO, and P. E. LILEY, Nat'l Stand. Ref. Data. Ser., Nat. Bur. Stand., No. 8 Part 1 (1966).

POWER, T. D., and M. D. DUB, Lancet 222 (1932) 338. 'The Leucopoietic Value of Sulfur.'

PRATT, C. J., Scientific Amer. 222 (1970) 63. 'Sulfur.'

PRICE, C. C., and S. OAE, "Sulfur Bonding," Ronald Press, New York, 1962.

PRINCIOTTA, F. T., and W. H. PONDER, Preprint Symp. Sulfur, Energy. Environ., LBL, Berkeley, Ca., 1974. 'Current Status of SO_2 Control Technology.'

PRINCIOTTA, F. T., Environ. Prot. Agency Rep. EPA-650/2-75-010a,b (1975) 'Sulfur Oxide Throwaway Sludge Evaluation Panel (SOTSEP), Volumes 1 and 2: Final Report– Executive Summary and Technical Discussion.'

PRINCIOTTA, F. T., Environ. Prot. Tech. Ser. EPA-600/2-76-152a (1976). 'Keynote Address at EPA's Stationary Source Combustion Symp., Atlanta, 1975.'

PROCTER, A. R., and H. M. APELT, Tappi 52 (1969) 1518. 'Reactions of Wood Components with Hydrogen Sulfide, III. The Efficiency of Hydrogen Sulfide Pretreatment Compared to Other Methods for Stabilizing Cellulose to Alkaline Degradation.'

PRONK, F. E., A. F. SODERBERG, and R. T. FRIZZELL, Proc. 1975 Ann. Conf. Can. Tech. Asphalt Assoc., Toronto, 1975. 'Sulphur Modified Asphaltic Concrete.'

PRYOR, W. A., "Mechanism of Sulfur Reactions," McGraw Hill, New York, 1962.

PRYOR, W. A., J. P. STANLEY, and T. H. LIN, Quart. Rep. Sulfur Chem. 4 (1970) 305. 'Radical Reactions and Radical Production from Sulfur Compounds.'

PUTNAM, A. A., E. L. KROPP, and R. E. BARRETT, U.S. NTIS, PB Rep., PB-248100 (1972). 'Evaluation of National Boiler Inventory.'

Q

QUARLES VAN UFFORD, J. J., and J. C. VLUGTER, Brennstoff Chem. 43 (1962) 173. 'Sulfur and Bitumen, I.'

QUARLES VAN UFFORD, J. J., and J. C. VLUGTER, Brennstoff Chem. 46 (1965) 1. 'Sulfur and Bitumen, II.'

QUECEDO, I., Proc. Symp. New Uses for Sulphur and Pyrites, Madrid, 1976, The Sulphur Institute, London, 1976. 'Use of Pyrites and Profitable Use of the Cinders.'

R

RABEN, I. A., Proc. Flue Gas Desulfurization Symp., 1973, Environ. Prot. Technol. Ser. EPA-650/2-73-038 (1973). 'Status of Technology of Commercially Offered Lime and Limestone Flue Gas Desulfurization Systems.'

RAHMAN, R., S. SAFE, and A. TAYLOR, Quart. Rev. 24 (1970) 208. 'The Stereochemistry of Polysulfides.'

RALL, H. T., C. J. THOMPSON, H. J. COLEMAN, and R. L. HOPKINS, "Sulfur Compounds in Crude Oil, " U.S. Dept. of the Interior, Bur. of Mines Bull. 659, 1972.

RAMMELSBERG, C., Pogg. Ann. 143 (1846) 245. 'Ueber die schwefligsauren Salze.'

RAMMELSBERG, C., Pogg. Ann. 143 (1846) 391. 'Ueber die schwefligsauren Salze.'

RAMOS, E., Proc. Symp. New Uses for Sulphur and Pyrites, Madrid, 1976, The Sulphur Institute, London, 1976. 'New Uses of SO_2.'

RANNEY, M. W., "Desulfurization of Petroleum 1975," Noyes Data Corp., Park Ridge, New Jersey, 1975.
RAO, S. T., P. J. SAMSON, and A. R. PEDDADA, Atmosph. Environ. 10 (1976) 375. 'Spectral Analysis Approach to the Dynamics of Air Pollutants.'
RASCH, R., Aufbereit. Tech. 16 (1975) 237. 'Bonding of Sulfur Dioxide in the Waste Gases of Combustion Installations.'
RASMUSSEN, H. E., R. C. HANSFORD, and A. N. SACHANEN, Ind. Eng. Chem. Int. Ed. 38 (1946) 376. 'Reactions of Aliphatic Hydrocarbons with Sulfur.'
RASMUSSEN, R. A., Tellus 26(1-2) (1974) 254. 'Emission of Biogenic Hydrogen Sulfide.'
RAU, H., T. R. N. KUTTY, and J. R. F. GUEDES DE CARVALHO, J. Chem. Thermodynam. 5 (1973) 291. 'High Temperature Saturated Vapour Pressure of Sulphur and the Estimation of its Critical Quantities.'
RAU, H., T. R. N. KUTTY, and J. R. F. GUEDES DE CARVALHO, J. Chem. Thermodynam. 5 (1973) 833. 'Thermodynamics of Sulphur Vapour.'
RAU, H., Philips Forschungslaboratorium Aachen Computer Program, 1976. 'Partial Sulphur Pressures and Vapour Densities.'
RAULIN, F., and G. TOUPANCE, Bull. Chim. Soc. 1976 (1976) 667. 'Etude Cinetique des Reactions en Solution Aqueuse de Nitriles Maloniques Avec l'Ethanethiol. Incorporation du Soufre dans les Syntheses Prebiotiques.'
RAVAGLIOLI, A., M. F. BALLI, and G. NEGRI, Proc. Symp. New Uses for Sulphur and Pyrites, Madrid, 1976, The Sulphur Institute, London, 1976. 'Sulphur Impregnation of Fired Ceramic Bodies: An Investigation into its Effects on Pore Size Distribution and Frost Resistance.'
RAY, D. L., U.S. Atomic Energy Comm. Report WASH-1281, 1973. 'The Nation's Energy Future.'
RAYMONT, M. E. D., and J. B. HYNE, Private Communication, 1974. 'Compilation of Thermodynamic Functions for Sulfanes.'
RAYMONT, M. E. D., Hydrocarb. Process. 54 (1975) 139. 'Make Hydrogen from Hydrogen Sulfide.'
REAVELL, J. A., Chem. Age 21 (1929) 103. 'Removal of Flue Gases Cannot be Achieved in an Economical Way.'
REED, T. B., and R. M. LERNER, Science 182 (1973) 1299. 'Methanol: A Versatile Fuel for Immediate Use.'
REHME, K. A., and F. P. SCARINGELLI, Anal. Chem. 47 (1975) 2474. 'Effect of Ammonia on the Spectrophotometric Determination of Atmospheric Concentrations of Sulfur Dioxide.'
REID, D. H., "Specialist Periodical Reports Series: Organic Compounds of Sulfur, Selenium, and Tellurium," Vol. 3, Chemical Society, London, 1975.
REID, E. E., "Organic Chemistry of Divalent Sulfur," Chemical Publishing Co., New York, 1960.
REID, G. L., K. ALLEN, and D. J. HARRIS, "The Nationalized Fuel Industries," Heinemann Educational Books, London, 1973.
RENNIE, W. J., D. DUNAY, and J. B. HYNE, Alberta Sulphur Res. Quart. Bull. 8(2-3) (1971) 13. 'The Effect of Small Quantities of Organic and Inorganic Additives, and Geometric Form on the Mechanical Strength and Resistance to Breakdown of Elemental Sulfur.'

RHEINGOLD, A., Ed., Proc. Int. Conf. Homonuclear Polyatomics, Plattsburgh, N. Y., 1976, Elsevier Scientific Publishers, 1977.
RICHARDSON, N. V., and P. WEINBERGER, J. Electron Spectros. Rel. Phenom. 6 (1975) 109. 'The Electronic Structure of S_8.'
RICHERT, H., Z. Anorg. Allg. Chem. 309 (1961) 171. 'Zur Struktur des monomeren Thionylimids.'
RICKLES, R. N., Chem. Eng. Progr. 64 (1968) 53. 'Desulfurization's Impact on Sulfur Markets.'
RICKS, J. M., and R. F. BARROW, Can. J. Phys 47 (1969) 2423. 'The Dissociation Energy of Gaseous Diatomic Sulfur.'
RIEBER, M., S. L. SOO, and J. STUKEL, U.S. NTIS, PB Rep., PB-248064 (1975). 'Flue Gas Desulfurization and Low BTU Gasification. A Comparison. Appendix G.'
RIEDIGER, B., Int. Chem. Eng. 16 (1976) 203. 'Demetallization, Detarring and Desulfurization of Distillation Residues.'
RIETH, H., and P. HANSEN, Med. Klinik 61 (1966) 510. 'Wiederbesinnung auf Schwefel bei der Behandlung von Fussmykosen und aehnlichen Krankheitsbildern.'
RILLING, J., and B. BALESDENT, Preprint, 1975. 'Thermogravimetrie Continue en Atmosphere de Soufre pur.'
RINCKHOFF, J. B., Adv. Chem. Ser. 139 (1975) 48. 'Sulfuric Acid Plants for Copper Converter Gas.'
RINKER, R. G., S. LYNN, D. M. MASON, and W. H. CORCORAN, Ind. Eng. Chem. Fundamentals 4 (1965) 282. 'Kinetics and Mechanism of the Thermal Decomposition of Sodium Dithionite in Aqueous Solution.'
RINKER, R. G., and S. LYNN, Ind. Eng. Chem. Prod. Res. Devel. 8 (1969) 338. 'Formation of Sodium Dithionite from Sodium Amalgam and Sulfur Dioxide in Nonaqueous Media.'
RITTER, R. D., and J. H. KRUEGER, J. Am. Chem. Soc. 92 (1970) 2316. 'Nucleophilic Substitution at Sulfur. Kinetics of Displacement Reactions Involving Trithionate Ion.'
ROBERTS, A. C., and D. J. MCWEENY, J. Fd. Technol. 7 (1972) 221. 'The Uses of Sulphur Dioxide in the Food Industry, A Review.'
ROBERTS, H. L., in "Inorganic Sulphur Chemistry," G. Nickless, Ed., Elsevier, Amsterdam, 1968. 'Compounds Containing Sulphur-Halogen Bonds.'
ROBERTS, P. T., and S. K. FRIEDLANDER, Atmosph. Environ. 10 (1976) 403. 'Analysis of Sulfur in Deposited Aerosol Particles by Vaporization and Flame Photometric Detection.'
ROBERTS, P. T., and S. K. FRIEDLANDER, Environ. Sci. Technol. 10 (1976) 573. 'Photochemical Aerosol Formation. Sulfur Dioxide, 1-Heptene, and NO_x in Ambient Air.'
ROBINSON, E., and R. C. ROBBINS, J. Air Pollut. Control Assoc. 20 (1970) 233. 'Gaseous Sulfur Pollutants from Urban and Natural Sources.'
ROBINSON, E., and R. C. ROBBINS, in "Global Effects of Environmental Pollution," AAAS Symp., Dallas, 1968, S. F. Singer, Ed., Springer, New York, 1970. 'Gaseous Atmospheric Pollutants from Urban and Natural Sources.'
ROBINSON, J. N., Int. J. Sulfur Chem. 7B (1972) 51. 'The History of Sulfur Dioxide Emission Control at COMINCO Ltd.'
ROCHE, R. S., Alberta Sulphur Res. Quart. Bull. 4(2) (1967) 4. 'Sulfur-Containing Polymers; A Review.'

ROCHLIN, G. I., Science 195 (1977) 23. 'Nuclear Waste Disposal.'
ROCKEFELLER BROTHERS FUND, "The Unfinished Agenda, The Citizen's Policy Guide to Environmental Issues," G. O. Barney, Ed., Crowell, Co., New York, 1977.
RODGERS, W. H., JR., "Environmental Law," West Publ. Co., St. Paul, Minn, 1977.
RODHE, H., Tellus 22 (1970) 137. 'Residence Time of Anthropogenic Sulfur in the Atmosphere.
RODHE, H., Tellus 24 (1972) 128. 'A Study of the Sulfur Budget for the Atmosphere over Northern Europe.'
ROESKY, H. W., in "Sulfur in Organic and Inorganic Chemistry," Vol. 1, A. Senning, Ed., Marcel Dekker, New York, 1971. 'The Sulfur-Nitrogen Bond.'
ROGERS, C. H., T. O. SOINE, and C. O. WILSON, " A Text-Book of Inorganic Pharmaceutical Chemistry," 4th Ed., Lea & Febiger, Philadelphia, 1948.
ROGERS, D. E., and G. NICKLESS, in "Inorganic Sulphur Chemistry," G. Nickless, Ed., Elsevier, Amsterdam, 1968. 'The Chemistry of the Phosphorus-Sulphur Bond.'
ROOF, R. B., Aust. J. Phys. 25 (1972) 335. 'Indexing Powder Patterns for High Pressure Sulphur, Phases I and II.'
ROSE, H., Pogg. Ann. 21 (1831) 431. 'Ueber die Verbindungen des Chlors mit dem Schwefel, dem Selen und dem Tellur.'
ROSE, H., Pogg. Ann. 27 (1833) 107. 'Sulphur Chlorides, Bromides and Iodides.'
ROSEN, E., and R. TEGMAN, Chemica Scripta 2 (1972) 221. 'Solid, Liquid and Gas Phase Equilibria in the System Sodium Monosuulfide-Sodium Polysulfide-Sulfur.'
ROSEN, H., and T. NOVAKOV, Lawrence Berk. Lab. Rep. 5228, 1976. 'Application of Raman Scattering to the Characterization of Atmospheric Aerosol Particles.'
ROSENBERG, H. S., R. B. ENGDAHL, J. H. OXLEY, and J. M. GENCO, Chem. Eng. Progr. 7(May) (1975). 'The Status of SO_2 Control Systems.'
ROSS, R. A., and M. R. JEANES, Ind. Eng. Chem. 13 (1974) 102. 'Oxidation of Hydrogen Sulfide over Cobalt Molybdate and Related Catalysts.'
ROSS, R. D., Ed., "Air Pollution and Industry," Van Nostrand Reinhold, New York, 1972.
ROSSINI, F. D., J. Chem. Thermodynam. 2 (1970) 447. 'Report on International Practical Temperature Scale of 1968.'
ROTHSCHILD, W. G., J. Chem. Phys. 45 (1966) 3594. 'Radiation Chemistry of Liquid Sulfur Dioxide. II. Radiation-Induced Oxidation-Reduction Reactions in Liquid Sulfur Dioxide Solutions of Sulfur Trioxide and Potassium Iodide.'
ROY, R. A., and P. A. TRUDINGER, "The Biochemistry of Inorganic Compounds of Sulphur," Cambridge Univ. Press, 1970.
RUBERO, P. A., J. Chem. Eng. Data 9 (1964) 481. 'Effect of Hydrogen Sulfide on the Viscosity of Sulfur.'
RULE, A., and J. S. THOMAS, J. Chem. Soc. 105 (1914) 177 'The Polysulphides of the Alkali Metals.'
RUSHTON, J. D., Chem. Eng. Progr. 69 (1973) 39. 'Handling Sulfur Compounds from Pulp Mills.'
RUSS, C. R., and I. B. DOUGLASS, in "Sulfur in Organic and Inorganic Chemistry," Vol. 1, A. Senning, Ed., Marcel Dekker, New York, 1971. 'The Sulfur-Chlorine Bond.'
RUTHERFORD, W. M., and W. J. ROOS, in "Isotope Ratios as Pollutant Source and Behavior Indicators," Int. Atomic Energy Agency, Vienna, 1975. 'Separation of Sulphur Isotopes by Liquid Thermal Diffusion and by Chemical Exchange.'

RYBCZYNSKI, W., and A. MORSE, "The Use of Elemental Sulphur in Building. Patent Survey 1859-1974," Minimum Cost Housing Group, McGill University, Montreal, April 1975.

RYBCZYNSKI, W., Proc. Symp. New Uses for Sulphur and Pyrites, Madrid, 1976, The Sulphur Instit., London, 1976. 'Sulphur Concrete and Low-Cost Construction.'

S

SAENGER, G., Zement 22 (1933) 566. 'Mit Schwefel getraenkter Zementmoertel.'

SAKAI, H., Geochem. J. 2 (1968) 29. 'Isotopic Properties of Sulfur Compounds in Hydrothermal Processes.

SAKAI, H., T. YAMABE, H. KATO, S. NAGATA, and K. FUKUI, Bull. Chem. Soc. Japan 48 (1975) 33. 'An MO Theoretical Investigation of the Electronic Spectra of Divalent Sulfur Compounds.'

SAKANISHI, J., and R. H. QUIG, Environ. Prot. Technol. Ser. EPA-650/2-73-038 (1973). 'One Year's Performance and Operability of the Chemico-Mitsui Carbide Sludge (Lime) Additive Sulfur Dioxide Scrubbing System at Ohmuta No. 1.'

SALANECK, W. R., N. O. LIPARI, A. PATON, R. ZALLEN, and K. S. LIANG, Phys. Rev. B 12 (1975) 1493. 'Electronic Structure of S_8.'

SALOMAA, P., A. VESALA, and S. VESALA, Acta Chem. Scand. 23 (1969) 2107. 'Solvent Deuterium Isotope Effects on Acid Base Reactions; II. First and Second Acidity Constants of Sulfurous Acid.'

SALTZMAN, R. S., Proc. Ann. Symp. Instrum. Process Ind. 30 (1975) 33. 'Continuous Photometric Analyses for Measurement and Control in Sulfur Recovery Operations.'

SARRIA, J. M., Amer. Perfumer Essen. Oil Refiner 47 (1945) 39. 'Colloidal Sulfur Earns its Place in Cosmetics.'

SAWYER, F. G., R. H. HADER, L. K. HERNDON, and E. MORNINGSTAR, Ind. Eng. Chem. 42 (1940) 1938. 'Sulfur from Sour Gas.'

SAYLAK, D., B. M. GALLAWAY, and H. AHMAD, Adv. Chem. 140 (1975) 102. 'Beneficial Use of Sulfur in Sulfur-Asphalt Pavements.'

SCARGILL, D., J. Chem. Soc 1971A (1971) 2461. 'Dissociation Constants of Anhydrous Ammonium Sulphite and Ammonium Pyrosulfite Prepared by Gas Phase Reactions.'

SCEP REPORT, (Study of Critical Environmental Problems) sponsored by MIT, "Man's Impact on the Global Environment; Assessment and Recommendations for Action," The Colonial Press Inc., U.S.A., 1970.

SCHAEFER, K., and W. KÖHLER, Z. Anorg. Allg. Chem. 104 (1918) 212. 'Optische Untersuchungen ueber die Konstitution der schwefligen Saeure, ihrer Salze und Ester.'

SCHAU, P., Alberta Sulphur Res. Quart. Bull. 1 (1966) 33. 'The Sulphur Trade.'

SCHAUG, J., and A. SEMB, Science 194 (1976) 646. 'Acid in Rain Water.'

SCHEELE, W., and M. CHERUBIM, Kaut. Gummi 13 (1960) 49. 'Zur Kenntnis der Vulkanisation Hochelastischer Polymerisate. Beitrag zur Kinetik der Konzentrations-Abnahme des Schwefels bei Vulkanisations-Vorgaengen.'

SCHEIB, R. M., and G. H. MCCLELLAN, Sulphur Institute J. 12(1) (1976) 2. 'Characteristics of Sulphur Texture on SCU.'

SCHENCK, R., Liebig's Ann. 290 (1896) 171. 'Ueber den Schwefelstickstoff.'

SCHENK, J., Physica 23 (1957) 325. 'Some Properties of Liquid Sulfur and the Occurrence of Long Chain Molecules.'

SCHENK, P. W., and R. STEUDEL, in "Inorganic Sulphur Chemistry," G. Nickless, Ed., Elsevier, Amsterdam, 1968. 'Oxides of Sulphur.'

SCHERTZ, W. W., E. C. GAY, F. J. MARTINO, and K. E. ANDERSON, unpublished, 1974. 'Metal-Sulfide and Metal-Sulfide-Lithium Sulfide Mixtures as Positive Electrode Materials for High Energy Density Batteries.'

SCHERTZ, W. W., A. A. CHILENSKAS, and V. M. KOLBA, Proc. 10th Intersoc. Energy Convers. Eng. Conf., Newark, 1975. 'Battery Design and Cell Testing for Electric-Vehicle Propulsion.'

SCHIMMEL, H., and T. J. MURAWSKI, J. Air Pollut. Control Assoc. 25 (1975) 939. 'SO_2–Harmful Pollutant or Air Quality Indicator?'

SCHIPPER, L, and A. J. LICHTENBERG, Science 194 (1976) 1001. 'Efficient Energy Use and Well-being: The Swedish Example.'

SCHLUPP, K. F., and H. WIEN, Angew. Chem. Int. Ed. 15 (1976) 341. 'Production of Oil by Hydrogenation of Coal.'

SCHMIDBAUR, H., M. SCHMIDT, and W. SIEBERT, Chem. Ber. 97 (1964) 3374. 'Die Protonenresonanzspektren der Sulfane, II.'

SCHMIDKUNZ, H., Ph.D. Thesis, Univ. of Frankfurt, 1963. 'Chemilumineszenz der Sulphitoxidation.'

SCHMIDT, K. H., Chem. Ind. 26 (1974) 737. 'Bleibt Schwefel knapp und teuer?'

SCHMIDT, M., Habilitationsschrift, Univ. of Munich, 1956. 'Zur Kenntnis einer neuen Klasse von Schwefelsaeuren.'

SCHMIDT, M., and G. TALSKY, Z. Analyt. Chem. 166 (1959) 274. 'Jodometrische und Colorimetrische Bestimmung von Sulfanen, Elementarem Schwefel und Sulfan-Schwefel-gemischen. Ueber Saeuren des Schwefels. XVI.'

SCHMIDT, M., and K. BLAETTNER, Angew. Chem. 71 (1959) 407. 'Tetramerer Thioformaldehyd.'

SCHMIDT, M., and B. WIRWOLL, Z. Anorg. Allg. Chem. 303 (1960). 'Kondensation von Thioschwefelsaeure mit Dischwefeldichlorid.'

SCHMIDT, M., and G. TALSKY, Ber. 94 (1961) 1352. 'Preparation of Thiosulfuric Acid.'

SCHMIDT, M., and M. WIEBER, Chem. Ber. 94 (1961) 1426. 'Vulkanisierung von Siliconen.'

SCHMIDT, M., in "Inorganic Polymers," Academic Press, New York, 1962. 'Sulfur Polymers.'

SCHMIDT, M., Oesterr. Chem. Zeit. 64 (1963) 236. 'Zur Problematik der Schwefel-Schwefel-Bindung.'

SCHMIDT, M., and T. SAND, J. Inorg. Nucl. Chem. 26 (1964) 1165, 1173, 1185 & 1189. 'Physikalische Charakterisierung der Sulfan-disulfonate $K_2S_3O_6$ bis $K_2S_6O_6$; Reindarstellung Mikroaufnahmen und Loeslichkeitskurven; UV Extinktionen Waessriger Sulfan-disulfonatloesungen; zur Polarographie von Sulandisulfonaten; & Notiz zur Bestimmung ihrer Dielektrizitaetskonstante.'

SCHMIDT, M., and D. EICHELSDÖRFER, Z. Anorg. Allg. Chem. 330 (1964) 113, 122 & 130. 'Untersuchungen zur Quantitativen Bestimmung von Schwefel einschliesslich seiner Oxydationsstufe; Ueber die Reaktion von Jodwasserstoff mit elementarem Schwefel; & Thiolyse von Sulfurylchlorid.'

SCHMIDT, M., and T. SAND, Z. Anorg. Allg. Chem. 330 (1964) 179. 'Zur Identitaet der aus Thioschwefelsaeure mit Dichlordisulfan und aus Thioschwefelsaeure mit salpetriger Saeure hergestellten Tetrasulfan-disulfonate und zur Existenz eines 'saueren' Tetrasulfan-disulfonats.'

SCHMIDT, M., and E. WILHELM, Inorg. Nucl. Chem. Letters 1 (1965) 39. 'Gezielte Synthese von Cyclohexa-Schwefel S_6 und Cyclodeka-Schwefel S_{10}.'

SCHMIDT, M., in "Elemental Sulfur," B. Meyer, Ed., Interscience, New York, 1965. 'Reactions of the Sulfur-Sulfur Bond.'

SCHMIDT, M., and E. WILHELM, Angew. Chem. 78 (1966) 1020. 'Cyclododecaschwefel, S_{12}–eine neue Verbindung des Schwefels mit sich selbst.'

SCHMIDT, M., and H. D. BLOCK, Angew. Chem. 79 (1967) 944. 'Auftreten von Cyclododecaschwefel in schwefelschmelzen.'

SCHMIDT, M., Ann. Genie Chim. 3 (1968) 16. 'Ueber die direkte Verbrennung von Schwefel zu Schwefeltrioxid.'

SCHMIDT, M., G. KNIPPSCHILD, and E. WILHELM, Chem. Ber. 101 (1968) 381. 'Notiz zur vereinfachten Darstellung von Cyclododecaschwefel, S_{12}.'

SCHMIDT, M., and H. SCHULZ, Z. Naturforsch. 23b (1968) 1540. '1,2-Dithioliumsalze aus Sulfanen und 1,2-Dialdehyden.'

SCHMIDT, M., B. BLOCK, H. D. BLOCK, H. KÖPF, and E. WILHELM, Angew. Chem. 80 (1968) 660. 'Cycloheptaschwefel, S_7, und Cyclodekaschwefel, S_{10}–zwei neue Schwefelringe.'

SCHMIDT, M., Inorg. Macromol. Rev. 1(2) (1970) 101. 'Sulfur-containing Polymers.'

SCHMIDT, M., and E. WILHELM, J. Chem. Soc. Dalton 1970 (1970) 17. 'Cyclonona Sulfur, S_9.'

SCHMIDT, M., and H. D. BLOCK, Z. Anorg. Allg. Chem. 385 (1971) 119. 'Loeslichkeiten von Cyclohexaschwefel und Cyclododekaschwefel in Schwefelkohlenstoff und Benzol.'

SCHMIDT, M., and E. WEISSFLOG, Int. J. Sulf. Chem. 2 (1972) 19. 'A Simple Synthesis of Substituted Thioformaldehydes.'

SCHMIDT, M., in "Chemistry of Organic and Inorganic Sulfur Compounds," Vol. 2, A. Senning, Ed., Marcel Dekker, New York, 1972. 'Oxyacids of Sulfur.'

SCHMIDT, M., Int. J. Sulf. Chem. 7B (1972) 11. 'Fundamental Chemistry of Sulfur Dioxide Removal.'

SCHMIDT, M., Angew Chem. Int. Ed. 12 (1973) 445. 'Elemental Sulfur–A Challenge to Theory and Practice.'

SCHMIDT, M., Chem. Zeit. 7 (1973) 11. 'Schwefelchemie.'

SCHMIDT, M., and W. SIBERT, in "Comprehensive Inorganic Chemistry," Vol. 15, A. F. Trotman-Dickenson, Ed., Pergamon Press, Oxford, 1973. 'The Chemistry of Sulfur.'

SCHMIDT, M., E. WILHELM, T. DEBAERDEMAEKER, E. HELLNER, and A. KUTOGLU, Z. Anorg. Allg. Chem. 405 (1974) 153. 'Darstellung und Kristallstruktur von Cyclooktadekaschwefel, S_{18}, und Cycloikosaschwefel, S_{20}.'

SCHMIDT, M., and H. P. KOPP, Angew. Chem. 87 (1975) 631. 'Chemistry of Flue Gas Desulfurization. 2. Selenium-Catalyzed Formation of Sulfate and Sulfur from Sulfite.'

SCHNELLER, S. W., Int. J. Sulf. Chem. 7B (1972) 155; 8 (1976) 579. 'Name Reactions in Sulfur Chemistry, I and II.'

SCHORA, F. C., JR., Ed., "Fuel Gasification," Adv. Chem. Ser. 69 (1967).

SCHUMACHER, R., VDI-Ber. 246 (1975) 119. 'Reaction Series of Sulfur with Oxygen and Water with Regard to Association Compounds.'

SCHUMANN, H., Z. Anorg. Allg. Chem. 23 (1900) 43. 'Ueber die Einwirkungsprodukte von Schwefeldioxyd auf Ammoniak.'

SCHUMANN, H., in "Sulfur in Organic and Inorganic Chemistry," Vol. 3, A. Senning, Ed., Marcel Dekker, New York, 1973. 'Reactions of Elemental Sulfur with Inorganic, Organic, and Metal-Organic Compounds.'

SCHWALM, W. J., Alberta Sulphur Res. Quart. Bull. 9(2) (1972) 1. 'Reactions of Sulphur with Hydrocarbons.'

SCHWALM, W. J., Alberta Sulphur Res. Quart. Bull. 9(3) (1972) 28. 'Plasticisation of Sulphur Using Polythiol Additives.'

SCHWALM, W. J., and M. E. D. RAYMONT, Alberta Sulphur Res. Quart. Bull. 12(1) (1975) 21. 'Methods for Analysis of H_2S in Sulphur.'

SCHWARTZ, M. A., and T. O. LLEWELLYN, Adv. Chem. Ser. 140 (1975) 75. 'Sulfur in Construction Materials.'

SCHWARZENBACH, G., and A. FISCHER, Helv. Chim. Acta 43 (1960) 1365. 'Die Aciditaet der Sulfane und die Zusammensetzung waesseriger Polysulfidloesungen.'

SCORER, R. S., "Air Pollution," Pergamon Press, Oxford, 1968.

SCOTT, R. L., in "Elemental Sulfur," B. Meyer, Ed., Interscience, New York, 1965. 'Liquid Solutions of Sulfur.'

SCOTT, W. D., D. LAMB, and D. DUFFY, J. Atmosph. Sci. 26 (1969) 727. 'The Stratospheric Aerosol Layer and Anhydrous Reactions Between Ammonia and Sulfur Dioxide.'

SCOTT, W. D., and D. LAMB, J. Am. Chem. Soc. 92 (1970) 3943. 'Two Solid Compounds Which Decompose into a Common Vapor. Anhydrous Reactions of Ammonia and Sulfur Dioxide.'

SEEHL, E. R., Phil. Trans. 43 (1744-45) 1. 'An early Method of procuring the Volatile Acid of Sulphur.'

SEEL, F., and H. D. GÖLITZ, Z. Anorg. Allg. Chem. 327 (1964) 32. 'Ueber Dischwefeldifluorid (Thio-thionylfluorid).'

SEEL, F., R. BUDENZ, W. GOMBLER, and H. SEITTER, Z. Anorg. Allg. Chem. 380 (1971) 262. 'Nachweis der Entstehung von Difluorpolysulfanen bei der Umsetzung von Schwefel mit Silberfluorid.'

SEEL, F., and G. SIMON, Z. Naturforsch. 27b (1972) 1110. 'Ueber die Natur der Farbtraeger der Loesungen von Polysulfiden in EPD-Loesungsmitteln und die Farbtraeger der Ultramarine.'

SEEL, F., in "Sulfur in Organic and Inorganic Chemistry," Vol. 2, A. Senning, Ed., Marcel Dekker, New York, 1972. 'Mixed Sulfur Halides.'

SEEL, F., and H. J. GÜTTLER, Angew. Chem. 85 (1973) 416. 'Polysulfide-Radikalionen.'

SEEL, F., Adv. Inorg. Nuclear Chem. 16 (1974) 300. 'Lower Sulfur Fluorides.'

SEIBERT, A., Z. Phys. Chem. Neue Folge 97 (1975) 22. 'Salzschmelzen von Thiocyanaten und Thiosulfaten. II. Untersuchungen ueber den Platzwechsel der Schwefelatome in Natriumthiosulfat.'

SEILER, H., and H. ERLENMEYER, Helv. Chim. Acta 47 (1964) 264. 'Anorganische Duennschicht-Chromatographie, 9. Mitteilung, Duennschichtchromatographische Trennung von Sulfaten und Polythionaten.'

SELLECK, F. T., L. T. CARMICHAEL, and B. H. SAGE, Ind. Eng. Chem. 44 (1952) 2219. 'Phase Behavior of the Hydrogen Sulfide-Water System.'

SEMLYEN, J. A., Trans. Faraday Soc. 63 (1967) 743. 'Unperturbed Dimensions of Polycatenasulphur and Polycatenaselenium.'

SEMLYEN, J. A., Trans. Faraday Soc. 63 (1967) 2342; 64 (1968) 1396. 'Rotational Isomeric State Models of Sulphur and Selenium Chains. Part 2. Calculation of Entropy Changes in Formation of Cyclooctasulphur and Coclooctaselenium. Part 3. Large Rings in Liquid Sulphur and the Nature of pi-S.'

SEMLYEN, J. A., Polymer Sci. 12 (1971) 383. 'Equilibrium Ring Concentrations and the Statistical Conformations of Polymer Chains: Part 6. Freezing Point of Liquid Sulfur.'

SEMRAU, K., Adv. Chem. Ser. 139 (1975) 1. 'Controlling the Industrial Process Sources of Sulfur Oxides.'

SENDIVOGIUS, M., "Tractatus de Sulphure," Cologne, 1616.

SENNING, A., Ed., "Sulfur in Organic and Inorganic Chemistry," Vol. 1-3, Marcel Dekker, New York, 1971-1973.

SEVERS, R. K., Texas Reps. Biol. Medicine 33 (1975) 45. 'Air Pollution and Health.'

SHANE, G., and R. A. BURGESS, Proc. Symp. New Uses for Sulphur and Pyrites, Madrid, 1976, The Sulphur Institute, London, 1976. 'The Thermopave Process.'

SHARMA, G. C., A. J. PATEL, and D. A. MAYS, J. Am. Soc. Hortic. Sci. 101(2) (1976) 142. 'Effect of Sulfur-Coated Urea on Yield, Nitrogen Uptake, and Nitrate Content in Turnip Greens, Cabbage, and Tomato.'

SHARMA, R. A., J. Electrochem. Soc. 119 (1972) 1439. 'Equilibrium Phases in the Lithium-Sulfur System.'

SHASKEY-ROSENLUND, C., and B. MEYER, Lawrence Berk. Lab. Rep. 4160, 1977. 'History of Sulfur Chemistry.'

SHAW, E. W., Alberta Sulphur Res. Quart. Bull. 4(3) (1967) 5. 'Some General Relationships between Sour Gas Occurrences and the Geological History of the Western Canada Basin.'

SHEEHAN, J. B., Preprint Symp. Sulfur, Energy, Environ., LBL, Berkeley, Ca., 1974. 'Desulfurization of Power Plant Flues with the POWERCLAUS System.'

SHEEHAN, W. C., T. B. COLE, and L. G. PICKLESIMER, J. Polymer. Sci. A3 (1965) 1443. 'Thiazole Polymers for Heat Resistant Film and Fibers.'

SHELL OIL CO., Report to U. S. Environ. Prot. Agency, 1972. 'Computer Aided Forecast for Sulfur Supply and Demand to 2010.'

SHEPPARD, W. L. JR., Sulphur Institute J. 11(3-4) (1975) 15. 'Sulphur Mortars: A Historical Survey.'

SHERR, A. E., and J. CARLIN, J. Appl. Polymer Sci. 11 (1967) 311. 'Foams Produced from Sulfur Dioxide and Various Polymers.'

SHERWOOD, T. K., Int. J. Sulfur Chem. 7B (1972) 1. 'Processes for Removal of Sulfur from Combustion Products: The Challenge to Chemists.'

SHIMOMURA, Y., K. NISHIKAWA, S. ARINO, T. KATAYAMA, Y. HIDA, and T. ISOYAMA, Tetsu To Hagane 62 (1976) 547. 'Report on the Dismantling of Blast Furnaces. 2. On the Inside State of the Lumpy Zone of Blast Furnaces.'

SHIRLEY, A. R., JR., and R. S. MELINE, Adv. Chem. Ser. 140 (1975) 33. 'Sulfur-Coated Urea from a 1-Ton-Per-Hour Pilot Plant.'

SHORE, D., J. J. O'DONNELL, and F. K. CHAN, U.S. NTIS PB Rep., PB 238263/8GA, 1974. 'Evaluation of Research and Development Investment Alternatives for Sulfur Oxides Air Pollution Control Processes.'

SHORT, J. N., and H. W. HILL, JR., Chem. Tech. (ACS) 2 (1972) 481. 'Polyphenylene Sulfide Coating and Molding Resins.'

SIEGEL, L. M., in "Metabolic Pathways. Vol. 7. Metabolism of Sulfur," D. M. Greenberg, Ed., Academic Press, New York, 1975. 'Biochemistry of the Sulfur Cycle.'

SIJ, J. W., E. T. KANEMASU, and S. M. GOLTZ, Trans. Kans. Acad. Sci. 76 (1973) 199. 'Some Preliminary Results in Sulfur Dioxide Effects on Photosynthesis and Yield in Field-Grown Wheat.'

SILVER, B. A., Ed., "Permian Basin Section Symposium and Guidebook, 1968 Field Trip," Soc. Econ. Paleontologists Mineralogists Publ. 68-11, 1968. 'Guadelupian Facies, Apache Mountains area, West Texas.'

SIMON, A., and H. KÜCHLER, Z. Anorg. Allg. Chem. 260 (1949) 161. 'Ueber die Struktur des S_2O_4-Ions.'

SIMON, A., and K. WALDMANN, Z. Phys. Chem. 204 (1955) 235. 'Zur Struktur des Sulfitions.'

SIMON, A., and H. KRIEGSMANN, Z. Phys. Chem. 204 (1955) 369. 'Ueber den Charakter der SO-Bindung.'

SIMON, A., and K. WALDMANN, Z. Anorg. Allg. Chem. 281 (1955) 113, 135; 284 (1956) 36, 47. 'Ueber die Konstitution der sauren Sulfite. I. Waessrige Disulfit-(Pyrosulfit-)Loesungen und ihr Verhalten in Abhaengigkeit von Konzentration und Alkalisierungsgrad. II. Die Gleichgewichte der sauren Sulfite in waessrigen Loesungen auf Grund der Raman-Spektren der in Wasser und in Methanol geloesten sowie der kristallisierten Alkalisalze. III. Die Systeme saures Sulfit (geloest)/Alkalidisulfit (fest) und saures Sulfit (geloest)/Schwefeldioxyd (geloest). IV. Das 'Zeitphaenomen' der sauren Sulfite.'

SIMON, A., K. WALDMANN, and E. STEGER, Z. Anorg. Allg. Chem. 288 (1956) 131. 'Ueber die Konstitution der sauren Sulfite. V. Die Struktur des Disulfit-($S_2O_5^{2-}$-)Ions und ueber den Reaktionsmechanismus der Disulfitbildung bei der Halbneutralisation der Schwefligen Saeure.'

SIMON, A., and H. KRIEGSMANN, Ber. 89 (1956) 2442 'Zu den Ionengleichgewichten in den waessrigen Loesungen der sauren Sulfite.'

SIMON, A., and K. WALDMANN, Z. Anorg. Allg. Chem. 283 (1956) 358. 'Ueber die Ionen der Schwefligen Saeure in waessriger Loesung.'

SIMON, A., and W. SCHMIDT, Z. Electrochem. 64 (1960) 737. 'Praeperative und spektroskopische Untersuchungen an den sauren Sulfiten des Rubidiums und Caesiums.'

SIMON, A., and A. PISCHTSCHAN, J. prakt. Chem. 14 (1961) 196. 'Ueber den Zustand des Wassers im fluessigen Schwefeldioxyd.'

SINENCIO, F. S., Ph.D. Thesis, Univ. of San Paulo, 1969. 'Imperfeicoes, Polarizacao e Fotoemissao Interna em Monocristails de Enxofre Ortorrohbico.'

SITTIG, M., "Sulfuric Acid Manufacture and Effluent Control," Noyes Data Corp., Park Ridge, N. J., 1971.

SKJERVEN, O., Z. Anorg. Allg. Chem. 291 (1957) 325. 'Entfernung von Kohlenstoff und Schwefelwasserstoff aus fluessigem Schwefel.'

SKJERVEN, O., Z. Anorg. Allg. Chem. 314 (1962) 206. 'Die Hochreinigung von Schwefel. Der 'Schwarze Niederschlag.' '

SLACK, A. V., Chem. Eng. 74(Dec.) (1967) 188. 'Air Pollution: The Control of SO_2 from Power Stacks. Part III. Processes for Recovering SO_2.'

SLACK, A. V., TVA Conceptual Design Cost Study for Nat'l Center Air Pollut. Control, 1968. 'Sulfur Oxide Removal from Power Plant Stack Gas; Sorption by Limestone or Lime Dry Process.'

SLACK, A. V., "Sulfur Dioxide Removal from Waste Gases," 2nd Edition, Noyes Data Corp., Park Ridge, N. J., 1975.
SLUTSKAYA, L. D., Agrokhimiya 1 (1972) 130. 'Sulfur as Fertilizer.'
SMIRNOV, M. V., and N. P. PODLESNYAK, Prikl. Khim. 43 (1970) 1463. 'Interaction of Lithium with Molten Lithium Chloride and with LiCl-KCl Eutectic Mixture.'
SMITH, A. R., "Air Pollution," Soc. Chem. Ind., Monograph 22 (1966).
SMITH, J. J., D. JENSEN, and B. MEYER, J. Chem. Eng. Data 15 (1970) 144. 'Liquid Hydrogen Sulfide in Contact with Sulfur.'
SMITH, K. A., J. M. BREMNER, and M. A. TABATABAI, Soil Sci. 116 (1973) 313. 'Sorption of Gaseous Atmospheric Pollutants by Soils.'
SMITH, N., D. PAZSINT, and J. KARALIUS, Preprint 4th Joint Chem. Eng. Conf. Can. Soc. Chem. Eng., Am. Inst. Chem. Eng., Vancouver, B.C., 1973. 'Laboratory Development and Field Testing of a Sulfur/Foamed Polystyrene Insulation Composite.'
SMITH, R., R. G. DE PENA, and J. HEICKLEN, J. Colloid Interface Sci 53 (1975) 202. 'Kinetics of Particle Growth. VI. Sulfuric Acid Aerosol from the Photooxidation of Sulfur Dioxide in Moist Oxygen-Nitrogen Mixtures.'
SNEED, M. C., J. L. MAYNARD, and R. C. BRASTED, "Comprehensive Inorganic Chemistry," Vol. 8, D. Van Nostrand Co., Princeton, N. J., 1953.
SNELL, R. E., and P. C. LUCHSINGER, Arch. Environ. Health 18 (1969) 693. 'Effects of Sulfur Dioxide in Expiration, Flow Rates, and Total Respiratory Resistance in Normal Human Subjects.'
SNYDER, R. N., and J. J. LANDER, Electrochem. Tech. 4 (1966) 179. 'On the Self-Discharge Phenomenon of Lithium and Chlorine in a Fused Salt Electrolyte.'
SOCIETE NATIONALE DES PETROLES D'AQUITAINE, 7th Int. Road Fed. World Meet., Munich, SNPA, Paris, France, 1973. 'Properties of Sulphur Bitumen Binders.'
SOLLMANN, T., "Manual of Pharmacology," Saunders, London, 1957.
SOONDERJI, L. G., Pesticides 6 (1972) 13. 'Sulphur as Fungicide.'
SÖRBO, B., in "Sulfur in Organic and Inorganic Chemistry," Vol. 2, A. Senning, Ed., Marcel Dekker, New York, 1972. 'The Pharmacology and Toxicology of Inorganic Sulfur Compounds.'
SORENSEN, T. S., Alberta Sulphur Res. Quart. Bull. 2(4) (1966) 1. 'Sulphuric Acid as a Solvent.'
SOROKIN, Y. I., and N. DONATO, Hydrobiologia 47 (1975) 241. 'Carbon and Sulfur Metabolism in the Meromictic Lake Faro (Sicily).'
SOUBEIRAN, E., Ann. Chim. Phys. 67 (1838) 71. 'Le Sulfure d'Azote et sur le Chloride de Soufre Ammoniacal.'
SPAITE, P. W., Int. J. Sulfur Chem. 7B (1972). 'Barriers to Development of Effective Control Technology.'
SPEAR, W. E., and A. R. ADAMS, in "Elemental Sulfur," B. Meyer, Ed., Interscience, New York, 1965. 'The Electrical and Photoconductive Properties of Orthorhombic Sulfur Crystals.'
SPEDDING, D. J., Atmosph. Environ. 6 (1972) 583. 'Sulphur Dioxide Absorption by Sea Water.'
SPENGLER, G., "Die Schwefeloxyde in Rauchgasen und in der Atmosphaere; Ein Problem der Lufteinhaltung," VDI-Verlag, Duesseldorf, 1965.

SPERANDIO, G. J., Diss. Abstr. 30 (1970) 4227B. 'The Formulation and Clinical Evaluation of Acne Preparations Containing Sulfur and Resorcinol.'

SPITZER, K., Ph.D. Thesis, Univ. of Washington, Seattle, Wa., 1972. 'Spectroscopic Studies of Small Molecules in Inert Gas Matrices and Extended Hueckel Calculations on the Color of Sulfur Chains and Rings.'

SPRINGLER, E., Plastica 19 (1966) 269. 'Polysulfone–A New Engineering Material.'

SQUIRES, A. M., Chem. Eng. 74(Nov. 6 & 20) (1967) 260 & 131. 'Air Pollution: The Control of SO_2 from Power Stacks. Part I. The Removal of Sulfur From Fuels. Part II. The Removal of SO_2 from Stack Gases.'

SQUIRES, A. M., Int. J. Sulfur Chem. 7B (1972). 'New Systems for Clean Power.'

SQUIRES, A. M., Science 184 (1974) 340. 'Clean Fuels from Coal Gasification.'

SQUIRES, A. M., Science 191 (1976) 689. 'Chemicals from Coal.'

SRB, I., and A. VASKO, J. Chem. Phys. 37 (1962) 1892. 'Infrared Spectrum of CS_2.'

STACHEWSKI, D., Chem. Technik 4 (1975) 269. 'Aktuelle Produktionsverfahren zur Gewinnung von stabilen Isotopen der Elemente C, N, O und S.'

STAMM, H., O. SEIPOLD, and M. GOEHRING, Z. Anorg. Allg. Chem. 247 (1941) 277. 'Zur Kenntnis der Polythionsaeuren und ihrer Bildung. 4. Mitteilung: Die Reaktionen zwischen Polythionsaeuren und schwefliger Saeure bzw. Thioschwefelsaeure.'

STANLEY, E., Acta Cryst. 9 (1956) 897. 'The Structure of Potassium Dithionate.'

STANTON, R. L., "Ore Petrology," McGraw-Hill, New York, 1972

STARKEY, R. L., Soil Sci. 101 (1966) 297. 'Oxidation and Reduction of Sulfur Compounds in Soils.'

STAUFFER CHEMICAL CO., Report F1524, Westport, Conn., 1967. 'Stauffer Industrial Sulfurs.'

STAUFFER CHEMICAL CO., Westport, Conn., Brochures: B-11138 (1974), 'Sulfuric Acid.' B-11143 (1974), 'Sulfur Dioxide.' B-11221 (1974), 'NaHS.' B-11166R (1975), 'Liquid Sulfur Trioxide.' B-1110R (1975), 'CS_2.' B-11317 (1975), 'Sulfur Chlorides,' and unnumbered (1975), 'Crystex, Insoluble Sulfur.'

STECHER, P. G., M. J. FINKEL, O. H. SIEGMUND, and B. M. SZAFRANSKI, Eds. "The Merck Index of Chemicals and Drugs," Merck & Co., Inc., Rahway, N. J., 1960.

STEIJNS, M., and P. MARS, J. Colloid Interface Sci. 57 (1976) 175. 'The Adsorption of Sulfur by Microporous Materials.'

STEIJNS, M., J. Colloid Interface Sci. 57 (1976) 181. 'Mercury Chemisorption by Sulfur Adsorbed in Porous Materials.'

STEINER, P., H. JUENTGEN, and K. KNOBLAUCH, Adv. Chem. Ser. 139 (1975) 180. 'Removal and Reduction of Sulfur Dioxides from Polluted Gas Streams.'

STEINHART, J. S., and C. E. STEINHART, Science 184 (1974) 307. 'Energy Use in the U.S. Food System.'

STEINLE, K., Ph.D. Thesis, Ludwig-Maximilians Univ., Muenchen, 1962. 'Ueber die Bestandteile der Wachenroderschen Fluessigkeit und ihren Bildungsmechanismus.'

STEPUKHOVICH, A. D., N. V. KOZHEVNIKOV, and Y. F. MYASNIKOV, Korroz. Zashch. Neftegazov. Prom-sti. (11) (1975) 6. 'Study of the Combined Action of Some Inhibitors of Hydrogen Sulfide Corrosion of Steel.'

STERN, A. C., J. Occup. Med. 18 (1976) 297. 'Air Pollution Standards: An Overview.'

STEUDEL, R., and M. REBSCH, Z. Anorg. Allg. Chem. 413 (1975) 252. 'Zur Reaction von Thionylchlorid mit Polysulfanen: Darstellung und Eigenschaften von S_8O.'

STEUDEL, R., Angew. Chem. 87 (1975) 683. 'Eigenschaften von S-S-Bindungen.'

STEUDEL, R., and T. SANDOW, Angew. Chem. 88 (1976) 24. 'Darstellung von S_7O.'

STEUNENBERG, R. K., Preprint Int. Conf. Liquid Metal Technol. Energy Prod., 1976. 'Liquid Metal Electrodes in Secondary Batteries.'

STEVENS, R. K., J. D. MULIK, A. E. O'KEEFE, and K. F. KROST, Analyt. Chem. 43 (1971) 827. 'Gas Chromatography of Reactive Sulfur Gases in Air at the Parts per Billion Level.'

STILES, D. A., W. J. R. TYERMAN, O. P. STRAUSZ, and H. E. GUNNING, Can. J. Chem. 44 (1966) 2149. 'Photolysis of Group VI Hydrides. I. Solid H_2S, H_2S_2 and D_2S.'

STILLE, J. K., and J. A. EMPEN, J. Polymer Sci. (A1) 5 (1967) 273. 'Polymerization of Cyclic Sulfides.'

STILLINGS, M., M. C. R. SYMONS, and J. G. WILKINSON, J. Chem. Soc. 1971A (1971) 3201. 'Unstable Intermediates. Electron Spin Resonance Studies of Solutions of Sulphur in Oleum: The S_8^{+} Cation.'

STOECKLI, F., and H. STOECKLI-EVANS, Helv. Chim. Acta 58 (1975) 2291. 'The Gas Solid Interface. The Interaction of Argon, Xenon and Sulfur Hexafluoride with Rhombic Sulfur.'

STOIBER, R. E., and W. I. ROSE, JR., Bull. Volcanol. 37 (1973) 454. 'Cl, F and SO_2 in Central American Volcanic Gases.'

STOPINSKI, A., Biul., Inst. Geol., Warsaw 286 (1975) 89. 'Changes in the Result of Mining of Sulfur at Grzybow by Underground Melting Based on Microgravimetric Studies.'

STRAUSS, H. L., and J. A. GREENHOUSE, in "Elemental Sulfur," B. Meyer, Ed., Interscience, New York, 1965. 'Vibrational Spectra of Elemental Sulfur.'

STRAUSZ, O. P., R. J. DONOVAN, and M. DESORGO, Ber. Buns. Phys. Chem. 72 (1968) 253. 'Electronically Excited S_2 in the Disproportionation and Abstraction Reactions of Sulfur Radicals.'

STRAUSZ, O. P., in "Sulfur in Organic and Inorganic Chemistry," Vol. 2, A. Senning, Ed., Marcel Dekker, New York, 1972. 'The Chemistry of Atomic Sulfur.'

STRAUSZ, O. P., R. K. GOSAVI, A. S. DENES, and I. G. CSIZMADIA, Theor. Chim. Acta (Berlin) 26 (1972) 367. 'Molecular Orbital Calculations on the Ethylene Episulfide Molecule and its Isomers.'

SUDWORTH, J. L., Sulphur Instit. J. 8(4) (1972) 12. 'The Sodium/Sulphur Battery.'

SULLIVAN, J. L., and P. WARNECK, Microchem. J. 8 (1964) 241. 'Microanalysis of SO_4^{2-} and SO_3^{2-}.'

SULLIVAN, T. A., W. C. MCBEE, and D. D. BLUE, Adv. Chem. Ser. 140 (1975). 'Sulfur in Coatings and Structural Materials.'

SULPHUR INSTITUTE, THE, Washington, D.C., and London, Tech. Bull. 19 (1972). 'Potential Uses for Sulphur Dioxide.'

SULPHUR INSTITUTE, THE, Brochure, 1975. 'Sulphur—The Essential Plant Nutrient.'

SULPHUR INSTITUTE, THE, London, Proc. Conf., Madrid, 1976. 'New Uses for Sulphur and Pyrites.'

SULPHURIC ACID ASSOC. JAPAN, Sulphuric Acid Ind. 29(11) (1976) 12. 'Properties and Information for Safe Handling of Sulphuric Acid.'

SUMMERS, C. M., Sci. Amer. 225(3) (1971) 148. 'The Conversion of Energy.'

SUNDHEIM, C., Proc. Symp. New Uses for Sulphur and Pyrites, Madrid, 1976, The Sulphur Institute, London, 1976. 'New Uses of Sulphur: Their Part in Future World Consumption of Sulphur.'

SURMA-SLUSARSKA, B., J. RUTKOWSKI, and W. REIMSCHUSSEL, Prezegl. Papier, 32(4) (1976) 118. 'Radioisotope Study of the Process of Sulfite Cooking Liquor Penetration into Wood.'

SUZUKI, E., E. OSHIMA, and S. YAGI, Kogyo Kagaku Zasshi 69 (1966) 1841. 'Kinetic Studies on the Production of Zinc Dithionite.'

SUZUKI, H., Y. OSUMI, M. NAKANE, and Y. MIYAKE, Bull. Chem. Soc. Japan 47 (1974) 757. 'Studies on Preparation of High Purity Sulfur. II. Removal of Trace Amounts of Selenium and Tellurium in Sulfur by Means of Distillation with Silver.'

SUZUKI, H., K. HIGASHI, and Y. MIYAKE, Bull Chem. Soc. Japan 47 (1974) 759. 'Studies on Preparation of High Purity Sulfur. IV. Effect of the Distillation with Adsorbent and Silver on the Removal of the Impurities in Sulfur.'

SWAIM, C. D., JR., Adv. Chem. Ser. 139 (1975) 111. 'The Shell Claus Offgas Treating (SCOT) Process.'

SWIFT, S. C., F. S. MANNING, and R. E. THOMPSON, Soc. Pet. Eng. J. 16(2) (1976) 57. 'Sulfur-Bearing Capacity of Hydrogen Sulfide Gas.'

SYMONS, M. C. R., and J. G. WILKINSON, Nat. Phys. Sci. 236 (1972) 126. 'Nature of the Paramagnetic Cation in Solutions of Sulphur in Oleum.'

SZMANT, H. H., in "Sulfur in Organic and Inorganic Chemistry," Vol. 1, A. Senning, Ed., Marcel Dekker, New York, 1971. 'The Sulfur-Oxygen Bond.'

T

TABATABAI, M. A., Sulphur Institute J. 10(2) (1974) 11. 'Determination of Sulphate in Water Samples.'

TABOR, B. J., E. P. MAGRE, and J. BOON, Europ. Polymer J. 7 (1971) 1127. 'The Crystal Structure of Poly-p-Phenylene Sulphide.'

TAMAKI, A., Chem. Eng. Progr. 7(May) (1976) 55. 'The Thoroughbred 101 Desulfurization Process.'

TAMATE, R., and F. OHTAKA, Ibaraki Daigaku Nogakuba Gakujutsu Hokoku 23 (1975) 29. 'Sulfur Content in Pork and the Differences due to Year, Season, Breed, Muscle, and Sex.'

TANAKA, T., M. KOIZUMI, and Y. ISHIHARA, Denryoku Chuo Kenkyusho Gijutsu Dai Ichi Kenkyusho Hokoku 73107 (1974). 'Mechanism of Sulfur Dioxide Absorption in Limestone Slurry.'

TANG, S. Y., and C. W. BROWN, J. Raman Spectrosc. 3 (1975) 387. 'Raman Spectra of Gaseous and Matrix Isolated SO_3.'

TAYLOR, R. B., P. R. GAMBARANI, and D. ERDMAN, Proc. Symp. Flue Gas Desulfurization, New Orleans, 1976, Environ. Prot. Technol. Ser. EPA-600/2-76-136 (1976). 'Summary of Operations of the Chemico-Basic MgO FGD System at the PEPCO Dickerson Generating Station.'

TEDER, A., Arkiv Kemi 31 (1969) 173. 'The Spectra of Aqueous Polysulfide Solutions, Part II: The Effect of Alkalinity and Stoichiometric Composition at Equilibrium.'

TEDER, A., and J. TIBERG, Acta Chem. Scand. 24 (1970) 991. 'Sulfur Pressure Over Liquid Polysulfides.'

TEDER, A., Acta Chem. Scand. 25 (1971) 1722. 'The Equilibrium between Elementary Sulfur and Aqueous Polysulfide Solutions.'

TEDER, A., and A. Wilhelmsson, Svensk Papperstich. 78 (1975) 480. 'The Kinetics of the Reaction between Sulfide and Sulfite in Aqueous Solution.'

TEMPLETON, L. K., D. H. TEMPLETON, and A. ZALKIN, Inorg. Chem. 15 (1976) 1999. 'Crystal Structure of Monoclinic Sulfur.'

TEXAS GULF SULPHUR CO., "Sulphur Manual," Texasgulf Co., New York, 1959.

TEXASGULF CO., New York, Brochures: 'Analysis of Sulphur,' 1959; 'Properties of Sulphur,' 1961.

THACKER, C. M., Hydrocarb. Process. 49 (1970) 124. 'What's Ahead for Carbon Disulfide.'

THACKRAY, M., in "Elemental Sulfur," B. Meyer, Ed., Interscience, New York (1976). 'Phase Transition Rate Measurements.'

THACKRAY, M., J. Chem. Eng. Data 15 (1970) 495. 'Melting Point Intervals of Sulfur Allotropes.'

THAULOW, N., Proc. Symp. New Uses for Sulphur and Pyrites, Madrid, 1976, The Sulphur Institute, London, 1976. 'Sulphur Impregnation of Concrete Pipes.'

THIELER, E., "Schwefel," Tech. Fortschrittsber. Vol. 38, DFG, Dresden, 1936.

THOMAS, R. L., V. DHARMARAJAN, G. L. LUNDQUIST, and P. W. WEST, Analyt. Chem. 48 (1976) 639. 'Measurement of Sulfuric Acid Aerosol, Sulfur Trioxide, and the Total Sulfate Content of the Ambient Air.'

THOMPSON, C. J., and C. S. ALLBRIGHT, in "Analytical Chemistry of Sulfur and its Compounds," Wiley-Interscience, New York, 1970. 'Total Sulfur.'

THOMPSON, C. J., E. M. SHELTON, and H. J. COLEMAN, Hydrocarb. Process. 2(1) (1976) 73. 'Sulfur in World Crudes.'

THOMPSON, J. F., in "Annual Review of Plant Physiology," Vol. 18, L. W. Machlis, R. Briggs, and R. B. Park, Eds., Annual Reviews, Inc., Palo Alto, Ca, 1967. 'Sulfur Metabolism in Plants.'

THOMPSON, R. C., in "Inorganic Sulphur Chemistry," G. Nickless, Ed., Elsevier, Amsterdam, 1968. 'Fluorosulphuric Acid.'

THOMPSON, S. D., D. G. CARROLL, F. WATSON, M. O'DONNELL, and S. P. MCGLYNN, J. Chem. Phys. 45 (1966) 1367. 'Electronic Spectra and Structure of Sulfur Compounds.'

TIEMESSE, H., Z. Anorg. Allg. Chem. 41 (1928) 1242. 'Socio-Economic Impact of Sulfur Recovery from Flu Gases.'

TIMMERHAUS, K. D., Chem. Eng. Progr. (Aug.) (1976) 9. 'The Future of Chemical Engineering.'

TIMSON, G. F., and J. HELEIN, Hydrocarb. Process. 1(Jan.) (1974) 115. 'How Amoco Controls H_2S Clean-up.'

TISDALE, S., Sulphur Instit Brochure, 1967. 'Sulphur–The Essential Plant Nutrient.'

TOBOLSKY, A. V., F. LEONARD, and G. P. ROESER, J. Polymer. Sci., 3 (1948) 604. 'Polymerizable Ring Compounds in Constant Volume Polymerizations.'

TOBOLSKY, A. V., and A. EISENBERG, J. Am. Chem. Soc. 82 (1960) 289. 'A General Treatment of Equilibrium Polymerization.'

TOBOLSKY, A. V., G. D. T. OWEN, and A. EISENBERG, J. Colloid. Sci. 17 (1962) 717. 'Viscoelastic Properties of S-Se-As Copolymers.'

TOBOLSKY, A. V., W. J. MACKNIGHT, R. B. BEEVERS, and V. D. GUPTA, Polymer 4 (1963) 423. 'The Glass Transition Temperature of Polymeric Sulphur.'

TOBOLSKY, A. V., and M. TAKAHASHI, J. Polymer. Sci. A2 (1964) 1987. 'Elemental Sulfur as a Plasticizer for Polysulfide Polymers and Other Polymers.'

TOBOLSKY, A. V., and W. J. MACKNIGHT, "Polymeric Sulfur and Related Polymers," Interscience, New York, 1965.

TOBOLSKY, A. V., J. Polymer. Sci. 12C (1966) 71. 'Polymeric Sulfur and Related Polymers.'
TOKUNAGA, J., J. Chem. Eng. Data 19 (1974) 162. 'Solubilities of Sulfur Dioxide in Aqueous Alcohol Solutions.'
TOLAND, W. G., J. B. WILKES, and F. J. BRUTSCHY, J. Am. Chem. Soc. 75 (1953) 2263. 'Reactions of Toluic Acids with Sulfur. I. Stilbenedicarboxylic Acids.'
TOMANY, J. P., "Air Pollution: The Emissions, The Regulations, and The Controls," Elsevier, Amsterdam, 1975.
TOMCZUK, Z., A. E. MARTIN, and R. K. STEUNENBERG, Unpublished. 'Investigation of High-Temperature Lithium/Metal Sulfide Cells.'
TONZETICH, J., and D. M. CATHERALL, Arch. Oral Biol. 21 (1976) 451. 'Metabolism of [35-S]-thiosulfate and [35-S]-thiocyanate by Human Saliva and Dental Plaque.'
TOSTEVIN, W. C., Preprint, 4th Joint Chem. Eng. Conf., Can. Soc. Chem. Eng., Am. Inst. Chem. Eng., Vancouver, 1973. 'Sulphur Removal from Athabasca Bitumen.'
TOURO, F. J., J. Phys. Chem. 70 (1966) 239. 'H_2S in Sulfur.'
TRAIN, R. E., Science 189 (1975) 748. 'Sulfur Dioxide Pollution.'
TREECE, L. C., R. M. FELDER, and J. K. FERRELL, Environ. Sci. Technol. 10 (1976) 457. 'Polymeric Interfaces for Continuous Sulfur Dioxide Monitoring in Process and Power Plant Stacks.'
TREIBMANN, H., Muenchen. Med. Woch. 79 (1931) 1723. 'Therapie der Akne mit Sulfoderm-Puder.'
TRILLAT, J. J., and J. FORESTIER, Compt. Rend. Acad. Sci., Paris 192 (1931) 559. 'Etude sur la Structure du Soufre mou.'
TROFIMOV, B. A., N. K. GUSAROVA, S. V. AMOSOVA, and M. G. VORONKOV, Tezisy Dokl. Nauchn. Sess. Khim. Tekhnol. Org. Soedin. Sery Sernistykh Neftei, 13th (1974) 119. 'New Reactions of Polysulfides and Elemental Sulfur with Acetylene.'
TRUDINGER, P. A., J. Bacteriol. 93 (1967) 550. 'Metabolism of Thiosulfate and Tetrathionate by Heterotrophic Bacteria from Soil'
TRUDINGER, P. A., Adv. Microbial Physiol. 3 (1969) 111. 'The Biochemistry of the Sulfur Cycle.'
TSUNASHIMA, S., T. YOKOTA, I. SAFARIK, H. E. GUNNING, and O. P. STRAUSZ, J. Phys. Chem. 79 (1975) 775. 'Abstraction of Sulfur Atoms from Carbonyl Sulfide by Atomic Hydrogen.'
TUCKER, J. R., and H. E. SCHWEYER, Ind. Eng. Chem. Prod. Res. Devel. 4 (1965) 51. 'Distribution and Reactions of Sulfur in Asphalt During Air Blowing and Sulfurizing Processes.'
TUINSTRA, F., Acta Crystallogr. 20 (1966) 341. 'The Structure of Fibrous Sulphur.'
TUINSTRA, F., "Structural Aspects of the Allotropes of Sulfur and other Divalent Elements," Waltman, Delft, 1967.
TUINSTRA, F., Physica 34 (1967) 113. 'The Structure of Insoluble Sulfur.'
TULLER, W. N., Ed., "The Sulfur Data Book," McGraw Hill, New York, 1954.
TULLER, W. N., "Analytical Chemistry of Sulfur and Its Compounds," Wiley-Interscience, New York, 1970.
TURNER, N. I., A. C. LUDWIG, J. M. DALE, and A. CORREDOR, "Techniques for Sulfur Surface Bonding for Low Cost Housing," Southwest Research Institute, San Antionio, Texas. 1975.

TUTTLE, J. H., and P. R. DUGAN, Can. J. Microbiol. 22 (1976) 719. 'Inhibition of Growth, Iron, and Sulfur Oxidation in Thiobacillus Ferrooxidans by Simple Organic Compounds.'

TVA and THE SULPHUR INSTITUTE, Symp. Marketing Fertilizer Sulphur, Memphis, 1971, Sulphur Institut. Bull. Y-35, 1971.

U

UCHIDA, S., C. Y. WEN, and W. J. MCMICHAEL, Ind. Eng. Chem. Process Des. Dev. 15 (1976) 88. 'Role of Holding Tank in Lime and Limestone Slurry Sulfur Dioxide Scrubbing.'

ULLMANN, F., "Encyklopaedie der Technischen Chemie," 3rd Ed., 1950-1968; 4th Ed., 1970-Present, Urban & Schwarzenberger, Munich.

UNSWORTH, M. H., P. V. BISCOE, and H. R. PICKNEY, Nature 239 (1972) 458. 'Stomatal Responses to SO_2.'

U.S. BUREAU OF CENSUS, "Historical Statistics of the U.S., Colonial Times to 1970,' Superintendent of Documents, Washington, D.C., 1975.

U.S. BUREAU OF MINES, Final Reports Texas A & M Res. Proj. RF-983, 1974-76. 'Beneficial Use of Sulphur in Sulphur-Asphalt Pavements.'

U.S. ENVIRONMENTAL PROTECTION AGENCY, "Environmental Protection Research Catalog," EPA-ORM-72-1, Part 1, Washington, D.C., January, 1972.

U.S. ENVIRONMENTAL PROTECTION AGENCY, Office of Research, Quarterly Releases of Environmental Monitoring and Support Lab. "List of Designated Reference and Equivalent Methods.'

U.S. NATIONAL AIR POLLUTION CONTROL ADMINISTRATION, "Air Quality Criteria for Sulfur Oxides," NAPCA (now EPA), Research Triangle, N. C., 1969.

U.S. PHARMACOPEIAL CONVENTION, INC., "The United States Pharmacopeia," 19th Revision, U.S. Pharmacopeial Convention, Rockville, Md., 1975.

USUNARIZ, U., Proc. Symp. New Uses for Sulphur and Pyrites, Madrid, 1976, The Sulphur Institute, London, 1976. 'Production and Consumption of Pyrites.'

V

VALENTOVA, M., M. UHROVA, and V. TYMAN, Scientific Papers Inst. Chem. Tech. Prague H 8 (1972) 65. 'Bestimmung von Schwefel und Dessen Verbindungen im impraegnierten Holz.'

VANCE, J. L., and L. K. PETERS, Ind. Eng. Chem. Fundam. 15 (1976) 202. 'Aerosol Formation Resulting from the Reaction of Ammonia and Sulfur Dioxide.'

VAN KREVELEN, D. K., "Coal," Elsevier, Amsterdam, 1961.

VAN WART, H. E., and H. A. SCHERAGA, J. Phys. Chem. 80 (1976) 1812. 'Raman Spectra of Cystine-Related Disulfides. Effect of Rotational Isomerism about Carbon-Sulfur Bonds on Sulfur-Sulfur Stretching Frequencies.'

VAN WART, H. E., and H. A. SCHERAGA, J. Phys. Chem. 80 (1976) 1823. 'Raman Spectra of Strained Disulfides. Effect of Rotation about Sulfur-Sulfur Bonds on Sulfur-Sulfur Stretching Frequencies.'

VAN WART, H. E., H. A. SCHERAGA, and R. B. MARTIN, J. Phys. Chem. 80 (1976) 1832. 'Agreement Concerning the Nature of the Variation of Disulfide Stretching Frequencies with Disulfide Dihedral Angles.'

VASAN, S., Chem. Eng. Progr. 71 (1975) 61. 'The Citrex Process for SO_2 Removal.'

VENNART, J., and P. J. N. D. ASH, Health Phys. 30 (1976) 291. 'Derived Limits for 35-S in Food and Air.'

VERNOTTE, J., J. M. MAISON, A. CHEVALLIER, H. HUCK, C. MIEHE, and G. WALTER, Phys. Rev. C 13 (1976) 984. 'Electromagnetic Properties of the 6621- and 7950-keV levels in 32-S.'

VINCENT, P., Proc. Symp. New Uses for Sulphur and Pyrites, Madrid, 1976, The Sulphur Institute, London, 1976. 'Sulphur in Road Construction Technology.'

VINEYARD, B. D., J. Org. Chem. 32 (1967) 3833. 'The Versatility and the Mechanism of the n-Butylamine-Catalyzed Reaction of Thiols with Sulfur.'

VINOGRADOV, V. I., Priroda (Moscow) (12) (1975) 50. 'How Old is the Ocean.'

VIRTANEN, A. I., Biolog. Chem. Vyz. Zvirat. 6(2) (1970) 127. 'Production of Milk by Cows fed Protein-Free Diets.'

VISSERS, D. R., Z. TOMCZUK, and R. K. STEUNENBERG, J. Electrochem. Soc. 121 (1974) 665. 'A Preliminary Investigation of High Temperature Lithium/Iron Sulfide Secondary Cells.'

VITA, N., and E. SALMOIRAGHI, Arch. Exp. Patholog. Pharmakol. 166 (1932) 519. 'Ueber die entgiftende Wirkung des Kolloidalschwefels bei der Kohlenoxydvergiftung.'

VIVIAN, J. E., Environ. Prot. Agency Rep. EPA-650/2-73-047 (1973). 'Absorption of SO_2 into Lime Slurries: Absorption Rates and Kinetics.'

VLADZIMIRSKA, O. V., Pharm. J. (USSR) 29(3 & 4) (1974) 73 & 43. 'Sulfur Compounds as Modern Medicinal Agents, Parts 1 & 2.'

VON DEINES (see also DEINES)

VON DEINES, O., Z. Anorg. Allg. Chem. 177 (1928) 13. 'Ueber die Zersetzung von Thiosulfat durch Salzsaeure.'

VON ETTINGSHAUSEN, O. G., and E. KENDRICK, Polymer 7 (1966) 469. 'Polythioacetone.'

VON HALASZ, S. P., and O. GLEMSER, in "Sulfur in Organic and Inorganic Chemistry," Vol. 1, A. Senning, Ed., Marcel Dekker, New York, 1971. 'The Sulfur-Fluorine Bond.'

VOS, A., and E. H. WIEBENGA, Acta Cryst. 8 (1955) 217. 'The Crystal Structures of P_4S_{10} and P_4S_7.'

VROOM, A. H., Report for Nat'l Res. Council of Can., NRC 12241, 2nd Ed., 1971. 'Sulphur Utilization, A Challenge and An Opportunity.'

VROOM, A. H., Hydrocarb. Process. 51(7) (1972) 79. 'New Uses for Sulfur: The Canadian Viewpoint.'

VROOM, A. H., in "Sulfur Concrete, A New Material for Arctic Constructions," M. B. Ives, Ed., ASM, Metals Park, Ohio, 1977. 'Materials Engineering in the Arctic.'

W

WACKENRODER, H., Arch. Pharm. 97 (1846) 272. 'Ueber eine neue Saeure des Schwefels.'

WACKENRODER, H., Arch. Pharm. 98 (1846) 140. 'Ueber eine neue Saeure des Schwefels.'

WADDINGTON, T. C., in "Non-Aqueous Solvent Systems," Academic Press, London, 1965 'Liquid Sulphur Dioxide.'

WAESER, B., Chem. Zeit. Chem. Apparat. 83 (1959) 602. 'Gibt es ausbaufaehige Moeglichkeiten zur Verwendung von Schwefel und Schwefeldioxyd?'

WAGGONER, A. P., A. J. VANDERPOL, R. J. CHARLSON, S. LARSEN, L. GRANAT, and C. TRAEGARDH, Nature 261 (1976) 120. 'Sulphate-light Scattering Ratio as an Index of the Role of Sulphur in Tropospheric Optics.'

WAGNER, G., H. BOCK, R. BUDENZ, and F. SEEL, Chem. Ber. 106 (1973) 1285. 'Photoelectron Spectra and Molecular Properties, 19. FSSF and SSF_2.'

WAGNER, H., and H. SCHREIER, J. Radioanal. Chem. 32 (1976) 511. 'High Voltage Radioionophoretic Studies on the Isotope Exchange Reactions between Polythionates and Sulfite.'

WAKISAKA, I., Nippon Eiseigaku Zasshi 30 (1975) 543. 'Sensory Irritation of the Upper Respiratory Tract by Sulfur Dioxide.'

WAKSMUNDZKI, A., Przem. Chem. 55(6) (1976) 317. 'Pilot Plant Testing of Emulsion Flotation of Sulfur.'

WALKER, J. F., "Formadehyde," 2nd Ed., Reinhold Publishing Co, New York, 1953.

WALSH, W. J., J. W. ALLEN, J. D. ARNTZEN, L. C. BARTHELME, H. SHIMOTAKE, H. C. TSAL, and N. P. YAO, Proc. 9th IECEC, San Francisco, Ca., 1974. 'Development of Prototype Lithium/Sulfur Cells for Application to Load-Leveling Devices in Electric Utilities.'

WALTON, R. K., "1968 Modern Plastics Encyclopedia," McGraw-Hill, New York, 1967.

WANG, W. C., Y. L. YUNG, A. A. LACIS, T. MO, and J. E. HANSEN, Science 194 (1976) 685. 'Greenhouse Effects due to Man-Made Perturbations of Trace Gases.'

WANNER, H. U., Soz. Praeventivmed. 21 (1976) 65. 'Air Pollution in Switzerland.'

WARD, A. T., J. Phys. Chem. 72 (1968) 744. 'Crystal-Field Splitting of Fundamentals in the Raman Spectrum of Rhombic Sulfur.'

WARD, A. T., J. Phys. Chem. 72 (1968) 4133. 'Raman Spectrosocpy of Sulfur, Sulfur-Selenium, and Sulfur-Arsenic Mixtures.'

WARD, A. T., and M. B. MYERS, J. Phys. Chem. 73 (1969) 1374. 'An Investigation of the Polymerization of Liquid Sulfur, Sulfur-Selenium, and Sulfur-Arsenic Mixtures Using Raman Spectroscopy and Scanning Differential Calorimetry.'

WARSON, H., Paint Mfr. 40(6) (1970) 41. 'Synthetic Resins.'

WASAG, T., J. GALKA, and M. FRACZAK, Ochr. Powietrza 9(3) (1975) 72. 'Effect of Organic Acids on the Kinetics of Sulfur Dioxide Absorption.'

WASSERBERG, F. A. v., "Chemische Abhandlung vom Schwefel, J. q. Kraus, Vienna, 1788.

WATANABE, Y., N. SAITO, and Y. INUISHI, J. Phys. Soc. Japan 25 (1968) 1081. 'Carrier Transport and Generation in Orthorhombic Sulphur Crystals by Pulsed X-Rays.'

WATANABE, Y., Acta Cryst. B 30 (1974) 1396. 'The Crystal Structure of Monoclinic gamma-Sulphur.'

WAYMAN, M., and W. J. LEM, Can. J. Chem. 48 (1970) 782. 'Decomposition of Aqueous Dithionite. Part II. A Reaction Mechanism for the Decomposition of Aqueous Sodium Dithionite.'

WAYMAN, M., and W. J. LEM, Can. J. Chem. 49 (1971) 1140. 'Comment: On the Decomposition of Aqueous Sodium Dithionite.'

WEBER, C. O., Chem. Zeit. 18 (1894) 837. 'Ueber die Vulcanisation des Kautschuks.'

WEBER, R., Pogg. Ann. Phys. Chem. 217 (1870) 432. 'Beobachtungen ueber amorphen Schwefel.'

WEBER, R., Bull. Mes. Soc. Chim. Paris 15-16 (1871) 34. 'Observations sur le Soufre.'

WEBER, R., Pogg. Ann. Phys. Chem. 6 (1875) 531. 'Ueber eine neue Sauerstoffverbindung des Schwefels.'
WEGLER, R., E. KÜHLE, and W. SCHÄFER, in "Neuere Methoden der Praeperativen Organischen Chemie," Vol. 3, W. Foerster, Ed., Verlag Chemie, Weinheim, 1961. 'Reaktionen des Schwefels mit arylaliphatischen sowie aliphatischen Verbindungen.'
WEIR, A., JR., D. G. JONES, and L. T. PAPAY, Preprint Int. Conf. Environmental Sensing and Assessment, Las Vegas, 1975. 'Measurement of Particle Size and Other Factors Influencing Plume Opacity.'
WEIR, A., JR., D. G. JONES, L. T. PAPAY, S. CALVERY, and S. C. YUNG, Environ. Sci. Technol. 10 (1976) 539. 'Factors Influencing Plume Opacity.'
WEISS, J., Forschr. Chem. Forsch. 5 (1966) 635. 'Metall-Schwefelstickstoff-Verbindungen.'
WEITKAMP, A. W., J. Am. Chem. Soc. 81 (1959) 3431. 'The Action of Sulfur on Terpenes. The Limonene Sulfides.'
WEITZ, E., and F. ACHTERBERG, Ber. 61 (1928) 399. 'Ueber hoehere Polythionsaeuren, I. Mitteil: Die Hexathionsaeure.'
WEITZ, E., K. SPOHN, J. SINGER, F. BECKER, and K. GIELES, Angew. Chem. 64 (1952) 166. 'Polythionsaeuren mit mehr als 6 Schwefelatomen und Odensche Schwefelsole.'
WEITZ, E., and K. SPOHN, Chem. Ber. 89 (1956) 2332. 'Ueber Hoehere Polythionsaeuren, II. Mitteil: Polythionsaeuren mit mehr als 6 Schwefelatomen.'
WEITZ, E., F. BECKER, and K. GIELES, Chem. Ber. 89 (1956) 2345. 'Ueber Hoehere Polythionsaeuren, II. Mitteil: Vergleich der Darstellungsmethoden fuer Kaliumhexthionat.'
WEITZ, E., F. BECKER, K. GIELES, and B. ALT, Chem. Ber. 89 (1956) 2353. 'Ueber hoehere Polythionsaeuren, IV. Mitteil: Polythionsaeuren mit 15-40 Schwefelatomen.'
WEITZ, E., K. GIELES, J. SINGER, and B. ALT, Chem. Ber. 89 (1956) 2365. 'Ueber hoehere Polythionsaeuren, V. Mitteil: Ueber die Polythionat-Natur der hydrophilen Odenschen Schwefelsole.'
WEN, C. Y., and L. S. FAN, U.S. NTIS, PB Rep. PB-247234, 1975. Absorption of Sulfur Dioxide in Spray Column and Turbulent Contacting Absorbers.'
WESOLOWSKI, H., and L. ROZEWICKA, Rocz. Pomor. Akad. Med. Szczecinie 21 (1975) 269. 'Effect of Large Doses of Powdered Sulfur on Enzyme Reactions and Mucous Substances in the Duodenum and Jejunum of the Guinea Pig.'
WEST, E. D., J. Am. Chem. Soc. 81 (1959) 29. 'Cryoscopic Constant of S.'
WEST, J. R., in "Encyclopedia Britannica," 15th Ed., 1974. 'Sulfur Properties.'
WEST, J. R., Ed., "New Uses of Sulfur," Adv. Chem. Ser. 140 (1975).
WHANGER, P. D., Sulphur Institute J. 6(3) (1970) 6. 'Sulphur-Selenium Relationships in Animal Nutrition.'
WHELPDALE, D. M., and R. W. SHAW, Tellus 26(1-2) (1974) 196. 'Sulphur Dioxide Removal by Turbulent Transfer over Grass, Snow, and Water Surfaces.'
WHITAKER, M. C., J. Ind. Eng. Chem. (Feb.) (1912) 131. 'Introduction of H. Frasch, Recipient of Perkin Medal 1912.'
WHITE, W. W., J. A. ANDERSON, D. L. BLUMENTHAL, R. B. HUSAR, N. V. GILLANI, J. D. HUSAR, and W. E. WILSON, JR., Science 194 (1976) 187. 'Formation and Transport of Secondary Air Pollutants: Ozone and Aerosols in the St. Louis Urban Plume.'

WHITEHEAD, H. C., and G. ANDERMANN, J. Phys. Chem. 77 (1973) 721. 'An Interpretation of the K-beta X-Ray Emission Spectra of Dibenzyl Sulfide and S_8.'
WHITKOP, P. G., and P. L. GOODFRIEND, Photochem. Photobiol. 24 (1976) 303. 'The Flash Photolysis-Kinetic Spectroscopy of Aerosols.'
WICKERT, K., Mitteil. der VGB 83 (1963) 74. 'Experiments on Desulfurization Before and After the Burner for Reducing the Release of Sulfur Dioxide.'
WIELAND, D. R., Ph.D. Thesis, Agricultural and Mechanical College of Texas, College Station, Texas, 1958. 'The Solubility of Elemental Sulfur in Methane, Carbon Dioxide and Hydrogen Sulfide Gas.'
WIESER, H., P. J. LRUEGER, E. MULLER, and J. B. HYNE, Can. J. Chem. 47 (1969) 1633. 'Vibrational Spectra and a Force Field Calculation for H_2S_3 and H_2S_4.'
WIEWIOROWSKI, T. K., Ph.D. Thesis, Tulane Univ., New Orleans, La., 1965. 'The Structure of Solid Intermediates of the Reaction between Carbon Black and Sulfur.'
WIEWIOROWSKI, T. K., J. Phys. Chem. 70 (1966) 3534. 'Phase Diagram of S_8-CS_2.'
WIEWIOROWSKI, T. K., and F. J. TOURO, J. Phys. Chem. 70 (1966) 234. 'The Sulfur-Hydrogen Sulfide System.'
WIEWIOROWSKI, T. K., and B. L. SLATEN, J. Phys. Chem. 71 (1967) 3014. 'Molten Sulfur Chemistry. IV. The Oxidation of Liquid Sulfur.'
WIEWIOROWSKI, T. K., and B. L. SLATEN, J. Chem. Eng. Data 13 (1968) 38. 'Ternary System, S-Naphth-n-Octadecane.'
WIEWIOROWSKI, T. K., A. PARTHASARATHY, and B. L. SLATEN, J. Phys. Chem. 72 (1968) 1890. 'Molten Sulfur Chemistry. V. Kinetics of Chemical Equilibration in Pure Liquid Sulfur.'
WIEWIOROWSKI, T. K., Alberta Sulphur Res. Quart. Bull. 6 (1969) 27. 'Sulphur Recovery from Sour Gas Via a Catalyzed Reaction in Molten Sulphur.'
WIEWIOROWSKI, T. K., Endeavour 29 (1970) 9. 'The Sulphur-Hydrogen Sulphide-Hydrogen Polysulfide System.'
WIEWIOROWSKI, T. K., in "MTP Int. Rev. of Sci., Inorganic Chem. Ser. 1," Vol. 2, 1972. 'Binary and Ternary Systems Involving Sulphur.'
WIGAND, A., Z. Phys. Chem. 65 (1909) 442. 'Statik des fluessigen Schwefels im Dunkeln und unter dem Einfluss des Lichtes.'
WIGAND, A., Ann. Phys. 29 (1911) 1. 'Statik des fluessigen Schwefels.'
WILCOXON, F., and S. E. A. MCCALLAN, Phytopath. 20 (1930) 391. 'The Fungicidal Action of Sulphur: I. The Alleged Role of Pentathionic Acid.'
WILD, R. B., Proc. Royal Soc. (Melbourne) 1(2) (1911) 13. 'On the Action and Uses of Sulphur and Certain of its Compounds as Intestinal Antiseptics.'
WILHELM, E., Ph.D. Thesis, Univ. of Marburg, 1966. 'Untersuchungen zur Synthese neuer Schwefel-Ringe.'
WILLGERODT, C., Ber. 20 (1887) 2467. 'Ueber die Einwirkung von gelben Schwefelammonium auf Ketone und Chinone.'
WINNEWASSER, M., and J. HAASE, Z. Naturf. 23a (1968) 56. 'Elektronenbeugunsmessungen zur Struktur des gasfoermigen Disulfans.'
WINNEWISER, M., Fortschr. Chem. Forsch. 44 (1974) 1. 'Interstellar Molecules.'
WOLFKOWITSCH, S. I., and D. L. ZIRLIN, Z. Anorg. Allg. Chem. 211 (1933) 257. 'Die Oxydation von Ammoniumsulfit und die Gewinnung von Ammoniumsulfat aus gasfoermigem Schwefeldioxyd, Ammoniak und Wasserdampf.'
WOO, G. L., and R. W. CAMPBELL, Preprint, Am. Inst. Chem. Eng. Meet., Vancouver, B.C., 1973. 'Sulfur Foam, A New Rigid Insulation.'

WOOD, H. C., J. P. REMINGTON, and S. P. SADTLER, "The Dispensatory of the United States of America," 16th Ed., J. B. Lippincott Co., Philadelphia, Pa., 1889.

WOODHOUSE, E. J., and T. H. NORRIS, Inorg. Chem. 10 (1971) 614. 'Complex Formation between Sulfur Dioxide and Halide Ions in Acetonitrile Solution.'

WORLD HEALTH ORGANIZ., "Air Pollution," Columbia Univ. Press, N.Y., 1961.

WORLD HEALTH ORGANIZ., Reg. Off. Europe, EURO 3103A, Copenhagen, 1971. 'Long-Term Programme in Environmental Pollution Control in Europe.'

WRATTEN, S. J. and D. J. FAULKNER, J. Org. Chem. 41 (1976) 2465. 'Cyclic Polysulfides from the Red Alga Chondria californica.'

WRIGHT, R. F., and A. HENRIKSEN, Science 194 (1976) 647. 'Acid in Rain Water.'

XYZ

YANG, R. T., P. T. CUNNINGHAM, W. I. WILSON, and S. A. JOHNSON, Adv. Chem. Ser. 139 (1975) 149. 'Kinetics of the Reaction of Half-Calcined Dolomite with SO_2.'

YAO, N. P., and J. R. BIRK, Proc. 10th IECEC, Newark, 1975. 'Battery Energy Storage for Utility Load Leveling and Electric Vehicles: a Review of Advanced Secondary Batteries.'

YOKOSUKA, F., A. OKUWAKI, T. OKABE, and T. KURAI, Nippon Kagaku Kaishi (10&11) (1975) 1722 & 1901. 'Chemical Behavior of Low Valence Sulfur Compounds. XI. Electrolytic Oxidation of Sodium Thiosulfate and Low-Valence Sulfur Compounds in Aqueous Solution. XII. Oxidation of Sodium Thiosulfate with Hydrogen Peroxide and Sodium Hypochlorite.'

YOKOTA, T., M. G. AHMED, I. SAFARIK, O. P. STRAUSZ, and H. E. GUNNING, J. Phys. Chem. 79 (1975) 1758. 'Reaction of Hydrogen Atoms with Thiirane.'

YOUNG, H. C., and R. WILLIAMS, Science 67 (1928) 19. 'Pentathionic Acid, The Fungicidal Factor of Sulphur.'

YUKEL'SON, I. I., L. V. FEDOTOVA, E. F. KOZYREVA, and T. M. EVSYUKOVA, Izv. Vyssh. Ucheb. Zaved., Khim. Khim. Tekhnol. 14 (1971) 1213. 'Structure of Phenol Sulfidation Products Studied by an IR-Spectroscopic Method.'

ZACHARIASEN, W. H., Phys. Rev. 40 (1932) 113. 'The Crystal Lattice of Potassium Pyrosulfite, $K_2S_2O_5$, and the Structure of the Pyrosulfite Group.'

ZAHRADNIK, R. L., Chem. Eng. Progr. (June) (1976) 25. 'Coal Conversion R and D: What the Government is Doing.'

ZAJIC, J. E., "Microbial Biogeochemistry," Academic Press, New York, 1969.

ZALLEN, R., Phys. Rev. B 9 (1974) 4485. 'Pressure-Raman Effects and Vibrational Scaling Laws in Molecular Crystals: S_8 and As_2S_3.'

ZHAKOVA, M. A., Tr. Perm. Farm. Inst. No. 3 (1969) 167. 'Rheological Studies of Suspension Type Ointments.'

ZIPPERT, L., Deutsch. Med. Woch. Zeit. 55 (1929) 484. 'Sulfoderm, eine neue Form der Schwefelapplikation.'

ZLOBIN, V. S., and O. V. MOKANU, Okeanologiya 15 (1975) 1018. 'Kinetics of the Reaction between Hydrogen Sulfide and Oxygen in Sea Water.'

ZOBELL, C. E., Prod. Month. 22 (1958) 12. 'Ecology of Sulfate-Reducing Bacteria.'

ZUCKERMAN, B., Master's Thesis, Univ. of Pennsylvania, Philadelphia, Pa., 1963. 'An X-Ray Diffraction Study of Fibrous Sulfur.'

Author Index

This index lists only authors quoted in Chapters 1 through 15. Thus, the bibliography, pages 346 through 419, should be consulted parallel with this index. If a name is followed by another name, rather than a page number, the quoted person is co-author of a paper listed in the bibliography. For several patent authors the initials are not listed. If the inventor is not identified in a patent, the private or corporate owner appears as author.

Subject Index

Chemical formulae are in alphabetical order in the sequence S, S_2, SO, SO_2; abbreviations are listed alphabetically as if they were proper names. Greek letters are listed according to the English spelling.